T0139947

IFIP Advances in Information and Communication Technology 583

Editor-in-Chief

Kai Rannenberg, Goethe University Frankfurt, Germany

Editorial Board Members

IFIP – The International Federation for Information Processing

IFIP was founded in 1960 under the auspices of UNESCO, following the first World Computer Congress held in Paris the previous year. A federation for societies working in information processing, IFIP's aim is two-fold: to support information processing in the countries of its members and to encourage technology transfer to developing nations. As its mission statement clearly states:

> IFIP is the global non-profit federation of societies of ICT professionals that aims at achieving a worldwide professional and socially responsible development and application of information and communication technologies.

IFIP is a non-profit-making organization, run almost solely by 2500 volunteers. It operates through a number of technical committees and working groups, which organize events and publications. IFIP's events range from large international open conferences to working conferences and local seminars.

The flagship event is the IFIP World Computer Congress, at which both invited and contributed papers are presented. Contributed papers are rigorously refereed and the rejection rate is high.

As with the Congress, participation in the open conferences is open to all and papers may be invited or submitted. Again, submitted papers are stringently refereed.

The working conferences are structured differently. They are usually run by a working group and attendance is generally smaller and occasionally by invitation only. Their purpose is to create an atmosphere conducive to innovation and development. Refereeing is also rigorous and papers are subjected to extensive group discussion.

Publications arising from IFIP events vary. The papers presented at the IFIP World Computer Congress and at open conferences are published as conference proceedings, while the results of the working conferences are often published as collections of selected and edited papers.

IFIP distinguishes three types of institutional membership: Country Representative Members, Members at Large, and Associate Members. The type of organization that can apply for membership is a wide variety and includes national or international societies of individual computer scientists/ICT professionals, associations or federations of such societies, government institutions/government related organizations, national or international research institutes or consortia, universities, academies of sciences, companies, national or international associations or federations of companies.

More information about this series at http://www.springer.com/series/6102

Ilias Maglogiannis · Lazaros Iliadis ·
Elias Pimenidis (Eds.)

Artificial Intelligence Applications and Innovations

16th IFIP WG 12.5 International Conference, AIAI 2020
Neos Marmaras, Greece, June 5–7, 2020
Proceedings, Part I

Springer

Editors
Ilias Maglogiannis ⓘ
Department of Digital Systems
University of Piraeus
Piraeus, Greece

Elias Pimenidis ⓘ
Department of Computer Science
and Creative Technologies
University of the West of England
Bristol, UK

Lazaros Iliadis ⓘ
Department of Civil Engineering,
Lab of Mathematics and Informatics (ISCE)
Democritus University of Thrace
Xanthi, Greece

ISSN 1868-4238   ISSN 1868-422X (electronic)
IFIP Advances in Information and Communication Technology
ISBN 978-3-030-49163-5   ISBN 978-3-030-49161-1 (eBook)
https://doi.org/10.1007/978-3-030-49161-1

This Springer imprint is published by the registered company Springer Nature Switzerland AG
The registered company address is: Gewerbestrasse 11, 6330 Cham, Switzerland

Preface

AIAI 2020

Artificial Intelligence (AI) is already affordable through a large number of applications that offer good services to our post-modern societies. Image-face recognition and translation of speech are already a reality. File sharing via Dropbox, Uber transportations, social interaction through Twitter, and shopping from eBay are employing Google's TensorFlow platform. AI has already developed high levels of reasoning. Respective applications like AlphaGo (released by Google DeepMind) have managed to defeat human experts in highly sophisticated demanding games, like *Go*. This is a great step forward, if we realize that in the *Go* game the number of potential moves is higher than the number of atoms in the entire Universe. *AlphaGo Zero* is a recent and impressive advance which is using Reinforcement Learning to teach itself. It started with no knowledge at all and in three days it bypassed the capabilities of *AlphaGo Lee*, which is the version that defeated one of the best *Go* human players in four out of five games in 2016. In 21 days, it evolved even further, and it reached Master level. More specifically, it defeated 60 top professional *Go* players online and the world champion himself.

Deep Learning has significantly contributed to the progress made during the last decades. Meta-Learning and intuitive intelligence are on the way, and soon they will enable machines to understand what learning is all about. The concept of Generative Adversarial Neural Networks will try to fuse "imagination" to AI. However historic challenges for the future of mankind will be faced. Potential unethical use of AI may violate democratic human rights and may alter the character of western societies.

The 16th conference on Artificial Intelligence Applications and Innovations (AIAI 2020) offers an insight to all timely challenges related to technical, legal, and ethical aspects of AI systems and their applications. New algorithms and potential prototypes employed in diverse domains are introduced. AIAI is a mature international scientific conference held in Europe and well established in the scientific area of AI. Its history is long and very successful, following and spreading the evolution of Intelligent Systems.

The first event was organized in Toulouse, France, in 2004. Since then, it has a continuous and dynamic presence as a major global, but mainly European, scientific event. More specifically, it has been organized in China, Greece, Cyprus, Australia, and France. It has always been technically supported by the International Federation for Information Processing (IFIP) and more specifically by the Working Group 12.5 which is interested in AI applications.

Following a long-standing tradition, this Springer volume belongs to the IFIP AICT Springer Series and contains the papers that were accepted to be presented orally at the AIAI 2020 conference. An additional volume comprises the papers that were accepted and presented at the workshops – held as parallel events.

The diverse nature of papers presented, demonstrates the vitality of AI algorithms and approaches. It certainly proves the very wide range of AI applications as well.

The event was held during June 5–7, 2020, as a remote LIVE event with live presentations, a lot of interaction, and live Q&A sessions. There was no potential for physical attendance due to the COVID-19 global pandemic.

Regardless of the extremely difficult pandemic conditions, the response of the international scientific community to the AIAI 2020 call for papers was overwhelming, with 149 papers initially submitted. All papers were peer reviewed by at least two independent academic referees. Where needed a third referee was consulted to resolve any potential conflicts. A total of 47% of the submitted manuscripts (70 papers) were accepted to be published as full papers (12-pages long) in these Springer proceedings. Due to the high quality of the submissions, the Program Committee decided additionally to accept five more papers to be published as short papers (10-pages long).

The following scientific workshops on timely AI subjects were organized under the framework of AIAI 2020:

9th Mining Humanistic Data Workshop (MHDW 2020)

We would like to thank the Steering Committee of MHDW 2020 Professors Ioannis Karydis and Katia Lida Kermanidis from Ionian University, Greece, and Professor Spyros Sioutas from the University of Patras, Greece, for their important contribution towards the organization of this high-quality and mature event. Also, we would like to thank Professor Christos Makris and Dr. Andreas Kanavos from the University of Patras, Greece, and Professor Phivos Mylonas for doing an excellent job as Senior Members of the Program Committee. MHDW is an annual event that attracts an increasing amount of high-quality research papers, under the framework of the AIAI conference.

5th Workshop on 5G-Putting Intelligence to the Network Edge (5G-PINE 2020)

We would like to thank Dr. Ioannis P. Chochliouros, Research Programs Section, Research and Development Department, Fixed & Mobile, Hellenic Telecommunications Organization (OTE). We really appreciate his great efforts in organizing this high-quality 5th 5G-PINE workshop, which is a well-established annual event. It presents timely and significant results from state-of-the-art research in the 4th industrial revolution area. This workshop connects the conference with the latest AI applications in the telecommunication industry.

The workshops organized under the framework of AIAI 2020, also followed the same review and acceptance ratio rules. More specifically the 5th 5G-PINE workshop accepted 11 full papers (50%) and one short paper out of 23 submissions, whereas the MHDW 2020 accepted 6 full (37.5%) and 3 short papers out of 16 submitted papers.

Four keynote speakers gave state-of-the-art lectures (after invitation) in timely aspects-applications of AI.

Professor Leontios Hadjileontiadis, Department of Electrical and Computer Engineering at Aristotle University of Thessaloniki, Greece, and Coordinator of i-PROGNOSIS, gave a speech on the subject of "Smartphone, Parkinson's and Depression: A new AI-based prognostic perspective".

Professor Hadjileontiadis has been awarded, among other awards, as innovative researcher and champion faculty from Microsoft, USA (2012), the Silver Award in Teaching Delivery at the Reimagine Education Awards (2017–2018), and the Healthcare Research Award by the Dubai Healthcare City Authority Excellence Awards (2019). He is a Senior Member of IEEE.

Professor Nikola Kasabov, FIEEE, FRSNZ, Fellow INNS College of Fellows, DVF RAE UK, Director, Knowledge Engineering and Discovery Research Institute, Auckland University of Technology, Auckland, New Zealand, Advisory/Visiting Professor SJTU and CASIA China, RGU UK, gave a speech on the subject of "Deep Learning, Knowledge Representation and Transfer with Brain-Inspired Spiking Neural Network Architectures".

Professor Kasabov has received a number of awards, among them: Doctor Honoris Causa from Obuda University, Budapest; INNS Ada Lovelace Meritorious Service Award; NN Best Paper Award for 2016; APNNA 'Outstanding Achievements Award'; INNS Gabor Award for 'Outstanding contributions to engineering applications of neural networks'; EU Marie Curie Fellowship; Bayer Science Innovation Award; APNNA Excellent Service Award; RSNZ Science and Technology Medal; and 2015 AUT Medal; and Honorable Member of the Bulgarian and Greek Societies for Computer Science.

Dr. Pierre Philippe Mathieu, European Space Agency (ESA) Head of the Philab (Φ Lab) Explore Office at the European Space Agency in ESRIN (Frascati, Italy).

Professor Xiao-Jun Wu Department of Computer Science and Technology Jiangnan University, China, gave a speech on the subject of "Image Fusion Based on Deep Learning".

Professor Xiao-Jun Wu has won the most outstanding postgraduate award by Nanjing University of Science and Technology. He has won different national and international awards for his research achievements. He was a visiting postdoctoral researcher at the Centre for Vision, Speech, and Signal Processing (CVSSP), University of Surrey, UK, from 2003–2004.

The following two tutorial sessions by experts in the AI field completed the program:

Professor John Macintyre Dean of the Faculty of Applied Sciences, Pro Vice Chancellor at the University of Sunderland, UK. During the 1990s he established the Center for Adaptive Systems – at the university, which became recognized by the UK government as a Center of Excellence for applied research in adaptive computing and AI. The Center undertook many projects working with and for external organizations in industry, science, and academia, and for three years ran the Smart Software for Decision Makers program on behalf of the Department of Trade and Industry.

Professor Macintyre presented a plenary talk on the following subject: "AI Applications during the COVID-19 Pandemic - A Double Edged Sword?".

Dr. Kostas Karpouzis Associate Researcher, Institute of Communication and Computer Systems (ICCS) of the National Technical University of Athens, Greece. Tutorial Subject: "AI/ML for games for AI/ML".

Digital games have recently emerged as a very powerful research instrument for a number of reasons: they involve a wide variety of computing disciplines, from databases and networking to hardware and devices, and they are very attractive to users regardless of age or cultural background, making them popular and easy to evaluate with actual players. In the fields of AI and Machine Learning (ML), games arc used in a two-fold manner: to collect information about the players' individual characteristics (player modeling), expressivity (affective computing), and playing style (adaptivity) and also to develop AI-based player bots to assist and face the human players as a test-bed for contemporary AI algorithms.

This tutorial discusses both approaches that relate AI/ML to games: starting from a theoretical review of user/player modeling concepts, it discusses how we can collect data from the users during gameplay and use them to adapt the player experience or model the players themselves. Following that, it presents AI/ML algorithms used to train computer-based players and how these can be used in contexts outside gaming. Finally, it introduces player modeling in contexts related to serious gaming, such as health and education.

Intended audience: researchers in the fields of ML and Human Computer Interaction, game developers and designers, as well as health and education practitioners.

The accepted papers of the AIAI 2020 conference are related to the following thematic topics:

- AI Constraints
- Classification
- Clustering – Unsupervised Learning
- Deep Learning Long Short Term Memory
- Fuzzy Algebra-Fuzzy Systems
- Image Processing
- Learning Algorithms
- ML
- Medical – Health Systems
- Natural Language
- Neural Network Modeling
- Object Tracking-Object Detection
- Ontologies/AI
- Sentiment Analysis/ Recommendation Systems
- AI/Ethics/Law

The authors of submitted papers came from 28 different countries from all over the globe, namely:

Algeria, Austria, Belgium, Bulgaria, Canada, Cyprus, Czech Republic, Finland, France, Germany, Greece, The Netherlands, India, Italy, Japan, Mongolia, Morocco, Oman, Pakistan, China, Poland, Portugal, Spain, Sweden, Taiwan, Turkey, the UK, and the USA.

June 2020

Ilias Maglogiannis
Plamen Angelov
John Macintyre
Lazaros Iliadis
Stefanos Kolias
Elias Pimenidis

Organization

Executive Committee

General Chairs

Ilias Maglogiannis	University of Piraeus, Greece (President of the IFIP WG12.5)
Plamen Angelov	University of Lancaster, UK
John Macintyre	University of Sunderland, UK (Dean of the Faculty of Applied Sciences and Pro Vice Chancellor of the University of Sunderland)

Program Chairs

Lazaros Iliadis	Democritus University of Thrace, Greece
Stefanos Kolias	University of Lincoln, UK

Advisory Chairs

Andreas Stafylopatis	Technical University of Athens, Greece
Vincenzo Piuri	University of Milan, Italy (IEEE Fellow (2001), IEEE Society/Council active memberships/services: CIS, ComSoc, CS, CSS, EMBS, IMS, PES, PHOS, RAS, SMCS, SPS, BIOMC, SYSC, WIE)

Honorary Chair

Robert Kozma	University of Memphis, USA

Liaison Co-chairs

Ioannis Kompatsiaris	IPTIL Research Institute, Greece
Ioannis Chochliouros	Hellenic Telecommunications Organization, Greece

Workshop Chairs

Christos Makris	University of Patras, Greece
Phivos Mylonas	Ionian University, Greece
Spyros Sioutas	University of Patras, Greece
Katia Kermanidou	Ionian University, Greece

Publication and Publicity Chairs

Antonis Papaleonidas	Democritus University of Thrace, Greece
Konstantinos Demertzis	Democritus University of Thrace, Greece
George Tsekouras	University of the Aegean, Greece

Special Sessions Chairs

Panagiotis Papapetrou	Stockholm University, Sweeden
Georgios Paliouras	National Center for Scientific Research NSCR Demokritos, Greece

Steering Committee Chairs

Ilias Maglogiannis	University of Piraeus, Greece
Plamen Angelov	University of Lancaster, UK
Lazaros Iliadis	Democritus University of Thrace, Greece

Program Committee

Michel Aldanondo	Toulouse University, IMT Mines Albi, France
Georgios Alexandridis	University of the Aegean, Greece
Serafín Alonso Castro	University of Leon, Spain
Ioannis Anagnostopoulos	University of Thessaly, Greece
Costin Badica	University of Craiova, Romania
Giacomo Boracchi	Politecnico di Milano, Italy
Ivo Bukovsky	Czech Technical University in Prague, Czech Republic
George Caridakis	University of the Aegean, Greece
Francisco Carvalho	Polytechnic Institute of Tomar, Portugal
Ioannis Chamodrakas	National and Kapodistrian University of Athens, Greece
Adriana Coroiu	Babeş-Bolyai University, Romania
Kostantinos Delibasis	University of Thessaly, Greece
Konstantinos Demertzis	Democritus University of Thrace, Greece
Sergey Dolenko	Lomonosov Moscow State University, Russia
Georgios Drakopoulos	Ionian University, Greece
Mauro Gaggero	National Research Council of Italy, Italy
Ignazio Gallo	University of Insubria, Italy
Angelo Genovese	Università degli Studi di Milano, Italy
Spiros Georgakopoulos	University of Thessaly, Greece
Eleonora Giunchiglia	Oxford University, UK
Foteini Grivokostopoulou	University of Patras, Greece
Peter Hajek	University of Pardubice, Czech Republic
Giannis Haralabopoulos	University of Nottingham, UK
Ioannis Hatzilygeroudis	University of Patras, Greece
Nantia Iakovidou	King's College London, UK
Lazaros Iliadis	Democritus University of Thrace, Greece
Zhu Jin	University of Cambridge, UK
Jacek Kabziński	Lodz University of Technology, Poland
Andreas Kanavos	University of Patras, Greece
Stelios Kapetanakis	University of Brighton, UK
Petros Kefalas	CITY College, International Faculty of the University of Sheffield, Greece

Katia Kermanidis	Ionio University, Greece
Niki Kiriakidou	University of Patras, Greece
Giannis Kokkinos	University of Macedonia, Greece
Petia Koprinkova-Hristova	Bulgarian Academy of Sciences, Bulgaria
Athanasios Koutras	Technical Educational Institute of Western Greece, Greece
Paul Krause	University of Surrey, UK
Florin Leon	Technical University of Iasi, Romania
Aristidis Likas	University of Ioannina, Greece
Ioannis Livieris	University of Patras, Greece
Doina Logofătu	Frankfurt University of Applied Sciences, Germany
Ilias Maglogiannis	University of Piraeus, Greece
Goerge Magoulas	Birkbeck College, University of London, UK
Christos Makris	University of Patras, Greece
Mario Malcangi	University of Milan, Italy
Francesco Marceloni	University of Pisa, Italy
Giovanna Maria Dimitri	University of Cambridge, UK and University of Siena, Italy
Nikolaos Mitianoudis	Democritus University of Thrace, Greece
Antonio Moran	University of Leon, Spain
Konstantinos Moutselos	University of Piraeus, Greece
Phivos Mylonas	Ionio University, Greece
Stefanos Nikiforos	Ionio University, Greece
Stavros Ntalampiras	University of Milan, Italy
Mihaela Oprea	Petroleum-Gas University of Ploieşti, Romania
Ioannis P. Chochliouros	Hellenic Telecommunications Organization, Greece
Basil Papadopoulos	Democritus University of Thrace, Greece
Vaios Papaioannou	University of Patras, Greece
Antonis Papaleonidas	Democritus University of Thrace, Greece
Daniel Pérez López	University of Leon, Spain
Isidoros Perikos	University of Patras, Greece
Elias Pimenidis	University of the West of England, UK
Panagiotis Pintelas	University of Patras, Greece
Nikolaos Polatidis	University of Brighton, UK
Bernardete Ribeiro	University of Coimbra, Portugal
Leonardo Rundo	University of Cambridge, UK
Alexander Ryjov	Lomonosov Moscow State University, Russia
Simone Scardapane	Sapienza University, Italy
Evaggelos Spyrou	National Center for Scientific Research – Demokritos, Greece
Antonio Staiano	University of Naples Parthenope, Italy
Andrea Tangherloni	University of Cambridge, UK
Azevedo Tiago	University of Cambridge, UK
Francesco Trovò	Polytecnico di Milano, Italy

Nicolas Tsapatsoulis   Cyprus University of Technology, Cyprus
Petra Vidnerová   Czech Academy of Sciences, Czech Republic
Paulo Vitor de Campos   CEFET-MG, Brazil
  Souza
Gerasimos Vonitsanos   University of Patras, Greece

Abstracts of Invited Talks

Smartphone, Parkinson's and Depression: A New AI-Based Prognostic Perspective

Leontios Hadjileontiadis

Khalifa University of Science and Technology, UAE, and Aristotle
University of Thessaloniki, Greece
leontios@auth.gr

Abstract. Machine Learning (ML) is a branch of Artificial Intelligence (AI) based on the idea that systems can learn from data, identify patterns, and make decisions with minimal human intervention. While many ML algorithms have been around for a long time, the ability to automatically apply complex mathematical calculations to big data – over and over, faster and faster, deeper and deeper – is a recent development, leading to the realization of the so called Deep Learning (DL). The latter has an intuitive capability that is similar to biological brains. It is able to handle the inherent unpredictability and fuzziness of the natural world. In this keynote, the main aspects of ML and DL will be presented, and the focus will be placed in the way they are used to shed light upon the Human Behavioral Modeling. In this vein, AI-based approaches will be presented for identifying fine-motor skills deterioration due to early Parkinson's and depression symptoms reflected in the keystroke dynamics, while interacting with a smartphone. These approaches provide a new and unobtrusive way for gathering and analyzing dense sampled big data, contributing to further understanding disease symptoms at a very early stage, guiding personalized and targeted interventions that sustain the patient's quality of life.

Deep Learning, Knowledge Representation and Transfer with Brain-Inspired Spiking Neural Network Architectures

Nikola Kasabov

Auckland University of Technology, New Zealand
nkasabov@aut.ac.nz

Abstract. This talk argues and demonstrates that the third generation of artificial neural networks, the spiking neural networks (SNN), can be used to design brain-inspired architectures that are not only capable of deep learning of temporal or spatio-temporal data, but also enabling the extraction of deep knowledge representation from the learned data. Similarly to how the brain learns time-space data, these SNN models do not need to be restricted in number of layers, neurons in each layer, etc. When a SNN model is designed to follow a brain template, knowledge transfer between humans and machines in both directions becomes possible through the creation of brain-inspired Brain-Computer Interfaces (BCI). The presented approach is illustrated on an exemplar SNN architecture NeuCube (free software and open source available from www.kedri.aut.ac.nz/neucube) and case studies of brain and environmental data modeling and knowledge representation using incremental and transfer learning algorithms These include predictive modeling of EEG and fMRI data measuring cognitive processes and response to treatment, AD prediction, BCI for neuro-rehabilitation, human-human and human-VR communication, hyper-scanning, and others.

The Rise of Artificial Intelligence for Earth Observation (AI4EO)

Pierre Philippe Mathieu

European Space Agency (ESA), Head of the Philab (Φ Lab), Explore Office
at the European Space Agency in ESRIN, Frascati, Italy

Abstract. The world of Earth Observation (EO) is rapidly changing as a result
of exponential advances in sensor and digital technologies.

The speed of change has no historical precedent. Recent decades have witnessed extraordinary developments in ICT, including the Internet, cloud computing and storage, which have all led to radically new ways to collect, distribute
and analyse data about our planet. This digital revolution is also accompanied by
a sensing revolution that provides an unprecedented amount of data on the state
of our planet and its changes.

Europe leads this sensing revolution in space through the Copernicus initiative and the corresponding development of a family of Sentinel missions. This
has enabled the global monitoring of our planet across the whole electromag-
netic spectrum on an operational and sustained basis. In addition, a new trend,
referred to as "New Space", is now rapidly emerging through the increasing
commoditization and commercialization of space.

These new global data sets from space lead to a far more comprehensive
picture of our planet. This picture is now even more refined via data from
billions of smart and inter-connected sensors referred to as the Internet of
Things. Such streams of dynamic data on our planet offer new possibilities for
scientists to advance our understanding of how the ocean, atmosphere, land and
cryosphere operate and interact as part on an integrated Earth System. It also
represents new opportunities for entrepreneurs to turn big data into new types of
information services.

However, the emergence of big data creates new opportunities but also new
challenges for scientists, business, data and software providers to make sense
of the vast and diverse amount of data by capitalizing on powerful techniques
such as Artificial Intelligence (AI). Until recently AI was mainly a restricted
field occupied by experts and scientists, but today it is routinely used in
everyday life without us even noticing it, in applications ranging from recommendation engines, language services, face recognition and autonomous vehicles.

The application of AI to EO data is just at its infancy, remaining mainly
concentrated on computer vision applications with Very High-Resolution
satellite imagery, while there are certainly many areas of Earth Science and big
data mining/fusion, which could increasingly benefit from AI, leading to entire
new types of value chain, scientific knowledge and innovative EO services.

This talk will present some of the ESA research/application activities and
partnerships in the AI4EO field, inviting you to stimulate new ideas and collaboration to make the most of the big data and AI revolutions.

Image Fusion Based on Deep Learning

Xiao-Jun Wu

Jiangnan University, China
wu_xiaojun@jiangnan.edu.cn

Abstract. Deep Learning (DL) has found very successful applications in numerous different domains with impressive results. Image Fusion (IMF) algorithms based on DL and their applications will be presented thoroughly in this keynote lecture. Initially, a brief introductory overview of both concepts will be given. Then, IMF employing DL will be presented in terms of pixel, feature, and decision level respectively. Furthermore, a DL inspired approach called MDLatLRR which is a general approach to image decomposition will be introduced for IMF. A comprehensive analysis of DL models will be offered and their typical applications will be discussed, including Image Quality Enhancement, Facial Landmark Detection, Object Tracking, Multi-Modal Image Fusion, Video Style Transformation, and Deep Fake of Facial Images, respectively.

Contents – Part I

Image Processing

Learning Algorithms

Neural Network Modeling

Object Tracking/Object Detection Systems

Ontologies/AI

Sentiment Analysis/Recommender Systems

Contents – Part II

Medical-Health Systems

Natural Language

Classification

An Adaptive Approach on Credit Card Fraud Detection Using Transaction Aggregation and Word Embeddings

Ali Yeşilkanat$^{(\boxtimes)}$, Barış Bayram, Bilge Köroğlu, and Seçil Arslan

Applied AI and R&D Department, Yapi Kredi Technology, Istanbul, Turkey
ali.yesilkanat@ykteknoloji.com.tr

Abstract. Due to the surge of interest in online retailing, the use of credit cards has been rapidly expanded in recent years. Stealing the card details to perform online transactions, which is called fraud, has also seen more frequently. Preventive solutions and instant fraud detection methods are widely studied due to critical financial losses in many industries. In this work, a Gradient Boosting Tree (GBT) model for the real-time detection of credit card frauds on the streaming Card-Not-Present (CNP) transactions is investigated with the use of different attributes of card transactions. Numerical, hand-crafted numerical, categorical and textual attributes are combined to form a feature vector to be used as a training instance. One of the contributions of this work is to employ transaction aggregation for the categorical values and inclusion of vectors from a character level word embedding model which is trained on the merchant names of the transactions. The other contribution is introducing a new strategy for training dataset generation employing the sliding window approach in a given time frame to adapt to the changes on the trends of fraudulent transactions. In the experiments, the feature engineering strategy and the automated training set generation methodology are evaluated on the real credit card transactions.

Keywords: Fraud detection · Imbalanced data · Concept drift · Decision system · Character-level word embedding

1 Introduction

With the advances in information technology and electronic commerce, the use of credit cards raises in recent years. According to the fifth report of The Single European Payments Area (SEPA) report in 2016 [1], the total value of the fraudulent transactions was €1.8 billion, which 73% of this value comes from Card-Not-Present (CNP) payments. Comparing to the ATM and POS frauds, CNP fraud is the one which increases most by 2.1% over four years. Therefore, CNP fraud is a considerably serious problem in the credit card business.

© IFIP International Federation for Information Processing 2020
Published by Springer Nature Switzerland AG 2020
I. Maglogiannis et al. (Eds.): AIAI 2020, IFIP AICT 583, pp. 3–14, 2020.
https://doi.org/10.1007/978-3-030-49161-1_1

While developing a fraud detection model, numerous problems are encountered such as:

- generation of training set from hundreds of millions of imbalanced transactions,
- selection of the most appropriate feature combination due to indistinguishable features between fraud and non-fraud instances,
- skewness of the data and cost of the false-positive samples,
- durability of the model being affected by trend variation called concept drift caused by changes on behaviors of the customers and fraudsters.

Data imbalance is defined as possessing an unequal distribution of classes within a dataset. The transactions from credit cards commonly form an imbalanced dataset, including very few fraud transactions comparing to legitimate ones. In order to solve this problem, various methods are provided, namely under-sampling [15] and oversampling [4]; however, these techniques are problematic because the dataset is massively imbalanced and instances of the dataset individually carry important information (such as transactions belonging to the same credit card). In this work, we introduce a card-based equalization method on fraud and non-fraud cards to form a training dataset by incorporating all transactions of the sampled cards to provide a solution to the mentioned problems.

Properties of the fraudulent credit card transactions usually alter over time. The main reason for this is that fraudsters try to bypass fraud detection systems. Moreover, purchasing trends of online markets vary over time, such as the establishment of a new payment channel, new merchant, or merchant category. Those trend shifts are called concept drift, where the distribution of a data stream, is not stationary. The conventional fraud detection systems are developed using previous fraud transactions and cannot adapt to concept drift. In this work, the proposed system is retrained by itself over time to prevent concept drift adaptively. In this study, we propose a sliding window-based automated training dataset generation technique to solve this issue.

The transactions of a credit card can be performed from various sources, named terminals. These terminals transfer the data related to the transaction to the owner bank of the card in ISO 8538 Standard [9]. Generally, the sender terminal feeds the data in a structural and standardized format; yet, some attributes on the transactions are not sent correctly which may cause problems for fraud detection systems. For instance, the same merchant name are retrieved differently from the related property of the transaction, like, *FACEBOOK* or *FACEBK*. Therefore, a character-level word embedding is required to map the name to a vector of real numbers. In this way, the name of the merchant can be used as a distinctive feature to detect fraudulent behaviour.

The online transactions are often real-time events, in which a fraud decision system has to determine if it is a fraud or not in milliseconds. In this work, with the purpose of designing a fast and reliable credit card fraud detection system, a Gradient Boosting Tree (GBT) based approach is developed with different types of features. Each transaction includes the typical numeric, hand-crafted numeric, categorical, and textual features.

The rest of the paper is organized as follows: In Sect. 2, the related work is discussed. The modules of the proposed credit card fraud detection system and the processing modules are described in Sect. 3. Section 4 is devoted to our results of the experiments to evaluate the importance of feature engineering and automated training set generation approach. Finally, the results are given in Sect. 5.

2 Related Work

For the detection of credit card frauds, various feature engineering strategies mostly based on an aggregation of transactions have been proposed in the literature. Also, Gradient Boosting Tree (GBT) has been investigated and compared with the other state-of-the-art algorithms.

Feature Engineering. To improve the performance of credit card fraud detection, Jha et al. [7] proposed a transaction aggregation strategy to derive certain aggregated features such as the total amount or the number of transactions initiated on the same day with the same merchant, currency or country for capturing the buying behaviors of customers. The features are created on real data of credit card transactions. In the study of Whitrow et al. [14], using a transaction aggregation process resulting in more robust to the problem of population drift, several features are derived at different aggregation periods. Afterwards, for the detection of credit card frauds, different algorithms are performed on transactional features including the derived ones. Also, a novel approach is proposed in this work [12] for the generation of a cardholder's profile by extracting the aggregated daily amounts of the cardholder's transactions. Comparison of the decision tree, artificial neural network, and logistic regression approaches for detecting credit card fraud transaction is studied by [13] on the combination of the raw features and eight aggregated features such as the total and average amount, the number of transactions and the failure number of transactions in the same day and the last five days, showing that neural network and logistic regression models over-perform decision tree. Bahnsen et al. [2,3] presented a transaction strategy for the capturing of the patterns for the periodic behavior of a transaction based on von Mises distribution and spending behavior belonging to a customer by exploiting the previous transactions of the customer. Also, it is revealed that spending patterns and time features improve the performance of the detection of fraudulent transactions. Moreover, Lim et al. [10] proposed a novel transaction-level based aggregation strategy using conditional weighted aggregation on the previous transactions. In this work, two weighting functions using the number of transactions and time are performed to assign more weights to the transactions.

In addition, the contribution of the aggregation on the detection of credit card frauds is investigated using various time-series based algorithms. Jurgovsky et al. [8] have used Long Short-Term Memory (LSTM) on the aggregated features extracted from the cardholder-present transactions. In another time-series based approach [11], Hidden Markov Model (HMM) is employed to extract descriptive aggregated features from a transaction sequence of a customer. For each transaction in the sequence, a likelihood value based on the previous transaction is computed, and then, this likelihood values are used as additional features to generate a Random Forest model.

Gradient Boosting Trees. Fang et al. proposed a Light Gradient Boosting Machine (LightGBM) for detecting credit card frauds [6]. In the work of [5], for the detection of credit card frauds, the boosting algorithms, Adaboost, Gradient Boost, and eXtreme Gradient Boosting (XGBoost) were implemented to evaluate the detection performances. It was found out that the XGBoost method outperforms the other algorithms.

3   Proposed Fraud Detection System

The proposed system for the detection of credit card frauds is composed of two main modules; offline training and real-time detection as shown in Fig. 1. There is a feature engineering step common in both of the components. The offline training phase covers the automatic generation of the training set to be used in the generation of a model for fraud detection. In the real-time detection phase, the feature vector is extracted from streaming transaction to decide whether it is fraudulent or not.

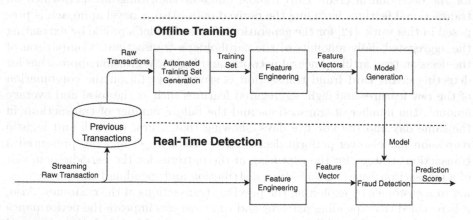

Fig. 1. The flowchart of the proposed system with offline training and credit card fraud detection in real-time.

3.1   Feature Engineering

A feature engineering process shown in Fig. 2 is required to improve the performance of model under the imbalanced class distributions, the noisy labels and features, skewness of the data, and overlapping transactions existing in both of the classes. Making combination of the different types of features, selection among the distinctive features, and generation of the additional hand-crafted features are used in feature engineering phase on the model generation component of our time-constrained fraud detection system.

Before the combination, the encoding of categorical features, generation of aggregated features, and extraction of word representations for the textual data are performed in the offline training and real-time detection stages. The categorical features are encoded in which a numeric value is set to a distinct categorical value. In addition, the vectors from the word embedding model of merchant names are included to compose the training instance for offline training component and to query the model on real time detection component.

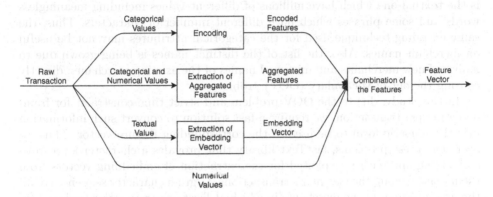

Fig. 2. The overview of the feature engineering process employed in the initial phases of the training and detection modules.

3.1.1 Encoding of Categorical Attributes

A few algorithms have been proposed to handle the data with categorical features. Therefore, an encoding process is required to transform them into numerical features. The values of some categorical attributes belonging to the transactions such as merchant category code, country code, etc. are directly encoded into a set of discrete integer values 0 through k where k is the number of unique categorical values.

3.1.2 Extraction of Aggregated Features

The properties of a credit card transaction do not represent the customer's financial behavior. In order to capture purchasing patterns to improve the model performance, an aggregation strategy is applied on raw transactional features such as currency, amount, merchant category. Utilizing the previous transactions of a credit card in a predefined time window to project the account's e-marketing behavior is the key point of our feature aggregation approach. Seeking the transaction history of a card in a time frame is needed because the recent transactions include much recent information about the customer. The day before the current day is decided as fixed time range. Various aggregated features per card, like the total amount of previous transactions, the total number of transactions which have the same merchant category code, total count of the same currency and country, etc., are computed in real-time from the transactions performed in the day before the current day.

3.1.3 Training a Word Embedding Model

For the use of the typical numeric, hand-crafted numeric, categorical, and textual attributes together, a set of a newly composed feature vectors are constructed to train a GBT model. Nevertheless, the attribute corresponding to merchant name is the textual one which have millions of different values including meaningless words and some phrases which have different number of characters. Thus, the same encoding technique used for the categorical attributes may not be useful on merchant names. Also, the list of the distinct names is being grown due to names of new merchants and mistyped names of the existing merchants, directly causing the out-of-vocabulary (OOV) problem.

In this study, due to the OOV problem and strict time constraint for fraud detection per transaction, we require a fast solution to convert such information into the numeric form to be used in the corresponding feature vector. Thus, to overcome these problems, fastText library that provides a character-level word embedding approach is exploited for the extraction of embedding vectors from the names. Among the last year's transactions, trigram character sequences of all the unique names of the merchants in which at least one transaction and contain only alphabetical characters, are considered to include the model. We train an embedding model on a corpus with millions of these filtered unique names. The model provides a 16-D embedding vector generated for each merchant name. The vector for the name of an incoming transaction is basically computed by summing all vectors extracted for each trigram sequence in the name, and included into the other features. Due to the restriction for the prediction time, retraining of the fastText model on the merchant names, including the new ones, is not applicable.

It is required to estimate how the embedding vector should be efficiently utilized. Combining the entire vector with the other features may make the name attribute dominate the other transactional features. To select more distinctive features and also reduce the dimensionality, Principled Component Analysis (PCA) is performed on the vectors of the names in training set like in Fig. 3. This technique removes the correlation and preserves the orthogonality between the word representations. Using PCA, the 16-D vectors are projected onto 4-D subspace determined by experiments. It is observed that the PCA based reduction prevents the dominancy of the vector on the other transactional features, enhances the fraud detection performance, and decreases the prediction time.

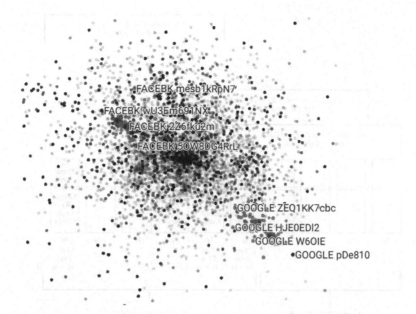

Fig. 3. A set of embedding vectors projected onto 2-D space generated on a set of merchant names in the training set.

3.2 Training of Credit Card Fraud Detection Model

In this study, the training set is constructed within several stages to create a data set in which classifiers can have stability against class skewness problems. It is desired to equalize fraudulent and legitimate credit card counts and to make sure that the transactions of each credit card covers the equal time range.

We identify credit cards as a fraudulent card if it has at least one fraud transaction in a given time range. If the credit card has no fraud transaction, then we denote it as a legitimate card. Instead of equalizing fraud and non-fraud transaction count, such as undersampling, we take the number of training instances the same for fraud and non-fraud ones. Although there is still massive class imbalance; however, we consider detecting fraud credit cards instead of the transaction more important.

The training set is constructed in a period of T, with m number of consecutive months. The last C number of months of this period is defined as the card selection zone, and the beginning B months denotes the transaction buffer zone. In Fig. 4, m is chosen as 10 to define the number of months in the period T. In the first place, containing the same count of the fraudulent credit cards, K, legitimate credit cards are sampled for each t_i month in the C card selection zone. This ensures that the dataset contains the same number of training instances for fraud and legitimate credit card transactions. Following the card selection process, all transactions of the selected cards in t_i, are obtained from the range of t_{i-n}, where n is the sliding window parameter. This operation needs reservation

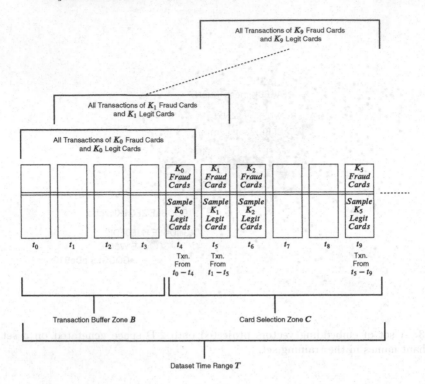

Fig. 4. The illustration of the automatic training set generation using transaction (txn) between T_i and T_{i+4}, where T denotes the month. The sliding window n parameter is chosen as 4 for this figure.

of the first n months only to fetch transactions of the t_{i+n} cards. As a result, the training set makes each card to have transactions from $(n + 1)$ months.

In addition, after a new training set is constructed, a PCA model for the embedding vectors of the merchant names is also updated using the vector representations of the merchant names. However, the same fastText model is utilized since the update of the word-embedding model like the fraud detection model may not enhance the vector representation.

4   Experiments

4.1   Experimental Setup

The experiments were performed using two training sets and test sets incorporating credit card transactions in a private bank. Training sets are formed using the approach described in Table 4. These two training sets contain transactions from 2018 October to 2019 July (Oct.'18–Jul.'19) and 2018 November to 2019 August (Nov.'18–Aug.'19), respectively. For the test sets, we use all CNP transactions in the next month for each training set that are August (Aug.'19) and

September (Sep.'19). Also, 30% of each training set has been utilized as a validation set, to be used in the training of the detection model. More details about the datasets can be found in Table 1.

The proposed system for credit card fraud detection was developed in Python 3.6.8. For the GBT model generation, XGBoost library 0.71 presenting an efficient Gradient Boosted Tree (GBT) implementation is used on these datasets. Also, Redis, an in-memory data structure store based on a server-client model with TCP sockets, is utilized for the real-time transaction aggregation.

4.2   Evaluation Metrics

Experiments are evaluated by the following metrics; the False-Positive Rate (FPR), recall, precision, and Area Under Curve (AUC). We also calculated the Equal Error Rate (EER), which determines the threshold values for its false acceptance rate and its false rejection rate when the rates are equal. Using the determined threshold values from EER calculation, we evaluated the performance of the credit card fraud detection model. Moreover, we experimentally found that 0.3 is decided as a fixed FPR to obtain the optimum threshold value.

Table 1. Overview of the training and test sets (Transactions = Txn).

Overview	Training set		Test set	
	Oct.'18–Jul.'19	Nov.'18–Aug.'19	Aug.'19	Sep.'19
# of txn.	1,3 M	1,4 M	20,2 M	21,3 M
# of cards	31,158	49,002	3 M	3,1 M
# of fraud txn.	61,130	52,610	6,767	7,045
# of fraud cards	15,579	24,501	3,626	3,997

4.3   Experiments of Feature Engineering

In the experiments regarding feature engineering process, only encoded features, encoded with aggregated ones, and all of them including word embedding features have been used to train a GBT model. In Table 2, the evaluation is shown on the test set Aug.'19 composed of all the CNP transactions on August, 2019. Also, on the test set, Sep.'19 composed of all the CNP transactions in September, 2019, the fraud detection results of these three models trained using three different feature sets are shown in the Table 3.

4.4   Experiments of Automated Training Set Generation

Two different datasets are formed by sliding window approach using transactions in Oct.'18–Jul.'19 & Nov.'18–Aug.'19. For each set, a brand new GBT model

Table 2. Success measurement results of our fraud detection system trained on Oct.'18–Jul.'19 training set and tested on Aug.'19 test dataset.

Features	AUC	Fixed FPR		EER			
		FPR	Recall	FPR	Recall	Error	Threshold
Encoded	0.947	0.3	0.962	0.120	0.879	12.08	0.064
Agg. + Encoded	0.951	0.3	0.962	0.115	0.883	11.68	0.056
Emb. + Agg. + Encoded	0.958	0.3	0.974	0.110	0.889	11.02	0.055

Table 3. Success measurement results of the fraud detection models which are trained on Nov.'18–Aug.'19 training sets and tested on Sep.'19 test set.

Features	AUC	Fixed FPR		EER			
		FPR	Recall	FPR	Recall	Error	Threshold
Encoded	0.960	0.3	0.984	0.104	0.895	10.48	0.018
Agg. + Encoded	0.964	0.3	0.960	0.099	0.900	9.90	0.017
Emb. + Agg. + Encoded	0.968	0.3	0.989	0.091	0.908	9.15	0.017

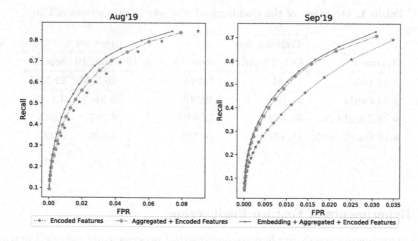

Fig. 5. Recall-FPR curve of the detection models trained on these features for test set composed of all CNP transactions in August (figure on the left), and test set composed of all CNP transactions in September (figure on the right)

Table 4. Results of the models trained on the sets generated by sliding window

Training sets	AUC	Fixed FPR		EER			
		FPR	Recall	FPR	Recall	Error	Threshold
Oct.'18–Jul.'19	0.968	0.3	0.989	0.091	0.908	14.16	0.038
Nov.'18–Aug.'19	0.940	0.3	0.960	0.141	0.858	9.15	0.017

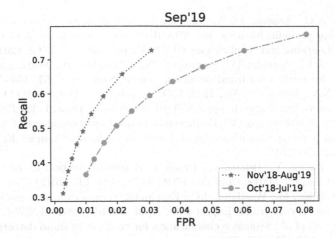

Fig. 6. The contribution of the sliding window based approach being employed for training set generation.

is trained and their detection performance are evaluated on the same test set Sep.'19 (Fig. 5).

Figure 6 and Table 4 reveals that, sliding the training set enhances the fraud detection performance in terms of AUC by 0.028%, and in fixed 0.3 FPR, Recall is increased by 0.029%. Also, we realized that joining recent transactions and decreasing more distant transactions to the training set increases the success of our credit card fraud detection model.

5 Conclusion

In this work, for the real-time detection of credit card fraud, a new approach is proposed for training dataset construction to make usable and valuable of different types of attributes of a transaction by combining numerical, hand-crafted numerical, categorical and textual features. Also, a character-level word embedding method is utilized to generate embedding vector for each merchant name. Also, there is a major problem affecting the durability of the model. Therefore, automatic generation of a transaction dataset has been developed to extract feature vectors used as training set to generate a Gradient Boosting Tree (GBT) model for the detection of the fraud transactions. In the future work, new types of aggregated features will be created and an ensemble method will be developed by combining of a number of models trained on the transactions of long- and short-term periods.

References

1. Fifth report on card fraud (2018). https://www.ecb.europa.eu/pub/cardfraud/html/ecb.cardfraudreport201809.en.html. Accessed 03 Feb 2020

2. Bahnsen, A.C., Aouada, D., Stojanovic, A., Ottersten, B.: Detecting credit card fraud using periodic features. In: 2015 IEEE 14th International Conference on Machine Learning and Applications (ICMLA), pp. 208–213. IEEE (2015)
3. Bahnsen, A.C., Aouada, D., Stojanovic, A., Ottersten, B.: Feature engineering strategies for credit card fraud detection. Expert Syst. Appl. **51**, 134–142 (2016)
4. Chawla, N.V., Bowyer, K.W., Hall, L.O., Kegelmeyer, W.P.: SMOTE: synthetic minority over-sampling technique. J. Artif. Intell. Res. **16**, 321–357 (2002)
5. Divakar, K., Chitharanjan, K.: Performance evaluation of credit card fraud transactions using boosting algorithms. Int. J. Electron. Commun. Comput. Eng. IJECCE **10**(6), 262–270 (2019)
6. Fang, Y., Zhang, Y., Huang, C.: Credit card fraud detection based on machine learning. Comput. Mater. Continua CMC **61**(1), 185–195 (2019)
7. Jha, S., Guillen, M., Westland, J.C.: Employing transaction aggregation strategy to detect credit card fraud. Expert Syst. Appl. **39**(16), 12650–12657 (2012)
8. Jurgovsky, J., et al.: Sequence classification for credit-card fraud detection. Expert Syst. Appl. **100**, 234–245 (2018)
9. Korman, B.R., Bergman, D.J.: Multi-transactional architecture. US Patent 6,308,887, 30 Oct 2001
10. Lim, W.-Y., Sachan, A., Thing, V.: Conditional weighted transaction aggregation for credit card fraud detection. In: Peterson, G., Shenoi, S. (eds.) DigitalForensics 2014. IAICT, vol. 433, pp. 3–16. Springer, Heidelberg (2014). https://doi.org/10.1007/978-3-662-44952-3_1
11. Lucas, Y., et al.: Multiple perspectives HMM-based feature engineering for credit card fraud detection. In: Proceedings of the 34th ACM/SIGAPP Symposium on Applied Computing, pp. 1359–1361 (2019)
12. Seyedhossein, L., Hashemi, M.R.: Mining information from credit card time series for timelier fraud detection. In: 2010 5th International Symposium on Telecommunications, pp. 619–624. IEEE (2010)
13. Shen, A., Tong, R., Deng, Y.: Application of classification models on credit card fraud detection. In: 2007 International Conference on Service Systems and Service Management, pp. 1–4. IEEE (2007)
14. Whitrow, C., Hand, D.J., Juszczak, P., Weston, D., Adams, N.M.: Transaction aggregation as a strategy for credit card fraud detection. Data Min. Knowl. Disc. **18**(1), 30–55 (2009)
15. Yen, S.J., Lee, Y.S.: Under-sampling approaches for improving prediction of the minority class in an imbalanced dataset. In: Huang, D.S., Li, K., Irwin, G.W. (eds.) Intelligent Control and Automation. LNCIS, vol. 344, pp. 731–740. Springer, Heidelberg (2006). https://doi.org/10.1007/978-3-540-37256-1_89

Boosted Ensemble Learning for Anomaly Detection in 5G RAN

Tobias Sundqvist[1]([⊠])[iD], Monowar H. Bhuyan[1]([⊠])[iD], Johan Forsman[2]([⊠]),
and Erik Elmroth[1]([⊠])[iD]

[1] Department of Computing Science, Umeå University, 901 87 Umeå, Sweden
{sundqtob,monowar,elmroth}@cs.umu.se
[2] TietoEvry, 907 36 Umeå, Sweden
johan.forsman@tietoevry.com

Abstract. The emerging 5G networks promises more throughput, faster, and more reliable services, but as the network complexity and dynamics increases, it becomes more difficult to troubleshoot the systems. Vendors are spending a lot of time and effort on early anomaly detection in their development cycle and majority of the time is spent on manually analyzing system logs. While main research in anomaly detection uses performance metrics, anomaly detection using functional behaviour is still lacking in depth analysis. In this paper we show how a boosted ensemble of Long Short Term Memory classifiers can detect anomalies in the 5G Radio Access Network system logs. Acquiring system logs from a live 5G network is difficult due to confidentiality issues, live network disturbance, and problems to repeat scenarios. Therefore, we perform our evaluation on logs from a 5G test bed that simulate realistic traffic in a city. Our ensemble learns the functional behaviour of an application by training on logs from normal execution time. It can then detect deviations from normal behaviour and also be retrained on false positive cases found during validation. Anomaly detection in RAN shows that our ensemble called BoostLog, outperforms a single LSTM classifier and further testing on HDFS logs confirms that BoostLog also can be used in other domains. Instead of using domain experts to manually analyse system logs, BoostLog can be used by less experienced trouble shooters to automatically detect anomalies faster and more reliable.

Keywords: Anomaly detection · AdaBoost · LSTM · Radio Access Network (RAN) · 5G · System logs · Functional area

1 Introduction

The 5G Radio Access Network (RAN) is a complex large-scale system where parts are both virtualized and distributed. Detection and diagnostics of system issues (e.g., faults, outages, degradation) are of great importance due to end-user experience and brand reputation. A key to success for companies is to find

© IFIP International Federation for Information Processing 2020
Published by Springer Nature Switzerland AG 2020
I. Maglogiannis et al. (Eds.): AIAI 2020, IFIP AICT 583, pp. 15–30, 2020.
https://doi.org/10.1007/978-3-030-49161-1_2

anomalies fast and early in the development cycle [8], this reduces the cost and time to market when new features are developed. Finding a way to speed up analytics and make it easier to analyze RAN traffic would therefore be very useful since it would both make the system more reliable and also earn money for the company. When it comes to anomaly detection in RAN, the main research has focused on using performance counters, e.g., network traffic and Key Performance Indicators [2,11,16,23]. A foreseen area is the functional domain where system logs can be used to determine the behaviour of the applications. System logs are today mainly used to troubleshoot systems when testing or when live traffic fails, in those cases it is very time consuming to manually analyze logs since they are huge.

In this paper, we use machine learning to develop a system that will allow trouble shooters in the Telecom area to find anomalies faster and with less system knowledge. By tracking and separating call sessions, we turn system logs into sequential time series that is suitable for anomaly detection. A model that has shown to be very useful in predicting time series is the Long Short Term Memory (LSTM) model [4,14,17,22]. One advantage of the LSTM model is that it can learn and remember long term dependencies which is useful when long sequences of events are analysed. By training on sequences of events in system logs, the LSTM model can also grasp the application behaviour without having any deep knowledge of the system. Recent research use the predictions made by the LSTM model to detect anomalies [5,6,18]. The common idea in these articles is that a LSTM model is used to detect single deviations from normal behaviour. The problem with these approaches are that they rely on single deviations and cannot detect multiple small deviations from the normal behaviour. Another aspect is that these anomaly detection methods are performed on numerical data, when it comes to analysing functional behaviour using system logs very little has been done using machine learning. In a recent survey of anomaly detection in system logs [1], the most matured method DeepLog [9], also the LSTM model and it shows the potential of the technique. In large scale, real time systems such as RAN, there are many concurrent threads that causes events to occur in different order from time to time. This makes it very difficult for a single LSTM model to predict the next state and as in the DeepLog case, one LSTM models are trained for each unique sequence. To overcome the real time challenges we propose an adaptive boosting ensemble [20], that uses LSTM models in a novel way to detect both single anomalies but also multiple deviations from the normal functional behaviour. The main contributions of this work are as follows:

- The design of an adaptive boosting ensemble, BoostLog, which outperforms single LSTM models and automatically detects anomalies in system logs.
- A novel, single LSTM classifier that can detect both multiple small deviations and also single larger ones in system logs.
- Evaluation of BoostLog and single LSTM classifier using real-time and benchmark data including the 5G test bed and HDFS log set.

This paper is organized as follows. Section 2 presents how machine learning can be used to find anomalies in system logs and why this is important in RAN.

In Sect. 3 we propose our method and in Sect. 4, we outline how our experiments were carried out. Finally our conclusions and how to proceed in future work are presented in Sect. 5.

2 Background

2.1 Recurrent Neural Network

Instead of using predefined log rules as in analysis tools such as Logsurfer [10], Swatch [13], or SEC [19], we want a model that can use a system log to learn the normal behavior without deep knowledge about the system. The state events in the log, are typically sequential and one model that has been proven to be useful for such time series is the Recurrent Neural Network (RNN) [9,27]. RNN has the advantage of possessing an internal memory and can remember its input. This has been found very useful in areas such as speech, text, financial data, audio, video, weather and much more. An extension of the RNN is the Long Short Term Memory (LSTM) model [14], which expands the memory of the RNN to be able to remember the inputs over a long period of time. Just like RNN, the LSTM also uses a chain of repeating modules, in which each module takes an input, produces an output, and also forward some information to the next module. Instead of having just one neural network in each module as in the normal RNN case, the LSTM uses four neural networks that interacts with each other. The main idea of the LSTM is to allow each module to select its input using an input gate, delete the information that is not important using a forget gate, and finally let the information in each module control the output using an output gate, see Fig. 1.

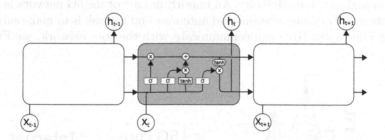

Fig. 1. LSTM architecture, X and h are input respectively output for a time step

2.2 Detecting Anomalies

The first part in detecting an anomaly using a LSTM model is to make a prediction of the next state, given previous data. The prediction is then compared with the state that followed and if they differs it can be classified as an anomaly, see also Sect. 3.1. The normal procedure when using time series to make predictions of future events is to re-frame the events into a supervised learning problem. By using previous events in the sequence, the next time step can be predicted and compared with the next event. The number of time steps to look back are very important when the data is sequential [9]. If several sequence chains are similar, more time steps are needed in order to predict the correct outcome and if too many time steps are used it will be difficult for the model to separate the sequence chains form each other. In Fig. 2, the importance of using many time steps, when predicting the events can be seen. In case A, only one time step is used and it is impossible to tell if 7, 8, or 9 is going to follow after event 6. In case B, two time steps are used and it is possible to tell the difference between 3-6-7 and 2-6-7 but it is still impossible to predict the 2-6-9 sequence. Only after looking at three time steps as in case C, we can tell the difference between all three sequences.

<table>
<tr><td>A</td><td>B</td><td>C</td></tr>
<tr><td>⑥→⑦</td><td>③→⑥→⑦</td><td>①→③→⑥→⑦</td></tr>
<tr><td>⑥→⑧</td><td>②→⑥→⑧</td><td>⓪→②→⑥→⑧</td></tr>
<tr><td>⑥→⑨</td><td>②→⑥→⑨</td><td>①→②→⑥→⑨</td></tr>
</table>

Fig. 2. Using previous time steps to predict the next, case A, B and C

2.3 RAN Metrics

The 5G generation of wireless telecom networks are designed to increase the speed, throughput, and reliability. An important part of the 5G network is RAN. It consists mainly of base stations and antennas and its task is to make sure that the User Equipment (UE) can communicate with the core network, see Fig. 3.

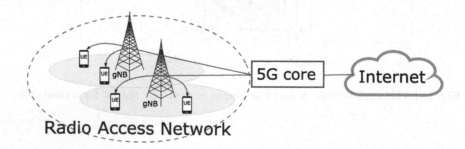

Fig. 3. 5G Radio Access Network (RAN)

In RAN there are three categories of metrics that are interesting to consider:

- Configurations, describes how the hardware and software are configured.
- Performance measurements, metrics that describes the performance of the system, often during a certain time period. Some performance measurements are used as quality indicators of the system and are often referred as Key Performance Indicators.
- System logs, contains time series of application specific events.

System logs contains most useful information when it comes to anomaly detection and root cause analytics. A faulty configuration can cause software to malfunction but logs are needed to find the configuration issues. Performance measurements can also hint that something is faulty during a certain time period but the logs are often needed to locate the fault. System logs are used to understand the behavior of the application, the vendor decides how often and what to log and it can contain start up sequences, important data, special events or alarms, etc. The best way to locate an anomaly in RAN is to use all three categories of data but we will limit our work in this article to look at how time series in system logs can be used to find the anomalies.

3   BoostLog: Proposed Method

3.1   LSTM Classifier

RAN system logs contains a mix of events from thousands of session calls that occurs in parallel of each other. Such concurrency causes a random distributions of sequence events and makes it impossible to learn the behavior for any model. As in [26] a better approach is to separate work flows or in this case sessions, into separate sequences and let the model train on each sequence instead of the whole list of events. In Fig. 4 it can be seen how metrics are collected from each Base Station and stored in the metric collector. The system log are then filtered and separated into UE sessions before the ensemble of LSTMs uses them for training or prediction.

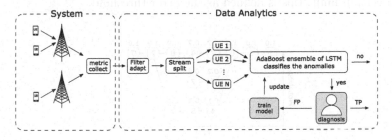

Fig. 4. BoostLog architecture

Even if it helps to separate the call sessions, all sessions will not look the same if sequence of events are analyzed. This is due to the nature of real time systems where several applications work in parallel to perform a task and depending on the load of applications, some events may occur in different order from time to time. This means that even if a LSTM model is trained on a large data set there will most likely be sequences that the model has not seen before. But even if the combined variations creates a sequence that has not been seen before, the single variations might be seen in other sequences that are used for training. This will also make the LSTM model more forgiving and sequences not seen before can be predicted normally if they only contains those small variations. Since sequences vary a lot we cannot use the most probable state from the LSTM model as the prediction, this would make the model to always choose the state that it has seen most times during training. Instead we let the model to output a vector with predictions for all states and look at the probability for the state it is trying to predict.

Figure 5 shows the output probabilities predicted by a LSTM model after it has trained on several different sequences. The probability for a state to occur is written above the state (1.0 means 100% probability), states receiving 0 probability are not shown. To determine if a particular sequence contains any anomaly or not, the probability for each state is first compared with the highest probable state:

$$\frac{p(act)_i}{p(max)_i} = c_i \qquad (1)$$

$p(act)_i$ is the probability for the state, $p(max)_i$ is the maximum probability among all states predicted, and c_i is the comparative value. Then all state comparatives is multiplied with each other resulting in a comparative value for the whole sequences:

$$c_1 \cdot c_2 \cdot \ldots \cdot c_n = C \qquad (2)$$

Where C is the product of all comparatives for the whole sequence, we call this the sequence error. The top sequence in Fig. 5 has the highest probability and would produce a product of $1 \cdot 1 \cdot 1 \cdot \ldots = 1$. If a sequence shown in the bottom of the Fig. 5 occurs then it would get a product of $1 \cdot 1 \cdot (0.1/0.7) \cdot (0.1/0.7) \cdot (0.1/0.7) \cdot 1 \cdot 1 = 0.0029$. The sequences can then be classified as anomaly if the sequence error is below a certain limit, this is defined as the error threshold.

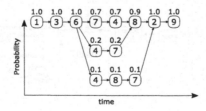

Fig. 5. Probabilities for a state to occur in different sequences detection.

The advantage of using an error threshold for the whole sequence is that it will both capture single events that are very unlikely and also sequences that has several minor deviations from the normal behaviour.

3.2   Filtering Data

The RAN data used in this article contains several thousands of sequences for calls between users. Many of these sequences are similar but small differences in order of events causes a large number of unique sequences. If a LSTM model would train on the whole data it would learn to predict the most probable state sequences. Sequences that occur very seldom would be treated as an anomaly since they deviate from the normal behaviour. In order to address this issue the data is filtered so that the LSTM model only train on sequences it has not seen before.

3.3   AdaBoost Ensemble of LSTM Classifiers

As the single LSTM model is trained on more and more unique sequences, the probability for a certain event to occur will decrease since there are more events for the LSTM model to choose between. In these cases it would be more beneficial to use several models that trained on sequences that are similar to each other. Sequences can be compared in many ways using the length of the sequence, number of matching events, how close the events are to each other, the time difference between them, etc. Even if we divided the sequences based on these parameters, the LSTM model would perhaps not treat the sequences as similar anyway. The best would be to let the LSTM model determine if the sequences are equal or not and divide the sequences into groups and then retrain the models. This is similar of how the AdaBoost ensemble works [12], where multiple weak learners are trained in sequence. The first learner trains and classifies the data used for training and then the next learner trains more on the data that were misclassified by the previous and so on until all learners have trained and classified the data. All the learners can then be used to classify new data by using a voting among the learners. The better a learner has been during training the more importance is given to the vote. AdaBoost in combination with LSTM models has also recently proven to be useful for predicting numerical time series [4,22,24]. The procedure we propose to train the AdaBoost LSTM ensemble for our purpose are the following:

1. Initialize an equal weight to all sequences used for training:

$$\omega_t(i) = \frac{1}{N} \tag{3}$$

2. The weights are used to select which sequences to train on, the higher number the more likely it is that the sequences are chosen for training.
3. A new LSTM model trains on the selected sequences.
4. The same model then predicts if there is an anomaly in any of the sequences.

5. If the accuracy is less then 0.55 then step 3–5 is repeated until the accuracy is larger than 0.55. If the accuracy does not reach 0.55 within 10 times the best model is chosen and we continue to step 6.
6. When all sequences is classified a value α is calculated based on the accuracy of the LSTM model, (t is the iteration of step 2–7):

$$\alpha_t = \frac{1}{2} \ln \frac{1 - \epsilon_t}{\epsilon_t} \tag{4}$$

where ϵ_t is the sum of the weighted error for the prediction:

$$\epsilon_t = \sum_{i=1}^{m} \omega_t(i)[y_i \neq h(x_i)] \tag{5}$$

7. α is the used to increase the weights for sequences classified as anomaly and decrease weights if classified as normal:

$$\omega_{t+1}(i) = \omega_t e^{-\alpha^t h^t(x)y(x)} \tag{6}$$

$\omega(i)$ is then normalised so that the sum of all $\omega(i)$ is 1:

$$\omega_t(i) = \frac{\omega_t(i)}{\sum_{i=1}^{N} \omega_t(i)} \tag{7}$$

8. Step 2–7 are repeated for all models selected to be part of the ensemble.
9. To detect anomalies, all models classifies a sequence and a voting procedure is used to determine the outcome. For each model, α is used to tell the importance of a vote, if the model had high accuracy during the training the vote will be more important.

During training of the ensemble it is also important that the training sequences does not contains any anomalies. If anomalies are present the LSTM models would also learn the anomaly behaviour and could not differ between normal and abnormal behaviour.

3.4   Anomaly Feedback Loop

During the continuous product development cycle, the software is constantly updated with new features and the behavior might change slightly for each software update. The update may cause the LSTM model to predict normal sequences as abnormal, this might also happen if the data contains a lot of sequences that has not been seen before (discussed in Sect. 3.1). The false positive (FP) sequences, can then easily be used to train the LSTM model and its weights will be adjusted during the training so that it recognizes the FP sequences. By analyzing all sequences that are classified as anomaly the FP cases can be separated and sent into the LSTM model for extra training, see Fig. 4.

4 Evaluation

4.1 RAN Data

RAN system logs can contain data that are sensitive for users and vendors and are hard to obtain. Instead of using logs from commercial live systems we used a 5G test bed created by TietoEvry, which can simulate a one day scenario for a small city with 10 000 inhabitants. The advantage in using a test bed is that we can control the scenarios, repeat them, and also choose what to include in the logs without disturbing any live system. Two data sets containing 7458 respectively 7456 session calls were collected from the one day scenario in which all traffic were normal. By letting the test bed skip or add extra events in the system log, a third data set containing 26 individual anomalies were created. The anomalies were: missing event, extra event, impossible state transitions, events not seen before, multiple events missing, and multiple extra events. The session calls in the logs were separated into event sequences and the key values for each event were extracted and encoded to be suitable for the machine learning problem.

To address the imbalanced data we discussed in Sect. 3.2 the two normal data sets were filtered to create two lists with unique sequences. The third data set was filtered to only contain the anomaly sequences. The three data sets will later be referred to as training (173 seq), test (45 seq) and error data set (26 seq).

4.2 Benchmark Data

To evaluate how BoostLog performs in other domains than RAN and also to compare our findings with recent research, a second data set was retrieved from a Hadoop Distributed File System (HDFS), in which 200 Amazon's EC2 nodes been running Hadoop-based map-reduce jobs. The block sequences in the data have been labeled by domain experts as anomaly or normal and contains 11 175 630 events in 575 061 sequence chains. This data is described more in detail in [28] and has been used for research in anomaly detection by [25] and [9]. In similar way as for the RAN data set only the unique sequences were extracted and split into train (12 406 seq), test (5317 seq) and anomaly data set (4375 seq). This data set is much larger than the RAN data set and in order to reduce time to tune the model and make it similar to the RAN log we only used 10% of the normal data while we kept all the anomalies.

4.3 Performance Measures

The main idea of this research was to find new ways to automatically analyze the system logs in order to find anomalies faster without being expert on the domain. One important aspect is also to be able to trust the model, for trouble shooters in RAN it are very important to not classify normal data as an anomaly. If the model keeps classifying normal data as faulty then the same thing as happens in the legendary Aesop's fable *The Boy Who Cried Wolf* will occur, the users will stop relying on the model. To make the model useful and trustworthy it need to fulfill two things, it should be able to find many anomalies and seldom classify normal behavior as faulty.

4.4   The LSTM Model

The single LSTM model used has 2 hidden layers with about 100 neurons, some models had fully connected layers and some layers used dropout. In the AdaBoost ensemble it is more beneficial to use many different classifiers and 15 LSTM models with a range from 2–3 hidden layers and 32–200 neurons were used. For all LSTM models, the keras [7] back-end with the implementation of the LSTM layer was used.

4.5   Parameter Tuning

Single LSTM Classifier. There are two important parameters to choose for the single LSTM classifier, firstly what error threshold that should be used and secondly how many states it should look back on in order to predict the next state.

To see the impact of the error threshold used for classifying the sequences we choose to set the look back to 4 and test error thresholds in range or 1 to $1e − 9$. The model is first trained on the training data set and then the accuracy is measured on the training, test and error data sets, the result can be seen

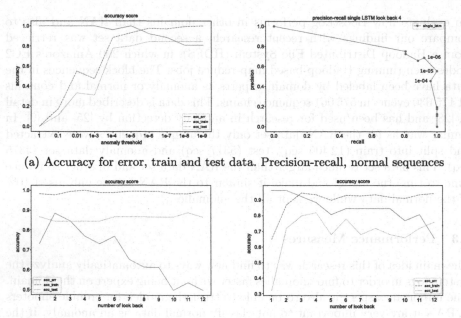

(a) Accuracy for error, train and test data. Precision-recall, normal sequences

(b) Left: Accuracy for error threshold $1e − 9$, Right: Accuracy for error threshold $1e − 5$

Fig. 6. Tuning of Single LSTM classifier

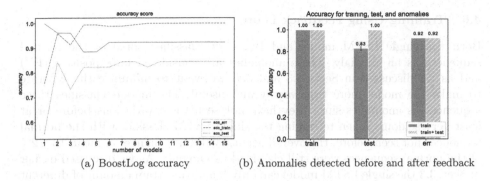

(a) BoostLog accuracy   (b) Anomalies detected before and after feedback

Fig. 7. BoostLog

in Fig. 6a. When the error threshold is high all sequences will be classified as anomalies and if the error threshold is very low it will miss some anomalies but it will classify the normal sequences more correctly. It seems that the threshold can be lowered down to $1e-5$ before the true positive (TP) rate starts to drop and we still get a quite high accuracy for the normal sequences. This can also be seen in the precision-recall curve in Fig. 6a.

Next parameter to tune is the number of look backs, ranging the look backs from 1 to 12, we tested both the error threshold at $1e-5$ where the true positive and true negative (TN) rate was high and also at $1e-9$ were TN rate was highest. For $1e-9$ the accuracy for the normal sequences stayed stable at the same level as in Fig. 6a but the accuracy kept decreasing with the increasing amount of look backs, see Fig. 6b. The drop in accuracy is explained by the increasing certainty of the model when the number of states to look back on are increased, a higher threshold is needed to compensate for the more precise model but then the accuracy would also drop for the normal sequences. The same tendency is seen for the $1e-5$ threshold and even if the accuracy for the training sequences increases the accuracy is not increased any more for the test sequences when the number of look backs are increased above 4, see Fig. 6b.

BoostLog. To start with, we used the same error threshold and look back parameter in BoostLog as for the single LSTM classifier. To see how the AdaBoost classifier improves with the amount of learners, the accuracy was calculated for the three data sets, see Fig. 7. As we can see the accuracy are getting stable at 10 learners. When comparing with the single LSTM model we can also see that the accuracy is better for both test and error sequences. This is very promising since the accuracy improved for those sequences not seen yet. Normally it is better to use weak learners that achieve an accuracy just above 0.5 in an AdaBoost ensemble. As expected the ensemble improved even more when increasing the error threshold and decreasing number of look back to get learners that were not equally good as the best single LSTM model. Final result can be see in Table 1.

4.6 Training Using Feedback Loop

Both the single LSTM model and BoostLog classified some of the normal sequences as an anomaly. These anomalies are considered false positive (FP) and as we discussed in Sects. 4.3 and 3.4, we want to minimize the FP rate to make the model more trustworthy and useful. The models classifies these sequences as anomalies since they have not seen these variations before or at least very seldom. When retraining the single LSTM classifier with the normal sequences not seen before, the overall accuracy did not change, in mean $13 \pm 2\%$ of the normal sequences were still classified as anomaly. As we discussed earlier in Sect. 3.3 the single LSTM model can only handle a certain amount of different sequences.

When the same procedure was performed for the BoostLog the accuracy stayed the same for the anomaly sequences and increased from 84% to 100% for the sequences it had not seen before, see Fig. 7.

Among the false negative cases, there were only two sequences that both the single LSTM and BoostLog classified as normal and both these cases missed one state in the sequence. These anomalies are difficult for the LSTM classifier to detect since the single missed state will not cause a very large drop in sequence error. If the LSTM predicts the probability for time step $t + 1$ it will also know what state that is likely to occur in time step $t + 2$. The probability for the state to occur in $t + 2$ will be slightly higher than other states and if the state for time step $t + 1$ is missing, the second highest probability is likely to be the state for $t + 2$.

4.7 Evaluation

In this work we have limit our models to only look at one feature from the system logs. The main focus have been to investigate how the anomaly detection can be improved in system logs using an ensemble of LSTM models. To do this we have compared the single LSTM model with the new BoostLog ensemble. Precision, Recall, and F-score was calculated for each model before and after training on RAN testbed, see Table 1, the models are called feedback model when training has been performed on normal sequences not seen before. The single LSTM model improved in detecting anomalies (the recall value), but instead it also classified more normal sequences as anomalies (the precision value). In total there was no improvement in the F-score. The BoostLog ensemble outperforms the single LSTM model and adapted very well to the new data while it kept the recall rate high. After training on test data it was very reliable and did not point out one single normal sequence as anomaly and still found 24 of the 26 anomalies.

Table 1. Precision, recall, and F-score for RAN data set

Model	Precision	Recall	F-score
Single LSTM	0.64	0.88	0.74
AdaBoost LSTM	0.77	0.92	0.84
Feedback single LSTM	0.59	1.0	0.74
Feedback AdaBoost LSTM	1.00	0.92	0.96

The same models were then tuned for the HDFS data set and tested in the same way as for the RAN data, see Table 2.

Table 2. Precision, recall, and F-score for HDFS data set

Model	Precision	Recall	F-score
Single LSTM	0.96	0.94	0.95
AdaBoost LSTM	0.94	0.98	0.96
Feedback single LSTM	0.96	0.93	0.94
Feedback AdaBoost LSTM	0.97	0.98	0.97

It seems that it is much easier for the single LSTM model to learn the sequence behaviour of the HDFS block sequences compared to the RAN call chains, it can also more easily find the anomalies. A quick check with fast Dynamic Time Warping (fastDTW) [21], confirms that the unique sequences in RAN data set is much more similar to each other than for the HDFS data. There is also a larger difference between the normal and anomaly sequences for HDFS data compared with the RAN data. Similar sequences will make it more difficult to predict the outcome for the single LSTM model and if the anomaly sequences also are more similar to normal sequences it will get even more difficult to detect the anomalies. This might explain why the single LSTM model can detect the anomalies easier in the HDFS data. We can also see that, as in the RAN data set, there is no improvement for the single LSTM model when training extra on normal sequences not seen before. Since the single LSTM model already performs very well on the HDFS data set, the results cannot be improved much using an ensemble, we can only see a slight improvement using the BoostLog ensemble.

Our results cannot be directly compared with previous research that has used the HDFS data set [9, 25], since they use multi features and also look at the whole data set. But the results are promising, in [9], the best model, also got a F-score of 0.96 when it analysed the HDFS data set.

5 Conclusion and Future Work

We have created a novel method that turns the LSTM model into a classifier for anomaly detection in system logs, in this case, logs from a large scale system such as 5G RAN. Furthermore we have also shown how the AdaBoost algorithm can be adopted to use our LSTM classifier to create an even better anomaly detection ensemble, the BoostLog.

Evaluation of performance in two different datasets shows that it is more beneficial to use BoostLog in domains such as RAN, where the sequences are similar to each other and the anomalies are difficult to distinguish from the normal behaviour.

BoostLog also adapts well when a feedback loop was used to train extra on false positive sequences and maximized the precision for the RAN data set. We expect that deep ensemble based networks such as BoostLog, can be used to find more anomalies early in RAN development cycle and will require less domain expertise in order to find anomaly behaviour in the system logs.

In this paper we investigated how one feature could be used to predict the anomalies and it will be interesting to see how our future work can be improved by selecting more features and also take into account the time that events occur. In our previous work we have examined how anomalies and security issues can be detected using performance metrics [3,15]. It will be interesting to see if the anomaly detection can be further improved using a combination of ensembles that analyze both functional and performance behaviour. We must also test our method on system logs from more different areas to see if the ensemble also is suitable to use in those domains.

Acknowledgement. This work was partially supported by the Wallenberg AI, Autonomous Systems and Software Program (WASP) funded by Knut and Alice Wallenberg Foundation.

References

1. Ajith, D.: A survey on anomaly detection methods for system log data. Int. J. Sci. Res. (IJSR) **8**, 23 (2019)
2. Al Mamun, S.M.A., Valimaki, J.: Anomaly detection and classification in cellular networks using automatic labeling technique for applying supervised learning. Procedia Comput. Sci. **140**, 186–195 (2018). https://doi.org/10.1016/j.procs.2018.10.328
3. Bhuyan, M.H., Bhattacharyya, D.K., Kalita, J.K.: Network Traffic Anomaly Detection and Prevention: Concepts, Techniques, and Tools, 1st edn. Springer, Cham (2017). https://doi.org/10.1007/978-3-319-65188-0
4. Bian, G., Liu, J., Lin, W.: Internet traffic forecasting using boosting LSTM method. In: 2017 International Conference on Computer Science and Application Engineering (CSAE 2017). DEStech Transactions on Computer Science and Engineering (2018). https://doi.org/10.12783/dtcse/csae2017/17517
5. Chalapathy, R., Chawla, S.: Deep learning for anomaly detection: a survey. CoRR abs/1901.03407 (2019). http://arxiv.org/abs/1901.03407

6. Chniti, G., Bakir, H., Zaher, H.: E-commerce time series forecasting using LSTM neural network and support vector regression. In: Proceedings of the International Conference on Big Data and Internet of Things, BDIOT 2017, pp. 80–84. ACM, New York (2017)

7. Chollet, F., et al.: Keras (2015). https://keras.io

8. Damm, L.O.: Early and cost-effective software fault detection: measurement and implementation in an industrial setting. Ph.D. thesis (2007)

9. Du, M., Li, F., Zheng, G., Srikumar, V.: DeepLog: anomaly detection and diagnosis from system logs through deep learning. In: Proceedings of the 2017 ACM SIGSAC Conference on Computer and Communications, pp. 1285–1298 (2017). https://doi.org/10.1145/3133956.3134015

10. Prewett, J.E.: Analyzing cluster log files using Logsurfer. In: Proceedings of Annual Conference on Linux Clusters (2003)

11. Fernández Maimó, L., Perales Gómez, L., García Clemente, F.J., Gil Pérez, M., Martínez Pérez, G.: A self-adaptive deep learning-based system for anomaly detection in 5G networks. IEEE Access 6, 7700–7712 (2018). https://doi.org/10.1109/ACCESS.2018.2803446

12. Freund, Y., Schapire, R.E.: A decision-theoretic generalization of on-line learning and an application to boosting. J. Comput. Syst. Sci. 55(1), 119–139 (1997)

13. Hansen, S.E., Atkins, E.T.: Automated system monitoring and notification with swatch. In: Proceedings of the 7th USENIX Conference on System Administration, LISA 1993, pp. 145–152. USENIX Association, Berkeley (1993)

14. Hochreiter, S., Schmidhuber, J.: Long short-term memory. Neural Comput. 9(8), 1735–1780 (1997)

15. Ibidunmoye, O., Rezaie, A., Elmroth, E.: Adaptive anomaly detection in performance metric streams. IEEE Trans. Netw. Serv. Manag. 15(1), 217–231 (2018). https://doi.org/10.1109/TNSM.2017.2750906

16. Iyer, A.P., Li, L.E., Stoica, I.: Automating diagnosis of cellular radio access network problems. In: Proceedings of the 23rd Annual International Conference on Mobile Computing and Networking. MobiCom 2017, pp. 79–87. ACM, New York (2017). https://doi.org/10.1145/3117811.3117813

17. Karevan, Z., Suykens, J.A.: Transductive LSTM for time-series prediction: an application to weather forecasting. Neural Netw. 125, 1–9 (2020)

18. Malhotra, P., Ramakrishnan, A., Anand, G., Vig, L., Agarwal, P., Shroff, G.: LSTM-based encoder-decoder for multi-sensor anomaly detection. CoRR abs/1607.00148 (2016). http://arxiv.org/abs/1607.00148

19. Rouillard, J.P.: Real-time log file analysis using the simple event correlator (SEC). In: Proceedings of LISA XVIII, pp. 133–150 (2004)

20. Polikar, R.: Ensemble based systems in decision making. IEEE Circuit Syst. Mag. 6, 21–45 (2006). https://doi.org/10.1109/MCAS.2006.1688199

21. Salvador, S., Chan, P.: Toward accurate dynamic time warping in linear time and space. Intell. Data Anal. 11(5), 561–580 (2007)

22. Sun, S., Wei, Y., Wang, S.: AdaBoost-LSTM ensemble learning for financial time series forecasting. In: Shi, Y., et al. (eds.) ICCS 2018. LNCS, vol. 10862, pp. 590–597. Springer, Cham (2018). https://doi.org/10.1007/978-3-319-93713-7_55

23. Szilagyi, P., Novaczki, S.: An automatic detection and diagnosis framework for mobile communication systems. IEEE Trans. Netw. Serv. Manag. 9(2), 184–197 (2012). https://doi.org/10.1109/TNSM.2012.031912.110155

24. Xiao, C., Chen, N., Hu, C., Wang, K., Gong, J., Chen, Z.: Short and mid-term sea surface temperature prediction using time-series satellite data and LSTM-AdaBoost combination approach. Remote Sens. Environ. 233, 111358 (2019)

25. Xu, W., Huang, L., Fox, A., Patterson, D., Jordan, M.I.: Detecting large-scale system problems by mining console logs. In: Proceedings of the ACM SIGOPS 22nd Symposium on Operating Systems Principles, SOSP 2009, pp. 117–132. ACM, New York (2009)
26. Yu, X., Joshi, P., Xu, J., Jin, G., Zhang, H., Jiang, G.: CloudSeer: workflow monitoring of cloud infrastructures via interleaved logs. In: ASPLOS 2016 (2016)
27. Zhang, K., Xu, J., Min, M.R., Jiang, G., Pelechrinis, K., Zhang, H.: Automated IT system failure prediction: a deep learning approach. In: 2016 IEEE International Conference on Big Data (Big Data), pp. 1291–1300 (2016). https://doi.org/10.1109/BigData.2016.7840733
28. Zhu, J., et al.: Tools and benchmarks for automated log parsing. CoRR abs/1811.03509 (2018). http://arxiv.org/abs/1811.03509

Machine Learning for Cognitive Load Classification – A Case Study on Contact-Free Approach

Mobyen Uddin Ahmed(⊠), Shahina Begum, Rikard Gestlöf, Hamidur Rahman, and Johannes Sörman

School of Innovation Design and Engineering (IDT), Mälardalen University, Västerås, Sweden
{mobyen.ahmed,shahina.begum,hamidur.rahman}@mdh.se

Abstract. The most common ways of measuring Cognitive Load (CL) is using physiological sensor signals e.g., Electroencephalography (EEG), or Electrocardiogram (ECG). However, these signals are problematic in situations e.g., in dynamic moving environments where the user cannot relax with all the sensors attached to the body and it provides significant noises in the signals. This paper presents a case study using a contact-free approach for CL classification based on Heart Rate Variability (HRV) collected from ECG signal. Here, a contact-free approach i.e., a camera-based system is compared with a contact-based approach i.e., Shimmer GSR+ system in detecting CL. To classify CL, two different Machine Learning (ML) algorithms, mainly, Support Vector Machine (SVM) and k-Nearest-Neighbor (k NN) have been applied. Based on the gathered Inter-Beat-Interval (IBI) values from both the systems, 13 different HRV features were extracted in a controlled study to determine three levels of CL i.e., S0: low CL, S1: normal CL and S2: high CL. To get the best classification accuracy with the ML algorithms, different optimizations such as kernel functions were chosen with different feature matrices both for binary and combined class classifications. According to the results, the highest average classification accuracy was achieved as 84% on the binary classification i.e. S0 vs S2 using k-NN. The highest F_1 score was achieved 88% using SVM for the combined class considering S0 vs (S1 and S2) for contact-free approach i.e. the camera system. Thus, all the ML algorithms achieved a higher classification accuracy while considering the contact-free approach than contact-based approach.

Keywords: Cognitive Load (CL) · Contact-free approach · k-Nearest-Neighbor (k-NN) · Support Vector Machines (SVM) · Machine Learning (ML)

1 Introduction

Driving is a complex task that requires high concentration with simultaneous skills and abilities [1]. Research shows that, most of the traffic accidents are caused by the human error. [1]. Most of the cases a road accident occurs when a driver loose concentration on

© IFIP International Federation for Information Processing 2020
Published by Springer Nature Switzerland AG 2020
I. Maglogiannis et al. (Eds.): AIAI 2020, IFIP AICT 583, pp. 31–42, 2020.
https://doi.org/10.1007/978-3-030-49161-1_3

driving due to look outside or thinking anything rather than driving task. Therefore, there is a connection between low concentration level and traffic accidents has been identified in [2]. Research shows that the number of road deaths increased by 28% in Sweden, by 14% in the Czech Republic and by 11% in the Netherlands in 2018 compared to 2017 [3]. Over 20% of those 90% are due to fatigue or CL (e.g. use of in-vehicle devices which can lead to cognitive overload for the driver) [4]. CL is also known as cognitive workload, which is a concept that describes the relationship between the cognitive demands from a task and the environment that influence the user's cognitive resources [5, 6].

To identify CL, psycho-physiological measures are mainly used, as they assess inter-actions between physiological and psychological states of a human. The most common psychophysiological measurement to detect CL is EEG, because it objectively identifies the cognitive cost of performing tasks [6]. Other parameters are also used to detect CL, such as: pupil dilation, HRV, galvanic skin response (GSR), and EEG etc. [7]. There are disadvantages with these contact-based approaches as most of them used electrodes and they should be connected to the human body, which can be cumbersome to use in situations such as driving. In many occasions the measurement signals are affected due to movements and artifacts [8]. That is why a contact-free way of measuring CL could be ideal for these situations. On the contrary, the contact-free approach e.g. camera systems have been incessantly growing in the research community [9]. Here, the camera system have developed based on color schemas such as RGB and the Lab color space to extract heart rate (HR) and IBI by detecting variation in facial skin-color caused by cardiac pulse [9]. McDuff et al. [10] has built a remote person independent classifier and predict CL with 85% classification accuracy, here, a contact-free camera-based system was used in order to measure HR, HRV and breathing rate (BR).

In order to detect or classify CL, HRV features are used, since the HR fluctuates with varying levels of CLs which can be detected with HRV metrics [11]. In medical diagnoses and stress detection HRV is a commonly used parameter, there are many different features in HRV that can be calculated and can be used as a parameter to estimate CL. There are three established ways to calculate the HRV, either by time-domain, frequency-domain or non-linear measurements [12]. CL detection is a very inefficient for humans to manually review large data sets because of the factors that occur naturally in everyday life situations [13]. This is why the ML algorithms are an essential factor in classifying CL.

The data collection for cognitive load research are generally conducted either in indoor environment using simulator driving or in outdoor environment using real road driving scenario [14–16]. The authors in [17] have focused on the feasibility of using visual attention features to classify CL in three different states i.e., low, medium and high cognitive load. Four different classification models such as k-NN, logistic regression, SVM, adaptive boosting and random forest have used and the k-NN received the highest accuracy of 81%. A study performed by Wang et al. [18] has received a 97.2% of classification accuracy in detecting high CL.

This paper presents a case study on contact-free approach for CL classification based on the features extracted through HRV analysis and using on ML. Here, the goal of the study is to compare the contact-free approach i.e., camera-based system with the contact-based approach i.e., a Shimmer GSR+ system to determine how well the contact-free approach performs in detecting CL. To classify the CL, two different ML algorithms, mainly, SVM and k-NN are used. This study includes a collection of data for both the systems in a controlled environment. Thirteen HRV features were extracted to determine three states of CL based on the gathered IBI values from both the systems. The three levels are, S0: reference point for low CL, S1: normal CL where the test subjects performed easy puzzles and drove normally in a video game and S2: high CL where the test subjects completed hard puzzles and drove on the hardest course of a video game while answering math questions [19]. To get the best classification accuracy with the ML algorithms, different optimizations such as kernel functions i.e., radial basis function (RBF), and sigmoid function were chosen with different feature matrixes both for binary and combined class classifications.

2 Materials and Methods

In the contact-free approach, a Logitech HD Webcam C615 camera with the resolution of 1920 × 1080, and a frame rate of 30fps equipped with auto focus was used to record faces of the test subjects. The recorded videos were saved in MP4 format. The camera was adjusted according to the height of the head of the test subject so that it can capture the facial image properly. The test subjects were sat roughly 70 cm in front of the camera during the different phases of the controlled experiment, thus the desired ROI of the test subjects and facial features were extracted [9]. During the data collection, the test subjects were asked to move their heads as normal but not too much (Fig 1).

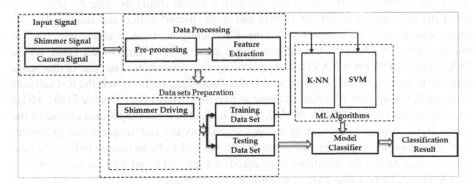

Fig. 1. Overview of the workflow for measuring CL

In the contact-based approach, a Shimmer GSR+ system was wired up to the subjects' body through a clamp to the earlobe, which provided the corresponding IBI values. The data recorded by the shimmer sensor was connected to a laptop via Bluetooth. The standard settings of Consensus were recorded as Unix_Timestamps, PPG_A13 as IBI and PPGtoHR. The IBI value was recorded by the shimmer system and saved in Excel sheets through a software called Consensus to monitor and manage the data gathered by the shimmersensing[1]. The shimmer had a sampling rate of roughly 40000 after a

---

[1] http://www.shimmersensing.com/products/consensys.

5-min test run. Here the frequency of the sensor reading was 133 Hz. An overview of the workflow used in this study is presented in Figure, that describes the different parts i.e., Data collection, Data Processing, Data Set Preparation, Classification Learner, ML Algorithms, Model Classifier, Classification Results.

2.1 Data collection

This study includes 11 test subjects with age between 21 and 33 years. Among them 9 were men with average age of 26 and two were women with average age of 27. Again, four of them had lowest education in math, three of them had highest math education and the rest had average math courses. The mean value of how good they see themselves in mental arithmetic were 5, 9 in a scale of 10. The mean value of time spent playing digital games per week was 13 h, but the time variation is high between the test subjects. The one with the highest h per week had forty h, and the lowest had one hour Two of the eleven test subjects only played on mobile, and the rest played on console/computer.

The data was collected through a controlled study consists of seven phases [19], between phases there were a minor pause to set the next phase up and in the fifth phase there were a 15-min break. The 1st phase of this experiment was a five-minute rest while sitting normal and watching TV silently. The 2nd Phase was to complete the two different easy metal puzzles while sitting normal. Here, the test subject played an easy puzzle first, and when they have dissembled the first one, they started over the next one. When both the puzzles were dissembled the test subject would start to reassemble both the puzzles in the same order. This process was repeated for five minute. The 3rd Phase was about to play both the metal puzzles with a higher difficulty. The 4th phase was about the test subject rested for 15 min and were offered coffee and snacks. The 5th phase was about to sit a normal watching TV silently same as in phase one, for five minute. In the 6th phase the test subjects play a game 'Mario Kart Double Dash' on the 150cc course BOWER'S CASTLE competing against seven bots. The time of this phase was five minute; however, the time was varied depending on how well the test subjects performed. Finally, the 7th phase is about to play the 'Mario Kart Double Dash' 150cc once again, but on RAINBOW ROAD, which is the final and the hardest course of the game. Here, instead of competing against seven bots the test subjects had to answer math questions while playing. The math questions had to be answered before the next question was asked, the questions were asked in every 30 s and five questions in total for each time. Table 1 illustrates the summary of the phases.

2.2 Data Pre-processing

For each phase of every test subject, the sample size was 40 000 on an average. A python program with the library OpenPyxl was created in order to handle samples, the program both transformed the UNIX time to 'date time' and removed the samples where the IBI was − 1.0. Thus, a smaller sample size is reduced to 250–500 depending on the recorded time as well as the precision of the device. The reason of this huge decrease was that the IBI was calculated approximately once per second, since the HR was around 60–80 BPM for all the test subjects. This resulted in roughly 300 recorded IBI's for each phase of every test subject, since the length of each experiment phase was five minute long.

Table 1. Summary of each phase during the data collection

Phases	1st	2nd	3rd	4th	5th	6th	7th
Duration (min)	5	5	5	15	5	5	5
Description	Normal sitting and watching TV	Normal sitting and solving easy puzzle	Normal sitting and solving hard puzzle	Rest and drinking coffee/tea	Normal sitting and watching TV	Playing driving game	Playing driving game and solve math
Goal	Low cognitive level	Medium high cognitive level	High cognitive level	Relax and recover state	Low cognitive level	Medium high cognitive level	High cognitive level
Datasets	S0	S1	S2		S0	S1	S2

For the camera data, IBI was extracted from the recorded video and saved in a separate file. From each of the recorded video, first, the face of the test subject was detected and then a region of interest (ROI) was selected from the detected face [20]. Each ROI was converted into three color frames which are red, green and blue which are transferred into Lab color space or Lab signal. Three signal processing algorithm First Fourier Transform (FFT) [21], Independent Component Analysis (ICA) [22] and Principle Component Analysis (PCA) [23] were used to extract IBI from the Lab signal by applying a band pass filter of 40–120 Hz [24]. An overview of the camera system that extracts IBI from a video recording is presented in Fig. 2.

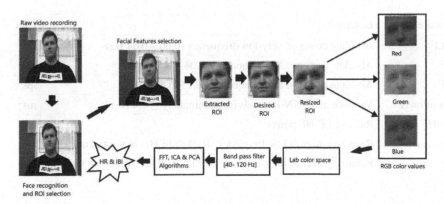

Fig. 2. Different steps of a camera system to extract IBI from camera data

2.3 Feature Extraction

Once the collected data were structured in a manner that would make it easier for a script to extract features from the recorded data the implementation of the feature extraction session started with the so-called trial and error phase in order to find a way of extracting all the HRV features properly. The purpose of the script was to reduce the total duration of the recorded time into three min and only consider the middle of the recording and remove the data from the beginning and end of the recording to avoid faulty recording. These three min were then be further divided into three parts of one min each. Then for each min of these three min data, the feature function in MATLAB was used to calculate the 13 features where six time domain features and seven frequency domain features as shown in Table 2 and Table 3 respectively. Then the features for each phase was calculated for each test subject. This was resulted in a 3 * 14 matrix linked to one class (in the controlled experiment phase), where SittingNormal = S0, Driving 1 or Puzzle 1 = S1, Driving 2 or Puzzle 2 = S2.

Table 2. List of different time-domain HRV features

Feature	Description	Unit
MeanNN	Is the average of all NN-intervals	ms
SDNN	Standard deviation of NN-intervals	ms
RMSSD	Root mean square successive RR interval differences	ms
SDSD	Standard deviation of successive differing between neighbor NN-intervals	ms
NN50	Number of pairs of successive NN-intervals differing by more than 50 ms	int
pNN50	Percentage of successive RR-intervals differing by more than 50 ms	%

Table 3. List of different frequency-domain HRV features

Feature	Description	Unit
VLF	Absolute power of very-low-frequency (0.003–0.04 Hz)	ms^2
LF	Absolute power of low-frequency (0.04–0.15 Hz)	ms^2
HF	Absolute power of high-frequency (0.15–0.40 Hz)	ms^2
TotalPower	Variance of all NN-intervals (approximately $\leq 0,4$ Hz)	ms^2
LF/HF ratio	Ratio of LF-HF power	%
LF (nu)	Relative power of low-frequency (0.04–0.15 Hz) in normal units	nu
HF (nu)	Relative power of high-frequency (0.15–0.40 Hz) in normal units	nu

2.4 Classification

For the classification using k-NN and SVM, the whole data sets were divided into four groups both for shimmer and camera data separately as follows:

- S0 vs S1, Normal sitting vs Driving normal
- S0 vs S2, Normal sitting vs Driving with distractions
- S1 vs S2, Driving normal vs Driving with distractions
- S0-S1-S2, Combined classification

First, 80% of the data for each group of data set was randomly selected for training and 20% for data was selected for testing. Binary classification was conducted for this paper works and two ML algorithms e.g. k-NN and SVM were considered an instance of supervised learning in which an algorithm learns to classify new observations from examples of labeled data [25]. SVM constructs a hyperplane or a set of hyperplane to analyses data to identify pattern and commonly used for classification and regression analysis [26]. K-NN is a nearest-neighbor classification model in which it can be altered both the distance metric and the number of nearest neighbors. Because a k-NN classifier stores training data, the model is used to compute re-substitution predictions. Alternatively, this model is used to classify new observations using the predict method [27].

Different kernel functions were investigated both for k-NN and SVM which were weighted, fine and medium kernel for k-NN and cubic, fine and medium kernel for SVM. The best cross-validation k-fold value was 10 for both k-NN and SVM. The binary SVM classifier was implemented according to [eq. 1], the fitcsvm[2] is a built-in function in MATLAB that returns a trained SVM model, based on the features of the training set x and the list with the corresponding class y. The parameters '*KernelFunction*' indicates that the kernel used in this model should be the variable k. The parameter '*KernelScale*' indicates that the kernel scale used in this model should be the variable s. '*BoxConstraint*' is the parameter indicating that the boxconstraint of the model should be the variable b.

$$trainedSVM = fitc(x, y', KernelFunction', k', kernelScale', s', BoxConstraint', b)$$

(1)

For k-NN model, [eq. 2] was implemented which is also a MATLAB built-in function. The function fitck-NN returns a trained K-NN model and based on the features of the training set x and the list with the corresponding class y. The parameter *NumNeighbours*' flags that the variable k should be used as the number of neighbors for this K-NN model. 'DisatnceWeight' is a parameter in the fitc k-NN that indicates that the variable w should be used as the distance weight.

$$trainedKNN = fitc(x, y,' NumNeighbors', k,' DistanceWeight', w)$$

(2)

The optimized classifiers were implemented for each training set of the class combinations based on the feature matrixes. The trained models were then given the features of the test sets and based on how they predicted the class of the features, the accuracy of each model could determine by how many correct class predictions it achieved. Each model executed 20 times for each data set and mean value of the result was considered.

3 Experimental Results

The classification learner revealed that the two best classification algorithms for the collected data were k-NN and SVM. Classification accuracy for the binary data sets (S0 vs S1), (S0 vs S2), (S1 vs S2) and (S0 vs S1andS2) and class classification accuracy for the combined data sets (S0, S1, S2) are calculated using ML algorithms i.e., k-NN and

---

[2] https://se.mathworks.com/help/stats/fitcsvm.html.

SVM. Also, F_1-score is calculated for each of the data set using the ML algorithms. A summary of the classification accuracy and F_1-score for the shimmer data is presented in Table 3 and Table 4 presents the classification accuracy for the camera data. The highest value of F_1-score both for the shimmer and camera are highlighted in gray color in the table (Table 5).

Table 4. Classification results using Shimmer system

Classification	Data set	Shimmer			
		k-NN		SVM	
		Mean accuracy with 20 runs	F_1	Mean accuracy with 20 runs	F_1
Binary	S0 vs S1	0.74	0.7	0.74	0.78
	S0 vs S2	0.78	0.74	0.66	0.79
	S1 vs S2	0.58	0.41	0.58	0.65
Combination	S0 vs S1 and S2	0.72	**0.86**	0.68	0.85

Table 5. Classification results using Camera system

Classification	Data set	Camera			
		k-NN		SVM	
		Mean accuracy with 20 runs	F_1	Mean accuracy with 20 runs	F_1
Binary	S0 vs S1	0.71	0.71	0.70	0.68
	S0 vs S2	0.81	0.84	0.74	0.74
	S1 vs S2	0.67	0.57	0.65	0.66
Combination	S0 vs S1 and S2	0.79	0.87	0.72	**0.88**

For k-NN and SVM, a graphical representation of average F_1-score between shimmer and camera is presented in Fig. 3 and Fig. 4. From the figures it can be seen that F_1-score for camera is larger than Shimmer both for k-NN and SVM considering binary and combined class classification.

4 Discussion and Conclusion

The experimental result shows that highest accuracy has been achieved in a controlled environment using the camera recording during the driving phase. Also, this shows that binary classification using the k-NN algorithm with the class combination of S0 vs S2 achieves an accuracy of 81% which was higher than the expected accuracy from the classification learner by one percent. With the same class combination and setup

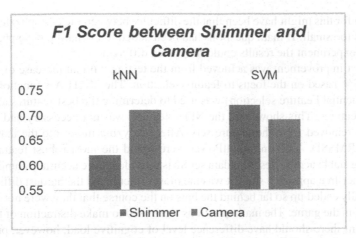

Fig. 3. Graphical representation of F_1-score between Shimmer and Camera data considering binary classification.

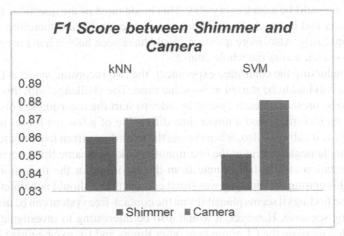

Fig. 4. Graphical representation of F_1-score between Shimmer and Camera data considering classification of combined class

recorded with the Shimmer sensor system only got 78% accuracy, which is higher than the expected accuracy 70%. For the class combination S0 vs (S1 and S2), the highest F_1 score 84% was achieved with medium gaussian SVM using the feature matrix recorded from the Camera system.

The results also demonstrates that the ML algorithms could classify CL in Driving using video games with equal or even higher accuracy than a traditional way (i.e., using metal puzzles). This is significant because controlled experiments that involve actual video games have not been extensively used in the state-of-the-art research. This research also reveals that the accuracy in classifying CL between driving normal and driving with distractions was greater than the accuracy between the Easy puzzle and Hard puzzle.

The reason for this might have been that the difficulty between the easy puzzle and hard puzzle was too small. If math questions had been added to the hard puzzle phase of the controlled experiment the results could have been different.

The best improvement was achieved from the testing with an increase of the accuracy with 3% based on the focus to feature selection. The MATLAB function sequentialfs (Sequential Feature selection) was used to determine the best features among the extracted features. This shows that the NFFT feature, was not necessary and therefore NFFT was removed from the feature sets. After analyzing more into the features, the MeanNN, RMSSD, VLF and TotalPower were found the most robust features while considering stable accuracies. The data set S2 is provided more accuracy compare to S0 using contact-free approach since it was the most difficult and the hardest difficulty, the driver usually ended up so far behind the bots on the course that they were not disturbed by the bots in the game. The math questions were used to make distraction of the driver which means there should have difference level of cognitive load; however, practically it was not found the difference between cognitive load levels due to less distraction of the drivers by math questions. The choice of an easier track could have made it possible to have the data differentiate even more than it did. Also, the time determined when a math question would be asked, every 30 s. This made most of the questions to be asked when the driver had a straight line without any worries and could therefore answer the question more easily. Also, more questions could have been held at hand in order for the questions to stretch across the whole course.

When conducting the controlled experiment, the two recording systems (i.e., Shimmer and Camera) had to be started at the same time. The challenge with this was that a button had to be pressed on each system in order to start the recordings. This made that the two system recordings had a minor time difference of a few ms for each recording for each of the test subject. Also, when cutting the recordings from five minute into three minute the implementation cut at the one-minute mark. The same thing occurred when the implementation cut the last minute from the recording at the three-minute mark. Because of this cutting the time gap was small enough that it should not affect the result.

With these findings it seems plausible that the contact-free system can be used reliably in real driving scenario. However, it would also be interesting to investigate if there is something that increase the CL more than other things and try to categorize how much it increased the CL. Also, do more experiments with the game over a longer period of time, to see how different test subjects will adapt to the Mario Kart game. Here, only the frequency and the time-domain features were considered. So, it would also be interesting to involve the non-linear HRV features in the future.

Acknowledgment. The authors would like to acknowledge Embedded Sensor Systems for Health Plus (ESS-H+) for their support of the research projects. The authors would also like to acknowledge all the participants for supporting in this study.

References

1. Kim, J.-P., Lee, S.-W.: Time domain EEG analysis for evaluating the effects of driver's mental work load during simulated driving, pp. 79–80 (2017)

2. Miyaji, M., Kawanaka, H., Oguri, K.: Study on effect of adding pupil diameter as recognition features for driver's cognitive distraction detection. In: 2010 7th International Symposium on Communication Systems, Networks & Digital Signal Processing (CSNDSP 2010), pp. 406–411 (2010)
3. Road Safety Annual Report 2019. OECD Publishing, Paris, OECD/ITF (2019)
4. Sena, P., d'Amore, M., Pappalardo, M., Pellegrino, A., Fiorentino, A., Villecco, F.: Studying the influence of cognitive load on driver's performances by a Fuzzy analysis of Lane Keeping in a drive simulation. IFAC Proc. **46**(21), 151–156 (2013)
5. Palinko, O., Kun, A.L., Shyrokov, A., Heeman, P.: Estimating cognitive load using remote eye tracking in a driving simulator. Presented at the Proceedings of the 2010 Symposium on Eye-Tracking Research & Applications, Austin, Texas (2010). https://doi.org/10.1145/174 3666.1743701
6. Kumar, N., Kumar, J.: Measurement of cognitive load in HCI systems using EEG power spectrum: an experimental study. Procedia Comput. Sci. **84**, 70–78 (2016)
7. Reimer, B., Mehler, B., Coughlin, J., Godfrey, K., Tan, C.: An on-road assessment of the impact of cognitive workload on physiological arousal in young adult drivers, pp. 115–118 (2009)
8. Zander, T.O.: Evaluation of a dry eeg system for application of passive brain-computer interfaces in autonomous driving. Front. Hum. Neurosci. **11**, 78 (2017). (in English)
9. Rahman, H., Ahmed, M.U., Begum, S.: Vision-based remote heart rate variability monitoring using camera. In: Ahmed, M.U., Begum, S., Bastel, J.-B. (eds.) HealthyIoT 2017. LNICST, vol. 225, pp. 10–18. Springer, Cham (2018). https://doi.org/10.1007/978-3-319-76213-5_2
10. McDuff, D., Gontarek, S., Picard, R.: Remote measurement of cognitive stress via heart rate variability. In: 2014 36th Annual International Conference of the IEEE Engineering in Medicine and Biology Society, EMBC 2014, vol. 2014, pp. 2957–2960 (2014)
11. Sahadat, M.N., Consul-Pacareu, S., Morshed, B.I.: Wireless ambulatory ECG signal capture for HRV and cognitive load study using the NeuroMonitor platform. In: 2013 6th International IEEE/EMBS Conference on Neural Engineering (NER), pp. 497–500 (2013)
12. Shaffer, F., Ginsberg, J.P.: An overview of heart rate variability metrics and norms. Front. Pub. Health **5**, 258 (2017). (in English)
13. Hussain, S., Chen, S., Calvo, R., Chen, F.: Classification of cognitive load from task performance & multichannel physiology during affective changes (2007)
14. Solovey, E.T., Zec, M., Garcia Perez, E.A., Reimer, B., Mehler, B.: Classifying driver workload using physiological and driving performance data: two field studies. In: Conference on Human Factors in Computing Systems – Proceedings (2014)
15. Engström, J., Johansson, E., Östlund, J.: Effects of visual and cognitive load in real and simulated motorway driving. Transp. Res. Part F: Traffic Psychol. Behav. **8**(2), 97–120 (2005)
16. Reimer, B., Mehler, B.: The impact of cognitive workload on physiological arousal in young adult drivers: a field study and simulation validation. Ergonomics **54**, 932–942 (2011)
17. Liu, A., Chen, C.: Towards practical driver cognitive load detection based on visual attention information. Masters of Science, University of Toronto, Canada (2017)
18. Wang, C., Guo, J.: A data-driven framework for learners' cognitive load detection using ECG-PPG physiological feature fusion and XGBoost classification. Procedia Comput. Sci **147**, 338–348 (2019)
19. Sörman, J., Gestlöf, R.: Contact-free cognitive load classification based on psycho-physiological parameters. Bachelor of Science, Mälardalen University, Sweden (2019)
20. Rahman, H., Begum, S., Ahmed, M.U., Funk, P.: Real time heart rate monitoring from facial RGB color video using Webcam. In: 29th Annual Workshop of the Swedish Artificial Intelligence Society (SAIS) 2016, Malmö, Sweden (2016)

21. Buijs, H., Pomerleau, A., Fournier, M., Tam, W.: Implementation of a fast Fourier transform (FFT) for image processing applications. IEEE Trans. Acoust. Speech Signal Process. **22**(6), 420–424 (1974)
22. Comon, P.: Independent component analysis, a new concept. Signal Process. **36**, 287–314 (1994)
23. Kramer, M.A.: Nonlinear principal component analysis using autoassociative neural networks. AIChE J. **37**(2), 233–243 (1991)
24. Rahman, H., Ahmed, M.U., Begum, S.: Non-contact heart rate monitoring using lab color space. In: 13th International Conference on Wearable, Micro & Nano Technologies for Personalized Health (PHealth 2016), Crete, Greece, 29–31 May 2016 (2016)
25. Alpaydin, E.: Introduction to Machine Learning. MIT Press (2010)
26. Cortes, C., Vapnik, V.N.: Support-vector networks. Mach. Learn. **20**(3), 273–297 (1995)
27. Tamrakar, P., Roy, S.S., Satapathy, B. Ibrahim, S.P.S.: Integration of lazy learning associative classification with k-NN algorithm. In: 2019 International Conference on Vision Towards Emerging Trends in Communication and Networking (ViTECoN), pp. 1–4 (2019)

Real-Time Prediction of Online Shoppers' Purchasing Intention Using Random Forest

Karim Baati[1(✉)] and Mouad Mohsil[2]

[1] Teolia Consulting, 12–14 Rond-Point des Champs Elysées, 75008 Paris, France
[2] Teolia D.A., 12–14 Rond-Point des Champs Elysées, 75008 Paris, France
{karim.baati,mouad.mohsil}@teolia.fr

Abstract. In this paper, we suggest a real-time online shopper behavior prediction system which predicts the visitor's shopping intent as soon as the website is visited. To do that, we rely on session and visitor information and we investigate naïve Bayes classifier, C4.5 decision tree and random forest. Furthermore, we use oversampling to improve the performance and the scalability of each classifier. The results show that random forest produces significantly higher accuracy and F1 Score than the compared techniques.

Keywords: Real-time online shopper behavior · Marketing offers in online stores · Random forest

1 Introduction

Nowadays, a large majority of businesses are supported or carried out online. In order to foster the generated virtual environments, marketing offers stand for one of the most valuable strategies which can be employed. Historically, these offers were indiscriminately suggested to the whole visitors of a given e-commerce website. Afterward, being aware about the necessity to orient their marketing actions to the right target, online stores opted for a near-real time analysis of visitors' information. The purpose is to contact the most relevant users (for instance by phone or e-mail) in order to suggest offers which are likely to induce them to go back to the website and achieve an effective purchase.

Recently, a new trend has emerged among virtual shopping environments so that potential visitors are identified at the time they are browsing the website. By contrast to the near-real time model, the advantage behind that is to avoid the high risk of losing users once disconnected from the online store. Indeed, in such a model, we imitate an experienced salesperson who struggles to retain potential visitors by providing a range of customized marketing actions which are likely to encourage them to buy. The latest study suggesting such a strategy in e-commerce websites could be found in [1] where authors proposed a system with two modules which predicts the purchasing intent of the visitor by using some

© IFIP International Federation for Information Processing 2020
Published by Springer Nature Switzerland AG 2020
I. Maglogiannis et al. (Eds.): AIAI 2020, IFIP AICT 583, pp. 43–51, 2020.
https://doi.org/10.1007/978-3-030-49161-1_4

session and user information along with aggregated pageview data kept track during the visit. The first module of this system is used to determine whether the user should be offered content and the second module is triggered only if the user is likely to abandon the site. Though the fact that the system proposed in [1] is appealing in terms of efficiency and scalability, the risk of abandon it implies stands for a problem which is not to be neglected.

In this paper, we do not aim to propose a new model that substitutes systems like the one of Sakar et al. [1]. On the contrary, our objective is to consolidate such systems by trying to retain the maximum number of potential visitors. In this scope, we suggest a system that allows to detect users with high purchasing intention as soon as they connect to an e-commerce website. For this purpose, we rest on the same data used in [1] but we only keep those pertaining to session and user information. As that will be detailed in this paper, by establishing our system, we aspire to be part of a global system that starts by proposing a first type of marketing offers to potential visitors once connected to the website. Later, that system calls a second subsystem (like the one of Sakar et al. [1]) to suggest more generous offers to visitors who did not carry out an effective purchase at the first stage but who displayed a high purchasing intention after a certain clickstream.

The remainder of the paper is structured as follows. Related work is reported in Sect. 2. Next, Sect. 3 details different functional and technical aspects of our proposed system. Afterward, experimentation results are communicated in Sect. 4. Lastly, Sect. 5 concludes the paper and suggests some directions for future research.

2   Literature Review

Literature encompasses many studies which are turned towards categorization of online visits in e-commerce websites.

A first study of Mobasher et al. [2] assessed two different clustering techniques based on user transactions and pageviews in order to find out useful aggregate profiles that can be used by recommendation systems to achieve effective personalization at early stages of user's visits in an online store.

Later, the study of Moe [3] prepared the ground for a system which can take customized actions according to the category of a visit. For this reason, the author proposed a system which makes use of page-to-page clickstream data from a given online store in order to categorize visits as a buying, browsing, searching, or knowledge-building visit. The proposed system rests on observed in-store navigational patterns (including the general content of the pages viewed) and a k-means algorithm for clustering.

In [4], Poggi et al. proposed a system which handles the loss of throughput in Web servers due to overloading, by assigning priorities to sessions on an e-commerce website according to the revenue that will generate. Data were formed of clickstream and session information and Markov chains, logistic linear regression, decision trees and naïve Bayes were investigated in order to measure the probability of users' purchasing intention [4].

In [5] and [6], authors designed the prediction of purchasing intention problem as a supervised learning problem and historical data collected from an online bookstore were used to categorize the user sessions as browsing and buyer sessions. In this scope, Support vector machines (SVMs) with different kernel types and k-Nearest Neighbor (k-NN) were respectively investigated in [5] and [6] to carry out classification.

In a more recent study [7], Suchacka and Chodak constructed a new approach to analyze historical data obtained from a real online bookstore. The proposed approach is based on association rule discovery in customer sessions and aims to evaluate the purchase probability in an online session.

In [8], the author proposed a system that identifies the website component that has the highest business impact on visitors. To build such a system, a data set based on the Google Analytics tracking code [9] has been created. Moreover, naïve Bayes and multilayer perceptron classifiers have been explored for classification.

In [1], authors set up a real-time user behavior analysis system for virtual shopping environment which is made up of two modules. In the first module, the purchasing intention of the visitor is predicted using aggregated pageview data kept track during the visit along with some session and user information. Further, oversampling and feature selection preprocessing algorithms were applied to improve the effectiveness and the scalability of a set of supervised machine learning techniques. The highest accuracy of 87.24% and F1 Score of 0.86 were obtained with a Multilayer Perceptron Network (MLP). In the second module, authors used a Long Short-Term Memory-based Recurrent Neural Network (LSTM-RNN) based on sequential clickstream data to produce the probability estimate of visitor's intention to leave the site without completing the transaction. Within the scope of the entire system proposed in [1], the first module is triggered only if the second module generates a greater value than a predetermined threshold and the final objective consists in deciding whether to offer a content to the online visitor.

3   Proposed System

In this section, we describe the functional aspects related to our proposed system. Next, we depict the dataset used for validation as well as the machine learning techniques which have been explored for classification.

3.1   Functional Description

Based on previous studies, we can notice that no work has addressed the purchasing intention of online visitors immediately after they connect to the website. By contrast to these studies, our system aims to detect users with high purchasing intention as soon as they connect to the e-commerce website and to offer content only to those who intend to complete a transaction. The advantage behind that is to avoid the risk of losing potential visitors who sometimes disconnect for trivial reasons (arrival of a guest at home, reception of a phone call, etc.). That risk should not be neglected since it naturally impacts the effectiveness of any system with a business objective (for instance each of the systems introduced respectively in [1,5,6] and [7]).

As mentioned earlier, through our proposed system, we do not aspire to replace the previous systems that have tackled the same problem by using click-stream data. On the contrary, we aim to reinforce that systems by trying to interest the maximum number of potential users. Indeed, our model could be used as a part of a more global structure as shown in Fig. 1. This global structure starts by proposing a first type of marketing offers to potential visitors once connected to the website and appeals later a second subsystem (like the one proposed in [1]) to suggest more generous offers to visitors who did not accomplish an effective purchase at the first stage but who showcased a high purchasing intention after a certain clickstream (Fig. 1). In our opinion, such a global system can be investigated for e-commerce websites as a new tool that could be useful in terms of effective use of time, purchase conversion rates and sales figures.

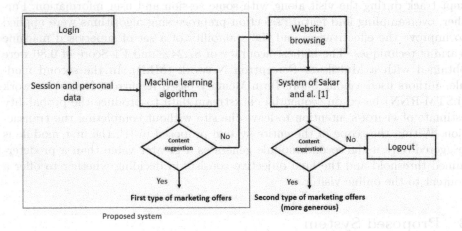

Fig. 1. Flowchart of the proposed system and its positioning within a more global desired system

3.2   Data Description

As mentioned earlier, we reposed on the same dataset used in [1] but we only keep the part pertaining to session and user information. Features related to the selected data are depicted in Table 1. As reported in [1], features belong to 12330 sessions and the data were constituted so that each session would belong to a different user in a 1-year period to avoid any tendency to a specific campaign, special day, user profile, or period. Moreover, among the 12330 sessions in the dataset, 84.5% (10422) were negative class samples that did not finalize a transaction and the rest (1908) were positive class samples ending with purchasing [1].

Based in Table 1, we can notice that, except the "Day" feature, all the rest of features are categorical. In order to homogenize the features types, we used a binning process in which the "Day" variable is discretized to 5 discrete levels [10].

Table 1. Features used for system validation

Feature	Feature description	Feature type
Day	Closeness of the site visiting time to a special day	Numerical
Operating Systems	Operating system of the visitor	Categorical
Browser	Browser of the visitor	Categorical
Region	Geographic region from which the session has been started by the visitor	Categorical
Traffic	Traffic source by which the visitor has arrived at the website	Categorical
Visitor	Visitor type as "New Visitor", "Returning Visitor" and "Other"	Categorical
Weekend	Boolean value indicating whether the date of the visit is weekend	Categorical
Month	Month value of the visit date	Categorical
Revenue	Class label indicating whether the visit has been finalized with a transaction	Categorical

3.3 Prediction

Technically, a challenging issue for the proposed system consists in finding out the most suitable machine learning technique for the prediction problem. In this context, specifications of input data are among main criteria which could be used to elect such a technique [11,12]. In our case, since features are categorical [13], three appropriate techniques for this type of features are selected, namely: naïve Bayes classifier, C4.5 classifier and random forest. In the following, fundamentals of each one of these techniques are briefly detailed.

- **Naïve Bayes classifier**
  Naïve Bayesian Classifier (NBC) is a probabilistic classifier which is based on Bayes fusion rule. It assumes the independence of the input features. That means that it considers each of these features to contribute independently to the probability of a class regardless of any possible correlations between

variables [14,15]. The final decision is assigned to the class with the highest probability. Despite its simplicity, NBC can often outperform more sophisticated classification methods [16].

- **C4.5 classifier**

  C4.5 is an extension of the earlier ID3 algorithm and stands for an algorithm used to generate a decision tree [17] which can be used for classification.

  At each node of the tree, C4.5 selects the attribute of the data that most effectively splits its set of samples into subsets enriched in one class or the other. The splitting criterion is the normalized information gain (difference in entropy). The attribute with the highest normalized information gain is chosen to make the decision. The C4.5 algorithm then recurses on the partitioned sublists.

- **Random forest classifier**

  Random Forest (RF) [18] is a well-known decision tree ensemble that is commonly used in classification. In many cases, RF showed a good performance that outstrips that of many other classification algorithms [19].

  As a decision tree ensemble, RF needs to construct several different decision trees. In order to accomplish that, each tree is set up considering a bootstrap sample set of the original training data. That consists in creating a new set sampling with replacement instances from the original set until getting the size of that original training data. Using one of the bootstrap sample sets of the training data, a random tree is obtained. In order to favor the diversity of the ensemble, each random tree follows a traditional top-down induction procedure with several modifications. At each step, when the best attribute is chosen, only a small subset of attributes from the dataset is considered. Considering only the subset of attributes chosen, we then compute the best attribute as in Classification And Regression Trees (CART) [20]. Each tree is built to its maximum depth and no pruning procedure is applied after the tree has been fully built. Lastly, to compute the predicted class for a sample, the predictions of the ensemble of decision trees are aggregated through majority voting.

4   Experiments and Results

To conduct experiments, data described in Sect. 3.2 are fed to naïve Bayes, C4.5 decision tree and random forest classifiers using 70% of dataset for training and the rest for validation [21,22]. Moreover, to ensure statistical significance, this procedure is repeated 100 times with random training/validation partitions.

Table 2 presents the average accuracy, sensitivity (true positive rate), specificity (true negative rate) and F1 Score for each classifier.

The results show that naïve Bayes classifier yields the highest accuracy rate on test data. However, a class imbalance problem emerges [23] since, for each classifier, the sensitivity is much lower than specificity (sensitivity is even null for naïve Bayes classifier). Indeed, it is clear that each classification technique tends to label the test samples as the majority class (negative class). This class imbalance problem is a natural situation for the addressed problem since most of the online visits do not end with purchasing [24]. However, this issue should be properly handled since correctly identifying directed buying visits which are represented with positive class in the dataset is as crucial as identifying negative class samples.

To deal with class imbalance problem, we considered the use of the widely used technique to balance the training set before the learning phase, namely the "Synthetic Minority Oversampling Technique" (SMOTE) methodology [25]. To do that, 30% of the data set consisting of 12330 samples is first excluded for testing and the oversampling SMOTE method is applied to the remaining 70% of the samples.

The results obtained on the balanced dataset are reported in Table 3. As it is seen, the highest accuracy of 86.78% and F1 Score of 0.60 is obtained with the random forest classifier.

Table 2. Results obtained on class imbalanced dataset

Classifier	Accuracy (%)	Sensitivity	Specificity	F1 score
Naïve Bayes	90.04	0.00	1.00	0.00
C4.5	86.27	0.11	0.94	0.13
Random forest	83.64	0.09	0.91	0.10

Table 3. Results obtained with oversampling

Classifier	Accuracy (%)	Sensitivity	Specificity	F1 score
Naïve Bayes	86.66	0.05	0.95	0.07
C4.5	86.59	0.55	0.92	0.56
Random forest	**86.78**	**0.62**	**0.91**	**0.60**

5   Conclusion and Prospects

The appeal of the proposed work is related to the need to set up a model that forecasts the visitor's shopping intent as soon as the e-commerce website is visited. The advantage behind that is to avoid the risk of abandon implied by each visit on the website. In this scope, the challenging issue consisted in finding out the most appropriate machine learning which could reach this purpose.

Three classification techniques have been investigated to resolve the addressed problem, namely naïve Bayes, C4.5 and random forest. Moreover, oversampling has been carried out to improve the performance and the scalability of each classifier. Based on experimentation and comparison results, we have proven the efficiency of the random forest classifier as a balanced classifier which is able to fit the requirements of our problem.

As aforementioned, it would be interesting to join a system like that of Sakar et al. [1] to our proposed classifier in order to form a more global system which could be useful for online stores. However, to confirm the effectiveness of a such global system, its exploration in online retailers in real-world and real-time settings should be achieved and its performance should be compared to that of competitive models.

References

1. Sakar, C.O., Polat, S.O., Katircioglu, M., Kastro, Y.: Real-time prediction of online shoppers' purchasing intention using multilayer perceptron and LSTM recurrent neural networks. Neural Comput. Appl. **31**(10), 6893–6908 (2019). https://doi.org/10.1007/s00521-018-3523-0
2. Mobasher, B., Dai, H., Luo, T., Nakagawa, M.: Discovery and evaluation of aggregate usage profiles for web personalization. Data Min. Knowl. Discov. **6**(1), 61–82 (2002). https://doi.org/10.1023/A:1013232803866
3. Moe, W.W.: Buying, searching, or browsing: differentiating between online shoppers using in-store navigational clickstream. J. Consum. Psychol. **13**(1–2), 29–39 (2003)
4. Poggi, N., Moreno, T., Berral, J.L., Gavaldà, R., Torres, J.: Web customer modeling for automated session prioritization on high traffic sites. In: Conati, C., McCoy, K., Paliouras, G. (eds.) UM 2007. LNCS (LNAI), vol. 4511, pp. 450–454. Springer, Heidelberg (2007). https://doi.org/10.1007/978-3-540-73078-1_63
5. Suchacka, G., Skolimowska-Kulig, M., Potempa, A.: Classification of e-customer sessions based on support vector machine. ECMS **15**, 594–600 (2015)
6. Suchacka, G., Skolimowska-Kulig, M., Potempa, A.: A k-nearest neighbors method for classifying user sessions in e-commerce scenario. J. Telecommun. Inf. Technol. **3**(64), 64–69 (2015)
7. Suchacka, G., Chodak, G.: Using association rules to assess purchase probability in online stores. Inf. Syst. e-Bus. Manag. **15**(3), 751–780 (2016). https://doi.org/10.1007/s10257-016-0329-4
8. Budnikas, G.: Computerised recommendations on e-transaction finalisation by means of machine learning. Stat. Transit. **16**(2), 309–322 (2015)
9. Clifton, B.: Advanced Web Metrics with Google Analytics. Wiley, Hoboken (2012)
10. Peng, H., Long, F., Ding, C.: Feature selection based on mutual information criteria of max-dependency, max-relevance, and min-redundancy. IEEE Trans. Pattern Anal. Mach. Intell. **27**(8), 1226–1238 (2005)
11. Kotsiantis, S.B., Zaharakis, I.D., Pintelas, P.E.: Supervised machine learning: a review of classification techniques. Front. Artif. Intell. Appl. **160**, 3–24 (2007)
12. Baati, K., Hamdani, T.M., Alimi, A.M., Abraham, A.: A new possibilistic classifier for mixed categorical and numerical data based on a bi-module possibilistic estimation and the generalized minimum-based algorithm. J. Intell. Fuzzy Syst. **36**(4), 3513–3523 (2019)
13. Baati, K., Hamdani, T.M., Alimi, A.M., Abraham, A.: A new classifier for categorical data based on a possibilistic estimation and a novel generalized minimum-based algorithm. J. Intell. Fuzzy Syst. **33**(3), 1723–1731 (2017)

14. Baati, K., Hamdani, T.M., Alimi, A.M., Abraham, A.: A modified Naïve Bayes style possibilistic classifier for the diagnosis of lymphatic diseases. In: Abraham, A., Haqiq, A., Alimi, A.M., Mezzour, G., Rokbani, N., Muda, A.K. (eds.) HIS 2016. AISC, vol. 552, pp. 479–488. Springer, Cham (2017). https://doi.org/10.1007/978-3-319-52941-7_47
15. Baati, K., Hamdani, T.M., Alimi, A.M., Abraham, A.: A modified Naïve possibilistic classifier for numerical data. In: Madureira, A.M., Abraham, A., Gamboa, D., Novais, P. (eds.) ISDA 2016. AISC, vol. 557, pp. 417–426. Springer, Cham (2017). https://doi.org/10.1007/978-3-319-53480-0_41
16. Langley P., Sage S.: Induction of selective Bayesian classifiers. In: Proceedings of 10th Conference on Uncertainty in Artificial Intelligence (UAI-94), pp. 399–406 (1994)
17. Quinlan, J.R.: C4.5: Programs for Machine Learning. Elsevier, Amsterdam (2014)
18. Breiman, L.: Random forests. Mach. Learn. 45(1), 5–32 (2001). https://doi.org/10.1023/A:1010933404324
19. Subudhi, A., Dash, M., Sabut, S.: Automated segmentation and classification of brain stroke using expectation-maximization and random forest classifier. Biocybern. Biomed. Eng. 40(1), 277–289 (2020)
20. Breiman, L., Friedman, J., Olshen, R., Stone, C.: Classification and Regression Trees. Wadsworth and Brooks, Belmont (1984)
21. Baati, K., Hamdani, T.M., Alimi, A.M., Abraham, A.: Decision quality enhancement in minimum-based possibilistic classification for numerical data. In: Abraham, A., Cherukuri, A.K., Madureira, A.M., Muda, A.K. (eds.) SoCPaR 2016. AISC, vol. 614, pp. 634–643. Springer, Cham (2018). https://doi.org/10.1007/978-3-319-60618-7_62
22. Baati, K., Kanoun, S.: Towards a hybrid system for the identification of Arabic and Latin scripts in printed and handwritten natures. In: Madureira, A.M., Abraham, A., Gandhi, N., Varela, M.L. (eds.) HIS 2018. AISC, vol. 923, pp. 294–301. Springer, Cham (2020). https://doi.org/10.1007/978-3-030-14347-3_28
23. Tian, J., Gu, H., Liu, W.: Imbalanced classification using support vector machine ensemble. Neural Comput. Appl. 20(2), 203–209 (2011). https://doi.org/10.1007/s00521-010-0349-9
24. Ding, A.W., Li, S., Chatterjee, P.: Learning user real-time intent for optimal dynamic web page transformation. Inf. Syst. Res. 26(2), 339–359 (2015)
25. Chawla, N.V., Bowyer, K.W., Hall, L.O., Kegelmeyer, W.P.: SMOTE: synthetic minority over-sampling technique. J. Artif. Intell. Res. 16, 321–357 (2002)

Using Classification for Traffic Prediction in Smart Cities

Konstantinos Christantonis[1], Christos Tjortjis[1(✉)] [iD], Anastassios Manos[2],
Despina Elizabeth Filippidou[2], Eleni Mougiakou[3], and Evangelos Christelis[2]

[1] International Hellenic University, 14th km Thessaloniki–Moudania, 57001 Thermi, Greece
c.tjortjis@ihu.edu.gr
[2] DOTSOFT SA, 3 Kountouriotou, 546 25 Thessaloniki, Greece
[3] Commonspace, 1-3 Akakiou & 60 Ipirou Street, 10439 Athens, Greece

Abstract. Smart cities emerge as highly sophisticated bionetworks, providing smart services and ground-breaking solutions. This paper relates classification with Smart City projects, particularly focusing on traffic prediction. A systematic literature review identifies the main topics and methods used, emphasizing on various Smart Cities components, such as data harvesting and data mining. It addresses the research question whether we can forecast traffic load based on past data, as well as meteorological conditions. Results have shown that various models can be developed based on weather data with varying level of success.

Keywords: Smart cities · Data mining · Prediction · Classification

1 Introduction

The deployment of modern and smart cities increasingly gains attention, as large urban centers over time present numerous challenges for citizens. Traffic is a stressful and time-consuming factor affecting citizens. Lately, many local authorities attempt to design and create smart infrastructures and tools in order to collect data and utilize models for better decision making and citizen support. Such data are often derived from sensors collecting information about Points Of Interest (POIs) in real time. Data mining techniques and algorithms can then support getting useful insights into the problem, whilst forming appropriate strategies to counter it.

This work focuses on analyzing different approaches regarding data manipulation in order to predict day-ahead traffic loads at random places around cities, based on weather conditions. Prediction efforts regard classification tasks aiming at highlighting factors that affect traffic prediction. This study utilizes weather data collected from sensor devices located in Athens and Thessaloniki, Greece. Three different day zones are introduced and compared while subsets of trimesters are also tested.

The remaining of the paper is structured as follows: Sect. 2 provides background information. Section 3 presents our approach for traffic prediction, including data selection as well as experimental results. Section 4 discusses and evaluates our findings while Sect. 5 concludes the paper with suggestions for future work.

I. Maglogiannis et al. (Eds.): AIAI 2020, IFIP AICT 583, pp. 52–61, 2020.
https://doi.org/10.1007/978-3-030-49161-1_5

2 Background

Traffic prediction is not a new subject; there are numerous scientific efforts that perform both classification and regression tasks in this domain [1]. However, few efforts attempted to predict traffic volumes based on low level data, such as weather data.

Nejad et al. examined the power of decision trees for classifying traffic loads on three levels, based only on time and temperature [2]. Results were positive and motivated our work. Xu et al. compared CART with k-NN and a direct Kalmann Filter [3]. Moreover, Wang et al. proposed the use of volume and occupancy data [4]. Such data, as well as speed, can be obtained by loop detectors. Loop detectors are sensors buried underneath highways which estimate traffic by observing information related to vehicles passing above them. Loop detectors along with rain predictions were also used in order to predict crashes [5].

Tree-based algorithms are widely used and justified, however even more sophisticated algorithms were used such as Support Vector Machines (SVMs) [6] and Neural Networks (NNs) [7]. A novel hybrid method for short-term traffic prediction under both typical and atypical traffic conditions. Theodorou et al. introduced an SVM based model that identifies atypical conditions [1]. We used ARIMA or k-NN regression to identify typical or atypical conditions, respectively. In addition to [4], Liu et al. introduced three binary variables, other than weather conditions, for holiday, special conditions and quality of roads [8].

However, in this work we chose not to include such features for several reasons. Firstly, holidays do not have the same effect on each season. For example, sunny holidays might result in lower levels of traffic in large urban centers whilst for winter holidays this is not the case. Special conditions, such as a major social events or protests are not easy to add to models since most of the times these occur unexpectedly. Abnormal conditions have a negative impact on such models, for instance when attempting to estimate the traffic flow based on Expressway Operating Vehicle [9].

Another significant decision on such predictions relates to the actual day of the week. As highlighted in [10, 11], weekdays tend to have different characteristics from weekends and should be carefully tested. Finally, most scenarios regard short-time prediction, meaning minutes to h ahead. Dunne et al. utilized neurowavelet models along with the rain expectation in the next hour [12].

3 Traffic Prediction

Weather data can be exploited in different smart cities problems and scenarios. Collecting accurate weather data can be beneficial for numerous daily problems which associate weather with human decisions. Traffic is affected by numerous factors, only some of which are predictable. As mentioned in Sect. 2, traffic is directly affected by weather conditions, however the scale differs across cities and cultures.

3.1 Problem Definition and Approach

This section tests the above intuition on traffic prediction. Traffic is a multi-dimensional problem; researchers focus on either predicting traffic loads on a certain time interval or selecting the optimal route based on real-time adaptations in order to minimize travel time. Accidents and social events can shortly disrupt normal traffic in specific areas, while seasonality and weather conditions can affect traffic in a larger scale. Based on that, we used weather data collected from sensors installed around carefully chosen specific city spots for predicting the day-ahead traffic volume.

To select the most appropriate locations to install sensors that either measure traffic loads or collect weather data, it is crucial to define their objective in advance. Our efforts focus on the question 'How can one exploit sensor data that are not personalized and create meaningful conclusions for the general public?' Deployment of smart city infrastructure requires a deep understanding of the traffic problem. Roads that are busier than others do not always provide more information in comparison to more isolated ones. The number of alternative roads or the location of busy buildings can affect the necessity of measuring traffic for a specific road.

Our approach, besides examining traffic predictability based on weather data, also aims to clarifying differences among locations. Moreover, we implement a series of tests regarding different monthly periods under the objective to understand which months contribute positively on the deployment of such models. The approach followed is explained in Sect. 3.2.

3.2 Dataset Description and Processing

For this task, our data source was the newly deployed ppcity.io, which is a set of platforms providing useful information derived from sensors to citizens and visitors of Athens and Thessaloniki, Greece. These two large urban centers host almost half the Greek population. The selected platform collects and analyses environmental, traffic and geospatial data. Those data are collected through several sensors located at central points around the two cities. In general, there are numerous city spots on both cities where the environmental and traffic conditions are measured. Initially, we chose locations about which we had information regarding both traffic and weather conditions. We focused on the ten most reliable spots that produce data, i.e. sensors with the most compact flow of data and wider range of recording.

Weather related attributes used for the research analysis involved the following: Humidity, Pressure, Temperature, Wind Direction, Wind Speed and Ultraviolet Radiation (UV). The platform provides additional weather data, but these were not included in the modeling process. Indicatively, such variables include measurements for ozone concentration, nitrogen dioxide etc.

The target variable (traffic load), named *Jam Factor*, represents the quality of travel. It ranges from 0 to 10, where 10 describes stopped traffic flow and 0 a completely empty road. Data have been collected for almost six months, a rather short period of time, but at least including August and December, two months when people tend to take holidays. August is widely considered in Greece as the month with the lowest traffic volume, since it is a period that most people go on holidays, away from large urban centers. Similarly, in December both cities face abnormalities in traffic loads, since many citizens move from and to the two large urban centers. The dataset comprises data collected between 29/7/19 and 3/2/20. Eight sensors are located in Athens and two in Thessaloniki. In addition, two sensors are located near school areas. More information about the sensors is available in Sect. 4.

The way data are processed is conceptually simple. Regarding the data structure, we defined three distinct time intervals within a day and examined their predictability. The first interval, named *Morning,* includes signals from sensors between 07:00 and 10:00. The second interval, named *Afternoon,* includes signals between 15:00 and 18:00 and the last one, named Evening, includes signals between 19:00 and 22:00.

Therefore, the average value for the given signals was computed. For example, if a sensor captures information every h(e.g. at 07:10, 08:10, 09:10 etc.), we computed and assigned the average value for each weather metric and the traffic load for that specific day period. The above strategy resulted in creating three different datasets consisting of 185 instances on average. Fortunately, the quality of the dataset was high, thus missing values were minimal. It is worth noting that our averaging strategy does not replace missing values. If there is a missing value in a weather averaging is performed using the remaining two available values, instead of averaging and labeling three values.

An important decision was to transform the problem into a classification task by categorizing the target variable into two and later into three classes. Initially, the split point was the median value for each dataset, aiming at a fully balanced task which would allow for safer conclusions. Therefore, the two classes named *High* and *Low* indicate if the traffic load on that specific day was higher than usual. In the second stage the target variable was split into three categories of equal size (High, Medium and Low). The metric used for classification evaluation was accuracy, since classes are balanced and of equal interest.

Finally, we split the dataset into three subsets consisting of data regarding different trimesters. The first one contains data for the period between 29-July and 25-October, the second for the period between 15-September and 20-December and the third one, between 25-October and 3-February. We aimed at distinguishing if any periods (seasonality) affect the models negatively.

3.3 Experiments

Different cities demonstrated different traffic patterns; however, in almost every case we observed a few common patterns. First, on weekends citizens tended to use vehicles much less than weekdays, while on holidays people tended to leave the urban centers. To further understand the case of Greece, Fig. 1 visualizes the mean value for all available sensors in order to explain phenomena beyond seasonality. The colors on all figures share the same palette and indicate the same subsets. More precisely, for Fig. 1, the blue line indicates the *Afternoon* values while red and green indicates *Morning* and *Evening* respectively.

For Figs. 2, 3, 4, 5, 6 and 7 blue indicates the subset of 29 July–3 February, red for 15 September–20 December, green for 29 July–25 October and purple for 25 October–3 February. As expected, a rapid introductory search on the data surface highlight that the volume of traffic was consistently higher during the Afternoon period, while both Morning and Evening periods seem to present similar amounts of traffic.

Regarding the predictability for each of these combinations between sensors and day periods, for all the experiments on this task, we used the Random Forest Classifier. It is an

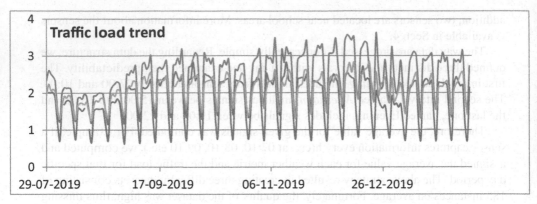

Fig. 1. Mean value for all sensors for each day period (Color figure online)

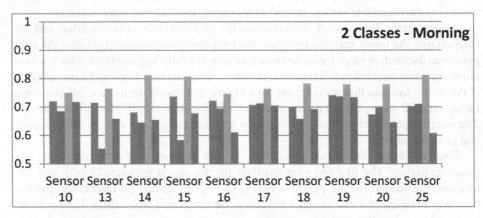

Fig. 2. Accuracy with 2-classes – Morning (Color figure online)

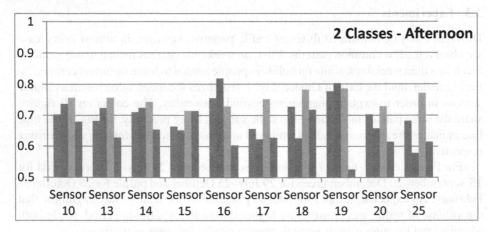

Fig. 3. Accuracy with 2-classes – Afternoon (Color figure online)

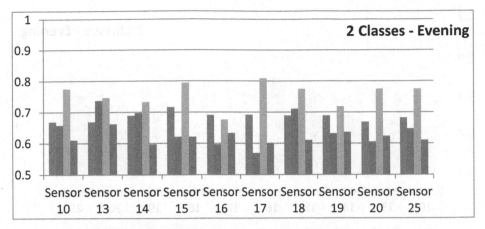

Fig. 4. Accuracy with 2-classes – Evening (Color figure online)

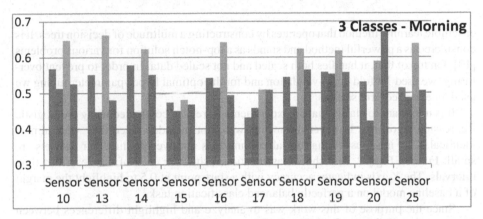

Fig. 5. Accuracy with 3-classes – Morning (Color figure online)

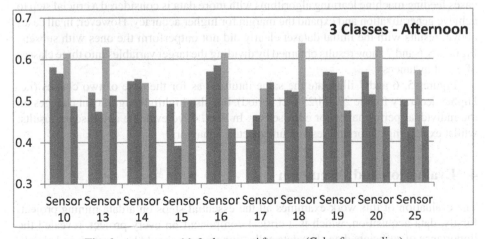

Fig. 6. Accuracy with 3-classes – Afternoon (Color figure online)

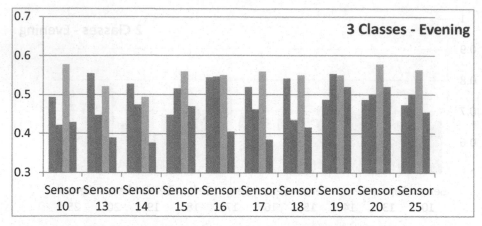

Fig. 7. Accuracy with 3-classes – Evening (Color figure online)

ensemble learning method that operates by constructing a multitude of decision trees. It is considered as a powerful method and stands as a top-notch solution for various problems [13]. On top of that, it handles both scaled and not scaled data. In order to prevent over-fitting, we used 10-fold cross-validation and for the optimal hyper-parameter tuning we used an extended grid search.

It is important to clarify that all experiments were also conducted using the Logistic Regression algorithm, however the results were not included since they were almost identical. That is to justify that our algorithm does not over-fit since our datasets are small. Further, Figs. 2, 3, 4 show classification results for each of the described time intervals. The Y-axis indicates accuracy with a start-point at 0.5 to highlight the margin of a baseline model on a perfectly balanced classification task.

Since the purpose of this work was to analyze and highlight differences between time and day periods on traffic prediction we did not focus on exact values. In most cases feeding machine learning algorithms with more data is considered a crucial step to achieve generalization and expand the margin for higher accuracy. However, in all cases above, results with the initial dataset clearly did not outperform the ones with subsets. Figures 5, 6 and 7 show results obtained by dividing the target variables into three classes of equal instances.

Figures 5, 6 and 7 illustrate the same intuition as for the case of two classes (i.e. Higher accuracy for the 29-Jul/25-Oct period), but also highlight significant changes on the individual performance for each sensor. In Sect. 4 we evaluate and discuss results, whilst explaining abnormalities and unexpected behavior.

4 Evaluation and Discussion

The evaluation of this work examines all the essential steps on a data mining project. Firstly, data acquisition, which is a critical component on every project, revealed the importance of sufficient data. The data collection process should be constant and clearly defined in advance. We dealt with a well-structured database that was recording sensor signals in a nearly perfect synchronization between traffic and weather data. Regarding the pre-processing step, we did not use all the available features, instead the models contained only weather metrics tested and introduced by the literature. For weather data exploitation, it is crucial to fully understand the correlation between variables, because the same level of information might be repeatedly captured. The rest of this section further discusses our findings per case.

The general problem in terms of real-time adjustment of routes in order to achieve traffic "congestions" is still most important for many cities. However, the day-ahead prediction of the volume was based on a major assumption that there would not be any accident or abnormal conditions in general. Starting from this point, the factor of environmental conditions was essential since many employees, students and tourists decide in advance the way they travel around the city on the upcoming day. The results of the approach presented in Sect. 3.3 are encouraging and justify what was stated above. All three-day periods on the binary classification achieved quite satisfying accuracy levels.

More precisely, the initial model, resulted in accuracy higher than 0.7 for most sensors on Morning and Afternoon. On the other hand, for the 3-classes experiments, results do not allow to reach safe conclusions. However, the initial model demonstrated robust performance achieving on average accuracy higher than 0.5. For the Morning period we observed predictability similar with the binary case. For Afternoon and Evening periods there were not significant gaps in accuracy even though the 29-Jul/25-Oct (light green line) period shows better performance than the rest. In addition, we observed that the 3-class experiments do not show similar patterns with the previous results.

Surprisingly, the middle period 15-Sep/20-Dec (red line) was expected to achieve much better results especially for the sensors 13 and 19 which are installed close to school areas. Indeed, the range of this period was selected to cover and adjust for the periods that schools operate, undisrupted. Results were contradicting, while Sensor 19 outperforms the rest on the Morning period for the binary case, Sensor 13 resulted in the lowest accuracy.

Figure 1 highlights the special case of August in Greece and the fact that the traffic load is highly reduced. In addition, as expected, on weekends traffic was heavily reduced while we observed that peaks on Afternoon and Evening periods for weekdays, happen mostly on Thursdays and Fridays.

Finally, it is worth noting that Sensors 15 and 17 are in Thessaloniki. For both, results were lower than the average for Athens. Based on these results, we observed that the range of traffic values for those two is amongst the highest.

5   Conclusions and Future work

Experiments for both stages produced some clear and informative results. The main conclusion is that the systematic recording and use of weather data can support decision making. The list below summarizes the conclusions.

- Standalone weather metrics can assist in building reliable prediction models regarding traffic volumes.

- Roads are busier in the afternoon and for most of the sensors even evenings have higher traffic volumes in comparison to mornings.
- Mornings do not return steady results for all sensors. As discussed, the results are stable only for sensors located near schools.
- For morning periods, the peaks of traffic happen in the beginning of the week. No earlier than the middle of September the load gets similar to that of Evenings and gets clearly lower again by the start of Christmas holidays.
- The first data subset regarding the period 29-Jul/25-Oct outperforms both other subsets and the initial data set consisting of all available data. That indicates that winter months introduce uncertainty and volatility, thus related models underperform significantly.
- Transforming the target variable into three classes resulted in admittedly good results, firmly better than the baseline model.

The above conclusions emerge from the detailed analysis of the models, weather metrics and traffic volumes. However, threats to their validity exist. We briefly summarize them in Sect. 5.1.

5.1 Threats to Validity

The biggest threat on approaches as such, is the fact that those day-ahead models rely on weather data which also are predicted. Thus, it is crucial that we have accurate predictions of weather conditions. Moreover, including the weekends that admittedly have different loads of traffic in the same models with weekdays may affect the validity on a negative way. Another threat could be sufficient deseasonilising; the factor of time could be possible analyzed into more explanatory variables. Not having available data for at least one year of recordings may lead to questionable conclusions about seasonal effects; however, this is not a rule. Finally, the sensors regard roads of different volume of traffic and even though traffic usually fluctuates uniformly that may conclude to misleading results.

Acknowledgements. The work is implemented within the co-funded project Public Participation City (ppCity - T1EΔK-02901 and MIS 5029727) by Action Aid "Research-Create-Innovate" implemented by General Secretariat of Research and Technology, Ministry of Development and Investments. The project (www.ppcity.eu) provides a set of tools and platforms to collect city environmental data and use these in an intelligent way in order to support informed urban planning. Key element in this process is the opinions and views of citizens which are collected in a crowd sourcing manner. The research depicted under this paper is based on Platform 3 (run from https://panel.ppcity.eu/platform3/) which provides an open data city portal from environmental data in Athens and Thessaloniki, Greece.

References

1. Theodorou, T.I., Salamanis, A., Kehagias, D., Tzovaras, D., Tjortjis, C.: Short-term traffic prediction under both typical and atypical traffic conditions using a pattern transition model. In: 3rd International Conference on Vehicle Technology & Intelligent Transport Systems, pp. 79–89 (2017)

2. Nejad, S.K., Seifi, F., Ahmadi, H., Seifi, N.: Applying data mining in prediction and classification of urban traffic. In: 2009 WRI World Congress on Computer Science and Information Engineering, pp. 674–678 (2009)
3. Xu, Y., Kong, Q., Liu, Y.: Short-term traffic volume prediction using classification and regression trees. In: 2013 IEEE Intelligent Vehicles Symposium (IV), pp. 493–498 (2013)
4. Wang, Y., Chen, Y., Qin, M., Zhu, Y.: Dynamic traffic prediction based on traffic flow mining. In: 2006 6th World Congress on Intelligent Control and Automation, Dalian, pp. 6078–6081 (2006)
5. Abdel-Aty, M.A., Pemmanaboina, R.: Calibrating a real-time traffic crash-prediction model using archived weather and ITS traffic data. IEEE Trans. Intell. Transp. Syst. **7**(2), 167–174 (2006)
6. Yan, H., Yu, D.: Short-term traffic condition prediction of urban road network based on improved SVM. In: 2017 International Smart Cities Conference, pp. 1–2 (2017)
7. Tang, J., Li, L., Hu, Z., Liu, F.: Short-term traffic flow prediction considering spatio-temporal correlation: a hybrid model combing type-2 fuzzy c-means and artificial neural network. IEEE Access **7**, 101009–101018 (2019)
8. Liu, Y., Wu, H.: Prediction of road traffic congestion based on random forest. In: 2017 10th International Symposium on Computational Intelligence and Design, pp. 361–364 (2017)
9. Ai, Y., Bai, Z., Su, H., Zhong, N., Sun, Y., Zhao, J.: Traffic flow prediction based on expressway operating vehicle data. In: 2018 11th International Conference on Intelligent Computation Technology and Automation, pp. 322–326 (2018)
10. Clark, S.: Traffic prediction using multivariate nonparametric regression. J. Transp. Eng. **129**(2), 161–168 (2003)
11. Christantonis, K., Tjortjis, C.: Data mining for smart cities: predicting electricity consumption by classification. In: IEEE 10th International Conference on Information, Intelligence, Systems and Applications, pp. 67–73 (2019)
12. Dunne, S., Ghosh, B.: Weather adaptive traffic prediction using neurowavelet models. IEEE Trans. Intell. Transp. Sys **14**(1), 370–379 (2013)
13. Tzirakis, P., Tjortjis, C.: T3C: improving a decision tree classification algorithm's interval splits on continuous attributes. Adv. Data Anal. Classif. **11**(2), 353–370 (2017). https://doi.org/10.1007/s11634-016-0246-x

Using Twitter to Predict Chart Position for Songs

Eleana Tsiara and Christos Tjortjis[(✉)] [iD]

The Data Mining and Analytics Research Group, School of Science and Technology,
International Hellenic University, 14th Km Thessaloniki–Moudania, 57001 Thermi, Greece
c.tjortjis@ihu.edu.gr

Abstract. With the advent of social media, concepts such as forecasting and now casting became part of the public debate. Past successes include predicting election results, stock prices and forecasting events or behaviors. This work aims at using Twitter data, related to songs and artists that appeared on the top 10 of the Billboard Hot 100 charts, performing sentiment analysis on the collected tweets, to predict the charts in the future. Our goal was to investigate the relation between the number of mentions of a song and its artist, as well as the semantic orientation of the relevant posts and its performance on the subsequent chart. The problem was approached via regression analysis, which estimated the difference between the actual and predicted positions and moderated results. We also focused on forecasting chart ranges, namely the top 5, 10 and 20. Given the accuracy and F-score achieved compared to previous research, our findings are deemed satisfactory, especially in predicting the top 20.

Keywords: Social media analytics · Prediction · Classification · Regression

1 Introduction

Social media have penetrated everyday life to the point that they constitute an integral part of our daily routines. The vast amount of data available through these services can be utilized in many domains including finance, marketing, and politics [8]. Social media can be exploited in a variety of cases, such as forecasting the commercial success of movies [1], election result predictions etc. [7, 9].

In this paper, we chose Twitter to generate predictions for the Billboard chart. After gathering chart data, including titles, artist names and rankings, as well as tweets related to the top 10 songs for each week, results showed a moderate correlation between the number of mentions of a song and its future performance, but no relation between the number of mentions of an artist and their imminent success.

This work has the following objectives: a) to acquire data from the Billboard chart, including rank, artist and song title of the top 100 songs at the current time. For that purpose we developed a method which extracts these parameters from the official site and saves them in a .json and a .csv file. b) to collect Twitter posts concerning the top 10 songs utilizing the Twitter Search API. c) to preprocess data into a homogenous, structured format removing redundant information. d) to perform sentiment analysis on

I. Maglogiannis et al. (Eds.): AIAI 2020, IFIP AICT 583, pp. 62–72, 2020.
https://doi.org/10.1007/978-3-030-49161-1_6

tweets, categorizing each post as positive, negative or neutral. e) to assess the contribution of the features, extracted from the Billboard chart and collected posts, to forecast the chart for the week to come.

The aim is to predict the top N songs for the following week and evaluate the efficiency of the process. This requires a classifier to generate predictions. The optimal number of N, the highest performing classification algorithm and the best feature combination can be determined by experimentation and result comparison.

The remaining of this paper reviews background in Sect. 2, describes our approach in Sect. 3, discusses results in Sect. 4 and concludes with directions for future work in Sect. 5.

2 Background

This section reviews forecasting related literature. Twitter was employed to predict the commercial success of movies [1]. The choice of film was based on the number of discussions and the difference of opinion, as well as by obtaining financial information about it. After collecting about 3 million tweets, the authors used a linear regression model and performed sentiment analysis to conclude that there is a correlation between the fame of a movie prior to its release and the revenue it produces.

Sales prediction was targeted in [3]. The aim was to estimate the impact of each variable such as posts, comments and likes on a group of Nike's Facebook page, examining each variable and each page individually, as well as a combination of all variables and all pages. Moreover, the predictive impact of search query data and the relation between Nike's events and the subsequent Facebook activity was explored. Simple regression scored as high as Bloomberg forecasts in terms of accuracy for predictions pertaining to the near future.

Another study that used Twitter data aimed at predicting the last presidential elections in the USA [7]. Its first objective was to retrieve data from Twitter, while organizing the timing of data collection according to important dates, such as debates. It then used sentiment analysis on 277,509 tweets. Each tweet was given a polarity and subjectivity score, depending on being negative, positive or neutral. Naïve Bayes was used for classification. The results were accurate and in fact predicted the right candidate, in contrast to most polls.

With regards to the prediction of the Billboard chart. Yekyung et al. focused on two tasks: a) the relationship between Twitter activity regarding music and forthcoming sales and b) predicting hit songs for the next Billboard chart [10]. The authors gathered over 30 million tweets, searching for the keywords #nowplaying, #np and #itunes, presumably because these hashtags are used to indicate the song a Twitter user is currently listening to. They also collected information from previous Billboard charts over the span of 10 weeks, which resulted in a dataset of songs, each with its own title, artist, rank and the time period it stayed on the chart. Altogether, the information retrieved related to 178 songs and 134 artists. 3 distinct metrics were established: a) song popularity, i.e. the number of tweets related to a specific song, b) artist popularity, i.e. the number of tweets mentioning this artist, and c) the number of weeks a song appeared on the Billboard chart. Forecasting targeted predictions for the top 10 songs, since this was the range with the highest accuracy achieved. Using the Pearson correlation, artist popularity and number of weeks on the chart, were not found to be strong predictors for a song's ranking. However, considering song popularity if all 3 metrics are combined, it is possible to predict the imminent success of a song quite accurately.

Koenigstein et al. approached Billboard chart prediction through the exploitation of Gnutella peer-to-peer network [6]. In total, 185,598,176 query strings, originating from the USA, were gathered during a 30-week period. They used both M5 and C4.5 algorithms to predict the Billboard Hot 100 and the Billboard Digital Songs charts. Moreover, in some cases, they also considered a song's debut rank on the chart, while predicted positions were usually either the top 10 or 20. They verified that queries used in peer-to-peer services are strong predictors for a song's success. For the Billboard Hot 100 chart, precision surpassed 86%, while for the Billboard Digital Songs accuracy exceeded 89%.

Finally, Zangerle et al. examined the relationship between song-related tweets and their ranking on the Billboard Hot 100 [11]. Their objective was to investigate the resemblance of the amounts of Twitter data referring to Billboard tracks to the state of the actual chart, along with the temporal offset between them, so that it can be determined if tweets have a predictive value for the chart and not vice versa. The authors used an existing dataset which consists of 111,260,925 tweets that included the term #nowplaying, gathered during 2014 and 2015. They also collected data from the Billboard Hot 100 chart for the same time period, corresponding to 886 distinct songs. They observed that a song remains on the chart for an average of 11.74 weeks and may stay on the top 100 for up to 58 weeks. To determine the correlation of rankings, they calculated three metrics per song, based on tweets: the median of play-counts per week, the mean play-counts per day and the total play-counts for a whole week. The median achieved the highest correlation (0.5), for 481 songs or 54.29% of the dataset.

The temporal relationship between tweets and charts was investigated through cross-correlation analysis that increased the mean correlation to 0.57 compared to the value of 0.5 that was previously found. 89.23% of all the examined tracks appear to have a temporal lag in relation to the Billboard chart, while 41.09% of all tracks have a negative lag. This means that the last percentage is exploitable for providing predictions about the chart's ranking for the weeks to come. Moreover, the authors followed the same process for songs that were first noticed in Twitter data and appeared on the chart at a later point (619 tracks in total). Similarly, 42.64% of them featured a negative lag thus facilitating future chart forecasts.

Regarding the predictive capability of Twitter data, they compared 3 models: one based exclusively on Billboard chart data, one relying only on tweets and the last one combining both. In terms of RMSE, the Twitter-based model had the worse score of 116.1, while the first one achieved an RMSE of 26.8. However, the multivariate model displayed a notable performance of 14.1. In conclusion a combination of data originating from both the Billboard chart and Twitter can significantly reduce forecasting error, thus song-related tweets can be useful for increasing the accuracy of ranking predictions. Summing up, about 41% of the collected tweets could be used for the prediction of the Billboard chart ranking if handled properly.

Having considered the literature, it can be hypothesized that the more attention a song gathers in Twitter, the more likely it is for people to listen to it and possibly end up purchasing it, further adding to its total popularity. Similarly, the whole publicity an artist gets and the image she presents, considering both her career and personal life, will probably motivate the public to check her work and increase her commercial success.

3   Approach

The methodology we followed is depicted in Fig. 1 and described below. The Search API was utilized for interacting with Twitter, using Python scripts for searching for tweets, storing posts and performing sentiment analysis. Database (DB) management tasks, including storing and retrieving Twitter data, were implemented using SQL Server. All useful information extracted from DB records was preprocessed and input to Weka [4] for mining.

3.1   Twitter Search API

Regarding song searching, tweets related to specific words were collected mainly to estimate
their airplay count, either from radio stations or
from individual users. These queries incorporate the term #nowplaying, which is joined via
the AND operator with the titles of all songs,
in brackets. Each title is placed in quotes and
brackets, if it consists of multiple words. Quotes
are necessary to ensure that the post contains all
the words in the same order, and they are not
just spread in the text. This would lead to the
collection of many irrelevant tweets and would
add noise to the dataset. Moreover, similarly to
the queries about artists, all words of each title
relate to the removal of spaces, and the hashtag symbol is inserted at the beginning of each
concatenated word.

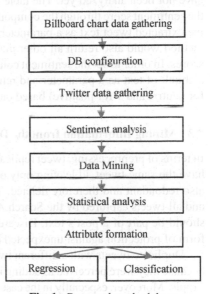

Fig. 1. Proposed methodology

The strings for searching for songs do not
include artist names, since they might exceed
the 500-character limit and unnecessary complexity would be added to queries. Instead, artist names and titles are matched, when retrieving data from the DB where all tweets are stored. This ensures that the posts refer to the songs that need to be analyzed.

3.2   Sentiment Analysis

Overall, the VADER lexicon comprises useful scripts [5]. Compared to VADER, senti-ment analysis with SentiWordNet [2] is not easy to implement, since it includes only the lexicon without any ready-to-use code. Therefore, it is up to the programmer to develop her own code to manipulate the lexicon and make the appropriate configurations to get the result she needs. The task is rather cumbersome because each word needs to be treated separately. It needs to be tagged based on the part of speech they represent, which can be noun, verb, adjective, adjective satellite (a subcategory of adjective according to WordNet) and adverb, and also each term has to be assigned with a score indicating its usage frequency.

For that reason, a handy script from [12] was utilized, featuring a class with methods for calculating different sentiment parameters for a given word or phrase. Most importantly, a procedure for estimating the total sentiment score of a sentence was used. This was the metric of interest we used to make comparisons with the respective result of VADER's sentiment analysis. The scoring method considers the following negation words: (not, n't, less, no, never, nothing, nowhere, hardly, barely, scarcely, nobody, none). Furthermore, there are 3 options for the generation of the sum of scores: average, geometric and harmonic.

The connection to the DB and the execution of queries was implemented in a similar way. However, in this script, there is a need for an extra query to fetch all the tweets that have not been analyzed yet. The table is updated with the compound column receiving the sentiment score through the compound dimension of the polarity_scores method with the extracted tweet text as a parameter. If the compound column was not specified, the method would also return all other metrics, meaning the positive, negative and neutral scores. In order to set the sentiment column, a new method has been defined, which takes a chunk of text as a parameter and returns its semantic orientation (-1 for negative, 0 for neutral and 1 for positive) based on the common thresholds of the compound value.

3.3 Mining Information from the Database

In terms of preprocessing, tweet replication is treated with the removal of all records that have the same tweet_id leaving only one copy. Any records with a black text field are also redundant and therefore deleted. Twitter typically does not allow tweeting blanks and all tweets captured by the Search API are retrieved using specific keywords, which should be part of a post's text. Discarding posts with an empty string is just an added form of protection against unexpected behavior.

Duplicate tweets with a different tweet_id field, but identical text, were not removed because they were perceived as adding information and depicting extra attention towards a topic. Moreover, especially in the case of songs that are gathered using the #nowplaying term, the content of the posts is mainly fixed, as it usually includes the artist and title of the track, sometimes followed by a link.

Twitter's Search API captures all tweets, including retweeted posts. These can be sometimes recognized from their textual content, which begins with the pattern "RT @[username]", where [username] refers to the user being quoted. Nevertheless, this is not an official feature and it is not necessarily inserted in every retweet, so there is no accurate way to determine that a post captured via the API is definitely a retweet.

In order to produce a tradeoff between the number of occurrences for each distinct text value (that is the number of posts that have the exact same text) and the number of retweets, the greatest value of the two is chosen. The idea behind this is to get the biggest sample possible and exclude the difference between the two parameters, which is probably tweets that coincide.

For retrieving song tweets from the DB, the song title is used with and without spaces among words. For most tracks, the artist must also appear in the text. This is a measure to filter out any records with different songs that happen to have the same title or phrases which are absolutely irrelevant to the tracks under examination.

3.4 Attribute Description

In order to use Weka, all the information extracted from the Billboard chart and gathered tweets should be organized in attributes and formatted properly to create an .arff file. The file consisted of 80 instances in total with no missing values. Each instance describes a specific track for one week, during which the track was at one of the top 10 positions of the chart. So, the data set consists of 10 instances per week.

Since our hypothesis is that most of the attributes are positively correlated to the success of a song, a song will get to a higher position when attribute values increase. The Billboard chart ranks songs in ascending order, starting with the most popular ones. So, mainly for harmonization with the rest of the attributes, as it was seen in [10], the position of each track is inverted by subtracting it from 101. For instance, number 1 song will be in position 100, while number 10 in 91 and so on. This also applies to previous-position and position attributes.

Each instance of the data represents a different song for a single week and it has to be compared to all other instances, which are the remaining 9 tracks from the top 10 positions of the chart for the same week and 10 more tracks for every week that has been monitored. Data gathering took place daily in a scheduled manner, but the total number of tweets captured on a weekly basis would differ each time, despite requesting a consistent number of items. Therefore, measuring every instance against all the others required some transformation on the specimen of tweets. For each week, the total number of posts for the top 10 songs was summed and the percentage of each track in comparison to the other 9 was calculated. Attributes song-play-count, artist-play-count and artist-tweets were tuned as described above.

The attributes selected are described below:

- Previous-position: The song position one week before.
- Chart-weeks: The number of weeks that the song has been in one of the 100 positions of the chart since its first entry. This value is set to 1 as soon as the song makes its chart debut.
- Top-weeks: The number of weeks that the song was in number 1.
- Song-play-count: The tuned number of song related tweets, including retweets, with the #nowplaying keyword.
- Artist-play-count: The tuned number of tweets, including retweets, which are related to the artist who sings this particular track, along with the #nowplaying keyword, not necessarily referring to this specific song. This is an estimation of the total number of play-counts for the artist, irrespective of the song being played. If a song belongs to more than one artist, the artist with the maximum number of tweets associated with her, represents the song.

- Artist-tweets: The tuned number of tweets, including retweets, which are related to the artist who sings this track. If a song belongs to more than one artist, the artist with the maximum number of tweets associated with her represents the song.
- Artist-sentiment-analysis-1: The average value of the score, produced by VADER via sentiment analysis on all tweets associated with the artist of this song. If there are many artists, the choice is made similarly to the previous cases.
- Artist-sentiment-analysis-2: This attribute is calculated like artist-sentiment-analysis-1 and the only difference is using SentiWordNet for score generation.
- Position: The position of the song for that week. This is the class attribute to be predicted.

Table 1 shows type and value restrictions for each attribute.

Table 1. Attribute type and values

Attribute	Type	Range
Previous-position	Integer	[0, 100]
Chart-weeks	Integer	≥ 1
Top-weeks	Integer	≥ 0
Song-play-count	Real	[0, 100]
Artist-play-count	Real	[0, 100]
Artist-tweets	Real	[0, 100]
Artist-sentiment-analysis-1	Real	[−1, 1]
Artist-sentiment-analysis-2	Real	[−1, 1]
Position	Integer	[0, 100]

4 Results and Discussion

This section presents and discusses results.

4.1 Results

Table 2 presents the attributes in Pearson's correlation value descending order.

Based on this output, the attributes can be divided in 3 categories with regards to their correlation (r) with the predicted class (position):

- **Moderately correlated attributes (0.2 < r < 0.7):** These are: *chart-weeks*, *song–play-count* and *previous-position*.
- **Weakly correlated attributes (0.1 < r < 0.2):** *Artist-play-count*, *artist-sentiment-analysis-2*, *artist-sentiment-analysis-1* and *top-weeks* are the attributes in this category.
- **Unrelated attributes (r ~ 0):** *Artist-tweets* is the only attribute in this category with its correlation coefficient close to 0.

Table 2. Pearson's correlation per attribute

Attribute	Pearson's correlation coefficient
Chart weeks	0.377
Previous position	0.370
Song playcount	0.292
Artist playcount	0.147
Artist sentiment analysis 2	0.134
Artist sentiment analysis 1	0.133
Top weeks	0.123
Artist tweets	0.013

Regarding regression analysis, the following results were generated using 10-fold cross-validation. Support Vector Regression had the smallest Mean Absolute Error (MAE): 4.0515 and Random Forest the lowest Root Mean Square Error (RMSE): 8.8117. Bagging also achieved decent results with RMSE: 8.961 and the highest correlation coefficient: 0.5697.

Hit prediction refers to forecasting whether a song is going to be within a specific range of positions or not. This is a classification problem, as each instance should be matched to one of the two available classes. For this purpose, a new attribute should be created through Weka's preprocessing functionality. After experimenting with all available algorithms using 10-fold cross-validation, J48 and PART appeared to be the classifiers with the best performance in terms of both accuracy and F1 scores.

The best scoring algorithms in the case of top 5 hits are Filtered Classifier and Decision Table, achieving 90% accuracy and the same for precision, recall and F1 score. F1-scores was 0.875 for hits and 0.917 for non-hits. For the prediction of the top 20 hits of the following week, the classifiers that achieved high accuracy and F1 values were LMT and Simple Logistic.

4.2 Discussion

This study investigated the forecasting strength of social media, focusing on the prediction of the Billboard's Hot 100, based on data extracted from the chart and music-related Twitter posts referring to artists and songs on the top ten each week. In total, more than one million tweets were gathered, and data collection lasted for about 2 months, during October and November in 2018.

Twitter data appear to be quite representative when it comes to the number of total play-counts for a song. Considering the generally accepted ranges, there is a moderate correlation (0.2917) between the number of tweets that include the title of a song and the #nowplaying terms as a subset and the imminent success of the song on the Billboard Hot 100 chart for the following week. On the contrary, tweets that provide an estimation for an artist's total play-counts in general are not adequate to give an accurate picture of the future performance of a particular song, as the correlation coefficient was shown to be weak.

There appears no relation between an artist's publicity, positive or negative, expressed by the total number of tweet mentions, and his ranking on the chart, since the value of the correlation coefficient was close to 0 (0.0134). Furthermore, there is no evidence that the positive attention an artist gets, represented by a value that estimates how favorable the posts related to her are, is significantly correlated to the positioning of her tracks on the Billboard chart, at least in the short term. Findings concerning sentiment analysis cannot be generalized for other domains other than this particular chart, as many other studies have ascertained [1, 7].

Features derived from tweets combined with chart data, can provide results of noteworthy accuracy, but this happens only in the investigation of specific aspects of the problem. Regarding regression analysis, the best performance was achieved by Support Vector Regression with MAE 4.0515 and Random Forest with RMSE 8.8117. MAE value means that the position predictions were on average 4 ranks away from the actual values. Nevertheless, RMSE is, in most cases, considered to be more reliable, as it penalizes outliers. According to RMSE, the average derivation is approximately 8 positions.

Result evaluation depends on the type of problem that needs to be addressed. For example, if the goal is to predict all the Billboard chart for the following week, the values could be considered useful. On the other hand, if it is desired to predict a particular range of chart positions, especially if this is limited, such as top 10 hits for instance, then this model would not be able to provide an accurate prediction. The highest correlation coefficient was 0.5697, with Random Forest, and the squared correlation coefficient was approximately 0.325, which is low compared to the best result found in [10], with a squared correlation of 0.57, using Support Vector Regression and 5-fold validation.

Hit prediction, implemented via classification, yielded some promising results: top 10 songs could be predicted with an accuracy of 85% utilizing J48 and PART. The F-scores for PART was 0.906 for hits and 0.625 for non-hits, while the same values for J48 were 0.909 and 0.571. In [10], Random Forest with 5-fold cross-validation, achieved 90% accuracy and 0.901 and 0.899 F-scores for hit and non-hits, respectively.

Experimenting with different ranges for hits, specifically the top 5 and top 20, scores were higher than for the top 10 range. Filtered Classifier and Decision Table achieved an accuracy of 90% and F-scores of 0.875 for the top 5 hits and 0.917 for non-hits. For the prediction of the songs in the 20 highest positions Simple Logistic and LMT achieved accuracy of 96.25%. Specifically, the F-scores for hits and non-hits were 0.980 and 0.727, respectively. In [10], the accuracy for the same range (with the Random Forest Classifier and a 5-fold validation) was 88.2% and the F-score for hit songs was 0.885. However, they achieved a better F-score for non-hits, which was 0.879.

5   Conclusions and Future Work

This study gathered data from two sources: titles, artist names and rankings from the Billboard Hot 100 chart and tweets related to songs in the top 10, in order to test how their integration could be used for forecasting future charts. Our findings suggest that there is a moderate correlation of 0.2917, between the number of mentions of a song and its chart performance. No significant relationship was observed between the attention an artist gets on Twitter, even via emotionally charged tweets, and their tracks' success.

Regarding regression analysis, the best score was a mean square error of 4.0515 and RMSE 8.8117. Both metrics show the average number of positions between the actual and predicted values, and their evaluation depends on the range of positions we want to predict. With the best scoring algorithm, hit prediction could be achieved with an accuracy of 80% and F-scores of 0.881 and 0.385 for hits and non-hits, for the 10 highest ranks. Similarly, for the top 5 hits these values were 90%, 0.917 and 0.875, respectively. Finally, for the top 20 songs, the results were competitive to previous research, achieving top accuracy of 96.25% and F-scores 0.980 for hits and 0.727 for non-hits.

Overall, the study has shown that mining Twitter data, extracting specific information and handling them properly can provide useful conclusions regarding the next Billboard chart, although it is not yet capable of perfect predictions.

5.1 Limitations and Future Research

This research gathers and examines Twitter and Billboard chart data for hit songs at the top 10. Tracks at lower ranks were not considered, thus it is impossible to predict their way up. So, future research could investigate the chart at a larger scale and consider all songs on the chart, and create a more comprehensive dataset, consisting of chart data for a period of one or more years, like the ones used in [10] and [11].

From a technical point of view, researchers are encouraged to try to capture tweets incorporating the #nowplaying term in general, without targeting specific songs or artists. Finding Twitter data referring to a set of 100 songs for each week would require a complicated methodology if the Search API is used, due to the search query limitations that were discussed in Sect. 3.1. On the other hand, a more generalized approach would demand a lot of space for storing data and would capture a lot of irrelevant tweets that would need to be excluded at a later stage. Thus, there should also be extra preprocessing effort. Nevertheless, collecting tweets indiscriminately ought to give a better insight regarding airplay trends and could probably aid the discovery of songs that are not at the chart at the given moment, but would be introduced in the weeks to come.

Undoubtedly, an exhaustive study of the optimized way to predict the Billboard chart should rely on a model that considers all parameters that affect song ranking, such as streaming activity and song sales. During data gathering and experimentation, it was observed that many tracks made their debut into the chart by occupying one of the first 10 positions, consequently they could not have been monitored even if all 100 songs were accounted for. They also tended to follow a specific pattern: usually they dropped significantly after their successful entrance into the chart and their play-counts seemed to be quite low at that time. It is, therefore, speculated that their steep rise to the top is basically owing to music sales and streaming data. In order to get a better idea about this phenomenon, research should be extended beyond the limits of the Billboard chart and the airplay counts and consider the other dimensions that contribute to the formation of the chart each week.

With regards to sentiment analysis, the degree to which tweets constitute a trust-worthy sample for determining the positive or negative disposition of the population towards an artist is still vague. There may be a relationship, but it may not be directly visible. This fact raises new questions about the predictive strength and limits of sentiment analysis in social media and is something worth investigating in the future. It is suggested that researchers explore the impact of sentiment analysis in a longer term context. This essentially means, monitoring sentiment orientation for an artist and track the performance of her songs for many weeks, possibly considering some time lag, as the public's response to a likeable or unlikeable person may be revealed gradually.

Different angles can also be investigated, for instance, the relationship between a positive or negative bias towards an artist and the number of tracks belonging to him that are on the chart at the present time or the rate at which these tracks ascend and descend in the chart. Researchers could, additionally, consider improving the mechanism of sentiment analysis and attempt to enhance existing lexicons in order to make them more suitable for music-related data. For instance, some terms referring to music, like radio, album, concert, listen, etc., which would otherwise be characterized as neutral, could infuse a positive tone into a tweet, when the name of an artist is included in the text.

References

1. Asur, S., Huberman, B.: Predicting the future with social media. In: 2010 IEEE/WIC/ACM International Conference on Web Intelligence and Intelligent Agent Technology (WI-IAT), vol. 1, pp. 492–499 (2010)
2. Baccianella, S., Esuli, A., Sebastiani, F.: SentiWordNet 3.0: an enhanced lexical resource for sentiment analysis and opinion mining. In: Proceedings of International Conference on Language Resources and Evaluation, LREC 2010, 17–23 May 2010
3. Boldt, L.C., Vinayagamoorthy, V., et al.: Forecasting nike's sales using facebook data. In:IEEE International Conference on Big Data (2016)
4. Witten, I.H., Frank, E., Hall, M.A., Pal, C.J.: Data Mining: Practical Machine Learning Tools and Techniques. Morgan Kaufmann, Burlington (2016)
5. Hutto, C.J., Gilbert, E.: VADER: a parisomonious rule-based model for sentiment analysis of social media text. In: Proceedings of 8th International AAAI Conference on Weblogs and Social Media (2014)
6. Koenigstein, N., Yuval, S., Zilberman, N.: Predicting billboard success using data-mining in P2P networks. In: 11th IEEE International Symposium on Multimedia (2009)
7. Oikonomou L., Tjortjis C.: A method for predicting the winner of the USA presidential elections using data extracted from Twitter. In: 3rd IEEE SE Europe Design Automation, Computer Engineering, Computer Networks, and Social Media Conference (2018)
8. Rousidis, D., Koukaras, P., Tjortjis, C.: Social media prediction a literature review. Multimed. Tools Appl. **79**(9), 6279–6311 (2020). https://doi.org/10.1007/s11042-019-08291-9
9. Beleveslis, D., Tjortjis, C., Psaradelis, D. Nikoglou, D.: A hybrid method for sentiment analysis of election related tweets. In: 4th IEEE SE Europe Design Automation, Computer Engineering, Computer Networks, and Social Media Conference (2019)
10. Yekyung, K., Bongwon, S., Kyogu, L.: #nowplaying the future billboard: mining music listening behaviors of twitter users for hit song prediction. In: Proceedings of 1st International Workshop on Social Media Retrieval and Analysis, pp. 51–56 (2014)
11. Zangerle, E., Pichl, M., Hupfauf, B., Specht, G.: Can microblogs predict music charts? An analysis of the relationship between #nowplaying tweets and music charts. In: Proceedings of 17th International Society for Music Information Retrieval Conference 2016
12. Sentiment analysis. github.com/anelachan/sentimentanalysis. Accessed 20 Feb

A Benchmarking of IBM, Google and Wit Automatic Speech Recognition Systems

Foteini Filippidou(✉) ⓘ and Lefteris Moussiades ⓘ

Department of Computer Science, International Hellenic University,
Agios Loukas, 65404 Kavala, Greece
{fnfilip,lmous}@cs.ihu.gr

Abstract. As the requirements for automatic speech recognition are continually increasing, the demand for accuracy and efficiency is also of particular interest. In this paper, we present most of the well-known Automated Speech Recognition systems (ASR), and we benchmark three of them, namely the IBM Watson, Google, and Wit, using the WER, Hper, and Rper error metrics. The experimental results show that Google's automatic speech recognition performs better among the three systems. We intend to extend the benchmarking both to include most of the available Automated Speech Recognition systems and increase our test data.

Keywords: Automatic speech recognition · WER · Hper · Rper

1 Introduction

Speech is probably the primary means of man communication. Therefore, acquiring speech from computers is reasonable to contribute to more effective man-machine communication. Two primary technologies have developed concerning speech: the Automatic Speech Recognition or ASR, for short, and the Text to Speech or TTS. ASR refers to the conversion of speech into text [1], while TTS, as its name suggests, is the reverse process [2]. Although man naturally acquires speech in early growth [3], speech production and recognition by computers is a complicated process that has extensively been addressed by the research community.

As early as 1952, the first system was built that could identify digits with high precision. This early system required users to pause after each digit [4]. Raj Reddy constructed the first recognition system of continuous speech as a student at Stanford University in the late 1960s [Wikipedia - Speech Recognition].

Shortly after, T.K. Vintsyuk presented an algorithm [7] that can recognize speech, creating a sequence of words that contained in continuous and connected speech.

Later, in 1981, Logica developed a real-time speech recognition system based on the original project of the Joint Speech Research Unit in the United Kingdom. This user-friendly system is one of the first steps in improving human-machine communication [5]. Today, ASR applications are not just confined to human-machine communication for personal use but include industrial machine guidance with voice commands [6, 7], automatic

I. Maglogiannis et al. (Eds.): AIAI 2020, IFIP AICT 583, pp. 73–82, 2020.
https://doi.org/10.1007/978-3-030-49161-1_7

telephone communication [8], communication with automotive systems, military vehicles, and other equipment, communication with health care, aerospace and other systems [9]. Also, ASR systems are utilized to provide more specialized services, such as training in pronunciation and vocabulary [10].

As the number of ASR systems is continuously increasing, it is quite challenging to select the most appropriate for a particular application. Our immediate plans are to build an artificial vocabulary learning assistant. In particular, this assistant will be able to contribute to the learning of vocabulary by entering into dialogues with the trainee. In this context, it is necessary to select the appropriate ASR. Thus, this work first attempts a presentation of known ASR systems and then proceeds to benchmark three well-known systems, namely IBM Watson, Google, and Wit. Therefore, we set the foundations for a more general assessment of most ASR systems with the ultimate goal of choosing the most appropriate vocabulary learning assistant.

The rest of this review is organized as follows. In section two, we briefly present analyze the Architecture of ASR systems. In section three, some of the most well-known ASR systems are introduced. In section four, we describe criteria and metrics used for system evaluation. Next, in section five, we analyze the data and the methodology we use for the experimental procedure. Finally, section six closes the review with final remarks and conclusions.

2 Architecture

An ASR application accepts the speech signal as input and converts it into a series of words in text form. During the speech recognition process, a list of possible texts is created, and finally, the most relevant text to the original sound signal is selected [4, 5]. A typical ASR consists of an acoustic front-end that process the speech signal to extract useful features [11]. Then, a feature vector is generated. Several feature extraction methods are used including, Principle Component Analysis (PCA), Linear Discriminant Analysis (LDA), Independent Component Analysis (ICA), Linear Predictive Coding (LPC), Cepstral Analysis, Mel-Frequency Scale Analysis, Filter-Bank Analysis, Mel-Frequency Cepstrum Coefficients (MFCC), Kernal Based Feature Extraction, Dynamic Feature Extraction, Wavelet-based features, Spectral Subtraction and Cepstral Mean Subtraction (CMS). Generally, some spectra temporal analysis of the signal generates features that usually transform into more compact and robust vectors [11]. At the processing step, the acoustic lexicon and language model are used by a decoder (search algorithm) to produce the hypothesized word or phoneme, as shown in Fig. 1. The acoustic model contains acoustic features for each of the distinct phonetic units and typically refers to the process of establishing the statistical representations for the feature vector sequences computed from the speech waveform. The Hidden Markov Model (HMM) is one of the most commonly used to build acoustic models. Other acoustic models include segmental and super segmental models, neural networks, maximum entropy models, and conditional random fields. An acoustic model is a file that contains statistical representations of each of the distinct sounds that makes up a word [11]. The lexicon includes terms of the vocabulary of the current application [11]. The language model consists of the limitations associated with the sequence of words that are accepted in a given language.

Two popular tools for language modelling are the CMU Statistical Language Modeling (SLM) Toolkit and the Stanford Research Institute Language Modeling Toolkit [11]. The decoder aims to find the most likely sequences of words that could match the audio signal by applying appropriate models. Most decoding algorithms produce a list of possible word sequences called the n-best list [4, 5].

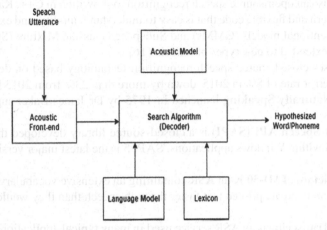

Fig. 1. Speech recognition architecture [11]

3 Well-Known ASR Systems

Many international organizations, such as Microsoft SAPI, Dragon-Natural-Speech, and Google, as well as research teams, are working in the area of ASR systems [3]. Next, we describe in short some of the most well-known ASR systems.

CMU-Sphinx was introduced in 1986 and is a set of open-source speech recognition libraries. It supports English but not Greek. Contains a series of packages for different tasks and applications:

Pocketsphinx: Lightweight library with C for written recognition [12]. It is fast, runs locally, and requires relatively modest computational resources. It provides nbest lists and lattices and supports incremental output. It provides voice activity detection functionality for continuous ASR and is also fully customizable and trainable [13].

Sphinxbase: Support for the libraries required by Pocketsphinx [12].

CMUclmtk: Language model tools [12].

Sphinxtrain: Acoustic model tool. It includes a set of programs and documentation for making and building audio models for any language [12].

Sphinx3: Voice recognition decoder is written in C [12].

Sphinx4: The latest Java-based speech recognition tool made by Carnegie Mellon University (CMU), Sun Microsystems Laboratories, and Mitsubishi

Electric Research Laboratories (MERL). It is flexible, supporting all types of acoustic models based on the Hidden Markov model (HMM), all basic types of language models, and multiple search strategies [14].

CMU also attempted to develop an Amazigh speech recognition system used in a vast geographical area of North Africa [15].

Kaldi is also an open-source speech recognition tool, written in C++. Kaldi's goal is to have a modern and flexible code that is easy to understand, modify, and expand. Kaldi supports conventional models (GMMs) and Subspace Gaussian Mixture (SGMMs) but can be easily extended to new types of models [16].

Google uses closed-source speech recognition technology based on deep learning achieving an error rate of 8% in 2015, down by more than 23% from 2013 [17].

Dragon Naturally Speaking launched in 1975 by Dr James Baker supports ASR and TTS [18].

Microsoft Speech API (SAPI) is a closed-source library developed that supports ASR and TTS within Windows applications. SAPI 5 is the latest major version released in 2000 [2].

DragonDictate (TM)-30 K an ASR containing an extensive vocabulary that allows interactive users to create printed text faster through speech than they would have done by hand [19].

Amazon Transcribe is an ASR service used in many typical applications, including phone customer service systems, and automatic subtitling for audio and video. Amazon Transcribe is continuously learning and improving to keep up with the development of language [20].

Microsoft Azure provided as a service, converts speech to text in real-time. It includes an API Speech SDK and a REST API. The latter can be used only for queries containing up to 10 s recorded audio [21].

Wit is free even for commercial use. It supports over 130 languages, and it applies the applicable EU data protection laws, including GDPR. Wit supports Node, Python, and Ruby programming languages [22].

Twilio includes vocabulary to support the industry. It recognizes 119 languages and dialects. It is paid but also has a free trial version [23].

Houndify is a platform that allows anyone to create intelligent communication interfaces with voice capability. It supports both ASR and TTS on Android, iOS, C++, Web, Python, Java, and C Sharp. It is available for a fee but also has a free version [24].

IBM® (Speech to Text) allows users to add up to 90 thousand out-of-vocabulary (OOV) words to a custom language model. The service utilizes machine learning to combine the knowledge of grammar and language structure as well as the synthesis of sound and voice signals to transform the human voice accurately. It continually renews and improves its function as it receives speech. It can support continuous speech. Programming languages that supports are Node, Java ™, Python, Ruby, Swift, and Go SDKs [25]. The service offers three interfaces: A WebSocket and synchronous HTTP interface, each being able to transfer up to 100 MB of audio data with one request. It also offers an asynchronous HTTP interface, which with one request can pass up to 1 GB of audio data [25].

AT&T Watson, created in 1995 by AT & T Research [26]. It allows users to create real-time multilingual voice and speech recognition applications [27]. It is a cloud-based service that can be accessed through HTML POST requests.

Web Speech API, developed by the W3C Speech API team, it is open but seems to be driven by two companies: Google and Openstream. To upload the audio data, the user needs to open an HTTPS connection with the network service through a POST request [28].

NaturalReader. It can read any text, such as Microsoft Word files, Web pages, PDFs, and emails, and convert them into audio files. It is available in free and paid versions [29].

iSpeech is an online free solution available for mobile applications. It has a unique multithreaded and multicore method for text-to-speech conversion, which allows for a simple text-to-speech approach with an unlimited number of processors simultaneously. Thus, it is speedy. It also provides an SDK for java applications [30].

WaveSurfer is an open-source model made in KTH's Center for Speech Technology. An essential feature of WaveSurfer is its versatile configuration [31].

Julius is open-source written in C with high performance. It incorporates significant cutting-edge speech recognition techniques, achieving LVCSR, real-time vocabulary, and real-time speech recognition. It works on Linux, Windows, Mac OS X, Solaris, and other Unix variants and also runs on the Apple iPhone. It supports several languages [32].

4 Evaluation Metrics

There are three types of errors that occur in speech recognition [33]:

If a word in the reference sequence is transcribed as a different word, it is called substitution. When a word is completely missing in the automatic transcription, it is characterized as deletion. The appearance of a word in the transcription that has no correspondent in the reference word sequence is called insertion.

The performance accuracy of a system is usually rated by the word error rate (WER) (1), a popular ASR comparison index [34]. It expresses the distance between the word sequence that produces an ASR and the reference series.

Defined as:

$$WER = (S + D + I)/N_1 = (S + D + I)/(H + S + D) \qquad (1)$$

where I = the total number of entries, D = total number of deletions, S = total number of replacements, H = total number of successes, and N_1 = total number of reference words [33].

Despite being the most commonly used, WER has some cons. It is not an actual percentage because it has no upper bound. When $S = D = 0$ and we have two insertions for each input word, then $I = N1$ (namely when the length of the results is higher than the number of words in the prompt), which means WER = 200%. Therefore it does not tell how good a system is, but only that one is better than another. Moreover, in noisy conditions, WER could exceed 100%, as it gives far more weight to insertions than to deletions [33].

Match Error Rate (MER) is the proportion of I/O word matches, which are errors, which means that is the probability of a given match being incorrect. The ranking behaviour of MER (2) is between that of WER and WIL [35].

$$MER = (S + D + I)/(N = H + S + D + I) = 1 - H/N \qquad (2)$$

WIL is a simple approximation to the proportion of word information lost, which overcomes the problems associated with the RIL (relevant information lost) measure that was proposed years ago [35].

The Relative Information Lost (RIL) measure was not widely taken up because it is not as simple as WER, and it measures zero error for any one-one mapping between input and output words, which is unacceptable [35].

The evaluation of an ASR system also correlates with PER (Position-Independent word Error Rate) automatic metric. PER compares the words in the two sentences (hypothesis and reference) and is always lower than or equal to WER. Hypothesis per (Hper) (similar to precision) (3) refers to the set of words in a hypothesis sentence which do not appear in the reference sentence. Reference PER (Rper) (similar to recall) (4), denotes the set of words in a reference sentence which do not appear in the hypothesis sentence. In other words, the main goal of Hper and Rper is to identify all words in the hypothesis which do not have a counterpart in the reference and vice versa [36].

$$WER = \frac{S + D + I}{N_1} = \frac{S + D + I}{H + S + D} \qquad (3)$$

$$MER = \frac{S + D + I}{N = H + S + D + I} = 1 - \frac{H}{N} \qquad (4)$$

5   Experimentation

5.1   Data and Methodology

From the tools described above, IBM Watson, Wit, and Google were selected to be tested and compared to the accuracy of the results they produce.

Three speakers recorded specific sentences in English. The recorded sentences are typical in dialogues between a teacher and a student while learning English as a second language. The used sentences contain words, numbers, and names.

Speakers are not native English speakers, and their native language is Greek. Speakers A and B are women, and speaker C is a man. We intend to evaluate ASR systems for an application that will train the student in vocabulary learning. Recording took place in an ideal environment with no noise except for speaker C, whose recording has a low level of noise. Each of the three speakers recorded twenty sentences.

Experimental measurements were made using audio files, recorded with Audacity 2.3.1. The WAV files were stereo with two recording channels, 44100 Hz, 32 bit, and they range in size from about 350 KB to 0.99 MB.

Generated texts were compared to the corresponding correct texts that had been given to the speakers. The Word error rate (WER), Hper, and Rper were measured using an Open Source Tool for Automatic Error Classification of Machine Translation Output available on GitHub [37].

6 Results

The text files were created with Uberi, available at https://github.com/Uberi/speech_
recognition. Uberi uses the Python Speech Recognition library and supports several
ASR engines.

Comparing the resulting texts with the reference texts, we calculated the WER
error measurements as a percentage with an accuracy of two decimal places. WER
measurements were made using the tools that are available on pages https://github.
com/belambert/asr-evaluation and https://github.com/cidermole/hjerson. In the end, the
average error of each tool per speaker was calculated and is presented in Table 1.

Table 1. Average WER (%) measurements

Speaker A			Speaker B			Speaker C		
IBM	Google	Wit	IBM	Google	Wit	IBM	Google	Wit
30.10	16.60	25.87	47.73	20.45	23.28	36.51	24.85	58.87

According to Table 1, we note that Google has the smallest average error in any
case. Specifically, Google system presents WER error 16.60%, 20.45%, and 24.85% for
Speaker A, B, and C respectively.

Taking into account the three error values, we calculate the average error for each
system. The overall mean error was 38.1% for IBM, 20.63% for Google, and 36% for
Wit. We observe that among the three, the smallest average error is reported by the
Google tool (20.63%). Based on these measurements, it seems that Google's system
responds better regardless of the speaker's type.

Then we recorded how many sentences from the ones processed by each system
were fully recognized, without any errors.

Table 2. Percent (%) accurate prediction of all 60 sentence

IBM	Google	Wit
26.67	43.33	36.67

As we see in Table 2, Google presents a higher percentage of accurate forecasts
(43.33%) than the other systems. Wit tool is the second in the rankings and the IBM
the last. These measurements reveal the ability of ASR systems to recognize a sentence
without any errors. This ability is essential if the ASR system is to be used in teaching
English as a foreign language, as the application we intend to build. Our app must be able
to fully recognize the pupil's speech to teach him the correct pronunciation, grammar,
and pension, so we do not want to have an approximation but an accurate recognition.

To get a better and more accurate estimate of the three systems, we measured the
Hper και Rper errors using the available tool on https://github.com/cidermole/hjerson.
In Table 3, we present the results of these measurements.

Table 3. Rper και Hper error percentage (%)

	Speaker A			Speaker B			Speaker C		
	IBM	Google	Wit	IBM	Google	Wit	IBM	Google	Wit
Rper	28.45	16.81	24.21	42.24	20.44	26,64	30.66	26.94	61.78
Hper	28.60	13.35	19.79	42.09	19.53	24,17	32.97	22.29	52.99

According to Table 2, the Google ASR system presents the lower Rper and Hper errors for each speaker. Knowing that Rper and Hper errors are associated with adding or removing words in recognized speech, the measurements confirm that the Google system presents the fewest distortions.

Finally, we calculated the total mean errors per tool, which are shown in Table 4.

Table 4. Total mean errors (%)

	IBM	Google	Wit
Rper	23,90	21,40	37,54
Hper	34,55	18,39	32,32

Observing the values in Table 4, we find out that Google's ASR shows the lower average error for both metrics (Rper and Hper).

7 Conclusions

This paper presents some of the most widely known speech recognition tools currently in use. Three of them were then selected to study and compare their performance. The tools selected are IBM Watson, Google API, and Wit. All three tools were tested on the same audio data files and not in a continuous speech stream. The audio files involved texts that occur very often when teaching English as a foreign language. The WER, Hper, and Rper error metrics were measured to evaluate the tools. The results showed that the Google tool is superior to the other two. The Google tool was able to predict most of the sentences accurately, and it presented the smallest error percent. In the future, the size of the data set could be increased by adding more sentences to increase the reliability of the measurements and conclusions. Other tools could also be tested to allow us to decide which one is most suitable for our application.

Acknowledgements. This work is partially supported by MPhil program "Advanced Technologies in Informatics and Computers", hosted by Department of Computer Science, International Hellenic University.

References

1. Anusuya, M.A., Katti, S.K.: Speech Recognition by Machine, A Review. ArXiv10012267 Cs, January 2010
2. Sharma, F., Wasson, S.G.: A Speech Recognition and Synthesis Tool : Assistive Technology for Physically Disabled Persons (2012)
3. Using Statistical Methods in a Speech Recognition System for Romanian Language - ScienceDirect. https://www.sciencedirect.com/science/article/pii/S1474667015373067. Accessed 05 May 2019
4. Automatic Recognition of Spoken Digits. J. Acoust. Soc. Am. **24**(6). https://asa.scitation.org/doi/abs/10.1121/1.1906946. Accessed 05 May 2019
5. Britain scores in easier man-machine communication. Sens. Rev. **1**(4), 172–173 (1981)
6. Upton, J.: Speech recognition for computers in industry. Sens. Rev. **4**(4), 177–178 (1984)
7. Applications and potential of speech recognition. Sens. Rev. 4 (1983)
8. US7363228B2 - Speech recognition system and method - Google Patents. https://patents.google.com/patent/US7363228B2/en. Accessed 11 May 2019
9. Alyousefi, S.: Digital Automatic Speech Recognition using Kaldi, Thesis (2018)
10. Srikanth, R., Salsman, J.: Automatic pronunciation scoring and mispronunciation detection using CMUSphinx. In: Proceedings of the Workshop on Speech and Language Processing Tools in Education, Mumbai, India, pp. 61–68 (2012)
11. Karpagavalli, S., Chandra, E.: A Review on Automatic Speech Recognition Architecture and Approaches (2016)
12. Shmyrev, N.: CMUSphinx Open Source Speech Recognition, CMUSphinx Open Source Speech Recognition. http://cmusphinx.github.io/. Accessed 05 May 2019
13. Morbini, F. et al.: Which ASR should I choose for my dialogue system? In: Proceedings of the SIGDIAL 2013 Conference, pp. 394–403 (2013)
14. Lamere, P., et al.: The CMU Sphinx-4 Speech Recognition System, p. 4 (2003)
15. Satori, H., ElHaoussi, F.: Investigation Amazigh speech recognition using CMU tools. Int. J. Speech Technol. **17**(3), 235–243 (2014)
16. Povey, D., et al.: The Kaldi Speech Recognition Toolkit (2011)
17. Këpuska, V.: Comparing speech recognition systems (Microsoft API, Google API And CMU Sphinx). Int. J. Eng. Res. Appl. **07**(03), 20–24 (2017). https://doi.org/10.1007/s10772-014-9223-y
18. Dragon NaturallySpeaking, Wikipedia, 17 April 2019
19. Baker, J.M.: DragonDictatetm-30K: natural language speech recognition with 30,000 words. In: Presented at the First European Conference on Speech Communication and Technology, Paris, France, September 1989. Accessed 26 Apr 2019
20. Amazon Transcribe – Automatic Speech Recognition - AWS, Amazon Web Services, Inc. https://aws.amazon.com/transcribe/. Accessed 26 Apr 2019
21. erhopf: Speech-to-text with Azure Speech Services - Azure Cognitive Services. https://docs.microsoft.com/en-us/azure/cognitive-services/speech-service/speech-to-text. Accessed 05 May 2019
22. Wit.ai. https://wit.ai/. Accessed 05 May 2019
23. Speech Recognition API - Twilio. https://www.twilio.com/speech-recognition. Accessed 05 May 2019
24. Houndify | Add voice enabled, conversational interface to anything. https://www.houndify.com/. Accessed 05 May 2019
25. IBM Watson | IBM. https://www.ibm.com/watson. Accessed 05 May2019
26. Sharp, R.D., et al.: The Watson speech recognition engine. In: 1997 IEEE International Conference Acoustics Speech Signal Process

27. Goffin, V., et al.: The AT&T WATSON Speech Recognizer. In: Proceedings of ICASSP 3905 IEEE International Conference Acoustics Speech Signal Process (2005)
28. Adorf, J.: Web Speech API (2013)
29. Text to speech online. https://www.naturalreaders.com/. Accessed 05 May 2019
30. Text to Speech I TTS SDK I Speech Recognition (ASR). https://www.ispeech.org/. Accessed 05 May 2019
31. Sjölander, K., Beskow, J.: WAVESURFER - AN OPEN SOURCE SPEECH TOOL, p. 4 (2001)
32. Lee, A., Kawahara, T.: Recent development of open-source speech recognition engine julius. In: Proceedings: APSIPA ASC 2009: Asia-Pacific Signal and Information Processing Association, 2009 Annual Summit and Conference, pp. 131–137, October 2009
33. Errattahi, R., El Hannani, A., Ouahmane, H.: Automatic speech recognition errors detection and correction: a review. Proc. Comput. Sci. **128**, 32–37 (2018)
34. Gaikwad, S.K., Gawali, B.W., Yannawar, P.: A review on speech recognition technique. Int. J. Comput. Appl. **10**(3), 16–24 (2010)
35. Morris, C., Maier, V., Green, P.: From WER and RIL to MER and WIL: improved evaluation measures for connected speech recognition, p. 4 (2004)
36. Seljan, S., Dunder, I.: Combined Automatic Speech Recognition and Machine Translation in Business Correspondence Domain for English-Croatian (2014)
37. Madl, D.: Hjerson: An Open Source Tool for Automatic Error Classification of Machine Translation Output: cidermole/hjerson (2018)

Clustering/Unsupervised Learning/Analytics

A Two-Levels Data Anonymization Approach

Sarah Zouinina[1,2,4]($\boxtimes$), Younès Bennani[1,2], Nicoleta Rogovschi[1,3], and Abdelouahid Lyhyaoui[4]

[1] LIPN UMR 7030 CNRS, Université Sorbonne Paris Nord, Villetaneuse, France
{Sarah.Zouinina,Younes.Bennani}@lipn.univ-paris13.fr,
nicoleta.rogovschi@parisdescartes.fr
[2] LaMSN, La Maison des Sciences Numériques, Université Sorbonne Paris Nord, Saint-Denis, France
[3] LIPADE, Université de Paris, Paris, France
[4] LTI-ENSA, Université Abdelmalek Essaadi, Tanger, Morocco
lyhyaoui@gmail.com

Abstract. The amount of devices gathering and using personal data without the person's approval is exponentially growing. The European General Data Protection Regulation (GDPR) came following the requests of individuals who felt at risk of personal privacy breaches. Consequently, privacy preservation through machine learning algorithms were designed based on cryptography, statistics, databases modeling and data mining. In this paper, we present two-levels data anonymization methods. The first level consists of anonymizing data using an unsupervised learning protocol, and the second level is anonymization by incorporating the discriminative information to test the effect of labels on the quality of the anonymized data. The results show that the proposed approaches give good results in terms of utility what preserves the trade-off between data privacy and its usefulness.

Keywords: Data anonymization · Learning vector quantization · Privacy utility tradeoff · Microaggregation

1 Introduction

Due to the saturation of cities with smartphones and sensors, the amount of information gathered about each individual is frightening. Humans are becoming walking data factories and third-parties are tempted to use personal data for malicious purposes. To protect individuals from the misuse of their precious information and to enable researchers to learn from data effectively, data anonymization is introduced with the purpose of finding balance between the level of anonymity and the amount of information loss. Data anonymization is

Supported by the ANR Pro-Text project. N° ANR-18-CE23-0024-01.

therefore defined as *it is the process of protecting individuals' sensitive information while preserving its type and format* [12,17].

Hiding one or multiple values or even adding noise to data as an attempt to anonymize data is considered inefficient because the reconstruction of the initial information is very probable [15]. Machine learning for data anonymization is still an underexplored area [3], although it provides some good assets to the field of data security. Inspired from the k-anonymization technique proposed by Sweeney [16], we aim to create micro clusters of similar objects that we code using the micro cluster's representative. In this way, the distortion of the data is minimal and the usefulness of the data is maximal. This can be achieved using supervised or unsupervised methods. For the unsupervised methods [8], the most used approach is the clustering that allows to open a new research direction in the field of anonymization i.e. create clusters of k elements and replace the data by the prototypes of the clusters (centroids) in order to obtain a good trade-off between the information loss and the potential data identification risk. However, usually, these approaches are based on the use of the k-means algorithm which is prone to local optima and may give biased results.

In this paper we answer the question of *how can the introduction of discriminative information affect the quality of the anonymized datasets*. To this purpose, we revisited all the previously proposed approaches, and we added a second level of anonymization by incorporating the discriminative information and using Adaptive Weighting of Features to improve the quality of the anonymized data. This aims to improve the anonymized data quality without compromising its level of privacy. The paper is organised into four sections: the first dresses the different approaches of privacy preserving using machine learning, the second sums up the previously proposed approaches, the third discusses the introduction of the discriminative information and the fourth validates the method experimentally on six different datasets.

2   Privacy Preservation Using Machine Learning

Anonymization methods for microdata rely on many mechanisms and *data perturbation* is the common technique binding them all. Those mechanisms modify the original data to improve data privacy but inevitably at cost of some loss in data utility. Strong privacy protection requires masking the original data and thus reducing its utility. Microaggregation is a technique for disclosure limitation aimed at protecting the privacy of data subjects in microdata releases. It has been used as an alternative to generalization and suppression to generate k-anonymous data sets, where the identity of each subject is hidden within a group of k subjects. Unlike generalization, microaggregation perturbs the data and this additional masking freedom allows improving data utility in several ways, such as increasing data granularity, reducing the impact of outliers and avoiding discretization of numerical data [4] microaggregation. Rather than publishing an original variable V_i for a given record, the average of the values of the group over which the record belongs is published. In order to minimize information loss, the groups should be as homogeneous as possible. The impact of microaggregation on the utility of anonymized data is quantified as the resulting accuracy of a machine learning model trained on a portion of microaggregated data and tested on the original data [13]. Microaggregation is measured in terms of *syntactic distortion*.

Achieving microaggregation might be done using machine learning models, like *clustering* and/or *classification*. LeFevre et al. [7] propose several algorithms for generating an anonymous data set that can be used effectively over pre-defined workloads. Workload characteristics taken into account by those algorithms include selection, projection, classification and regression. Additionally, LeFevre et al. consider cases in which the anonymized data recipient wants to build models over multiple different attributes. Nearest neighbor classification with generalization has been investigated by [11]. The main purpose of generalizing exemplars (by merging them into hyper-rectangles) is to improve speed and accuracy as well as inducing classification rules, but not to handle anonymized data. Martin proposes building non-overlapping, non-nested generalized exemplars in order to induce high accuracy. Zhang et al. discuss methods for building naive Bayes and decision tree classifiers over partially specified data [6, 18]. Partially specified records are defined as those that exhibit nonleaf values in the taxonomy trees of one or more attributes. Therefore generalized records of anonymous data can be modeled as partially specified data. In their approach classifiers are built on a mixture of partially and fully specified data. Inan et al. [5] address the problem of classification over anonymized data. They proposed an approach that models generalized attributes of anonymized data as uncertain information, where each generalized value of an anonymized record is accompanied by statistics collected from records in the same equivalence class. They do not assume any probability distribution over the data. Instead, they propose collecting all necessary statistics during anonymization and releasing these together with the anonymized data. They show that releasing such statistics does not violate anonymity.

3 Clustering for Data Anonymization

3.1 k-TCA and Constrained TCA

In previous articles we introduced an approach of k-anonymity using Collaborative Multi-view Clustering [22] and a k-anonymity through Constrained Clustering [21]. The two models propose an algorithm that relies on the classical Self Organizing Maps (SOMs) [10] and collaborative Multiview clustering in purpose to provide useful anonymous datasets [9]. They achieve anonymization in two-levels, the pre-anonymization step and the anonymization step. The pre-anonymization step is similar for both algorithms and it consists of horizontally splitting data so each observation is described in different spaces and then using the collaborative paradigm to exchange topological information between collaborators. The Davies Bouldin index (DB) [2] is used in this case as a clustering validity measure and a stopping criterion of the collaboration. When DB decreases, the collaboration is said to be positive, but if it increases, the collaboration is clearly negative, since it is degrading the clustering quality and therefore the utility of the provided anonymous data. The topological collaborative multiview clustering outputs homogeneous clusters after the clustering, the individuals contained in each view are coded using the Best Matching Units of each neuron in the case of k-TCA and using the linear mixture of models in the case of Constrained TCA. The pre-anonymized views are then gathered to be reconstructed in the same manner as the original dataset.

The anonymization step of the algorithms is totally different between the two. In the k-TCA, the pre-anonymized dataset will be fine-tuned using a SOM model with a map size determined automatically using the Kohonen heuristic. Each individual of the dataset is then coded using the BMU of the cluster and the level of k-anonymity is evaluated. In those model we have the advantage of determining the k-anonymity level automatically. In the second algorithm, Constrained TCA (C-TCA), the k level of anonymity is fixed ahead, before starting the experiments. A SOM is created using the pre-anonymized dataset as an input. Each node is examined to determine if it respects the constraint of k element in each cluster. Respectively the elements captured by the neurons that don't respect the predefined constraint are redistributed on the closest units. By using this technique, we design clusters of at least k elements and we code the objects using the BMUs in order to have k-anonymized dataset. We then evaluate the best k level that gives a good tradeoff between anonymity and utility.

3.2 Attribute Oriented Kernel Density Estimation for Data Anonymization

Another method that we proposed to anonymize a dataset was the Attribute Oriented Kernel Density Estimation [20]. The choice of 1 dimensional KDE was motivated by the ability of the model to determine where data is grouped together and where it is sparse relying on its density. KDE is a non parametric model that uses probability density to investigate the inner properties of a given dataset. The algorithm that we propose clusters the data by determining the points where density is the highest (local maximas) and the points with the smallest density (local minimas): those local minimas refer to the clusters' boarders and the local maximas are the clusters' prototypes. KDE is a non-parametric approach to approximate the distribution of a dataset and overcome the inability of the histograms to achieve this estimation because of the discontinuity of the bins. Each object that falls between two minimas is recoded using the corresponding local maxima. Doing this at a one dimensional level helps preserving the characteristics of each feature in the dataset and thus doesn't compromise its utility.

4 Incorporating Discriminative Power During Anonymization Process

After evaluating the different results of data anonymization using the methods in the previous works, we asked the question *What if data was labelled?* and

How the supervision can influence the obtained utility results? To answer to those questions we used the Learning Vector Quantization approach (LVQ). We applied it to enhance the clustering results of each of our proposed methods. LVQ is a pattern recognition model that takes advantage of the labels to improve the accuracy of the classification. The algorithms learns from a subset of patterns that best represent the training set.

The choice of the Learning Vector Quantization (LVQ) method was motivated by the simplicity and rapidity of convergence of the technique, since it is based on the hebbian learning. This is a prototype-based method that prepares a set of codebook vectors in the domain of the observed input data samples and uses them to classify unseen examples. Kohonen presented the self organizing maps as an unsupervised learning paradigm that he improves using a supervised learning technique, called the learning vector quantization. It is a method used for optimizing the performances of a trained map in a reward-punishment scheme.

Learning Vector Quantization was designed for classification problems that have existing data sets that can be used to supervise the learning by the system. LVQ is non-parametric, meaning that it does not rely on assumptions about that structure of the function that it is approximating. Euclidean distance is commonly used to measure the distance between real-valued vectors, although other distance measures may be used (such as dot product), and data specific distance measures may be required for non-scalar attributes. There should be sufficient training iterations to expose all the training data to the model multiple times. The learning rate is typically linearly decayed over the training period from an initial value until it is close to zero. Multiple passes of the LVQ training algorithm are suggested for more robust usage, where the first pass has a large learning rate to prepare the codebook vectors and the second pass has a low learning rate and runs for a long time (perhaps 10-times more iterations).

In the Learning Vector Quatization model, each class contains a set of fixed prototypes with the same dimension of the data to be classified. LVQ adaptively modifies the prototypes. In the learning algorithm, data is first clustered using a clustering method and the clusters' prototypes are moved using LVQ to perform classification. We chose to supervise the results of the clustering by moving the center clusters' using the wLVQ2 proposed in Algorithm 1 for each of the approaches. We use the wLVQ2 [1] since this upgraded version of the LVQ respects the characteristics of each features and adapts the weighting of each feature according to its participation to the discrimination. The system learns using two layers: the first layer calculates the weights of the features and then it is presented to the LVQ2 algorithm.

The cost function of this algorithm can be written as follows:

$$R_{wLVQ2}(x, m, W) = \begin{cases} \|Wx - m_j\|^2 - \|Wx - m_i\|^2, & If \ C_k = C_j \\ 0, & otherwise \end{cases}$$

Where $x \in C_k$ and W is the weighting coefficient matrix; m_i is the nearest codeword vector to Wx and m_j is the second nearest codeword vector to

Algorithm 1. Adaptive Weighting of Pattern Features During Learning

Initialization :

Initialize the matrix of weights W according to :

$$w_j^i = \begin{cases} 0, & when \ i \neq j \\ 1, & when \ i = j \end{cases}$$

The codewords **m** are chosen for each class using the k-means algorithm.

Learning Phase:

1. Present a learning example x.
2. Let $m_i \in C_i$ be the nearest codeword vector to x.
   - if $x \in C_i$, then go to 1
   - else then
     - let $m_j \in C_j$ be the second nearest codeword vector
     - if $x \in C_j$ then
       * a symmetrical window win is set around the mid-point of m_i and m_j.
       * if x falls within win, then
         Codewords Adaptation:

         * m_i is moved away from x according to the formula

         $$m_i(t+1) = m_i(t) + \alpha(t)[Wx(t) - m_j(t)]$$

         * m_j is moved closer x according to the formula

         $$m_j(t+1) = m_j(t) - \alpha(t)[Wx(t) - m_j(t)]$$

         * for the rest of the codewords

         $$m_k(t+1) = m_k(t)$$

         Weighting Patterns features:

         * adapt w_k^k according to the formula:

         $$w_k^k(t+1) = w_k^k(t) - \beta(t)x^k(t)(m_i^k(t) - m_j^k(t))$$

         * go to 1.

Where $\alpha(t)$ and $\beta(t)$ are the learning rates

Wx. The wLVQ2 with the Collaborative Paradigm enhances the utility of the anonymized data by the k-TCA and the Constrained TCA (C-TCA) models, the use of wLVQ2 is done after the collaboration between cluster centers' to improve the results of the Collaboration at the pre-anonymization and the anonymization steps.

The experimental protocol of using wLVQ2 with Attribute-oriented data anonymization and Kernel Density Estimation, takes in account the labels of the dataset and improves the found prototypes and then represents the micro-clusters using them.

5   Experimental Validation

5.1   Datasets

Six datasets from the UCI machine learning repository are used in the experiment. The Table 1 below presents the main characteristics of these databases.

Table 1. Some characteristics of datasets

Datasets	#Instances	#Attributes	#Class
Ecoli	336	8	8
Electrical	10000	14	2
Glass	214	10	7
Page blocks	5473	10	5
Waveform	5000	21	3
Yeast	1484	8	10

5.2   Quality Validity Indices

Cluster validity consists of techniques for finding a set of clusters that best fits natural partitions without any a priori class information. The outcome of the clustering process is validated by a cluster validity index. Internal validation measures reflect often the compactness, the connectivity and the separation of the cluster partitions. We choose to validate the results of the proposed methods using Silhouette Index and Davies Bouldin Index. The results are given in the Tables 2 and 3.

Table 2. Silhouette Index

	Ecoli	Electrical	Glass	Page blocks	Yeast	Waveform
k-TCA	0.26	−0.05	0.42	−0.40	0.13	0.18
k-TCA^{++}	0.89	0.08	0.59	−0.49	0.84	0.24
C-TCA	0.24	−0.05	0.43	−0.34	0.07	0.13
C-TCA^{++}	0.84	0.08	0.45	−0.43	0.81	0.25
KDE	0.26	0.069	−0.19	−0.54	0.28	−0.27
KDE^{++}	0.99	0.99	0.57	0.98	0.99	1

As illustrated, the Attribute oriented microaggregation using wLVQ2 (++: Discriminative version of each approach, KDE^{++}, $k\text{-}TCA^{++}$, $C\text{-}TCA^{++}$) outperforms by far the Attribute Oriented microaggregation in both Silhouette and Davies Bouldin indices.

Table 3. Davies Bouldin Index

	Ecoli	Electrical	Glass	Page blocks	Yeast	Waveform
k-TCA	2.68	2.28	0.40	3.23	2.31	1.51
k-TCA^{++}	0.59	3.38	0.40	3.11	0.24	1.37
C-TCA	1.61	2.58	0.55	3.04	2.95	1.92
C-TCA^{++}	0.14	3.38	0.51	3.10	0.26	1.35
KDE	0.57	3.96	4.99	3.83	2.43	6.96
KDE^{++}	9.91E-08	0.02	1.32	0.52	4.20E-08	3.58E-06

5.3   Combined Utility Measure

Separability Utility. To measure the utility of the anonymized datasets we propose a test on the original and the anonymized data. The test consists of comparing the accuracy of a decision tree model with 10 folds cross validation before and after microaggregation to evaluate the practicality of the proposed anonymization. We call it separability utility since it measures the separability of the clusters. We give the results of this measure in Table 4, we also provide a comparison between the separability utility measures of the original and the anonymized datasets.

The separability measure was improved after LVQ for 83% of the tests done on the datasets, this can be explained by the tendency of microaggregation to remove non decisive attributes from the dataset in order to gather together elements that are similar. The $^{++}$ in the name of the methods refers to discriminant version.

Table 4. Separability utility: Glass, Ecoli, Electrical, Page blocks, Waveform & Yeast

Datasets	Glass	Ecoli	Electrical	Page blocks	Waveform	Yeast
Before anonymization	0.692	0.821	0.995	0.966	0.748	0.812
k-TCA	0.943	0.845	0.999	0.905	0.830	0.862
k-TCA $^{++}$	0.944	0.988	0.735	0.919	0.884	1
C-TCA	0.747	0.848	0.999	0.915	0.816	0.876
C-TCA $^{++}$	0.859	0.863	0.745	0.918	0.884	0.887
KDE	0.701	0.801	0.988	0.955	0.755	0.834
KDE $^{++}$	0.743	0.806	0.982	0.962	0.758	0.845

Structural Utility Using the Earth Mover's Distance. We believe that measuring the distance between two distributions is the way to evaluate the difference between the datasets. The amount of utility lost in the process of anonymization can be see as the distance between the anonymized dataset and the original one.

The Earth Mover's distance (EMD) also known as the Wasserstein distance [14], extends the notion of distance between two single elements to that of a distance between sets or distributions of elements. It compares the probability distributions P and Q on a measurable space (Ω, Ψ) and is defined as follows (We are using the distance of order 1):

$$W_1(P,Q) = \inf_{\mu} \left\{ \int_{\Omega \times \Omega} |x - y| d\mu(x,y) \middle| \begin{array}{l} \mu : \text{prob. measure on } (\Omega \times \Omega, \Psi \otimes \Psi) \\ \textit{with marginals} : P, Q \end{array} \right\} \quad (1)$$

where $\Omega \times \Omega$ is the product probability space. Notice that we may extend the definition so that P is a measure on a space (Ω, Ψ) and Q is a measure on a space (Ω', Ψ').

Let us examine how the above is applied in the case of discrete sample spaces. For generality, we assume that P is a measure on (Ω, Ψ) where $\Omega = \{x_i\}_{i=1}^{n}$ and Q is a measure on (Ω', Ψ') where $\Omega' = \{y_i\}_{j=1}^{n'}$ - the two spaces are not required to have the same cardinality.

Then, the distance between P and Q becomes:

$$W_1(P,Q) = \inf_{\{\lambda_{i,j}\}, i,j} \left\{ \sum_{i=1}^{n} \sum_{j=1}^{n'} \lambda_{i,j} |x_i - y_j| : \sum_{i=1}^{n} \lambda_{i,j} = q_j, \sum_{j=1}^{n'} \lambda_{i,j} = p_i, \lambda_{i,j} \geq 0 \right\}$$

EMD is the minimum amount of work needed to transform a distribution to another. In our case we measure the EMD between the anonymized and the original datasets, attribute by attribute, to get an idea about the distortion of the anonymized datasets. We then normalize all distances between 0 and 1, then we define the utility by $1 - W_1(P, Q)$. The smaller the distance W_1 is, the more the data utility is preserved.

Preserving Combined Utility. To choose the anonymization method which best addresses the separability-Structural utility Trade-off, we propose to combine the two types of utility structural and separability in a combined form while $\alpha = \frac{1}{2}$:

$$Comb\_Utility = \alpha.Separability + (1 - \alpha).Structural$$

Table 5 summarize the clustering results of the proposed approaches in terms of combined utility ($Comb\_Utility$). As it can be seen, our approach Attribute-oriented generally performs best on all the datasets. To further evaluate the performance, we compute a measurement score by following [19]:

$$Score(A_i) = \sum_{j} \frac{Comb\_Utility(A_i, D_j)}{\max_i Comb\_Utility(A_i, D_j)}$$

where $Comb\_Utility(A_i, D_j)$ refers to the combined Utility value of A_i method on the D_j dataset. This score gives an overall evaluation on all the

Table 5. Combined separability and structural utility Comb_Utility

	Ecoli	Electrical	Glass	Page blocks	Waveform	Yeast	Score
k-TCA	0.63	0.74	0.71	0.60	0.66	0.51	**4.96**
k-TCA^{++}	0.78	0.62	0.74	0.82	0.69	0.92	**5.18**
C-TCA	0.74	0.75	0.54	0.70	0.65	0.51	**4.92**
C-TCA $^{++}$	0.62	0.62	0.76	0.71	0.70	0.87	**5.20**
KDE	0.60	0.54	0.44	0.71	0.63	0.73	**4.98**
KDE $^{++}$	0.83	0.95	0.91	0.77	0.75	0.79	**5.17**

datasets, which shows our approach Attribute-oriented outperforms the other methods substantially in most cases.

As shown in the Table 5, the introduction of the discriminant information improves the utility of the anonymized datasets for all of the methods proposed.

6   Conclusion

In this paper we studied the impact of incorporating the discriminative information to improve data anonymization level and to preserve its usefulness. The anonymization is achieved in two levels process. The first, uses one of these three methods: k-TCA or Constrained TCA (C-TCA) or Attribute Oriented KDE, that we introduced for data anonymization through microaggregation approach. And the second, through the use of labels and the learning of the vectors weights adaptively using the weighted LVQ. The experimental investigation shown above prove the efficiency of the methods and illustrate its importance. The main contributions of the article are the addition of the supervised learning layer to improve utility of the model without compromising its anonymity. The separability utility reflects the usefulness of the data and the structural utility shows its level of anonymity. The combined utility is a weighted measure that combines both measures, we can change the weight of the utility tradeoff depending on which side we want to emphasise on.

References

1. Bennani, Y.: Adaptive weighting of pattern features during learning. In: IJCNN 1999. International Joint Conference on Neural Networks. Proceedings. vol. 5, pp. 3008–3013. IEEE Service Center, Piscataway, NJ (1999)
2. Davies, D., Bouldin, D.: A cluster separation measure. IEEE Trans. Pattern Anal. Mach. Intell. **PAMI–1**(2), 224–227 (1979)
3. Domingo-Ferrer, J., Soria-Comas, J., Mulero-Vellido, R.: Steered microaggregation as a unified primitive to anonymize data sets and data streams. IEEE Trans. Inf. Forensics Secur. **14**(12), 3298–3311 (2019)
4. Hundepool, A., et al.: Statistical Disclosure Control. John Wiley, Hoboken (2012)

5. Inan, A., Kantarcioglu, M., Bertino, E.: Using anonymized data for classification. In: IEEE 25th International Conference on Data Engineering, pp. 429–440 (2009)
6. Zhang, J., Kang, D., Silvescu, A., Honavar, V.: Learning accurate and concise Naive Bayes classifiers from attribute value taxonomies and data. Knowl. Inf. Syst. **9**, 157–179 (2006). https://doi.org/10.1007/s10115-005-0211-z
7. LeFevre, K., DeWitt, D.J., Ramakrishnan, R.: Workload-aware anonymization. In: KDD 2006, pp. 277–286 (2006)
8. Khan, S., Iqbal, K., Faizullah, S., Fahad, M., Ali, J., Ahmed, W.: Clustering based privacy preserving of big data using fuzzification and anonymization operation. ArXiv abs/2001.01491 (2019)
9. Kim, S., Chung, Y.D.: An anonymization protocol for continuous and dynamic privacy-preserving data collection. Future Gener. Comput. Syst. **93**, 1065–1073 (2019)
10. Kohonen, T.: Self-organizing Maps. Springer, Berlin (1995)
11. Martin, B.: Instance-based learning: Nearest neighbour with generalisation. Master's Thesis, Computer Science Department, Hamilton, New Zealand, University of Waikato (1995)
12. Raghunathan, B.: The Complete Book of Data Anonymization: From Planning to Implementation. CRC Press, Boca Raton (2013)
13. Rodríguez-Hoyos, A., Estrada-Jiménez, J., Rebollo-Monedero, D., Parra-Arnau, J., Forné, J.: Does k-anonymous microaggregation affect machine-learned macrotrends? IEEE Access **6**, 28258–28277 (2018)
14. Rubner, Y., Tomasi, C., Guibas, L.J.: The earth mover's distance as a metric for image retrieval. Int. J. Comput. Vision **40**(2), 99–121 (2000)
15. Sharma, A., Singh, G., Rehman, S.: A review of big data challenges and preserving privacy in big data. In: Kolhe, M.L., Tiwari, S., Trivedi, M.C., Mishra, K.K. (eds.) Advances in Data and Information Sciences. LNNS, vol. 94, pp. 57–65. Springer, Singapore (2020). https://doi.org/10.1007/978-981-15-0694-9_7
16. Sweeney, L.: K-anonymity: a model for protecting privacy. Int. J. Uncertain. Fuzziness Knowl.-Based Syst. **10**(5), 557–570 (2002)
17. Venkataramanan, N., Shriram, A.: Data Privacy: Principles and Practice. Chapman & Hall/CRC, Boca Raton (2016)
18. Zhang, J., Honavar, V.: Learning from attribute value taxonomies and partially specified instances, Washington, DC, USA, pp. 880–887 (2003)
19. Zhao, H., Fu, Y.: Dual-regularized multi-view outlier detection. In: IJCAI (2015)
20. Zouinina, S., Bennani, Y., Ben-Fares, M., Lyhyaoui, A., Rogovschi, N.: Preserving utility during attribute-oriented data anonymization process. Aust. J. Intell. Inf. Process. Syst. **16**(3), 25–35 (2019)
21. Zouinina, S., Grozavu, N., Bennani, Y., Lyhyaoui, A., Rogovschi, N.: Efficient k-anonymization through constrained collaborative clustering. In: IEEE Symposium Series on Computational Intelligence, SSCI 2018 (2018)
22. Zouinina, S., Grozavu, N., Bennani, Y., Lyhyaoui, A., Rogovschi, N.: A topological k-anonymity model based on collaborative multi-view clustering. In: Kůrková, V., Manolopoulos, Y., Hammer, B., Iliadis, L., Maglogiannis, I. (eds.) ICANN 2018. LNCS, vol. 11141, pp. 817–827. Springer, Cham (2018). https://doi.org/10.1007/978-3-030-01424-7_79

An Innovative Graph-Based Approach to Advance Feature Selection from Multiple Textual Documents

Nikolaos Giarelis(iD), Nikos Kanakaris(iD), and Nikos Karacapilidis(✉)(iD)

Industrial Management and Information Systems Lab, MEAD, University of Patras,
26504 Rio, Patras, Greece
giarelis@ceid.upatras.gr, nkanakaris@upnet.gr, karacap@upatras.gr

Abstract. This paper introduces a novel graph-based approach to select features from multiple textual documents. The proposed solution enables the investigation of the importance of a term into a whole corpus of documents by utilizing contemporary graph theory methods, such as community detection algorithms and node centrality measures. Compared to well-tried existing solutions, evaluation results show that the proposed approach increases the accuracy of most text classifiers employed and decreases the number of features required to achieve 'state-of-the-art' accuracy. Well-known datasets used for the experimentations reported in this paper include *20Newsgroups*, *LingSpam*, *Amazon Reviews* and *Reuters*.

Keywords: Feature selection · Graph-based text representation · Document clustering · Text mining · Natural Language Processing

1 Introduction

Graph-based text representations are widely used in various Natural Language Processing, Text Mining and Information Retrieval tasks (Vazirgiannis et al. 2018). These representations exploit concepts and techniques inherited from graph theory (e.g. node centrality and subgraph frequency) to address limitations of the classical *bag-of-words* representation (Aggarwal 2018); in this way, they are able to capture structural and semantic information of a text, mitigate the effects of the 'curse-of-dimensionality' phenomenon, identify the most important terms of a text, and seamlessly incorporate information coming from external knowledge sources. However, existing graph-based representations concern a single document each time. In cases where one needs to analyze a corpus of documents, these approaches demonstrate a series of weaknesses, the main of them being that they are incapable to assess the importance of a word for the whole set of documents.

Recently, graph-based text representations have been used to facilitate and augment the feature selection process, i.e. the process of selecting a subset of relevant features when constructing a model. These approaches combine statistical tests and graph algorithms to uncover hidden correlations between terms and document classes. However,

© IFIP International Federation for Information Processing 2020
Published by Springer Nature Switzerland AG 2020
I. Maglogiannis et al. (Eds.): AIAI 2020, IFIP AICT 583, pp. 96–106, 2020.
https://doi.org/10.1007/978-3-030-49161-1_9

while they take into account the co-occurrences between terms to identify the most representative features of a single document (something that is not the case in traditional statistical methods), they are not able to assess the importance of a term in a corpus of documents. To remedy the above weakness, this paper builds on a graph-based text representation model to introduce a novel approach to feature selection from multiple textual documents, namely *GraFS*. Contrary to existing approaches, the one introduced in this paper (i) enables the investigation of the importance of a term into a whole corpus of documents, (ii) incorporates the relationships between terms (co-occurrences) into the feature selection process, (iii) achieves state-of-the-art accuracy in ML tasks such as text classification using fewer features, and (iv) mitigates the effects of the 'curse-of-dimensionality' phenomenon. GraFS has been evaluated by using five datasets and five classifiers. Compared to four well-tried existing feature selection approaches, our initial experimental results show that GraFS increases the accuracy of most text classifiers and decreases the number of features required to achieve 'state-of-the-art' accuracy.

The remainder of the paper is organized as follows. Section 2 discusses related work issues. The proposed feature selection approach is presented in Sect. 3. Section 4 reports on the experiments carried out to assess the proposed approach against previous ones. Limitations of our approach, future work directions and concluding remarks are outlined in Sect. 5.

2 Background Work

The proposed feature selection approach builds on a graph-based representation of multiple textual documents and exploits advantages of contemporary graph databases. This section highlights related background work issues.

2.1 Graph-Based Text Representations

Graph-of-words is a well-known graph-based text representation method. Being similar to the bag-of-words approach that has been widely used in the NLP field, it enables a sophisticated keyword extraction and feature engineering process. In a graph of words, each node represents a unique term (i.e. word) of a document and each edge represents the co-occurrence between two terms within a sliding window of text. The utilization of a small sliding window size, due to the fact that larger ones produce heavily interconnected graphs where the valuable information is cluttered with noise, has been proposed in (Nikolentzos et al. 2017). In this direction, work described in (Rousseau et al. 2015) suggests that a window size of four is generally considered as the appropriate value, since it does not sacrifice either the performance or the accuracy of their approach.

2.2 Graph-Based Feature Selection

Several interesting graph-based feature selection approaches have been already proposed in the literature. For instance, (Rousseau et al. 2015) proposes various combinations and configurations of popular frequent subgraph mining techniques - such as *gSpan* (Yan and Han 2002), *Gaston* (Nijssen and Kok 2004) and *gBoost* (Saigo et al. 2009) - to perform unsupervised feature selection exploiting the k-core subgraph. In particular, aiming to increase performance, Rousseau and his colleagues rely on the concept of k-core subgraph to reduce the graph representation to its densest part. The experimental results show a significant increment of the accuracy compared to common classification approaches. The work reported in (Henni et al. 2018) applies centrality algorithms (such as PageRank) to calculate the centrality score of a graph's features and accordingly identify the most important ones. The approach presented in (Fakhraei et al. 2015) builds on combinations of several types of graph algorithms to discover highly connected features of a graph. Such algorithms include the Louvain Algorithm for community detection and the PageRank algorithm to discover influential nodes and other user-defined graph measures. This last approach combines PageRank and Coloring algorithms with the custom graph measures of in-degree and out-degree.

Other already proposed approaches rely on the recursive filtering of the existing feature space; for instance, one of them re-applies PageRank to find the most influential features (Ienco et al. 2008). These approaches use graph-connected features to include contextual information, as modelled implicitly by a graph structure, using relationships that describe connections among real data. They aim to reduce ambiguity in feature selection and improve accuracy in traditional Machine Learning methods.

2.3 Graph Databases

Compared to relational databases, graph databases provide a more convenient and efficient way to natively represent and store highly interlinked data. Moreover, they allow the retrieval of multiple relationships and entities with a single operation, thus avoiding the utilization of rigid join operations which are heavily used in relational databases (Miller 2013). An in-depth review of graph databases can be found in (Rawat and Kashyap 2017).

3 GraFS: Graph-Based Feature Selection

3.1 Graph-of-Docs Text Representation

To select the most representative features of a corpus of documents, we build on the *graph-of-docs* text representation, first proposed in (Giarelis et al. 2020). Aiming to represent multiple documents in a single graph, the graph-of-docs representation expands the well-known 'graph-of-words' model that produces a single graph for each individual document (Rousseau et al. 2013). Graph-of-docs allows diverse types of nodes and edges to co-exist in a graph, including nodes with types such as `document` and `word`, and edges with types such as `is_similar`, `connects`, `includes`, and `feature` (see Fig. 1).

Briefly, according to the graph-of-docs representation:

- each unique word node is connected to all the document nodes where it belongs to using edges of the `includes` type;
- each unique word node selected as a feature is connected to document nodes using edges of the `feature` type;

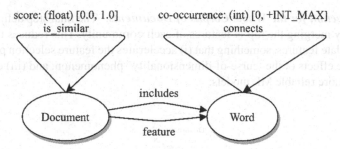

Fig. 1. The schema of the graph-of-docs representation model.

- edges of `connects` type are only applicable between two word nodes and denote their co-occurrence within a specific sliding text window;
- edges of the `is_similar` type link a pair of document nodes and indicate their contextual similarity.

Graph-of-docs enables us to investigate the importance of a term not only within a single document but also within a whole corpus of documents, which in turn augments the quality of the overall feature selection process.

3.2 Feature Selection

Our approach consists of four steps: (i) creation of a document similarity subgraph; (ii) detection of document communities; (iii) feature selection for each community, and (iv) feature selection for the whole corpus of documents.

Creation of a Document Similarity Subgraph. We argue that subgraphs from the graph-of-docs graph describing similar documents share common word nodes as well as similar structural characteristics. This enables us to calculate the similarity between two documents by using typical data mining similarity measures, which in turn facilitates the production of a similarity subgraph. The similarity subgraph consists of document nodes and edges of the `is_similar` type.

Detection of Document Communities. By exploiting the document similarity subgraph, we detect communities of contextually similar documents using the `score` property of the `is_similar` type edges as a distance value. A plethora of community detection algorithms can be found in the literature, including *Louvain*, *Label Propagation* and *Weakly Connected Components*.

Feature Selection for Each Community. Since documents that are in the same community are contextually similar, we assume that it is also more likely that they share common features (see Fig. 2). Aiming to find the top-N most representative features of each community, GraFS ranks the terms of each community by their document frequency and their PageRank score.

Feature Selection for the Whole Corpus of Documents. The final step defines the feature space by merging the top-N features of each community. This reduces the number of the candidate features, something that (i) accelerates the feature selection process, (ii) mitigates the effects of the 'curse-of-dimensionality' phenomenon, and (iii) enables the training of more reliable ML models.

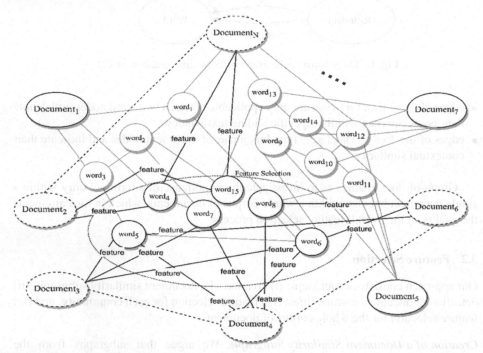

Fig. 2. Feature selection using the graph-of-docs text representation model. The selected features, shown within the circle, are linked to documents with edges of 'feature' type. Relationships between documents are denoted with dotted lines.

4   Experiments

For the implementation and evaluation of our approach, we used the Python programming language and the scikit-learn ML library (https://scikit-learn.org). The Neo4j graph database (https://neo4j.com) has been utilized for the needs of the graph-of-docs representation. The full code, the documentation and the evaluation results of our experiments are freely available at https://github.com/NC0DER/GraphOfDocs.

4.1   Baseline Methods

This subsection presents the benchmarks used to evaluate the performance of GraFS. For the implementation of these methods, we used the scikit-learn ML library (implementation details can be found at https://scikit-learn.org/stable/modules/feature_selection.html).

Low Variance Feature Selection (LVAR). The first benchmark removes the features that do not meet a predefined variance threshold (Aggarwal 2018). This method is referred to as *LVAR* in the remainder of this paper (scikit-learn library, class: `sklearn.feature_selection.VarianceThreshold`).

Univariate Feature Selection (KBEST). The second benchmark relies on univariate statistical tests to select the k-best features (Aggarwal 2018). In particular, it attempts to find correlations between an individual feature and a document class. In this paper, we adopt the x^2 test as our main statistical test. This method is referred to as *KBEST* in the remainder of this paper (scikit-learn library, classes: `sklearn.feature_selection.SelectKBest` and `sklearn.feature_sele-ction.chi2`).

Feature Selection Using a Meta-Transformer Model (META). The third benchmark uses a meta-transformer model to retain only the features with significant importance. It is assumed that a statistical model (e.g. logistic regression) provides importance metrics for each feature to be considered as a candidate meta-transformer model. Available meta-transformer models include logistic regression, linear SVM and neural networks, as well as more sophisticated methods such as word embeddings (e.g. *word2vec* (Mikolov et al. 2013)). In this paper, we use the linear SVM model, since it performs well regardless of the number of samples or the number of unique features of a dataset. In the remainder of the paper, this method is referred to as *META* (scikit-learn library, classes: `sklearn.feature_selection.SelectFromModel` and `sklearn.svm.LinearSVC`).

4.2 Datasets

This subsection describes the datasets used in our experiments to evaluate the performance of GraFS. These datasets are available at https://github.com/imis-lab/aiai-2020-datasets.

20Newsgroups. We tested the proposed model by utilizing an already preprocessed version of the well-known *20Newsgroups* dataset, which is a collection of approximately 20,000 newsgroup documents, partitioned evenly across 20 different newsgroups. As far as the multi-class text classification task is concerned, this dataset fits well to the purposes of our experimentations since it provides a large volume of different documents on the same subjects.

Reuters. We tested the proposed model on a preprocessed version of the Reuters dataset, which includes 21,578 news stories; since almost half of them lack the class field, we used only the ones that came along with their class (i.e. 10,377). For each news story, certain attributes were retained; for instance, the 'title' attribute that contains the title of the story and the 'body' attribute that contains the main text of the news story. In this paper, we used this dataset to execute experiments related to the multi-class text classification task.

Amazon Reviews. We also tested the proposed model on a preprocessed version of the Amazon Reviews dataset, which contains labeled (positive or negative) reviews of products belonging to different categories (e.g. automotive, electronics, grocery etc.). We picked four product categories (i.e. books, DVD, electronics, kitchen), each having 1000 positive and 1000 negative reviews. We utilized this dataset to conduct experiments related to the opinion mining task.

LingSpam. The LingSpam dataset (Androutsopoulos et al. 2000) contains 2,893 email messages, which are classified either as 'spam' or 'not spam'. We utilized this dataset to conduct experiments related to the spam detection task.

JiraIssues. The JiraIssues dataset, concerns the development of 168 software projects including 'Hadoop', 'Spark' and 'Airflow'. It contains information related to 228,969 Jira issues. Each Jira issue in this dataset has the attributes 'description', and 'assignee'. The set of the document classes of the dataset corresponds to the names of the available employees ('assignee' attribute). This dataset was retrieved from the publicly accessible Jira instance of Apache Software Foundation (https://issues.apache.org/jira). We utilized it to execute experiments related to the multi-class text classification task.

Table 1. The hyper-parameters of each feature selection method per dataset.

Method	Dataset	Hyper-parameter	Values
LVAR	20Newsgroups, Reuters, Amazon, LingSpam, JiraIssues	Variance threshold	[0.0005, 0.001, 0.0015, 0.002, 0.003, 0.004, 0.005, 0.01]
GraFS	20Newsgroups, Reuters, Amazon, LingSpam, JiraIssues	top-N	[5, 10, 15, 20, 25, 50, 100, 250, 500]
KBEST	20Newsgroups	k	[1000, 2000, 3000, 5000, 10000, 15000, 20000, 25000, 30000]
KBEST	Reuters, Amazon, JiraIssues	k	[1000, 2000, 3000, 5000, 6000, 7000, 8000, 10000, 14000]
KBEST	LingSpam	k	[250, 500, 1000, 1500, 2000, 2500, 3000, 4000, 5000]

4.3 Experimental Setup

To identify the most important words in the entire corpus of documents, we selected to use the *PageRank* algorithm, since it performs well regardless of the topics of the documents. To identify similar documents needed for the generation of the document similarity subgraph, we used the *Jaccard* similarity measure since it deals only with the absence or the presence of a word, ignoring its document frequency. To form communities of similar documents, we used the *Louvain* community detection algorithm. Finally, we executed several experiments with different hyperparameter values for the LVAR, KBEST and GraFS feature selection methods. Table 1 summarizes the values given to these hyperparameters per dataset.

4.4 Evaluation

To evaluate the effectiveness of our approach, we assess the contribution of GraFS in the accuracy of widely used text classifiers against the bag-of-words (BOW) text representation and the three domain-agnostic feature selection techniques described in Sect. 4.1 (see Table 2). The text classifiers considered are: *naive Bayes* (NB), *k-nearest neighbors* (5NN), *logistic regression* (LR), *neural networks* (NN100x50) and *linear support vector machines* (LSVM). It is noted that in the case of BOW, none of the feature selection techniques has been applied to the specific experiment. Results obtained show that GraFS (i) increases the accuracy in most cases, and (ii) decreases the number of features required to achieve 'state-of-the-art' accuracy (Fig. 3 – right part). Figure 3 illustrates the accuracy of the LSVM classifier per number of selected features for the GraFS, KBEST and LVAR feature selection techniques (additional comparisons can be retrieved from https://github.com/imis-lab/aiai-2020-datasets).

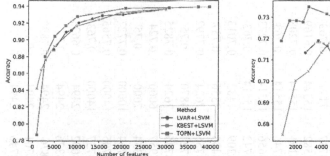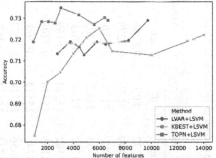

Fig. 3. Accuracy of the LSVM classifier per number of selected features for the GraFS (TOPN), KBEST and LVAR feature selection techniques on *20Newsgroups* (left) and *JiraIssues* (right) datasets.

Our approach differs from the existing ones in that it considers the whole corpus of documents (instead of each document separately) and the associated relationships between the words. Thus, the feature set selected using GraFS contains the most influential features of a document corpus. Hence, GraFS reduces the number of the selected features, which in turn mitigates the effects of the 'curse-of-dimensionality' phenomenon, i.e. the production of over-fitted ML models and sparse feature vectors. Contrary to our approach, common feature selection methods that are based on statistics ignore the interconnections between the terms (both within a single document and across the documents of a corpus), which has as effect that more features are required from the text classifiers to perform equally well (see the left graph in Fig. 3).

Table 2. Accuracy score (acc) and number of features (|f|) per text classifier for each feature selection technique on the five selected textual datasets. * and bold font highlights the best method for a specific dataset as far as the accuracy score and the number of features are concerned.

Method	20Newsgroups		Reuters		Amazon		LingSpam		JiraIssues	
	acc	|f|	acc	|f|	acc	|f|	acc	|f|	acc	|f|
GraFS+5NN	0.7192	20942	0.8272	809	0.6786	1065	0.9926	120	0.682	905
GraFS+NB	0.9421	20942	0.8399	7171	0.7273	1940	0.9963	2274	0.6958	6404
GraFS+LR	0.9402	37281	0.8782*	7171	0.776	4897	1.0*	120	0.7517	2598
GraFS+NN100x50	0.9575*	30793	0.8733	8187	0.7403	1940	1.0	758	0.7542*	6404
GraFS+LSVM	0.9392	39694	0.8737	7171	0.763	4897	0.9963	120	0.7348	3045
BOW+5NN	0.643	62384	0.7582	15514	0.6169	9771	0.8333	16695	0.6486	14539
BOW+NB	0.9361	62384	0.8191	15514	0.7208	9771	0.9963	16695	0.6989	14539
BOW+LR	0.9387	62384	0.8746	15514	0.763	9771	1.0	16695	0.7461	14539
BOW+NN100X50	0.9546	62384	0.8656	15514	0.7273	9771	0.9963	16695	0.741	14539
BOW+LSVM	0.9408	62384	0.8742	15514	0.763	9771	1.0	16695	0.7304	14539
LVAR+5NN	0.6942	4880	0.8209	1482	0.7013	1719	0.8926	5464	0.6493	2771
LVAR+NB	0.94	29992	0.8403	3356	0.737	2906	0.9963	8234	0.7373	5833
LVAR+LR	0.9384	29992	0.8755	7624	0.7792	3637	1.0	8234	0.7442	9706
LVAR+NN100X50	0.9541	29992	0.8724	4870	0.75	1719	1.0	11058	0.7398	6489
LVAR+LSVM	0.9368	29992	0.8746	7624	0.7825	1719	1.0	16695	0.7291	9706
KBEST+5NN	0.721	5000	0.7957	6000	0.724	350	0.9778	5000	0.6644	4000
KBEST+NB	0.9374	25000	0.8354	7000	0.75	500	0.9963	3000	0.6952	14000
KBEST+LR	0.9389	30000	0.8764	10000	0.7727	3000	1.0	1000	0.7461	6000
KBEST+NN100X50	0.9564	30000	0.8705	14000	0.7565	1000	1.0	1000	0.7423	10000
KBEST+LSVM	0.9366	30000	0.8737	14000	0.763	6000	1.0	1000	0.7253	6000
META+5NN	0.6542	14907	0.8002	2494	0.6623	2731	0.8704	2509	0.6329	2942
META+NB	0.9387	14907	0.8376	2494	0.737	2731	0.9963	2509	0.6989	2942
META+LR	0.9376	14907	0.876	2494	0.789*	2731	1.0	2509	0.7461	2942
META+NN100X50	0.952	14907	0.8701	2494	0.75	2731	1.0	2509	0.7467	2942
META+LSVM	0.9408	14907	0.8746	2494	0.7727	2731	1.0	2509	0.731	2942

5 Conclusions

This paper introduces a new approach for graph-based feature selection, namely *GraFS*. To test the proposed approach, we benchmarked GraFS against classical feature selection techniques. The evaluation outcome was very promising; state-of-the-art accuracy has been achieved in the classification of five well-known datasets using fewer features. In any case, our approach demonstrates two limitations: (i) it is unable to select features for outlier documents, i.e. documents that are not similar to any other document, and (ii) it requires significant time to generate the corresponding graph of documents in a disk-based graph database.

Aiming to address the above limitations as well as to integrate our approach into existing works on knowledge management systems, future work directions include: (i) the utilization and assessment of an in-memory graph database in combination with Neo4j; (ii) the exploitation of link prediction algorithms to deal with outlier documents; (iii) the application of graph and word embedding techniques, and (iv) the integration of our approach into collaborative argumentation environments where the underlying knowledge is structured through semantically-rich discourse graphs (e.g. integration with the approaches described in (Kanterakis et al. 2019) and (Karacapilidis et al. 2009)).

Acknowledgments. The work presented in this paper is supported by the OpenBio-C project (www.openbio.eu), which is co-financed by the European Union and Greek national funds through the Operational Program Competitiveness, Entrepreneurship and Innovation, under the call RESEARCH – CREATE – INNOVATE (Project id: T1EDK- 05275).

References

Aggarwal, C.C.: Machine Learning for Text. Springer, Cham (2018). https://doi.org/10.1007/978-3-319-73531-3

Androutsopoulos, I., Koutsias, J., Chandrinos, K., Paliouras, G., Spyropoulos, C.: An evaluation of Naïve Bayesian anti-spam filtering. In: Proceedings of the Workshop on Machine Learning in the New Information Age, 11th European Conference on Machine Learning, pp. 9–17 (2000)

Fakhraei, S., Foulds, J., Shashanka, M., Getoor, L.: Collective spammer detection in evolving multi-relational social networks. In: Proceedings of the 21 ACM SIGKDD International Conference on Knowledge Discovery and Data Mining, pp. 1769–1778 (2015)

Giarelis, N., Kanakaris, N., Karacapilidis, N.: On a novel representation of multiple textual documents in a single graph. In: Czarnowski, I., Howlett, R.J., Jain, L.C. (eds.) Intelligent Decision Technologies 2020 – Proceedings of the 12th KES International Conference on Intelligent Decision Technologies (KES-IDT-20), Split, Croatia, 17–19 June 2020. Springer (2020)

Henni, K., Mezghani, N., Gouin-Vallerand, C.: Unsupervised graph-based feature selection via subspace and PageRank centrality. Expert Syst. Appl. **114**, 46–53 (2018)

Ienco, D., Meo, R., Botta, M.: Using PageRank in feature selection. In: SEBD, pp. 93–100 (2008)

Kanterakis, A., et al.: Towards reproducible bioinformatics: the OpenBio-C scientific workflow environment. In: Proceedings of the 19th IEEE International Conference on Bioinformatics and Bioengineering (BIBE), Athens, Greece, pp. 221–226 (2019)

Karacapilidis, N., et al.: Tackling cognitively-complex collaboration with CoPe_it! Int. J. Web-Based Learn. Teach. Technol. **4**(3), 22–38 (2009)

Mikolov, T., Sutskever, I., Chen, K., Corrado, G.S., Dean, J.: Distributed representations of words and phrases and their compositionality. In: Advances in Neural Information Processing Systems (NeurIPS), pp. 3111–3119 (2013)

Miller, J.J.: Graph database applications and concepts with Neo4j. In: Proceedings of the Southern Association for Information Systems Conference, vol. 2324, no. 36, Atlanta, USA (2013)

Nijssen, S., Kok, J.N.: A quickstart in frequent structure mining can make a difference. In: Proceedings of the Tenth ACM SIGKDD International Conference on Knowledge Discovery and Data Mining, pp. 647–652. ACM Press (2004)

Nikolentzos, G., Meladianos, P., Rousseau, F., Stavrakas, Y., Vazirgiannis, M.: Shortest-path graph kernels for document similarity. In: Proceedings of the 2017 Conference on Empirical Methods in Natural Language Processing, pp. 1890–1900 (2017)

Rawat, D.S., Kashyap, N.K.: Graph database: a complete GDBMS survey. Int. J. **3**, 217–226 (2017)

Rousseau, F., Kiagias, E., Vazirgiannis, M.: Text categorization as a graph classification problem. In: Proceedings of the 53rd Annual Meeting of the Association for Computational Linguistics and the 7th International Joint Conference on Natural Language Processing, vol. 1, pp. 1702–1712 (2015)

Rousseau, F., Vazirgiannis, M.: Graph-of-word and TW-IDF: new approach to ad hoc IR. In: Proceedings of the 22nd ACM International Conference on Information & Knowledge Management, pp. 59–68. ACM Press (2013)

Saigo, H., Nowozin, S., Kadowaki, T., Kudo, T., Tsuda, K.: gBoost: a mathematical programming approach to graph classification and regression. Mach. Learn. **75**(1), 69–89 (2009). https://doi.org/10.1007/s10994-008-5089-z

Vazirgiannis, M., Malliaros, F., Nikolentzos, G.: GraphRep: boosting text mining, NLP and information retrieval with graphs. In: Proceedings of the 27th ACM International Conference on Information and Knowledge Management, pp. 2295–2296 (2018)

Yan, X., Han, J.: gSpan: graph-based substructure pattern mining. In: Proceedings of the IEEE International Conference on Data Mining, pp. 721–724. IEEE Press (2002)

k-means Cluster Shape Implications

Mieczysław A. Kłopotek[1]([⊠]), Sławomir T. Wierzchoń[1],
and Robert A. Kłopotek[2]

[1] Institute of Computer Science, Polish Academy of Sciences, Warsaw, Poland
{mieczyslaw.klopotek,slawomir.wierzchon}@ipipan.waw.pl
[2] Faculty of Mathematics and Natural Sciences, School of Exact Sciences,
Cardinal Stefan Wyszyński University in Warsaw, Warsaw, Poland
r.klopotek@uksw.edu.pl

Abstract. We present a novel justification why k-means clusters should
be (hyper)ball-shaped ones. We show that the clusters must be ball-
shaped to attain motion-consistency. If clusters are ball-shaped, one can
derive conditions under which two clusters attain the global optimum of
k-means. We show further that if the gap is sufficient for perfect separation, then an incremental k-means is able to discover perfectly separated
clusters. This is in conflict with the impression left by an earlier publication by Ackerman and Dasgupta. The proposed motion-transformations
can be used to the new labeled data for clustering from existent ones.

Keywords: Cluster shape · Motion-consistency · Outer-consistency ·
Incremental clustering · Perfect cluster separation · Clusterability

1 Introduction

Sufficiently diverse corpora of test data for the development and implementation of algorithms, in particular of clustering algorithms, constitute an imma-
nent challenge [9]. While various efforts like crowdsourcing produce considerable
resources, tasks like fine-tuning or prevention of overfitting require still bigger
sets of labeled data. Therefore, ways are sought on how to derive a new tagged
set from an existent one without violating clustering algorithm assumptions.

Therefore, the axiomatization of clustering algorithm properties are of interest. Kleinberg [5] introduced interesting clustering invariant properties like
scaling-invariance and consistency. *Scaling invariance* means that the cluster
structure should be preserved if all distances between data points are multiplied
by a constant. *Consistency* means that the clustering should be preserved if
distances between data points in the cluster are not increased, and distances
between data points from distinct clusters are not decreased. Regrettably, the
consistency transform (CT) cannot be applied to data being subject of k-means
clustering to generate a new labeled data set because this algorithm does not
have the consistency property [8]. Hence other properties like the inner or outer

© IFIP International Federation for Information Processing 2020
Published by Springer Nature Switzerland AG 2020
I. Maglogiannis et al. (Eds.): AIAI 2020, IFIP AICT 583, pp. 107–118, 2020.
https://doi.org/10.1007/978-3-030-49161-1_10

consistency [1] need to be investigated. The *inner-consistency* transform is identical with CT except for distances between data points from distinct clusters are unchanged. The *outer-consistency* transform is identical with CT except for distances between data points within each cluster are unchanged. But inner-consistency property is not possessed by k-means [1] and what is worse, in a fixed-dimensional space the inner CT reduces typically to identity transform [7] which is of no value for test data generation. The outer consistency, on the other hand, with respect to k-means is of little value under continuous setting because it requires synchronized motion of all clusters – one cluster alone cannot be moved. This paper investigates how the outer-consistency constraints can be relaxed to what we call motion-consistency. It shall be defined as follows:

Definition 1. Cluster area *is any solid body containing all cluster data points. Gap between two clusters is the minimum distance between the cluster areas, i.e., Euclidean distance between the closest points of both areas.*

Definition 2. *Given a clustered data set embedded in a fixed dimensional Euclidean space, the* motion-transformation *is any continuous transformation of the data set in the fixed dimensional space that (1) preserves the cluster areas (the areas may only be subject of isomorphic transformations) and (2) keeps the minimum required gaps between clusters (the minimum gaps being fixed prior to transformation). By* continuous *we mean: there exists a continuous trajectory for each data point such that the conditions (1) and (2) are kept all the way.*

Definition 3. *A clustering method has the property of* motion-consistency, *if it returns the same clustering after motion-transformation.*

Compared to outer-consistency, or even the consistency, we weaken the constraints imposed on the distances between points, because the distances between data points of different data sets do not need to be increased, but only the distances between cluster areas (gaps) should not be decreased below certain values.

We demonstrate for k-means that it is advantageous to define the cluster area as a ball centered at its gravity center and encompassing all the data points of a cluster. *Wherever we speak about a ball, we mean a hyper-ball that is the region enclosed by a hyper-sphere, that is an n-ball for n-dimensional Euclidean space* $\mathbb{R}^n$. k-means, one of the most popular algorithms, exists in a multitude of versions. For an extensive overview of the general concept of k-means and versatile versions of it, see e.g., [11]. We refer here to the following ones: (1) random-seed k-means, that is one with random initial seeding of cluster centers, (2) random-set k-means, that is one with the random initial assignment of data points to clusters, (3) k-means++, that is one with seeding of clusters according to a heuristic minimizing the distance to the closest cluster (4) k-means-ideal that is an "oracle" algorithm that finds the clustering minimizing absolutely the k-means objective. We consider so-called batch versions.

2   Moving Clusters – Motion-Consistency

As was observed in [4], clustering algorithms make implicit assumptions about the clusters' definition, shape, and other characteristics and/or require some predetermined free parameters. The shape of the clusters may constitute the foundation for choosing the right number of clusters to split the data into [3]. In this section, let us ask what should be the shape of the area covered by the *k*-means clusters. The usual way to look at the *k*-means clusters is one of the so-called Voronoi regions [10]. These regions are polyhedrons such that any point within the area of the polyhedron is closer to its cluster center than to any other cluster center. Obviously, the *outer* polyhedrons (at least one of them) can be moved away continuously from the rest without overlapping any other region so that at least the motion-transformation is applicable non-trivially. However, does the motion-consistency hold? A closer look at the issue tells us that it is not. As *k*-means terminates, the neighboring clusters' polyhedra touch each other via a hyperplane such that the straight line connecting centers of the clusters is orthogonal to this hyperplane. This causes that points on the one side of this hyperplane lie more closely to the one center, and on the other to the other one. But if we move the clusters in such a way that both touch each other along the same hyperplane, then it happens that some points within the first cluster will become closer to the center of the other cluster and vice versa.

Generally, moving the clusters will change their structure (points switch clusters) unless the points lie actually not within the polyhedrons but rather within *paraboloids* with appropriate equations. Then moving along the border, hyperplane will not change cluster membership (locally, that is, the data points of the two considered clusters will not switch cluster membership given that we fixed all other clusters and consider reclustering of these two clusters only). But the intrinsic cluster borders are now *paraboloids*. The problem will occur again if we relocate the clusters allowing for touching along the *paraboloids*.

Hence the question can be raised: What shape should the *k*-means clusters have in order to be (locally) immune to movement of whole clusters?

Assume that only one cluster would move. Let us consider the problem of susceptibility to class membership change within a 2D plane containing the two cluster centers and the motion vector of the moving cluster. Let the one cluster center be located (for simplicity) at the point (0,0) in this plane and the other at $(2x_0, 2y_0)$ for some x_0, y_0. Let further the border of the first cluster be characterised by a (symmetric) function $f(x)$ and let the shape of the border of the other one $g(x)$ be the same, but rotated by 180° around (x_0, y_0): $g(x) = 2y_0 - f(2x_0 - x)$. Let both have a touching point (we excluded already a straight line and want to have convex smooth borders). From the symmetry conditions one easily sees that the touching point must be (x_0, y_0). As this point lies on the surface of $f()$, $y_0 = f(x_0)$ must hold. Any point $(x, f(x))$ of the border of the first cluster must be closer to its centre $(0,0)$ than to the centre $(2x_0, 2y_0)$ of the other:

$$(x - 2x_0)^2 + (f(x) - 2f(x_0))^2 - x^2 - f^2(x) \geq 0 \tag{1}$$

That is

$$-x_0(x - x_0) - f(x_0)\,(f(x) - f(x_0)) \geq 0$$

Let us consider only positions of the center of the second cluster below the X axis ($y_0 < 0$). In this case $f(x_0) < 0$. Further let us concentrate on $x > x_0$. We get $\frac{f(x)-f(x_0)}{x-x_0} \geq \frac{x_0}{-f(x_0)}$. In the limit, when x approaches x_0, $f'(x_0) \geq \frac{x_0}{-f(x_0)}$. By analogy for $x < x_0$ in the limit $x \to x_0$ we get: $f'(x_0) \leq \frac{x_0}{-f(x_0)}$ This implies

$$f'(x_0) = \frac{-1}{\frac{f(x_0)}{x_0}} \tag{2}$$

$\frac{f(x_0)}{x_0}$ is the directional tangent of the straight line connecting both cluster centres. $f'(x_0)$ is tangential of the borderline of the first cluster at the touching point of both clusters. The equation above means both are orthogonal. But this property implies that $f(x)$ must define (a part of) a circle centred at $(0,0)$. As the same reasoning applies at any touching point of the clusters, a k-means cluster would have to be (hyper)ball-shaped in order to allow the movement of the clusters without elements switching cluster membership.

We know that most k-means versions tend to stick at local minima. We see here immediately that some kind of local minima is preserved under motion-consistency transform.

Theorem 1. *If random-set k-means has a local minimum in ball form that is such that the clusters are enclosed into equal radius balls centered at the respective cluster centers, and gaps are fixed at zero, then the motion-transform preserves this local minimum.*

Proof. For $k = 2$, this is obvious from the above consideration. For $k > 2$ consider just each pair of clusters to see that no cluster change occurs.

The tendency of k-means to recognize best ball-shaped clusters has been known long ago, but we are not aware of presenting such an argument for it.

3   Motion-Consistency Property for Two Clusters

The preservation of local minima does not guarantee the preservation of the global minimum even if the global minimum has the above-mentioned ball form.

A sufficient separation between the enclosing balls is needed, as we will show again for $k = 2$.

Let us consider, under which circumstances a cluster C_1 of radius r_1 containing n_1 elements would take over n_{21} elements (i.e. subcluster C_{21}) of a cluster C_2 of radius r_2 of cardinality n_2, if we perform the motion-consistency transform. As only (sub)cluster centres are of interest in our investigation, we can concentrate on the plane spanned by the gravity centres $c1, c21, c22$ of C_1, C_{21}, C_{22}, see left most Fig. 1. The enclosing (hyper)balls of both clusters C_1, C_2 intersect with this plane as circles, indicated as black lines. In worst case, either $c21$ or $c22$ would

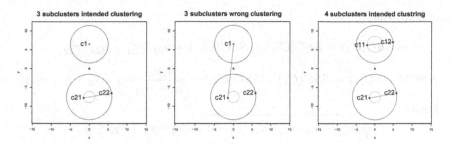

Fig. 1. Possible alternatives to the basic clustering into $\{C_1, C_2\}$. Left figure: one cluster is split into two clusters. Central figure: upon this split, one cluster takes over a subcluster of the other cluster. Right figure: both clusters are split into subclusters that form new clusters, as indicated in Fig. 2.

lie on the respective circle, which implies the other centre lying on a circle with smaller radius, drawn in grey. Generally, both will lie closer to C_2 gravity centre. Let $n_{22} = n_2 - n_{21}$ be the number of the remaining elements (subcluster C_{22} of the second cluster. Let the enclosing balls of both clusters be separated by the distance (gap) g. Let us consider the worst case that is that the center of the C_{21} subcluster lies on a straight line segment connecting both cluster centers. The centre of the remaining C_{22} subcluster would lie on the same line but on the other side of the second cluster centre. Let r_{21}, r_{22} be the distances of centros of n_{21} and n_{22} from the centre of the second cluster. The relations

$$n_{21} \cdot r_{21} = n_{22} \cdot r_{22}, \; r_{21} \leq r_2, \; r_{22} \leq r_2$$

must hold. Let us denote with $SSC(C)$ the sum of squared distances of elements of the set C to the center of this set.

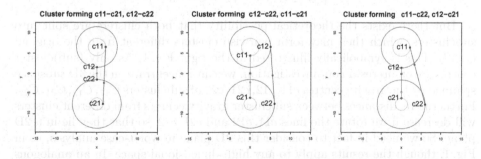

Fig. 2. Extreme cases to be considered when looking for a sufficient gap to avoid alternative clustering with a split of both clusters.

So in order for the clusters C_1, C_2 to constitute the global optimum

$$SSC(C_1) + SSC(C_2) \leq SSC(C_1 \cup C_{21}) + SSC(C_{22})$$

must hold. But

$$SSC(C_2) = SSC(C_{21}) + SSC(C_{22}) + n_{21} \cdot r_{21}^2 + n_{22} \cdot r_{22}^2$$

$$SSC(C_1 \cup C_{21}) = SSC(C_1) + SSC(C_{21}) + \frac{n_1 n_{21}}{n_1 + n_{21}}(r_1 + r_2 + g - r_{21})^2$$

Hence

$$\sqrt{(r_{21}^2 + r_{21} \cdot r_{22})(n_{21}/n_1 + 1)} - r_1 - r_2 + r_{21} \le g$$

As $r_{22} = \frac{r_{21} n_{21}}{n_2 - n_{21}}$

$$r_{21} \sqrt{\frac{n_2}{n_1} \frac{n_1 + n_{21}}{n_2 - n_{21}}} - r_1 - r_2 + r_{21} \le g$$

Let us consider the worst-case when the elements to be taken over are at the *edge* of the cluster region ($r_{21} = r_2$). Then

$$r_2 \sqrt{\frac{n_2}{n_1} \frac{n_1 + n_{21}}{n_2 - n_{21}}} - r_1 \le g$$

The lower limit on g will grow with n_{21}, but $n_{21} \le 0.5 n_2$, because otherwise r_{22} would exceed r_2. Hence in the worst case

$$r_2 \sqrt{\frac{n_2}{n_1} \frac{n_1 + n_2/2}{n_2/2}} - r_1 \le g$$

$$r_2 \sqrt{2(1 + 0.5 n_2/n_1)} - r_1 \le g \tag{3}$$

In case of clusters with equal sizes and equal radius this amounts to

$$g \ge r_1(\sqrt{3} - 1) \approx 0.7 r_1$$

But there exists the theoretical possibility that both clusters are split into subclusters, which then may form pairwise clusters different from the original C_1, C_2. This is symbolically illustrated in the right Fig. 1. As only (sub)cluster centres are of interest in our investigation, we can concentrate on the 3D subspace spanned by the gravity centres $c11, c12, c21, c22$ of subclusters $C_{11}, C_{12}, C_{21}, C_{22}$. Furthermore, distances between subcluster gravity centers from different clusters will decrease if we rotate the lines $c11, c12$ and $c21, c22$, so that they lie in a 2D plane. So we need in fact to consider this 2D plane in worst-case analysis, as in Fig. 1, though the results apply to any high-dimensional space. In an analogous way as above, we can derive (as a simple exercise) explicit requirements on minimum gap g needed in order to ensure that such a re-clustering will not happen. The worst cases of subcluster center positions to be considered are depicted in Fig. 2. In each case, the 50%–50% split of a cluster into subclusters turns out to be requiring the biggest gap. We conclude

Theorem 2. *k-means algorithm with $k = 2$ possesses the property of motion-consistency if (1) the k-means clustering global minimum Γ has the property that each cluster can be enclosed in a ball and (2) the gaps between balls fulfil the condition of taking the maximum of the gaps derived from Fig. 2, Fig. 1 and (3) the gap between clusters would not be decreased below the gap value from (2) during the motion.*

Note that under *k*-means objective, the globally optimal clustering is also pairwise optimal, but the inverse does not hold.

This means that Motion-Consistency transform can turn an optimal clustering to an unoptimal one for $k > 2$.

It should be emphasized that we consider here about the local optimum of k-means. With the aforementioned gap size, the global *k*-means minimum may lie elsewhere, in a clustering possibly without gaps. Also, the motion-transformation preserves as a local minimum the partition it is applied to. Other local minima and global minimum can change.

Note that the motion-consistency (applicable for $k = 2$ in *k*-means) is more flexible for the creation of new labeled data sets than outer-consistency.

4 Perfect Ball Clusterings

The problem with *k*-means (-random and ++) is the discrepancy between the theoretically optimized function (*k*-means-ideal) and the actual approximation of this value. It appears to be problematic even for *well-separated* clusters.

First, let us point to the fact that *well-separatedness* may keep the algorithm in a local minimum.

It is commonly assumed that a good initialization of a *k*-means clustering is one where the seeds hit different clusters. It is well-known that under some circumstances, the *k*-means does not recover from poor initialization, and as a consequence, a natural cluster may be split even for *well-separated* data.

Hitting each cluster may not be sufficient as neighboring clusters may be able to shift the cluster center away from its cluster. Hence let us investigate what kind of well-separability would be sufficient to ensure that once clusters are hit by one seed each, they would never lose the cluster center.

Let us investigate the working hypothesis that two clusters are well separated if we can draw a ball of some radius ρ around true cluster center of each of them, and there is a gap between these balls. We claim (see [6]) that

Theorem 3. *If the distance between any two cluster centres A, B is at least $4\rho_{AB} + \epsilon$, $\epsilon > 0$, where ρ_{AB} is the radius of a ball centred at A and enclosing its cluster (that is cluster lies in the interior of the ball) and it also is the radius of a ball centred at B and enclosing its cluster, then once each cluster is seeded the clusters cannot loose their cluster elements for each other during k-means-random and k-means++ iterations.*

Before starting the proof, let us introduce related definitions.

Definition 4. *We shall say that clusters centred at A and B and enclosed in balls centred at A, B and with radius ρ_{AB} each are* nicely ball-separated, *if the distance between A, B is at least $4\rho_{AB} + \epsilon$, $\epsilon > 0$. If all pairs of clusters are nicely ball separated with the same ball radius, then we shall say that they are* perfectly ball-separated.

Obviously, if there exists a perfect ball clustering into k-clusters in the data set, then after invariance transform as well as after consistency transform, there exists a perfect ball clustering into k-clusters in the data set. Let us restrict ourselves to clusterings with at least three data points in a cluster (violation of the most general richness, but nonetheless a reasonable richness, let us call it reachness-3++. In this case, it is obvious that if the perfect ball clustering exists then, it is unique. This means automatically that if k-means would be able to detect the perfect ball clustering into k clusters, then it would be consistent in the sense of Kleinberg. Therefore it is worth investigating whether or not k-means can detect a perfect ball clustering.

If the data set had a perfect ball clustering into k clusters (of at least 2 elements), but not into $k - n_1$ nor into $k + n_2$ clusters, where $1 \leq n_1 \leq k - 2, 1 \leq n_2 \leq n/2 - k$ are natural numbers, then under application of Kleinberg's consistency transform, the new data set can both have a perfect ball clustering into $k - n_1$ and into $k + n_2$ clusters. Hence the Kleinberg's impossibility theorem holds also within the realm of perfect ball clusterings.

Proof. For the illustration of the proof see Fig. 3.

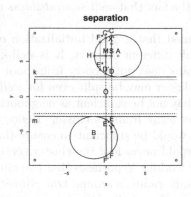

Fig. 3. An illustrative figure for proof of 4 radius distance ensuring good separability.

Consider two points A, B be two ball centers and two points, X, Y, one being in each ball (presumably the cluster centers at some stage of the k-means algorithm). To represent their distances faithfully, we need at most a 3D space.

Let us consider the plane established by the line AB and parallel to the line XY. Let X' and Y' be projections of X, Y onto this plane. Now let us establish that the hyperplane π orthogonal to X, Y, and passing through the middle of the line segment XY, that is the hyperplane containing the boundary between clusters centered at X and Y does not cut any of the balls centered at A and B. This hyperplane will be orthogonal to the plane of the Fig. 3 and so it will manifest itself as an intersecting line l that should not cross circles around A and B, being projections of the respective balls. Let us draw two solid lines k, m between circles $O(A, \rho)$ and $O(B, \rho)$ tangential to each of them. Line l should lie between these lines, in which case the cluster center will not jump to the other ball. Let the line $X'Y'$ intersect with the circles $O(A, \rho)$ and $O(B, \rho)$ at points C, D, E, F as in the figure. It is obvious that the line l would get closer to circle A, if the points X', Y' would lie closer to C and E, or closer to circle B if they would be closer to D and F.

Therefore, to show that the line l does not cut the circle $O(A, \rho)$, it is sufficient to consider $X' = C$ and $Y' = E$. (The case with ball $Ball(B, \rho)$ is symmetrical).

Let O be the center of the line segment AB. Let us draw through this point a line parallel to CE that cuts the circles at points C', D', E' and F'. Now notice that centric symmetry through point O transforms the circles $O(A, \rho)$, $O(B, \rho)$ into one another, and point C' into F' and D' into E'. Let E^* and F^* be images of points E and F under this symmetry. In order for the line l to lie between m and k, the middle point of the line segment CE shall lie between these lines.

Let us introduce a planar coordinate system centered at O with $\mathcal{X}$ axis parallel to lines m, k, such that A has both coordinates non-negative, and B non-positive. Let us denote with α the angle between the lines AB and k. As we assume that the distance between A and B equals 4ρ, then the distance between lines k and m amounts to $2\rho(2\sin(\alpha) - 1)$. Hence the $\mathcal{Y}$ coordinate of line k equals $\rho(2\sin(\alpha) - 1)$. So the $\mathcal{Y}$ coordinate of the center of the line segment CE shall be not higher than this. Let us express this in vector calculus:

$$4(y_{OC} + y_{OE})/2 \leq \rho(2\sin(\alpha) - 1)$$

Note, however that

$$y_{OC} + y_{OE} = y_{OA} + y_{AC} + y_{OB} + y_{BE} = y_{AC} + y_{BE} = y_{AC} - y_{AE^*} = y_{AC} + y_{E^*A}$$

So let us examine the circle with center at A. Note that the lines CD and E^*F^* are at the same distance from the line C' D'. Note also that the absolute values of direction coefficients of tangentials of circle A at C' and D' are identical. The more distant these lines are, as line CD gets closer to A, the y_{AC} gets bigger, and y_{E^*A} becomes smaller. But from the properties of the circle, we see that y_{AC} increases at a decreasing rate, while y_{E^*A} decreases at an increasing rate. So the sum $y_{AC} + y_{E^*A}$ has the biggest value when C is identical with C' and we need hence to prove only that

$$(y_{AC'} + y_{D'A})/2 = y_{AC'} \leq \rho(2\sin(\alpha) - 1)$$

Let M denote the middle point of the line segment $C'D'$. As point A has the coordinates $(2\rho\cos(\alpha), 2\rho\sin(\alpha))$, the point M is at distance of $2\rho\cos(\alpha)$ from A. But $C'M^2 = \rho^2 - (2\rho\cos(\alpha))^2$.

So we need to show that $\rho^2 - (2\rho\cos(\alpha))^2 \leq (\rho(2\sin(\alpha) - 1))^2$. In fact we get from the above $0 \leq 1 - \sin(\alpha)$ which is an obvious trigonometric relation.

5   Incremental k-means

It is not hard to demonstrate that if the perfect-ball separation was found, it does not suffice to state that we reached a global minimum of k-means. Moreover, even if the global minimum of k-means is a perfectly ball separated set of clusters, it does not mean that motion-transformation keeping this property will yield a new clustering being optimal for k-means.

However, there exists a version of k-means for which a perfectly ball separated set of clusters is the global optimum. Hence the motion-transform keeping the perfect ball separation keeps the optimum. We will introduce this algorithm in this section and demonstrate the respective property.

Ackerman and Dasgupta [2] study clusterability properties of incremental clustering algorithms. They introduce an incremental version of a very popular k-means algorithm (for an extensive overview of k-means versions see [11]).

They introduced the *perfect clustering* with the property that the smallest distance between elements of distinct clusters is larger than the distance between any two elements of the same cluster. They demonstrate that there exists an incremental algorithm discovering the *perfect clustering* that is linear in k with respect to space. But their incremental (*sequential*) k-means fails to do so.

Their case study is interesting because it demonstrates that the cluster shape plays a role - each cluster has to be enclosed into a convex envelope. The problem of incremental k-means is caused by the fact that this envelope is not ball-shaped.

Data: the data points $\mathbf{x}_i$, $i = 1, \ldots, m$, the required number of clusters k
Result: T - the set of cluster centres
Set $T = (t_1, \ldots, t_k)$ to the first k data points;
Initialize the counts $n_1, n_2, \ldots, n_k$ to 1;
while *any data point unvisited* **do**

> Acquire the next example, t_{k+1}. Set $n_{k+1} = 1$;
> Find $i, j \in \{1, \ldots, k+1\}, i < j$ such the distance between t_i and t_j is the smallest one among distances between $t_1, \ldots, t_{k/+1}$.
> Replace $t_i = (t_i n_i + t_j n_j)/(n_i + n_j)$, thereafter $n_i = n_i + n_j$;
> **if** $j \neq k+1$ **then**
>> replace $t_j = t_{k+1}$, $n_j = n_{k+1}$
>
> **end**

end

Algorithm 1: Sequential (incremental) k-means, our modification

Let us discuss at this point a bit the notions of *perfect separation*. In their Theorem 4.4. Ackerman and Dasgupta [2] show that the incremental k-means algorithm, as introduced in their Algorithm 2.2, is not able to cluster correctly data that is *perfectly clusterable* (their Definition 4.1). The reason is quite simple. The perfect separation refers only to separation of data points, and not to points in the convex hull of these points. But during the clustering process, the candidate cluster centres are moved in the convex hulls, so that they can occasionally get too close to data points of the other cluster. To avoid this effect, we will use the just introduced concept of perfect ball separation (see Definition 4)

Under the *perfect-ball-separation* as introduced here their incremental *k*-means Algorithm 2.2. (Sequential *k*-means) will discover the structure of the clusters after a modification (Algorithm 1):

Data: $T = (t_1, \ldots, t_k)$ be the resulting set of cluster centres from the Algorithm 1.

Result: Clusterability decision

Initialize the furthest neighbours $f_1, f_2, \ldots, f_k$ with $t_1, t_2 \ldots, t_k$ respectively;

while *any data point unvisited* **do**

    Acquire the next example, x; **if** t_i *is the closest centre to* x *and* x *is further away from* t_i *than* f_i **then**

        | Replace f_i with X;

    end

end

Compute distances between corresponding t_i and f_i, pick the highest one;

Compute distances between each pair t_i, t_j and pick the lowest one;

if *the latter is 4 times or more higher than the former one* **then**

    | We got a perfect ball clustering

else

    | Perfect ball clustering was not found

end

Algorithm 2: Sequential *k*-means, our modification – second pass

The reason is as follows. Perfect ball separation ensures that there exists an r of the enclosing ball such that the distance between any two points within the same ball is lower than $2r$, and between them is bigger than $2r$. So whenever Ackerman's incremental *k*-mean merges two points, they are the points of the same ball. Upon merging, the resulting point lies again within the ball.

Theorem 4. *The incremental k-means algorithm will discover the structure of perfect-ball-clustering.*

Proof. If t_i, t_j are points within the ball enclosing a single cluster, then also $(t_i n_i + t_j n_j)/(n_i + n_j)$ will lie within the same ball. If t_{k+1} stems from a cluster not represented by $t_1, \ldots, t_k$, then $k + 1$ will not be in the pair (i, j) of closest elements, because their distance is more than 2ρ, while those within a cluster at most 2ρ. On the other hand, t_i, t_j would not stem from two different clusters because $t_1, \ldots, t_k$, because the distance within a cluster is at most 2ρ, while the distance between elements from distinct clusters is at most 2ρ. In this way, no t_l, $l = 1, \ldots, k$ will lie outside of balls representing clusters. Furthermore, its position will be calculated only based on data points from the same cluster. Furthermore, it will be the average position of those points, so finally, after the full pass, all t_l will represent the k different cluster centers.

The incremental k-means algorithm returns only a set of cluster centers without stating whether or not we got a perfect ball clustering. However, if we are allowed to inspect the data for the second time, such information can be provided. See Algorithm 2: A second pass for other algorithms from Ackerman and Dasgupta Sect. 2 would not yield such a decision.

6   Conclusions

We derived in this paper the intended shape of a k-means cluster (a ball centered at cluster gravity center) as a necessary condition of clustering preserving motion-transformation of the dataset. This shape preserves ball-shaped local minima for k-means algorithm with random initial partition. We have also derived, for ball-shaped clusters for $k = 2$, gap condition for motion-transformation that preserves the global minimum of k-means. We have also shown that incremental k-means is able to find the perfect ball-shaped clustering. Therefore the motion-transform keeping the perfect ball separation will preserve the incremental-k-means clustering. Thus we have discovered a couple of transformations that preserve various aspects of clustering, suitable for deriving new labeled datasets from existent ones, as implied by our Theorems 1, 3, 2, 4.

References

1. Ackerman, M., Ben-David, S., Loker, D.: Towards property-based classification of clustering paradigms. In: Proceedings NIPS 2010, pp. 10–18. Curran Associates, Inc. (2010)
2. Ackerman, M., Dasgupta, S.: Incremental clustering: the case for extra clusters. In: Proceedings NIPS 2014, pp. 307–315. Curran Associates, Inc. (2014)
3. Chiang, M., Mirkin, B.: Intelligent choice of the number of clusters in k-means clustering: an experimental study with different cluster spreads. J. Classif. **27**, 3–40 (2010). https://doi.org/10.1007/s00357-010-9049-5
4. Everitt, B.S., Landau, S., Leese, M., Stahl, D.: Cluster Analysis, 5th edn. Wiley, Chichester (2011)
5. Kleinberg, J.: An impossibility theorem for clustering. In: Proceedings NIPS 2002, pp. 446–453 (2002). http://books.nips.cc/papers/files/nips15/LT17.pdf
6. Kłopotek, R., Kłopotek, M.: On the discrepancy between Kleinberg's clustering axioms and k-means clustering algorithm behavior. arXiv:1702.04577 (2017)
7. Kłopotek, M.A., Kłopotek, R.: Towards continuous consistency axiom (submitted)
8. Kłopotek, M.A., Kłopotek, R.: Clustering algorithm consistency in fixed dimensional spaces. In: Proceedings ISMIS2020 (2020, to appear)
9. Pei, Y., Zaïane, O.: Synthetic data generator for clustering and outlier analysis. Technical report, January 2006
10. Schreiber, T.: A Voronoi diagram based adaptive k-means-type clustering algorithm for multidimensional weighted data. In: Bieri, H., Noltemeier, H. (eds.) CG 1991. LNCS, vol. 553, pp. 265–275. Springer, Heidelberg (1991). https://doi.org/10.1007/3-540-54891-2_20
11. Wierzchoń, S.T., Kłopotek, M.A.: Modern Algorithms of Cluster Analysis. Studies in Big Data, vol. 34. Springer, Cham (2018). https://doi.org/10.1007/978-3-319-69308-8

Manifold Learning for Innovation Funding: Identification of Potential Funding Recipients

Vincent Grollemund[1,2(✉)], Gaétan Le Chat[2], Jean-François Pradat-Peyre[1,3], and François Delbot[1,3]

[1] Sorbonne Université, 75005 Paris, France
vincent.grollemund@lip6.fr
[2] FRS Consulting, 75009 Paris, France
[3] Nanterre Université, 92014 Nanterre, France

Abstract. finElink is a recommendation system that provides guidance to French innovative companies with regard to their financing strategy through public funding mechanisms. Analysis of financial data from former funding recipients partially feeds the recommendation system. Financial company data from a representative French population are reduced and projected onto a two-dimensional space with Uniform Manifold Approximation and Projection, a manifold learning algorithm. Former French funding recipients' data are projected onto the two-dimensional space. Their distribution is non-uniform, with data concentrating in one region of the projection space. This region is identified using Density-based Spatial Clustering of Applications with Noise. Applicant companies which are projected within this region are labeled potential funding recipients and will be suggested the most competitive funding mechanisms.

Keywords: Dimension reduction · Manifold learning · Clustering

1 Introduction

Given the diversity and quantity of unstructured information available on existing French funding mechanisms, innovative companies need guidance with regard to their financing strategy. finElink [4] is a recommendation system that meets this need. Developed by FRS Consulting, a French consulting firm specialized in public innovation funding, it was initially based on business knowledge of FRS Consulting associates. Analysis of financial data from former French funding recipients, using machine learning, helped identify applicant companies with a high potential and further enhance finElink's recommendation.

However, relevance of applicant companies cannot be solely assessed on former funding recipient data, as these data suffer from significant data sparsity,

© IFIP International Federation for Information Processing 2020
Published by Springer Nature Switzerland AG 2020
I. Maglogiannis et al. (Eds.): AIAI 2020, IFIP AICT 583, pp. 119–127, 2020.
https://doi.org/10.1007/978-3-030-49161-1_11

bias and missing data constraints. Funding recipients data were obtained through cross-checking information from funding mechanism websites and the French national company registry where companies' financial statements are available. Financial information was frequently missing, specifically for newly created companies which are the targeted recipients of numerous funding mechanisms. Data collected had a significant number of missing features. Moreover, most funding mechanisms did not communicate on their recipients, especially for small funding mechanisms. As such, available funding recipient data were strongly biased towards well-known funding mechanisms. Supervised learning on data with these limitations would have easily led to overfitting. These limitations were avoided using unsupervised learning and another larger dataset of French companies obtained using a proprietary software. This other dataset was representative of all French companies and suffered from fewer missing data.

These representative company data were reduced and projected into a two-dimensional space with Uniform Manifold Approximation and Projection (UMAP), a manifold learning algorithm. Former funding recipient data were then projected into the new space. Funding recipient projections showed an uneven distribution pattern with funding recipients concentrating in one projection space area. This area was identified using a density-based clustering method: Density-based Spatial Clustering on Applications with Noise (DBSCAN) [3].

This study presents our approach to use this target population of funding recipients in order to isolate a sub-population of potential funding recipients within a large representative population. This approach is neighborhood-based and combines a manifold learning algorithm with a density-based clustering method. Section 2 will present the data used and the data processing steps. Section 3 will focus on data reduction results. The conclusion will be addressed in Sect. 4.

1.1 Previous Work

Dimension reduction intends to represent high-dimensional data in a low-dimensional space while preserving data structure. Linear dimension reduction algorithms, namely Principal Component Analysis (PCA), strive to preserve global input data structure but describe poorly the true geometry of nonlinear data [7]. In this study, former funding recipient and representative populations had similar spatial distributions in the low-dimensional space, hence PCA was unable to isolate the target population from the representative population. Nonlinear dimension reduction algorithms, also referred to as manifold learning algorithms, can describe a wider range of variable interactions [14]. They are usually divided into two categories based on whether they focus on local or global data structure preservation. Global nonlinear dimension reduction algorithms such as Kernel PCA [10] or Isometric feature Mapping (ISOMAP) [17] strive to preserve input data geometry at all scales: neighborhood and remoteness are preserved between the input and output spaces. Local nonlinear dimension reduction algorithms such as Locally Linear Embedding (LLE) [11] or t-Stochastic Neighbor Embedding (t-SNE) [8] focus primarily on local geometry preservation in small neighborhoods of the manifold [14]. Recently developed, UMAP [9] falls into this last category. The local approach has two advantages over the global approach: first, lower computational complexity as computations involve sparse matrix manipulation, second, an enhanced ability to represent a wider range of manifolds, specifically when geometry is Euclidean at a local scale but is non-Euclidean at a global scale. t-SNE has proven to balance well local and global data structures on real life data giving t-SNE a competitive edge. This was not the case for LLE and its other nonlinear counterparts [8]. But t-SNE suffers from several drawbacks [13,18], which do not affect UMAP, such as:

- inability to scale computationally when working with widely used python libraries;
- non-convexity of its cost function, leading to potential initialization-based results;
- non-preservation of density and distances between the input and output spaces (neighborhood is nonetheless preserved).

Due to computational scaling constraints and the inability to use distances in the output space, UMAP was preferred to t-SNE. UMAP has already been successfully used in various medical contexts such as survival prognosis estimation of Amyotrophic Lateral Sclerosis (ALS) patients [6], gene co-expression analysis [5] and infection risk prediction of newly diagnosed B-cell chronic lymphocytic leukemia (B-CLL) patients [1].

Applying UMAP in the context of public innovation funding is original with regard to both testing this recent manifold learning algorithm and experimenting on novel data in a field where public data is sparse.

2   Methods

2.1   Data

The first dataset was our target population of French funding recipients which included 3,350 samples. The second dataset was our representative population of French companies which contained 152,899 instances, randomly sampled from the Amadeus database [2]. As such, companies sampled from that database were selected with less bias than funding recipient data. Features selected for this study were limited to turnover, net income, equity and headcount over a three-year period. Data were not processed as time-series. Feature selection was based upon finElink's use case: information asked to users needed to be easily available to improve user-friendliness. Age, business sector and location information was excluded as these features were not continuous. Age was discretized in years. Business sector and location were categorized with respectively NACE codes and region names. When categorical or discretized features are included in a UMAP projection, the algorithm primarily learns how to represent these different categories or bins without providing additional information on the data. As such, these UMAP projections were unable to isolate the target population from the representative population when these features were included.

2.2 Missing Data Analysis

Missing data were imputed using MissForest [15], a multiple imputation method based on a random forest model. Multiple imputation methods preserve input data distribution better than single imputation methods. MissForest has a good tolerance for high missing data rates and can handle Non Missing At Random (NMAR) schemes [16]. Multiple Imputation by Chained Equations (MICE) [12] is another multiple imputation method based on regression. Both MICE and MissForest deal with mixed data types (categorical and/or continuous). However, MissForest is non-parametric and, as such, can handle non-linearity and variable interactions in data, which MICE cannot. Initial missing data rates were 58% and 34% for respectively the target and representative populations. Given the high missing data rates, data imputation on the overall available population would have been inappropriate. Data with up to 7 missing features, on a total of 15 (age, business sector and location were included for missing data imputation) were selected. As such, initial feature distributions were not significantly altered after data imputation. Data were normalized prior to missing data imputation. After processing, the two datasets included 1,413 and 114,628 samples for respectively the target and representative populations.

2.3 Dimension Reduction

Data reduction was carried out using UMAP. The representative population was projected into a two-dimensional space. UMAP is neighborhood-based and works in two steps. First, a compressed embedding of the input space is built through topological analysis of the data structure using simplexes[1]. Second, a low-dimensional (in our case two-dimensional) data embedding is found through a cross-entropy[2] optimization process. UMAP preserves data neighborhoods, distances and density. The initial modeling step depends on whether the algorithm should focus on preserving the local or global input data structure. Data structure is estimated according to the size of the neighborhood investigated. The second compression step is mainly defined using two parameters which are the output dimension size and the minimum distance permitted between two points in the output space, i.e. how compact the output projection can be. In our study, UMAP parameters were set as follows:

- output dimensionality was set to 2, as adding an additional dimension did not provide more insight;

---

[1] In geometry, a simplex is defined as a set of points, where none is a barycentre of the remaining points. The convex hull of these points corresponds to the face of the simplex. In simpler terms, a n-simplex can be thought of as the generalization of a triangle in the nth dimension.

[2] In machine learning, cross entropy is frequently used as a cost function to compare two probability distributions (p,q): p is optimized to approximate q the fixed target distribution.

- neighborhood size was set as high as possible given computational time (6,500) in order to obtain a global overview of data structures, funding recipients were not isolated when the focus was made on local data structure;
- no minimum distance in the output space was set to allow overlapping.

2.4   Clustering

The UMAP projection space was divided into a grid and density differences within that grid were examined using the ratio of funding recipient samples within each cell over the total cell samples. Centroid-based clustering methods are not relevant given the data distribution as they are unable to deal with noise. Density-based clustering methods, such as DBSCAN, manage noise through density analysis which meets our problem's constraints. In DBSCAN, cluster identification is carried out by assessing the neighborhood density of each sample, i.e. evaluating the number of neighbors within an ϵ radius of that sample. Provided the number of neighbors is above the user-defined threshold, that sample is said to be a cluster core point. If the sample does not have enough neighbors within an ϵ radius while having at least one core point as a neighbor, then that point is assigned to the core point's cluster. Otherwise, that point is labeled as noise. Projections from the target population were fed into DBSCAN to isolate the projection space area with a high density of target population samples. The remaining samples were labeled as noise. DBSCAN tuning led to the following setup:

- the ϵ distance was set to the first percentile of the target population distance distribution;
- the minimum number of points within a ϵ radius required to form a cluster was set to 20.

3   Results

3.1   Input Feature Distribution Analysis

As UMAP is a non-linear dimension reduction method, projection features cannot be analyzed to provide any interpretability. Output dimension analysis, as commonly performed for PCA, cannot be carried out. Nonetheless, analysis of input feature distribution in the UMAP projection space is an alternative as it gives a broad overview of variable importance with regard to the projection. Plot analysis can help identify strong correlations between projections and input features. This was the case for turnover and headcount variables for year N-1, presented in respectively Fig. 1a and Fig. 1c. These variables appeared to have an impact on the overall data projection pattern. Net income and equity variables did not show a high degree of correlation with the projection, as shown respectively in Fig. 1b and Fig. 1d, as feature distribution appeared to be random in some projection space areas. Results were plotted for year N-1, but the patterns were similar for the two other years (N-2 and N-3). Turnover and headcount

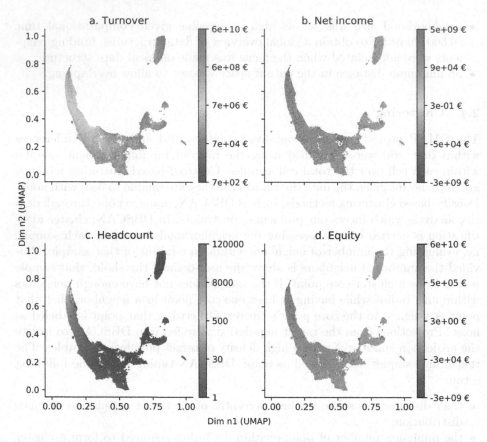

Fig. 1. Input feature distribution for samples from the representative population of French companies: turnover(*a*), net income (*b*), headcount (*c*) and equity (*d*) for the year N-1. Axes are dimensionless and come from UMAP dimension reduction (*a*, *b*, *c*, *d*).

appeared to be the variables that mattered the most distance-wise in the output space. Net income and equity, which showed a weaker or limited impact on the overall UMAP projection distribution, might have had a more local impact distance-wise in the output space.

3.2   Funding Recipient Distribution Analysis

Funding recipient samples were then projected onto the low-dimensional space. Distribution patterns for funding recipients were different from those observed for the representative population as shown in Fig. 2a. Funding recipients were prone to concentrate primarily in the left region of the projection in the shape of a curve. The curve went from the projection's upper left side to the lower center region, in the shape of a "backbone". Projection space division into a grid, presented in Fig. 1b, helped understand the projection space density distribution.

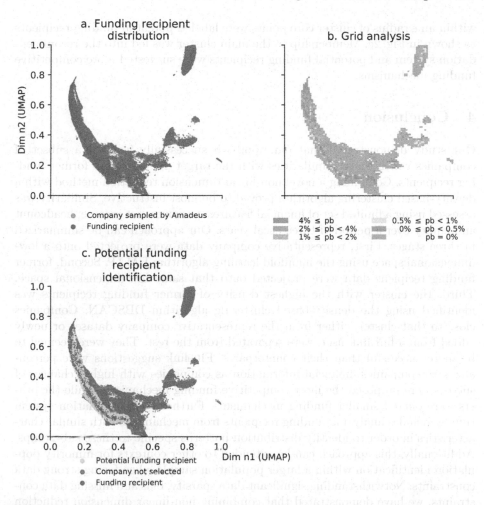

Fig. 2. Funding recipients were projected onto the two-dimensional plane with a non-uniform distribution pattern (*a*). The projection space was divided in a grid to analyze density of funding recipients within each cell(*b*). DBSCAN identified the main cluster of funding recipients. Companies close to the funding recipient cluster belonged to the potential funding recipient cluster (*c*). Axes are dimensionless and come from UMAP dimension reduction (*a, b, c*).

Density analysis confirmed the shape identification. Funding recipient samples were mainly located within the "backbone" shape as 74% of funding recipients belonged to it (i.e. 1,041 out of the 1,413 funding recipient samples). Funding recipient concentration within a specific projection space area confirmed that our similarity-based approach on financial features for potential funding recipient identification was relevant. DBSCAN was then applied on the target population and its main cluster was identified. The remaining funding recipients were labeled as noise. Company samples from the representative population that were within an ϵ radius of cluster core points were labeled potential funding recipients as shown in Fig. 2c. Membership to the main cluster was fed into the recommendation system and potential funding recipients were suggested more competitive funding mechanisms.

4 Conclusion

Our study demonstrated that our approach successfully isolated a subset of companies which shared similarities with the target population of former funding recipients. Combining a novel non-linear dimension reduction method with a density-based clustering algorithm proved to be most instructive. Similarity was assessed using a limited set of financial features: turnover, net income, headcount and equity over a period of three fiscal years. Our approach can be summarized in three stages. First, representative company data were projected onto a low-dimensional space using the manifold learning algorithm UMAP. Second, former funding recipient data were projected onto that same low-dimensional space. Third, the cluster with the highest density of former funding recipients was identified using the density-based clustering algorithm DBSCAN. Companies close to that cluster, either from the representative company dataset or newly added from a finElink user, were separated from the rest. They were deemed to be more successful than their counterparts. Finelink suggestions were personalized to companies' financial information as companies with higher chances of success were proposed the most competitive funding mechanisms while the others were offered smaller funding mechanisms. Further recommendation system tuning includes analyzing funding recipients from mechanisms with similar characteristics in order to identify distribution patterns specific to these sub-groups. Additionally, this approach can be extended to other contexts for minority population identification within a larger population sample when facing strong data constraints. Notwithstanding significant data sparsity, bias and missing data constraints, we have demonstrated that combining non-linear dimension reduction with density-based clustering, important correlations can be unraveled.

References

1. Agius, R.: Machine learning can identify newly diagnosed patients with CLL at high risk of infection. Nat. Commun. **11**(1), 1–17 (2020)
2. Amadeus.   https://www.bvdinfo.com/en-gb/our-products/data/international/amadeus. Accessed 25 Feb 2020
3. Ester, M., Kriegel, H., Sander, J., Xu, X.: A density-based algorithm for discovering clusters in large spatial databases with noise. In: KDD, vol. 96, no. 34, pp. 226–231 (1996)
4. finElink. https://www.finelink.eu. Accessed 25 Feb 2020
5. Fornito, A., Arnatkevičiūtė, A., Fulcher, B.: Bridging the gap between connectome and transcriptome. Trends Cogn. Sci. **23**(1), 34–50 (2019)

6. Grollemund, V., Pradat, P.F., Delbot, F., Le Chat, G., Pradat-Peyre, J.F., Bede, P.: Manifold learning for ALS prognosis: development and validation of a prognosis model. Scientific Reports (submitted manuscript)
7. Lee, J., Verleysen, M.: Nonlinear Dimensionality Reduction. Springer, Heidelberg (2007). https://doi.org/10.1007/978-0-387-39351-3
8. Maaten, L., Hinton, G.: Visualizing data using t-SNE. J. Mach. Learn. Res. **9**, 2579–2605 (2008)
9. McInnes, L., Healy, J., Melville, J.: UMAP: uniform manifold approximation and projection for dimension reduction. arXiv, preprint arXiv:1802.03426 (2018)
10. Mika, S., Schölkopf, B., Smola, A., Müller, K., Scholz, M., Rätsch, G.: Kernel PCA and de-noising in feature spaces. In: 11th International Conference on Neural Information Processing Systems, pp. 546–542. MIT Press, Denver (1998)
11. Roweis, S.: Nonlinear dimensionality reduction by locally linear embedding. Science **290**(5500), 2323–2326 (2000)
12. Schafer, J.: Analysis of Incomplete Multivariate Data. Chapman and Hall/CRC, London (1997)
13. Schubert, E., Gertz, M.: Intrinsic t-stochastic neighbor embedding for visualization and outlier detection. In: Beecks, C., Borutta, F., Kröger, P., Seidl, T. (eds.) SISAP 2017. LNCS, vol. 10609, pp. 188–203. Springer, Cham (2017)
14. Silva, V., Tenenbaum, J.: Global versus local methods in nonlinear dimensionality reduction. In: Advances in Neural Information Processing System, pp. 721–728 (2003)
15. Stekhoven, D.J., Bühlmann, P.: MissForest - non-parametric missing value imputation for mixed-type data. Bioinformatics **28**(1), 112–118 (2012)
16. Tang, F., Ishwaran, H.: Random forest missing data algorithms. Stat. Anal. Data Min.: ASA Data Sci. J. **10**(6), 363–377 (2017)
17. Tenenbaum, J.: A global geometric framework for nonlinear dimensionality reduction. Science **290**(5500), 2319–2323 (2000)
18. Wattenberg, M., Viégas, F., Johnson, I.: How to use t-SNE effectively. Distill **1**(10), e2 (2016)

Network Aggregation to Enhance Results Derived from Multiple Analytics

Diane Duroux[1(✉)], Héctor Climente-González[2,3,4], Lars Wienbrandt[5],
and Kristel Van Steen[1,6]

[1] BIO3 - GIGA-R Medical Genomics, University of Liège, Liège, Belgium
diane.duroux@uliege.be
[2] Institut Curie, PSL Research University, 75005 Paris, France
[3] INSERM, U900, 75005 Paris, France
[4] MINES ParisTech, PSL Research University,
CBIO-Centre for Computational Biology, 75006 Paris, France
[5] Institute of Clinical Molecular Biology, Kiel University, Kiel, Germany
[6] University of Liège, Liège, Belgium

Abstract. The more complex data are, the higher the number of possibilities to extract partial information from those data. These possibilities arise by adopting different analytic approaches. The heterogeneity among these approaches and in particular the heterogeneity in results they produce are challenging for follow-up studies, including replication, validation and translational studies. Furthermore, they complicate the interpretation of findings with wide-spread relevance. Here, we take the example of statistical epistasis networks derived from genome-wide association studies with single nucleotide polymorphisms as nodes. Even though we are only dealing with a single data type, the epistasis detection problem suffers from many pitfalls, such as the wide variety of analytic tools to detect them, each highlighting different aspects of epistasis and exhibiting different properties in maintaining false positive control. To reconcile different network views to the same problem, we considered 3 network aggregation methods and discussed their performance in the context of epistasis network aggregation. We furthermore applied a latent class method as best performer to real-life data on *inflammatory bowel disease (IBD)* and highlighted its benefits to increase our understanding about IBD underlying genetic architectures.

Keywords: Networks · Aggregation · Latent class methods · Epistasis

1 Introduction

Analyses carried out with different analytic tools often lead to inconsistent conclusions that are difficult to unify. In biology, integrative analyses usually aim at identifying the driving factors of a biological process by the joint exploration of several datasets, possibly reduced in dimension, or by obtaining a single solution

© IFIP International Federation for Information Processing 2020
Published by Springer Nature Switzerland AG 2020
I. Maglogiannis et al. (Eds.): AIAI 2020, IFIP AICT 583, pp. 128–140, 2020.
https://doi.org/10.1007/978-3-030-49161-1_12

per dataset prior to aggregation. All of these settings often involve a single analytical modelling framework to address the main question of interest.

Several aggregation methods exist and have been discussed in different contexts within human complex genetics [21]. Restricting attention to omics data,
we mention the context of multi-omics analyses with supervised methods [13] for
association or for prediction [17], and unsupervised methods for disease subtyp-
ing [29]. A returning common approach is the exploitation of network representations of the data. Here, nodes either represent samples (individuals) or biological
features and edges represent interactions. Features may be directly measured or
synthetic (modules); edges may be functional, biological or analytically derived
via statistical and machine learning models.

In *genome-wide association interaction studies (GWAIS)* thousands of individuals, typed for genome-wide sets of genetic variants, are mined to identify
interacting loci in association with a characteristic, such as disease state. The
most popular genetic variants in these studies are *Single Nucleotide Polymorphisms (SNPs)*. In this paper, the main question of interest is how to derive unified conclusions from GWAIS with SNPs that have been typed out on the same
dataset, yet with different analytic tools or protocols. The motivation for this
question is multi-fold. In Bessonov et al. [2], it was demonstrated that slightly
different GWAIS analysis protocols may lead to highly different analysis results.
At the same time, different analytics are believed to highlight only particular
aspects of the genetic architecture underlying complex traits under investigation.
Hence, in order to aid in generating robust genetic interaction findings, that can
be used as input to replication and experimental validation studies, there is a
need for novel approaches to prioritize interactions obtained by different analytic
workflows [32]. To our knowledge, the presented study is the first that explores
the utility of network aggregation in deriving an aggregated statistical epistasis
network across different epistasis detection analysis protocols.

The remainder of this paper is organized as follows. We present in silico data
and a case study on inflammatory bowel disease in Sect. 2. In Sect. 3 we outline
the aggregation methods included in a comparative study. We report results in
Sect. 4. Finally, in Sect. 5 we discuss and conclude this work.

2   Synthetic and Real-Life Data

2.1   In Silico Data

We created several imperfect networks with binary edges (i.e., an edge is present
or not) that partially represented a true network. In particular, we used the
function *huge.generator* from the package *huge* [39] in R [26] to generate data
with random graph structures. Essentially, we applied the number of observations
($n = 200$) and the number of variables ($d = 50$) as parameters. The adjacency
matrix θ with probability $3/d$ that a pair of nodes is connected was computed via
huge.generator. In other words, each pair of off-diagonal elements were randomly
set to $\theta[i, j] = \theta[j, i] = 1$ for $i \neq j$ with probability $3/d$, and 0 otherwise. Then,
a precision matrix was calculated from the adjacency matrix and was used to

compute a covariance matrix in order to create the generating data. It led to a true baseline binary network with 50 nodes and a random graph structure, and the associated generating data.

Next, we created 5 so-called partial networks in the following way. We first applied the graphical lasso estimator (*glasso* option in the function *huge*) on the data of 200 samples and 50 nodes that we previously generated. The employed function carries out undirected graph estimation using a lambda sequence of size 10 to control the regularization. It returns a list of precision matrices corresponding to the lambdas. Second, the function *huge.select* was applied to select the regularization parameter. We applied the *stability approach to regularization selection (stars)*, which selects the optimal network by variability of subsamplings and gives a supplementary estimated network by merging the corresponding subsampled networks using the frequency counts. Then, to actually build a partial network, we randomly selected 50% of the edge values of the estimated graph and kept them as is. The remaining edge values were set to zero, i.e. representing the lack of interaction between corresponding nodes. This selection of 50% of the edges was performed five times, to give rise to 5 partial networks. Several variations to the baseline network were considered as detailed in Fig. 2. For each of the considered configurations we created 1,000 replicates. We highlight that the partial networks constructed in this way are in line with the hypothesis that *statistical epistasis networks (SENs)* derived from multiple analytics only partially reflect a true underlying interaction network (Fig. 1).

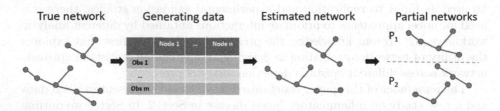

Fig. 1. Generation of the data

2.2 Inflammatory Bowel Disease Data

IBD defines several chronic idiopathic inflammatory conditions. *Crohn's disease (CD)* and *ulcerative colitis (UC)* are the two main forms of IBD. UC is related to the colon, whereas CD affects the whole gastrointestinal tract and especially the terminal ileum and colon [9]. To date, identified independent loci associated with human complex diseases such as IBD only explain a small part of the disease heritability. As previous studies indicate, genetic interactions may have a significant role in this missing heritability [40], yet only a handful of replicable and clinically actionable interactions have been discovered [32].

Using data as part of the International IBD Consortium, we performed a first data *quality control (QC)* check as in Ellinghaus et al. [6]. Then, additional QC measures were taken, specifically related to large-scale GWAIS, as motivated in Gusareva and Van Steen [10]. In particular, only common variants (MAF $> 5\%$) and those in Hardy-Weinberg equilibrium (p-value > 0.001) were considered. Also, we pruned out SNPs that were in Linkage Disequilibrium (LD $r^2 > 0.75$) with the option "--indep-pairwise 50 5 0.75" in PLINK [25]. Since this LD filtering is based on sliding windows, LD was not tested exhaustively among all possible pairs. This may induce redundant epistasis signals due to LD and requires taking additional measures post interaction analysis (see Sect. 4.1). Lastly, to enrich the data for known risk loci, all risk SNPs described in Liu et al. [20] were included. In addition, we adjusted phenotypes to correct for population structure using the top 7 principal components. These adjusted phenotypes were obtained as residuals from a logistic regression model by subtracting model-fitted values from observed phenotype values. Submitting the phenotype adjusted traits to analytic tools may reduce power but is a pragmatic choice when the analytics do not accept covariates or explanatory variables other than the SNPs under investigation. Overall, the obtained dataset contained 38,225 SNPs and 66,280 individuals, partitioned in 32,622 cases and 33,658 controls.

3   Comparative Study – Towards Network-Based Aggregation Methods for Statistical Epistasis Networks

In this project, we compared three unsupervised network aggregation methods. Given a set of edges and several networks, aggregation process was used to partition edges into two clusters [7], edge present or not, based on edge similarity across partial networks. The variables to assess similarity are the value of the edges in the different partial networks. The input matrix for clustering takes edges for rows and partial networks for columns. Matrix entries are 1 when an edge is present and 0 otherwise. First we selected one of the most popular unsupervised learning algorithms, k-means, using the function *kmeans* in R [26]. In particular, the *kmeans* function was applied to group the edges such that edges within the same cluster were as similar as possible, whereas edges in different clusters were as different as possible in order to maximize intra-cluster similarity, and minimize inter-cluster similarity. The algorithm of Hartigan and Wong [11] was applied. The total within-cluster variation was set as the sum of squared Euclidean distances between edges and the mean of edges in this cluster, called center. Each edge was associated with a cluster so the total within-cluster variation is minimized. In practice, two edges were picked randomly as cluster center. Then, each edge was assigned to their closest center based on the Euclidean distance. For each cluster, the center was updated by computing the mean values of all the edges in the cluster. The process was repeated 10 times to iteratively minimize the total within-cluster variation.

Second, we used the *Latent Class Modelling (LCM)* approach for clustering with the R [26] function *poLCA* [19]. It allows a dataset to be partitioned into exclusive groups called *latent classes*. The main latent class model is $P(y_n|\theta) = \sum_{j=1}^{S} \pi_j P_j(y_n|\theta_j)$ where y_n is the observation n (edge pair) of the variables (partial networks), S is the number of clusters (2), and π_j is the prior probability (random) of belonging to cluster j. P_j is the cluster specific probability of y_n given the cluster specific parameters θ_j. Expectation-Maximization algorithm was used to maximize the latent class mode log-likelihood function with $poLCA$. The output included a vector of predicted cluster memberships for each edge.

We also adapted the *Similarity Network Fusion (SNF)* approach from Wang et al. [35] to handle unweighted graphs. Note, that SNF was originally created for aggregating data types on a genomics scale so as to create an aggregated similarity matrix between individuals and that aggregation was based on normalized similarity matrices with continuous values. In our approach no normalization was performed and the partial networks were iteratively updated with information from the other networks to build an aggregated graph via the R library *SNFtool* [34] and the function *SNF* therein. We then set the diagonal of the aggregated adjacency matrix to 0. Since the outputted consensus network was continuous, it was binarized again by testing a variety of thresholds ranging from 0 to 1 with a step of 0.01 and selecting the threshold with maximal performance (simulation setting dependent). An edge was considered to be present in the final aggregated network if and only if the optimal threshold was surpassed.

Because most edges in the synthetic and real-life data are absent, we chose the F1-score to evaluate the performance of aggregation methods. It is defined as $F_1 = 2 \times \frac{\text{precision} \times \text{recall}}{\text{precision} + \text{recall}}$ and seeks a balance between precision (true positives divided by the number of true positives and false positives) and recall (true positives divided by the number of true Positives and false negatives). The F1-score ranges from 0 to its best value 1. First, we measured the initial F1-scores for each partial network compared to the true base network and selected the partial network with the highest score ($max(F1_{\text{Inital}})$). Then we computed the gain in F1 score ($F1_{\text{Gain}}$) defined as $F1_{\text{Gain}} = F1_{\text{AggregatedNetwork}} - max(F1_{\text{Inital}})$. The aggregation method for which $F1_{\text{Gain}}$ was the highest was our best performer.

Using the real-life data of Sect. 2.2, and by means of illustration, we applied 3 analytic methods to identify pairwise genetic interactions (hereafter referred to as epistasis). For each of these we subsequently constructed a statistical epistasis network with connected nodes representing SNPs involved in a significant interaction. The best performing network aggregation method was applied to obtain a single network comprising epistasis results from the 3 analytic methods.

The first method was regression-based and belongs to the most popular methods used in this context, as it is easy to implement and interpret. With PLINK 1.9 we fitted the linear regression model $E[Y|A, B] = \beta_0 + \beta_1 g_A + \beta_2 g_B + \beta_3 g_A g_B$, where Y is the phenotype adjusted for population structure as described in Sect. 2.2 and is assumed to follow a normal distribution, with g_A (g_B) representing genotype information for SNP A (B) under an additive encoding scheme, and with $\beta_i, i = 0, \ldots, 3$ the regression coefficients. The null hypothesis tested in PLINK [25] was H0: $\beta_3 = 0$ versus H1: $\beta_3 \neq 0$, i.e. a 1 degree of freedom test. Multiple testing corrected significance was assessed by creating permutation null samples while permuting Y values 400 times. We then produced a "top p-value" distribution with the smallest p-value of each permutation and we set an overall threshold at 5% of these top p-values. From that, we defined an overall p-value threshold with guarantee of 5% *Family Wise Error Rate (FWER)*, as in Hemani et al. [12].

The second method was a non-parametric dimensionality reduction method. In particular, we fitted *Model-Based Multifactor Dimensionality Reduction (MB-MDR)* [31] with default options that exhaustively explores the association between each SNP pair and Y adjusted for population structure as before. The method is non-parametric in the sense that no assumptions are made regarding the modes of interaction inheritance. Unlike the regression method above, MB-MDR is fairly robust to deviations from the normal distributions for Y, even though the final MB-MDR test for non-binary traits is by default the result of a sequence of t-tests. The Model-Based part of MB-MDR assumes the default of adjusting two-locus testing for main effects (SNP A, SNP B) and thus the considered MB-MDR alternative hypothesis was H1: the joint effect of SNP A and SNP B goes beyond additivity. Significance assessment with multiple testing correction was achieved by the default MB-MDR options of carrying out 999 permutations and gammaMAXT at a FWER of 5%.

The third method we considered was epiHSIC [16], as implemented in R's gpuEpiScan [15]. It searches for genomic interactions in a regression framework by efficiently scanning high-dimensional datasets. Efficiency is based on prescreening by HSIC, a statistical measure of non-independence between two variables: e.g. the larger HSIC value, the more likely it is that the correlation between SNP A and SNP B is independent from Y. Such independencies are believed to be indicative for potential epistasis. Significance was assessed via comparing obtained Bonferroni corrected p-values to 0.05.

PLINK analyses were performed on a cluster running Scientific Linux release 7.2 (Nitrogen), using 6 threads and the total runtime was 2 h 35 m. MBMDR analysis were implemented on the same computing system, with 100 threads and the total runtime was 2 h 05 m. EpiHSIC analysis was performed on CentOS Linux release 7.7.1908 (Core) cluster with 1 GPU (V100 GPU 2 × 12-Core Intel Xeon Gold 6126 2.6 GHz 192 GB RAM) and the total runtime was 20 min.

4 Results

4.1 Simulation Study

The average F1 gain is 0.18 for LCA, 0.13 for k-means and 0.01 for SNF with the baseline simulation scenario. Also, the more knowledge the partial network contains (i.e. percentage of edges overlapping with the estimated base network), the higher the initial F1 scores and the less beneficial the aggregation (Fig. 2A), which is in line with intuition.

In addition, when the number of observations in the generating data exceeds 500, then the F1 gain stabilizes (Fig. 2B). A too small number of nodes (here below 50) or a too high number of nodes (here above 100) shows to be suboptimal for both LCA and k-means. For SNF, the less nodes, the more beneficial the aggregation is, although its F1 gain is still the smallest of the 3 considered aggregation methods (Fig. 2C). The results of Fig. 2C may have repercussions for "true" epistasis networks that would be too large in terms of numbers of SNPs. Part of the problem can be alleviated by deriving gene-based SENs from SNP-based SENs. In fact, to date, there is little evidence that the number of gene-based interactions would be extremely large, especially when ruling out spurious interactions due to major gene effects. The situation may be different for other interactome networks such as protein-protein interaction networks.

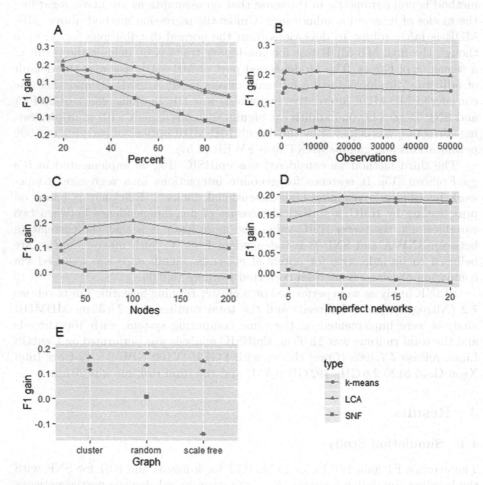

Fig. 2. Average F1 gain per simulation scenario. [A] Variation of the percentage of nodes of the estimated network used to create each partial network; [B] Number of observations in the generating dataset; [C] Number of vertices in the true network. [D] Number of partial networks aggregated to estimate the true network; [E] Graph type

Furthermore, with the SNF algorithm, F1 gain decreases with the inclusion of more partial networks in the aggregation process (Fig. 2D). In contrast, with k-means, the F1 gain increases as the number of partial networks varies from

5 to 20, whereas it remains stable with LCA when at least 10 networks are aggregated. Such information is relevant to have an idea about the number of epistasis networks (e.g. derived from different analytic protocols) to include in the aggregation process, with minimal loss of information compared to the "true" (unknown) underlying epistasis network. Notably, in real-life it is expected that non-random heterogeneity exists between partial epistasis networks. Not properly accounting for this may jeopardize the reliability of the aggregated network. Unfortunately, intrinsic differences between epistasis detection tools are often hard to assess based on the supporting literature that underlies each tool: to date there is no consensus about sufficiently advanced gold standard in silico datasets on human interactomes. Hence, a pragmatic way to deal with different forms of heterogeneity is to act at the level of the epistasis networks themselves, and includes accommodating potential scale differences in SEN edge weights across networks. We are currently working on a strategy around a notion of statistically significant differences between (groups of) SENs and clustering that combines the ideas of consensus clustering with meta clustering [4].

Finally, k-means and LCA, are quite stable across network structures, whereas SNF performs extremely poor on scale-free networks (Fig. 2E). The future will show what the implications are for the aggregation of SENs. Indeed, whether or not SENs or genetic interaction networks are scale-free is still under debate [3].

Based on all of the above, we selected LCA as SEN aggregation method of choice and compared the two LCA-derived clusters on the synthetic data, in more detail. In particular, we computed the average distance between and within clusters using the Manhattan distance for each of the 1000 runs. Overall, the average distance between clusters is 2.7 (standard error 0.005) and the average distance within groups is 0.15 (standard error 0.0001). Also, for each partial network and cluster, we calculated the frequency of 1's (i.e. edges present), to generate 1000 times two 5-dimensional vectors. Permutational multivariate analysis of variance with 1000 permutations (using R library *vegan* [24] and function *adonis*) shows that the clustering is significantly associated to edge abundance across partial networks (p-value of 0.001).

4.2 Inflammatory Bowel Disease Aggregated Statistical Epistasis Network

Here, the aim is to use knowledge derived from our simulation study to uncover the "true" epistasis network underlying inflammatory bowel disease, via multiple partial epistasis networks that are obtained from different analytic protocols on the same real-life data. As LCA performed best in Sect. 4.1, we applied it to combine 3 statistical epistasis networks for IBD, after further manipulation of the networks. We reduced the size of the networks while minimizing spurious edges. In particular, SNP pairs where both SNPs resided in the HLA region were deleted, as for this region it is notoriously hard to distinguish between main and additional non-additive effects [30]. Significant SNP pairs exhibiting strong LD ($r^2 > 0.75$) were eliminated as well. The resulting SENs are depicted in Fig. 3. The LCA aggregated SEN counts 193 nodes, 203 interactions and 12 modules. The size of the largest connected component (LCC) is 163.

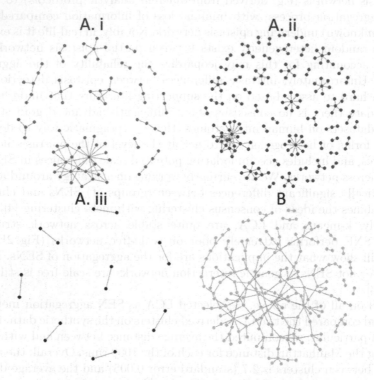

Fig. 3. SNP-based statistical epistasis networks (SENs). IBD SEN derived from [A.i] linear regression, [A.ii] MB-MDR, [A.iii] epiHSIC, [B] LCA aggregation.

To address the question whether the aggregated network gives added value over the contributing SENs to understanding underlying genetic architectures of IBD, we carried out several pathway enrichment analyses. To this end, we first mapped all SNPs of the LCA aggregated network to genes. This was done location-wise with FUMA [36] and its function *SNP2GENE*: SNPs were mapped to a gene whenever the SNP was located in that gene's region, i.e. including 10kb before and after the gene. Second, we ran pathway over-representation analyses of LCC containing at least 3 SNPs in R [26] using the library *clusterProfiler* [38] and the function *enrichKEGG*. FDR was controlled using the Benjamini-Hochberg procedure [1].

Since the LCC obtained with epiHSIC contained only 2 SNPs, no enriched pathway is obtained. The pre-screening approach from epiHSIC combined with stringent Bonferroni correction for multiple testing may not be a good choice since too much information gets lost, as we also saw with the small size of the associated network compared to the two other networks. For linear regression and MB-MDR the same 5 significant pathways were detected. This larger overlap between MB-MDR and linear regression is not surprising as neither of these methods involved a pre-screening, in contrast to epiHSIC. Also, filtering or not increases the heterogeneity in epistasis results [2]. The 5 pathways referred to cytokine-cytokine receptor interaction, JAK/STAT signaling pathway, Inflammatory Bowel Disease, Th17 cell differentiation and Th1 and Th2 cell differentiation and were already linked to IBD in earlier work [5,8,23]. Pathway enrichment analysis applied to the LCA aggregated SEN identified 12 significant pathways, including the 5 mentioned before. Therefore, in this case study, aggregation highlighted more pathways than the union of the pathways detected with each epistasis detection method. Note, that epiHSIC network contributed to the aggregated network. In fact, without including it, the LCA aggregated SEN lost 5 nodes, 1 module and 2 enriched pathways. The 7 unique pathways to the LCA aggregated SEN were viral protein interaction with cytokine, C-type lectin receptor, TNF, Yersinia infection, allograft rejection, intestinal immune network for IgA production and autoimmune thyroid disease. They seemed to be coherent with earlier work in relation to IBD [14,18,27,33].

5   Conclusion

Genetic interactions, beyond effects of independent SNPs or genes, can further unravel the genetic underpinnings of human complex diseases. Such interactions contribute to epistasis, which has grown into a more general theory and applications framework for the analysis of interactions across and between multiple omics data. Many methods have been created to understand the true role of these interactions but findings are often inconsistent. This is in part due to different analytic protocols for epistasis detection giving rise to, at best, partially overlapping results. To this end, we first summarized the results of epistasis analyses in networks with nodes representing SNPs and edges representing binary evidence for a statistically significant interaction between corresponding SNPs. We second investigated the utility of network aggregation methods built on unsupervised machine learning to reconstruct the "true" disease underlying epistasis network. Unsupervised machine learning techniques have been used before in different contexts to unravel disease associated biological knowledge, for instance to derive multimodal biomarker signatures of disease risk [28], to identify subphenotypes for asthma [37], or to provide a molecular reclassification of Crohn's Disease [22]. Here, we used it to predict epistasis network links via the aggregation of partial networks. Our simulations revealed that *Latent Class Analysis (LCA)* outperformed k-means and a customized version of Similarity Network Fusion. We furthermore applied LCA to data for inflammatory bowel disease and underlined the benefits of an aggregated network via pathway enrichment analyses performed on the largest connected component of aggregated and contributing networks. These enrichment analyses revealed 7 pathways that could not be detected with either of the 3 considered statistical epistasis detection models. This pilot study suggests the potential of network aggregation in epistasis research and the need to investigate the added value of between-network heterogeneity in advanced network aggregation algorithms.

Acknowledgements. Data collections and processing were supported by funds to the International IBD Genetics Consortium. DD acknowledges the European Union's Horizon 2020 research and innovation programme under grant agreement No. 813533.

References

1. Benjamini, Y., Hochberg, Y.: Controlling the false discovery rate: a practical and powerful approach to multiple testing. J. Roy. Stat. Soc.: Ser. B (Methodol.) **57**(1), 289–300 (1995)
2. Bessonov, K., Gusareva, E.S., Van Steen, K.: A cautionary note on the impact of protocol changes for genome-wide association SNP × SNP interaction studies: an example on ankylosing spondylitis. Hum. Genet. **134**(7), 761–773 (2015). https://doi.org/10.1007/s00439-015-1560-7
3. Broido, A.D., Clauset, A.: Scale-free networks are rare. Nat. Commun. **10**(1), 1–10 (2019)
4. Caruana, R., Elhawary, M., Nguyen, N., Smith, C.: Meta clustering. In: Sixth International Conference on Data Mining (ICDM 2006), pp. 107–118. IEEE (2006)
5. Coskun, M., Salem, M., Pedersen, J., Nielsen, O.H.: Involvement of JAK/STAT signaling in the pathogenesis of inflammatory bowel disease. Pharmacol. Res. **76**, 1–8 (2013)
6. Ellinghaus, D., et al.: Analysis of five chronic inflammatory diseases identifies 27 new associations and highlights disease-specific patterns at shared loci. Nat. Genet. **48**(5), 510 (2016)
7. Faber, V.: Clustering and the continuous k-means algorithm. Los Alamos Sci. **22**(138144.21), 67 (1994)
8. Gálvez, J.: Role of Th17 cells in the pathogenesis of human IBD. ISRN Inflamm. **2014**, 14 (2014)
9. Geboes, K., Dewit, O., Moreels, T.G., Faa, G., Jouret-Mourin, A.: Inflammatory bowel diseases. In: Jouret-Mourin, A., Faa, G., Geboes, K. (eds.) Colitis, pp. 107–140. Springer, Cham (2018). https://doi.org/10.1007/978-3-319-89503-1_8
10. Gusareva, E.S., Van Steen, K.: Practical aspects of genome-wide association interaction analysis. Hum. Genet. **133**(11), 1343–1358 (2014). https://doi.org/10.1007/s00439-014-1480-y
11. Hartigan, J.A., Wong, M.A.: Algorithm as 136: a k-means clustering algorithm. J. Roy. Stat. Soc. Ser. C (Appl. Stat.) **28**(1), 100–108 (1979)
12. Hemani, G., Shakhbazov, K., Westra, H.J., Esko, T., Henders, A.K., McRae, A.F., et al.: Detection and replication of epistasis influencing transcription in humans. Nature **508**(7495), 249–253 (2014). 00162
13. Huang, S., Chaudhary, K., Garmire, L.X.: More is better: recent progress in multi-omics data integration methods. Front. Genet. **8**, 84 (2017)
14. Hütter, J., et al.: Role of the C-type lectin receptors MCL and DCIR in experimental colitis. PLoS One **9**(7), e103281 (2014)
15. Jiang, B.: gpuEpiScan: GPU-Based Methods to Scan Pairwise Epistasis in Genome-Wide Level (2019). r package version 0.0.1
16. Kam-Thong, T., Putz, B., Karbalai, N., Muller-Myhsok, B., Borgwardt, K.: Epistasis detection on quantitative phenotypes by exhaustive enumeration using GPUs. Bioinformatics **27**(13), i214–i221 (2011). 00026
17. Kim, M., Tagkopoulos, I.: Data integration and predictive modeling methods for multi-omics datasets. Mol. Omics **14**(1), 8–25 (2018)

18. Koelink, P.J., Bloemendaal, F.M., Li, B., Westera, L., Vogels, E.W., van Roest, M., et al.: Anti-TNF therapy in IBD exerts its therapeutic effect through macrophage IL-10 signalling. Gut **69**, 1053–1063 (2019)
19. Linzer, D.A., Lewis, J.: poLCA: polytomous variable latent class analysis version 1. 4. J. Stat. Softw. **42**, 1–29 (2011)
20. Liu, J.Z., et al.: Association analyses identify 38 susceptibility loci for inflammatory bowel disease and highlight shared genetic risk across populations. Nat. Genet. **47**(9), 979 (2015)
21. Lópezde Maturana, E., Pineda, S., Brand, A., Van Steen, K., Malats, N.: Toward the integration of omics data in epidemiological studies: still a "long and winding road". Genet. Epidemiol. **40**(7), 558–569 (2016)
22. Maus, B., Jung, C., John, J.M.M., Hugot, J.P., Génin, E., Van Steen, K.: Molecular reclassification of Crohn's disease: a cautionary note on population stratification. PloS One **8**(10), e77720 (2013)
23. Nemoto, Y., Watanabe, M.: The Th1, Th2, and Th17 paradigm in inflammatory bowel disease. In: Baumgart, D. (ed.) Crohn's Disease and Ulcerative Colitis, pp. 183–194. Springer, Boston (2012). https://doi.org/10.1007/978-1-4614-0998-4_15
24. Oksanen, J., et al.: Package 'vegan'. Community Ecol. Package Version **2**(9), 1–295 (2013)
25. Purcell, S., Neale, B., Todd-Brown, K., Thomas, L., Ferreira, M.A., Bender, D., et al.: PLINK: a tool set for whole-genome association and population-based linkage analyses. Am. J. Hum. Genet. **81**(3), 559–575 (2007)
26. R Core Team: R: A Language and Environment for Statistical Computing. R Foundation for Statistical Computing, Vienna, Austria (2017). https://www.R-project.org/
27. Saebo, A., Vik, E., Lange, O.J., Matuszkiewicz, L.: Inflammatory bowel disease associated with yersinia enterocolitica O: 3 infection. Eur. J. Intern. Med. **16**(3), 176–182 (2005)
28. Shomorony, I., et al.: An unsupervised learning approach to identify novel signatures of health and disease from multimodal data. Genome Med. **12**(1), 1–14 (2020)
29. Tini, G., Marchetti, L., Priami, C., Scott-Boyer, M.P.: Multi-omics integration-a comparison of unsupervised clustering methodologies. Brief. Bioinform. **20**(4), 1269–1279 (2019)
30. Traherne, J.: Human MHC architecture and evolution: implications for disease association studies. Int. J. Immunogenet. **35**(3), 179–192 (2008)
31. Van Lishout, F., Gadaleta, F., Moore, J.H., Wehenkel, L., Van Steen, K.: gamma-MAXT: a fast multiple-testing correction algorithm. BioData Min. **8**(1), 36 (2015)
32. Van Steen, K., Moore, J.H.: How to increase our belief in discovered statistical interactions via large-scale association studies? Hum. Genet. **138**(4), 293–305 (2019). https://doi.org/10.1007/s00439-019-01987-w
33. Wadhwa, V., Lopez, R., Shen, B.: Crohn's disease is associated with the risk for thyroid cancer. Inflamm. Bowel Dis. **22**(12), 2902–2906 (2016)
34. Wang, B., et al.: SNFtool: similarity network fusion. Cran 2014 (2014)
35. Wang, B., et al.: Similarity network fusion for aggregating data types on a genomic scale. Nat. Methods **11**(3), 333 (2014)
36. Watanabe, K., Taskesen, E., van Bochoven, A., Posthuma, D.: Functional mapping and annotation of genetic associations with FUMA. Nat. Commun. **8**(1) (2017). https://doi.org/10.1038/s41467-017-01261-5. 00139

37. Woodruff, P.G., Modrek, B., Choy, D.F., Jia, G., Abbas, A.R., Ellwanger, A., et al.: T-helper type 2-driven inflammation defines major subphenotypes of asthma. Am. J. Respir. Crit. Care Med. **180**(5), 388–395 (2009)

38. Yu, G., Wang, L.G., Han, Y., He, Q.Y.: clusterProfiler: an R package for comparing biological themes among gene clusters. Omics: J. Integr. Biol. **16**(5), 284–287 (2012)

39. Zhao, T., Liu, H., Roeder, K., Lafferty, J., Wasserman, L.: The huge package for high-dimensional undirected graph estimation in R. J. Mach. Learn. Res. **13**(Apr), 1059–1062 (2012)

40. Zuk, O., Hechter, E., Sunyaev, S.R., Lander, E.S.: The mystery of missing heritability: genetic interactions create phantom heritability. Proc. Natl. Acad. Sci. **109**(4), 1193–1198 (2012)

PolicyCLOUD: Analytics as a Service Facilitating Efficient Data-Driven Public Policy Management

Dimosthenis Kyriazis[1]($\boxtimes$), Ofer Biran[2], Thanassis Bouras[3], Klaus Brisch[4], Armend Duzha[5], Rafael del Hoyo[6], Athanasios Kiourtis[1], Pavlos Kranas[7], Ilias Maglogiannis[1], George Manias[1], Marc Meerkamp[4], Konstantinos Moutselos[8], Argyro Mavrogiorgou[1], Panayiotis Michael[8], Ricard Munné[9], Giuseppe La Rocca[10], Kostas Nasias[11], Tomas Pariente Lobo[9], Vega Rodrigálvarez[6], Nikitas M. Sgouros[1], Konstantinos Theodosiou[3], and Panayiotis Tsanakas[8]

[1] University of Piraeus, Piraeus, Greece
{dimos,kiourtis,imaglo,gmanias,margy,sgouros}@unipi.gr
[2] IBM Research, Haifa, Israel
biran@il.ibm.com
[3] Ubitech, Athens, Greece
{bouras,ktheodosiou}@ubitech.eu
[4] DWF Rechtsanwaltsgesellschaft mbH, Cologne, Germany
{klaus.brisch,marc.meerkamp}@dwf.law
[5] Maggioli SpA, Santarcangelo di Romagna, Italy
armend.duzha@maggioli.it
[6] Instituto Tecnológico de Aragón, Saragossa, Spain
{rdelhoyo,vrodrigalvarez}@itainnova.es
[7] LeanXcale, Madrid, Spain
pavlos@leanxcale.com
[8] National Technical University of Athens, Athens, Greece
kmouts@gmail.com, panayiotismichael@mail.ntua.gr,
panag@cs.ntua.gr
[9] Atos Spain, Madrid, Spain
{ricard.munne,tomas.parientelobo}@atos.net
[10] EGI Advanced Computing Services for Research, Amsterdam, The Netherlands
giuseppe.larocca@egi.eu
[11] OKYS, Sofia, Bulgaria
knasias@okys.eu

Abstract. While several application domains are exploiting the added-value of analytics over various datasets to obtain actionable insights and drive decision making, the public policy management domain has not yet taken advantage of the full potential of the aforementioned analytics and data models. Diverse and heterogeneous datasets are being generated from various sources, which could be utilized across the complete policies lifecycle (i.e. modelling, creation, evaluation and optimization) to realize efficient policy management. To this end, in this paper we present an overall architecture of a cloud-based environment that facilitates

© IFIP International Federation for Information Processing 2020
Published by Springer Nature Switzerland AG 2020
I. Maglogiannis et al. (Eds.): AIAI 2020, IFIP AICT 583, pp. 141–150, 2020.
https://doi.org/10.1007/978-3-030-49161-1_13

data retrieval and analytics, as well as policy modelling, creation and optimization. The environment enables data collection from heterogeneous sources, linking and aggregation, complemented with data cleaning and interoperability techniques in order to make the data ready for use. An innovative approach for analytics as a service is introduced and linked with a policy development toolkit, which is an integrated web-based environment to fulfil the requirements of the public policy ecosystem stakeholders.

Keywords: Data analytics as a service · Public policies · Cloud computing · Policy modelling

1   Introduction

The ICT advances as well as the increasing use of devices and networks, and the digitalisation of several processes is leading to the generation of vast quantities of data. These technological advances have made it possible to store, transmit and process large amounts of data more effectively than before in several domains of human activity of public interest [1]. It is undeniable that we are inundated with more data than we can possibly analyse [2]. This rich data environment affects decision and policy making: cloud environments, big data and other innovative data-driven approaches for policy making create opportunities for evidence-based policies, modernization of public sectors and assistance of local governance towards enhanced levels of trust [4]. During the traditional policy cycle, which is divided into 5 different stages (agenda setting, policy formulation, decision making, policy implementation and policy evaluation), data is a valuable tool for allowing policy choices to become more evidence-based and analytical [3]. The discussion of data-driven approaches to support policy making can be distinguished between two main types of data. The first is the use of open data (administrative - open - data and statistics about populations, economic indicators, education, etc.) that typically contain descriptive statistics, which are used more intensively and in a linked way, shared through cloud environments [3]. The second main type of data is from any source, including data related to social dynamics and behaviour that affect the engagement of citizens (e.g. online platforms, social media, crowd-sourcing, etc.). These data are analysed with novel methods such as sentiment analysis, location mapping or advanced social network mining. Furthermore, one key challenge goes beyond using and analysing big data, towards the utilization of infrastructures for shared data in the scope of ethical constraints both for the citizens and for the policy makers. These ethical constraints include "data ownership" ones, which determine the data sharing rules, as well as data localisation constraints, which may unjustifiably interfere with the "free flow of data". As for policy makers, the dilemma, to what extent big data policy making is in accordance to values elected governments promote, is created. This is a problem deriving from the fact that it is difficult to point to the scope of the consent citizens may give to big data policy analysis [5]. Furthermore, such policies provide a broad framework for how decisions should be made regarding data, meaning that they are high-level statements and need more detail before they can be operationalized [6].

Big data enabled policy making should answer modern democratic challenges, considering facts about inequality and transparency both on national and local level, involving multi-disciplinary and multi-sectoral teams.

In this context, this paper presents the main research challenges and proposes an architecture of an overall integrated cloud-based environment for data-driven policy management. The proposed environment (namely PolicyCLOUD) provides decision support to public authorities for policy modelling, implementation and simulation through identified populations, as well as for policy enforcement and adaptation. Additionally, a number of technologies are introduced that aim at optimizing policies across public sectors by utilizing the analysed inter-linked datasets and assessing the impact of policies, while considering different properties (i.e. area, regional, local, national) and population segmentations. One of the key aspects of the environment is its ability to trigger the execution of various analytics and machine learning models as a service. Thus, the implemented and integrated service can be executed over different datasets, in order to obtain the results and compile the corresponding policies.

2   Related Work

In terms of managing diverse data sources, the evolution of varieties of data stores (i.e. SQL and NoSQL), where each variety has different strengths and usage models, is linked with the notion of "polyglot persistence" [7]. The latter emphasizes that each application and workload may need a different type of data store, tailored for its needs (e.g. graph, time series). Moreover, the field of data warehousing addresses creating snapshots of Online Transactional Processing (OLTP) databases for the purposes of Online Analytical Processing (OLAP). This often requires copying the data and preparing it for analytics by transforming its structure (i.e. Extract Transform Load process) and building the relevant indexes for the analytical queries. This costly process is performed in order to achieve fast query times for analytical queries of interest, and in order to support data mining. However, there is an increasing trend to adopt a "just in time data warehouse" model, where data are federated on the fly according to runtime parameters and constraints [8]. As a result, data analytics frameworks increasingly strive to cater for data regardless of the underlying data store. Apache Spark [9] is an open source framework for analytics, which is designed to run in a distributed setting within a data centre. Spark provides a framework for cluster computing and memory management, and invents the notion of a Resilient Distributed Dataset (RDD) [10] that can be stored persistently using any storage framework that implements the Hadoop File System interface, including the Hadoop Distributed Filesystem (HDFS), Amazon S3 and OpenStack Swift. The Spark SQL component additionally defines a DataFrame as an RDD having a schema, and provides a SQL interface over DataFrames [11]. Built in support is provided for data sources with a JDBC interface, as well as for Hive, and the Avro, Parquet, ORC and JSON formats. Moreover, there is an external data sources API, where new data sources can be added by implementing a driver for the data source that implements the API. Many such drivers have been implemented, for example for Cassandra, MongoDB and Cloudant. These data sources can be queried, joining data across them and thus provide the ability to run batch queries across multiple data sources and formats. Spark also integrates batch processing with real time processing in the form of Spark Streaming [12] that allows real-time processing using the same underlying framework and programming paradigm as for batch computations. In Spark 2.0, streaming Spark SQL computations are also planned [13]. With the advent of IoT and the increasing capabilities available at the edge, applications may store and process data locally [14]. In the PolicyCLOUD architecture, data are managed whether in flight or at rest and are federated across multiple frameworks, data sources, locations and formats.

Another key aspect is data interoperability given the diversity of the data sources. Among the main value propositions of the PolicyCLOUD environment and tools for policy development and management will be its ability to integrate, link and unify the datasets from diverse sources, while at the same time enabling analytics over the unified datasets. As a key prerequisite to providing this added-value, the interoperability of diverse datasets should be ensured. A wide array of data representation standards in various domains have emerged as a means of enabling data interoperability and data exchange between different systems. Prominent examples of such standards in different policy areas include: (i) the INSPIRE Data Specifications [15] for the interoperability of spatial data sets and services, which specify common data models, code lists, map layers and additional metadata on the interoperability to be used when exchanging spatial datasets, (ii) the Common European Research Information Format (CERIF) [16] for representing research information and supporting research policies, (iii) the Internet of Things ontologies and schemas, such as the W3C Semantic Sensor Networks (SSN) ontology [17] and data schemas developed by the Open Geospatial Consortium (e.g., SensorML) [18], (iv) the Common Reporting Standard (CRS) that specifies guidelines for obtaining information from financial institutions and automatically exchanging that information in an interoperable way, and (v) standards-based ontologies appropriate for describing social relationships between individuals or groups, such as the "The Friend Of A Friend" (FOAF) ontology [19] and the Socially Interconnected Online Communities (SIOC) ontology [20]. These standards provide the means for common representation of domain specific datasets, towards data interoperability (including in several cases semantic interoperability) across diverse databases and datasets. Nevertheless, these standards are insufficient for delivering what PolicyCLOUD promises for a number of reasons. Initially, there is a lack of semantic interoperability in the given domain. For example, compliance to ontologies about IoT and sensor data fails to ensure a unified modelling of the physics and mathematics, which are at the core of any sensing task. Hence, in several cases there is a need for extending existing models with capabilities for linking/relating various quantifiable and measurable (real-world) features to define, in a user understandable and machine-readable manner the processes behind single or combined tasks in the given domain. Furthermore, there is a lack of semantic interoperability across datasets from different sectors. There is not easy way to link related information elements stemming from datasets in different sectors, which typically comprise different schemas. In this context, environmental datasets and transport datasets for instance contain many related elements, which cannot however be automatically identified and processed by a system due to the lack of common semantics. Finally, one needs to consider the lack of process interoperability. PolicyCLOUD deals with data-driven policy development and management, which entails the simulation and validation of entire processes. Especially in the case of multi-sectoral considerations (e.g., interaction and trade-offs between different policies) process interoperability is required in order to assess the impact of one policy on another. PolicyCLOUD proposes a multi-layer framework for interoperability across diverse policy related datasets, which will facilitate semantic interoperability across related datasets both within a single sector and across different policy sectors. Within a specific sector of each use case, semantic interoperability will enable adhering to existing standards-based representations for the sector data and other auxiliary data (e.g. sensor data, social media data). Across different use cases, PolicyCLOUD proposes a LinkedData approach [21] to enable linking of interrelated data across different ontologies.

3 Main Challenges and Proposed Approach

3.1 Main Challenges and Objectives Addressed by PolicyCLOUD

A Data-Driven Approach for Effective Policies Management
The challenge is to provide a scalable, flexible and dependable methodology and environment for facilitating the needs of data-driven policy modelling, making and evaluation. The methodology should aim at applying the properties of policy modelling, co-creation and implementation across the complete data path, including data modelling, representation and interoperability, metadata management, heterogeneous datasets linking and aggregation, analytics for knowledge extraction, and contextualization. Moreover, the methodology should exploit the collective knowledge out of policy "collections"/clusters combined with the immense amounts of data from several sources (e.g. sensor readings, online platforms, etc.). These collections of policies should be analysed based on specific Key Performance Indicators (KPIs) in order to enable the correlation of these KPIs with different potential determinants of policies impact within and across different sectors (e.g. environment, radicalisation, migration, goods and services, etc.).

Compilation, Assessment and Optimization of Multi-domain Policies
Another challenge refers to holistic policy modelling, making and implementation in different sectors (e.g. environment, migration, goods and services, etc.), through the analysis and linking of KPIs of different policies that may be inter-dependant and inter-correlated (e.g. environment). The goal is to identify (unexpected) patterns and relationships between policies (through their KPIs) to improve policy making. Moreover, the approach should enable evaluation and adaptation of policies by dynamically extracting information from various data sources, community knowledge out of the collections of policies, and outcomes of simulations and evidence-based approaches. Policies should be evaluated to identify both their effective KPIs to be re-used in new/other policies, and the non-effective ones (including the causes for not being effective) towards their improvement. Thus, developed policies should consider the outcomes of strategies in other cases, such as policies addressing specific city conditions.

Data Management Techniques Across the Complete Data Path
Data-driven policy making highlights the need for a set of mechanisms that address the data lifecycle, including data modelling, cleaning, interoperability, aggregation, incremental data analytics, opinion mining, sentiment analysis, social dynamics and

behavioural data analytics. In order to address data heterogeneity from different sources, modelling and representation technologies should provide a "meta-interpretation" layer, enabling the semantic and syntactic capturing of data properties and their representation. Another key aspect refers to techniques for data cleaning in order to ensure data quality, coherence and consistency including the adaptive selection of information sources based on evolving volatility levels (i.e. changing availability or engagement level of information sources). Mechanisms to assess the precision and correctness of the data, correct errors and remove ambiguity beyond limitations for multidimensional processing should be incorporated in the overall environment, while taking into consideration legal, security and ethical aspects.

Context-Aware Interoperability
The specific challenge refers to the design and implementation of a semantic layer that will address data heterogeneity. To this end, the challenge is to research on techniques and semantic models for the interoperable use of data in different scenarios (and thus policies), locations and contexts. Techniques for interoperability (such as OSLC - Open Services for Lifecycle Collaboration) with different ontologies (as placeholders for the corresponding information) should be combined with semantic annotations. Semantic models for physical entities/devices (i.e. sensors related to different policy sectors), virtual entities (e.g. groupings of such physical entities according to intrinsic or extrinsic, permanent or temporary properties) and online platforms (e.g. social media, humans acting as providers) should be integrated in data-driven policy making environments. These models should be based on a set of transversal and domain-specific ontologies and could provide a foundation for high-level semantic interoperability and rich semantic annotations across policy sectors, online systems and platforms. These will be turned into rich metadata structures providing a paradigm shift towards content-based storage and retrieval of data instead of data-based, given that stakeholders and applications target and require such content based on different high-level concepts. Content-based networks of data objects need to be developed, allowing retrieval of semantically similar contents.

Social Dynamics and Incentives Management
One of the main barriers to public bodies experimenting with big data to improve evidence-based policy making is citizens' participation since they are lacking awareness of the extend they may influence policy design and the ways these will be feasible. The challenge is to raise awareness about policy consultations and enable citizens to take direct action to participate, thus ensuring higher levels of acceptance and of trust. A potential solution could be to follow a living lab approach and implement an engagement strategy based on different incentives mechanisms. Furthermore, a data-driven policy management environment should allow social dynamics and behaviour to be included in the policy lifecycle (creation, adaptation, enforcement, etc.) through the respective models and analytical tools. These will allow policy makers to obtain the relevant crowd-sourcing data and the knowledge created by the closed groups (i.e. communities evaluating proposed policies) and the engaged citizens to analyse and propose social requirements that will be turned into policy requirements. On top of this, incentives management techniques will identify, declare and manage incentives for citizens' engagement, supporting different types of incentives (e.g. social, cultural, political, etc.),

with respect to information exchange, contributions and collaboration aspirations. The environment should also provide strategies and techniques for the alignment participation incentives, as well as protocols enabling citizens to establish their participation.

Analytics as a Service Reusable on Top of Different Datasets
Machine and deep learning techniques, such as classification, regression, clustering and frequent pattern mining algorithms, should be realized in order to infer new data and knowledge. Sentiment analysis and opinion mining techniques should determine whether the provided contributor's input is positive or negative about a policy, thus developing a "contributor graph" for the contributors of the opinions that happen to be themselves contributors to ongoing policy making projects. In the same context, social dynamics and behavioural data analytics should provide insights regarding which data needs to be collected and aggregated in a given case (e.g. time window addressed by the policy, location of populations, expected impact, etc.), taking into consideration the requirements of the engaged citizens to model the required policies. Moreover, a main challenge refers to technologies that allow analytics tasks to be decoupled from specific datasets and thus be triggered as services and applied to various cases and datasets.

Transferable Methods and a Unique Endpoint to Exploit Analytics in Different Cases
A key challenge refers to an overall system, acting as an endpoint that will allow stakeholders (such as policy makers and public authorities) to trigger the execution of different models and analytical tools on their data (e.g. to identify trends, to mine opinion artefacts, to explore situational and context awareness information, to identify incentives, etc.) and obtain the results. Based on these results, the modelled policies (through their KPIs) will be realized/implemented and monitored against these KPIs. Moreover, the endpoint should allow stakeholders and public administration entities to express in a declarative way their analytical tasks and thus perform/ingest any kind of data processing. Another need is for an adaptive visualization environment, enabling policy monitoring to be visualized in different ways while the visualization can be modified on the fly. The environment should also enable the specification of the assets to be visualized: which data sources and which meta-processed information. The goal is to enable the selection of sources based on the stakeholders' needs. Incremental visualization of analytics outcomes should also be feasible enabling visualization of results as they are generated.

3.2 Architecture Overview

As a complete environment, the proposed architectural approach includes a set of main building blocks to realize the corresponding functionality as depicted in the following figure (Fig. 1).

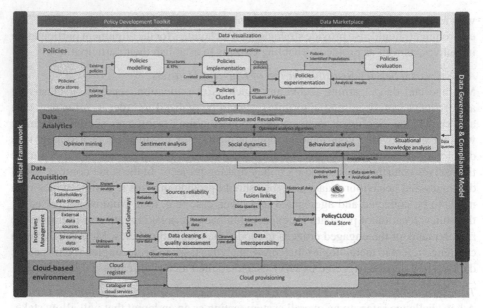

Fig. 1. Proposed functional architecture.

The overall flow is initiated from various data sources, as depicted in the figure through the respective *Data Acquisition* block. Data sources can be data stores from public authorities or external data sources (e.g. mobile devices, IoT sensors, etc.) that contribute data following the provision of incentives, facilitated through the *incentives management* mechanism. A set of APIs incorporated in a gateway component, enables data collection by applying techniques to identify the reliable sources exploiting the *sources reliability* component and for these sources obtain the data and perform the required *data quality assessment and cleaning*. *Semantic and syntactic interoperability* techniques are utilized over the cleaned data providing the respective interoperable datasets to the Policy CLOUD datastore following the required *data linking and aggregation* processes. The datastore is accessible from a set of machine learning models represented through the *Data Analytics* building block. Machine learning models incorporate opinion mining, sentiment and social dynamic analysis, behavioural analysis and situational/context knowledge acquisition. The data store and the analytics models are hosted and executed in a *cloud-based environment* that provides the respective services obtained from a catalogue of cloud infrastructure resources. Furthermore, all the analytics models are realized as services, thus enabling their invocation through a proposed policy development toolkit – realized in the scope of the *Policies* building block of the proposed architecture as a single point of entry into the PolicyCLOUD platform. The toolkit allows the compilation of *policies as data models*, i.e. structural representations that include key performance indicators (KPIs) as a means to set specific parameters (and their target values) and monitor the implementation of policies against these KPIs along with the list of analytical tools to be used for their computation. According to these analytics outcomes, the values of the KPIs are specified resulting to *policies implementation/creation*. It should be noted that PolicyCLOUD also introduces the concept of *policies clusters* in order to interlink different policies, and identify the KPIs and parameters that can be optimized in such policy collections. Across the complete environment, an implemented *data governance and compliance model* is enforced, ranging from the provision of cloud resources regarding the storage and analysis of data to the management of policies across their lifecycle.

4 Conclusions

The vast amounts of data that are being generated by different sources highlight an opportunity for public authorities and stakeholders to create, analyse, evaluate and optimize policies based on the "fresh" data, the information that can be continuously collected by citizens and other sensors. To this end, what is required refers to techniques and an overall integrated environment that will facilitate not only data collection but also assessment in terms of reliability of the data sources, homogenization of the datasets in order to make them interoperable (following the heterogeneity in terms of content and formats of the data sources), cleaning of the datasets and analytics. While several analytical models and mechanisms are being developed a key challenge relates to approaches that will enable analytics to be triggered as services and thus applied and utilized in different datasets and contexts. In this paper, we have presented the aforementioned challenges and the necessary steps to address them. We also introduced a conceptual architecture that depicts a holistic cloud-based environment integrating a set of techniques across the complete data and policy management lifecycles in order to enable data-driven policy management. It is within our next steps to implement the respective mechanisms and integrate them based on the presented architecture, thus realizing the presented environment.

Acknowledgment. The research leading to the results presented in this paper has received funding from the European Union's funded project PolicyCLOUD under Grant Agreement no 870675.

References

1. How can social media data be used to improve services? https://www.theguardian.com/local-government-network/2013/oct/03/social-media-improve-services-data
2. Data Growth, Business Opportunities, and the IT Imperatives. https://www.emc.com/leadership/digital-universe/2014iview/executive-summary.htm
3. Anderson, J.E.: Cases in Public Policy-Making. Praeger, New York (1976)
4. Setting Data Policies, Standards, and Processes. https://www.mcpressonline.com/analytics-cognitive/business-intelligence/setting-data-policies-standards-and-processes
5. Data for Policy: A study of big data and other innovative data-driven approaches for evidence-informed policymaking. http://media.wix.com/ugd/c04ef4_20afdcc09aa14df38fb646a33e6 24b75.pdf
6. Big Data: Basics and Dilemmas of Big Data Use in Policy-making. http://www.policyhub. net/en/experience-and-practice/153
7. What is Polyglot Persistence? http://www.jamesserra.com/archive/2015/07/what-is-polyglot-persistence

8. Building a Just-In-Time Data Warehouse Platform with Databricks. https://databricks.com/blog/2015/11/30/building-a-just-in-time-data-warehouse-platform-with-databricks.html
9. Apache Spark Homepage. http://spark.apache.org
10. Zaharia, M., et al.: Resilient distributed datasets: a fault-tolerant abstraction for in-memory cluster computing. In: Proceedings of the 9th USENIX conference on Networked Systems Design and Implementation. USENIX Association, Berkeley (2012)
11. Armbrust, M., et al.: Spark SQL: relational data processing in spark. In: Proceedings of the 2015 ACM SIGMOD International Conference on Management of Data (SIGMOD 2015), pp. 1383–1394. ACM, New York (2015)
12. Zaharia, M., Das, T., Li, H., Hunter, T., Shenker, S., Stoica, I.: Discretized streams: fault-tolerant streaming computation at scale. In: Proceedings of the Twenty-Fourth ACM Symposium on Operating Systems Principles, pp. 423–438. ACM, New York (2012)
13. Structuring Spark: DataFrames, DataSets and Streaming. https://databricks.com/session/structuring-spark-dataframes-datasets-and-streaming
14. Bonomi, F., Milito, R., Zhu, J., Addepalli, S.: Fog computing and its role in the internet of things. In: Proceedings of the first edition of the MCC workshop on Mobile cloud computing (MCC 2012), pp. 13–16. ACM, Helsinki (2012)
15. Data Specifications. http://inspire.ec.europa.eu/data-specifications/2892
16. Cordis. https://cordis.europa.eu/about/archives
17. Incubator. https://www.w3.org/2005/Incubator/ssn/ssnx/ssn
18. Open Geospatial Standards. https://www.ogc.org/standards/sensorml
19. FOAF Homepage. http://www.foaf-project.org
20. SIOC project. http://sioc-project.org
21. Heath, T., Bizer, C.: Linked Data: Evolving the Web into a Global Data Space, 1st edn. Morgan & Claypool, San Rafael (2011)

Demand Forecasting of Short Life Cycle Products Using Data Mining Techniques

Ashraf A. Afifi[1,2]([📧])

[1] Department of Engineering, Design and Mathematics, Faculty of Environment
and Technology, University of the West of England, Bristol, UK
Ashraf.Afifi@uwe.ac.uk
[2] Industrial Engineering Department, Faculty of Engineering,
Zagazig University, Zagazig, Egypt

Abstract. Products with short life cycles are becoming increasingly common in
many industries due to higher levels of competition, shorter product development
time and increased product diversity. Accurate demand forecasting of such products is crucial as it plays an important role in driving efficient business operations
and achieving a sustainable competitive advantage. Traditional forecasting methods are inappropriate for this type of products due to the highly uncertain and
volatile demand and the lack of historical sales data. It is therefore critical to
develop different forecasting methods to analyse the demand trend of these products. This paper proposes a new data mining approach based on the incremental
k-means clustering algorithm and the RULES-6 rule induction classifier for forecasting the demand of short life cycle products. The performance of the proposed
approach is evaluated using real data from one of the leading Egyptian companies
in IT ecommerce and retail business, and results show that it has the capability to
accurately forecast demand trends of new products with no historical sales data.

Keywords: Demand forecasting · Short life cycle products · Data mining ·
Clustering · Rule induction

1 Introduction

Forecasting is an important and necessary tool that helps managers make decisions about
future resourcing of their organisations. It plays a pivotal role in the effective planning
of various operations in an organisation including production, inventory, budget, sales,
personnel and facilities. Accurate forecasts can lead to significant cost savings, reduced
inventory levels, improved customer satisfaction and increased competitiveness.

The fast pace of new product introduction continually drives product life cycles
shorter. Products with life cycles of few weeks to few months are very common in fashion (e.g., toys and clothing) and high technology (e.g., mobile phones, computers and
consumer electronics) retail industries. A typical demand pattern for such products is
characterised by rapid growth, maturity, and decline phases [1]. The demand is also

© IFIP International Federation for Information Processing 2020
Published by Springer Nature Switzerland AG 2020
I. Maglogiannis et al. (Eds.): AIAI 2020, IFIP AICT 583, pp. 151–162, 2020.
https://doi.org/10.1007/978-3-030-49161-1_14

highly uncertain and volatile, particularly in the introduction stage. An additional problem is the inadequacy of historical data due to the short period of sales. In case of new products, there is complete unavailability of any previous data related to the sales of such products. These characteristics make it difficult to forecast the demand of short life cycle products.

Traditionally, demand forecasting is accomplished by statistical methods such as moving averages, exponential smoothing, Bayesian analysis, regression models, Holt-Winters and Box-Jenkins methods, and autoregressive integrated moving average (ARIMA) models [2]. Statistical methods are popular because of their simplicity and fast speed, and they provide satisfactory results in many forecasting applications. However, these methods are not designed for application in the short life cycle environment, especially most of them require large historical data to ensure accurate estimation of their parameters and they are limited to linear relations [3, 4].

Artificial intelligence (AI) methods such as neural networks, evolutionary algorithms, support vector machines, and fuzzy inference systems are widely used in forecasting activities [5, 6]. AI methods cope well with complexity and uncertainty, and they have better forecasting accuracy compared to statistical methods. However, they usually require a long computational time which makes them less appealing to the fast changing market of fashion and high technology products [7, 8].

Recently, various hybrid methods such as neural fuzzy systems are proposed in the literature to enhance demand forecasting [9, 10]. Hybrid methods utilise the strengths of different models to form a new forecasting method. They can learn complex relations in an uncertain environment and many of them are considered to be more accurate and efficient than the pure statistical and AI models [11].

In an attempt to cope with the lack of historical data of short life cycle products, some references use the data of similar products for which sufficient history is available to forecast the demand of new products [12–16]. For example, a hybrid method based on the k-means clustering technique and a decision tree classifier was developed to estimate the demand of textile fashion products [12]. A similar approach that uses the self-organizing map and the neural networks techniques was introduced in [13]. This paper reports on a new forecasting method using alternative data mining techniques to improve demand forecasting in the context of a large retail company. In particular, the incremental k-means clustering technique [17] and the RULES-6 classification rule learning algorithm [18] are employed. These methods are simple, effective and computationally efficient, which make them powerful and practical tools for retail sales forecasting.

The outline of this paper is as follow. Section 2 reviews the basic concepts of data mining with particular focus on clustering and rule induction methods. Section 3 presents the proposed forecasting approach. Section 4 reports and analyses the experimental results obtained with a real retail data. Section 5 concludes the paper and provides suggestions for future work.

2 Overview of Data Mining

Data mining, or knowledge discovery in databases (KDD), aims at processing large data into knowledge bases for better decision making. It has been successfully used in many applications to uncover hidden patterns and predict future trends and behaviours. There are three main steps in data mining, namely, data preparation, data modelling, and post processing and model evaluation [19–21]. This section gives a brief description of the data modelling techniques employed in this research.

2.1 Clustering Techniques

Clustering techniques are concerned with partitioning of data sets into several homogeneous clusters. These techniques assign a large number of data objects to a relatively small number of groups so that data objects in a group share the same properties while, in different groups, they are dissimilar. Many clustering techniques have been proposed over the years from different research disciplines [22, 23]. K-means is one of the best known and commonly used clustering algorithms. The algorithm forms k clusters that are represented by the mean value of the data points belonging to each of them. This is an iterative process that searches for a division of data objects into k clusters to minimise the sum of Euclidean distances between each object and its closest cluster centre.

The k-means algorithm is relatively scalable and efficient in clustering large data sets because its computational complexity grows linearly with the number of data objects. However, it is sensitive to the initial selection of cluster centres and requires the number of clusters k to be specified before the clustering process starts. The algorithm has been improved to address many of its deficiencies [17, 24, 25]. In particular, a new version called incremental k-means was introduced to reduce the dependence of the k-means algorithm on the initialisation of cluster centres [17]. To validate the robustness of the new algorithm it has been tested on a number of artificial and real datasets. The results showed clearly that incremental k-means consistently outperforms the original algorithm [17]. Therefore, this algorithm is applied in this research to search for interesting and natural clusters in the retail data.

The incremental k-means algorithm is summarised in Fig. 1. The algorithm starts initially with one cluster with the number of clusters k being incremented by 1 at each step thereafter. With each increase of k, a new cluster centre is inserted into the cluster with the highest distortion, and the objects are reassigned to clusters until the centres stop "moving". The process is repeated until k reaches the specified number of clusters.

2.2 Classification Learning Techniques

Classification learning employs a set of pre-categorised data objects to develop a model that can be used to classify new data objects from the same population or to provide a better understanding of the data objects' characteristics. Among the various classification learning techniques developed, inductive learning may be the most commonly used in real world applications [26]. The inductive learning techniques are relatively fast compared to other techniques. Another advantage is that they are simpler and the models that they generate are easier to understand.

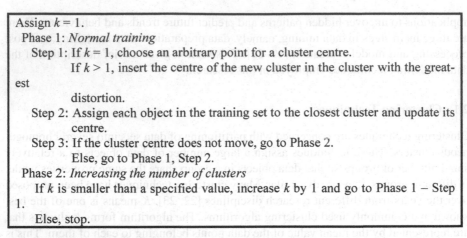

Assign $k = 1$.

Phase 1: *Normal training*

Step 1: If $k = 1$, choose an arbitrary point for a cluster centre.

If $k > 1$, insert the centre of the new cluster in the cluster with the greatest distortion.

Step 2: Assign each object in the training set to the closest cluster and update its centre.

Step 3: If the cluster centre does not move, go to Phase 2.

Else, go to Phase 1, Step 2.

Phase 2: *Increasing the number of clusters*

If k is smaller than a specified value, increase k by 1 and go to Phase 1 – Step 1.

Else, stop.

Fig. 1. A pseudo-code description of the incremental k-means algorithm [17].

In this study, a simple inductive learning algorithm called RULES-6 (RULe Extraction System – Version 6) [18] is used. RULES-6 extracts a set of classification rules from a collection of examples, each belonging to one of a number of given classes. The examples together with their associated classes constitute the set of training examples from which the algorithm generates the rules. Every example is described in terms of a fixed set of attributes, each with its own set of possible values.

In RULES-6, an attribute-value pair constitutes a condition. If the number of attributes is N_a, a rule may contain between one and N_a conditions, each of which must be a different attribute-value pair. Only conjunction of conditions is permitted in a rule and hence the attributes must all be different if the rule comprises more than one condition.

RULES-6 works in an iterative fashion. In each iteration, it takes a "seed" example not covered by previously created rules to form a new rule. Having found a rule, RULES-6 marks those examples that are covered by it and appends the new rule to its rule set. The algorithm stops when all examples in the training set are covered. In order to avoid the overfitting problem and to generate simple rules, the rules are pruned with a post pruning strategy [27]. A simplified description of RULES-6 is given in Fig. 2.

To form a rule, RULES-6 performs a general-to-specific beam search for a rule that optimises a given quality criterion. It starts with the most general rule and specialises it in steps considering only conditions extractable from the selected seed example. The aim of specialisation is to construct a rule that covers as many examples from the target class and as few examples from the other classes as possible, while ensuring that the seed example remains covered. As a consequence, simpler rules that are not consistent, but are more accurate for unseen data, can be learned. RULES-6 uses effective search-space pruning rules to avoid useless specialisations and to terminate a non-productive search during rule construction. This substantially increases the efficiency of the learning process. It deals with continuous-value attributes using a pre-processing discretisation method. With this method, the range of each attribute is split into a number of smaller

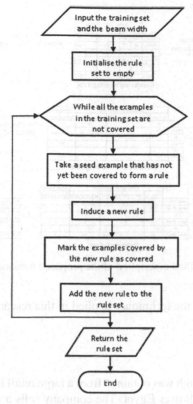

Fig. 2. A simplified description of RULES-6 [18].

intervals that are then regarded as nominal values. A detailed description of the process by which RULES-6 induces a rule can be found in [18].

3 Proposed Forecasting Approach

In this paper, it is proposed to forecast the demand of short life cycle products by applying data mining techniques. This forecasting approach includes the following steps (Fig. 3):

1. Preparing and transforming the available sales data into a format suitable for data mining techniques.
2. Using the incremental k-means algorithm to identify the sales profiles of historical products by grouping together the related time series of the products sales.
3. Applying the RULES-6 algorithm to build a classification model that links the descriptive attributes of the historical products and the sales profiles of the discovered groups in Step 2.
4. Using the classification model generated in Step 3 to forecast the sales profiles of the new products based on their descriptive attributes.

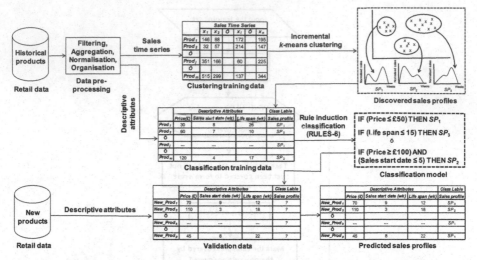

Fig. 3. A simplified description of the proposed forecasting approach.

This section discusses the techniques applied in this research to realise the steps of the proposed approach.

3.1 Data Preparation

The data used in this research was obtained from a large retail Egyptian company called 2B that has 37 branches all over Egypt. The company sells a wide range of high technology products (e.g., laptops, mobiles, tablets, gaming, networking, cables, software, laptop and mobile accessories) that have a short life cycle ranging from a few weeks to a few months. Two sets of data were extracted. The first contains 507 products corresponding to year 2015 and it is used for the learning process. The second is composed of 253 products for year 2016 and it is used for testing. The data is available weekly and covers a period of 26 weeks. The selected descriptive attributes of the products are the price, the starting date of the sales and the life span of each product.

High quality data is a prerequisite for applying any data mining technique [28]. Prior to data modelling, the data needs to be prepared. The objective at this stage is two-fold: to convert the data into a format required by the data mining algorithms and also to expose as much information as possible to data modelling. To prepare the available data for further analysis the following pre-processing steps are implemented.

1. Aggregate the sales for each product into weekly time buckets in order to cope with high uncertainty and volatility in sales patterns.
2. Filter the data by removing the negative sales numbers for customer returns.
3. Normalise the sales volume of all products during the sales period to enable comparison between the time series of products having different sales volumes. The normalised sales at period i, y_i, is computed by dividing the sale at period i, x_i, by the sum of sales during the sales period: $y_i = x_i / \sum_{k=1}^{n} x_k x_k$ where n is the number of periods.

4. Normalise the life span of all products to 26 weeks to allow comparison between the time series of products having different life spans. Normalisation of the life span is computed using homothetic transformation [12].
5. Organise the time series sales and descriptive attributes of the products into a data object format that is suitable for data mining algorithms (examples of the structure of the training and validation data sets created in this way are shown in Fig. 3).

3.2 Data Modelling

This section discusses the data mining techniques that were used to analyse the retail data. The analysis is performed in two stages. First, the time series sales of historical products are grouped into several homogeneous clusters. Second, the sales profiles of the new products are predicted by mapping them to the clusters using the descriptive attributes.

Sales Profiles Identification for Historical Products. The incremental k-means clustering algorithm is applied on the pre-processed retail data to discover groups of products with similar time series sales. Each cluster centre defines a sales profile characterising the sales behaviour of the products belonging to the cluster. The distortion error was used to evaluate the clustering results [17], with a lower value of this measure indicative of better quality of clustering.

The incremental k-means algorithm requires users to specify a number of parameters, namely, the number of clusters and the termination conditions for stopping the clustering process. To find a satisfactory clustering result, a number of iterations were conducted where the algorithm was executed with different values of k, the number of clusters, and the k value producing the lowest distortion error was selected [17]. In this work, k values in the range of 2 to $\sqrt{n}$, where n is the size of the training data set [29], were considered and the optimal value was found to be 13. The clustering process could be stopped by specifying termination conditions such as a predefined number of iterations and the percentage reduction of the distortion errors in one iteration being smaller than a given value ε. In this work, these two termination criteria were used. In particular, the maximum number of iterations was set to 50 and ε to 10^{-7} to stop the search process when one of these conditions is satisfied.

Sales Profiles Prediction for New Products. After clustering the time series data of the historical products, labels such as SP_1, SP_2, ..., SP_l, ..., SP_q representing the sales profiles of the formed groups, are assigned to the discovered clusters. A training set for the classification algorithm is then constructed using the descriptive attributes and the clusters labels. Each product is considered as a data point in the training set and it is described by the descriptive attributes as inputs and the label of the cluster to which it belongs is taken as its output.

The RULES-6 algorithm is used to extract if-then rules from the training data set. These rules provide a comprehensive insight into the data and are used to predict the sales profiles of the new products based on their descriptive attributes. RULES-6 has a number of parameters whose values determine the quality of the induced rule sets. In this research, the default settings were used [18].

4 Experimental Results

4.1 Clustering Results

The incremental k-means algorithm was applied to the training time series data and thirteen distinct clusters were created. Figure 4 shows four examples of the clustering results that have the maximum distortion errors. As can be seen in the figure, the incremental k-means clustering procedure is effective in producing accurate groups of products with similar time series sales. The sales behaviour of the products belonging to the different clusters are accurately described by the associated sales profiles (indicated in bold). It can then be concluded that a robust identification of the sales profiles has been achieved.

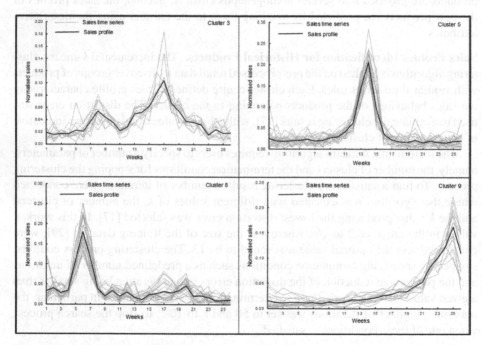

Fig. 4. Examples of the sales profiles created by the incremental k-means clustering algorithm.

4.2 Rule Induction Results

The RULES-6 algorithm was applied on the classification training data created from the descriptive attributes and the discovered sales profiles. Table 1 illustrates the produced set of rules to describe the training data. It is clear from the figure that the number of rules generated is significantly lower than the number of data points in the training set. Also, the number of features describing each rule is drastically reduced. The generated rule set is a compressed description of the training data that could be used to predict the sales profiles of new products.

Table 1. Rules derived by RULES-6 from the retail data.

R01: IF (Price <= £44) THEN Sales profile = SP1
R02: IF (Sales start date = 4) THEN Sales profile = SP4
R03: IF (Sales start date >= 14) THEN Sales profile = SP2
R04: IF (£101 <= Price <= £147) AND (Life span >= 22) THEN Sales profile = SP3
R05: IF (Price <= £98) AND (Sales start date <= 2) THEN Sales profile = SP7
R06: IF (Price <= £99) AND (Sales start date <= 11) THEN Sales profile = SP11
R07: IF (Life span <= 11) THEN Sales profile = SP6
R08: IF (Life span = 25) THEN Sales profile = SP9
R09: IF (Price <= £92) THEN Sales profile = SP10
R10: IF (Price >= £300) THEN Sales profile = SP12
R11: IF (£200 <= Price <= £249) THEN Sales profile = SP5
R12: IF (Price >= £250) THEN Sales profile = SP13
R13: IF (Sales start date = 1) AND (11 < Life span <= 14) THEN
     Sales profile = SP8

To evaluate its performance, the RULES-6 algorithm was applied on the validation data to predict the sales profiles of the new products. Figure 5 shows a comparison between the actual sales of the new products and the sales profiles predicted by the RULES-6 algorithm. As can be seen in the figure, RULES-6 assigns the correct profiles to most products. The actual sales associated with profile SP9 of cluster 9 are quite different in some weeks. Also, the actual sales accompanied with profile SP8 of cluster 8 are relatively similar but are slightly different from the predicted sales profile. This could arise from the failure of the descriptive attributes used to build the rule set to explain the sales behaviour of these products.

To further test its performance, the prediction accuracy of the RULES-6 algorithm was compared with that of five different well-known classifiers. The classification methods are the OneR algorithm [30], the Naïve Bayes method [31], the k-nearest-neighbours classifier [32], the RIPPER rule induction classifier [33], and the C4.5 decision tree [34]. These methods are representative of the different classification techniques and widely used in other comparative studies. Each of the classification methods was first applied on the classification training data and the generated classification models were then tested on the validation data to predict the sales profiles of the new products. The default parameters of the tested methods were used. The forecasting errors of the predicted sales profiles were analysed using three of the most popular measures, namely, Mean Absolute Error (MAE), Mean Absolute Percentage Error (MAPE), and Root Mean Squared Error (RMSE) [35].

The forecasting error values of the predicted sales profiles for the different classification methods are given in Table 2. As shown in the table, RULES-6 is the most accurate algorithm. It achieved the smallest error values for the three measures and the values were lower by 39.2% for the MAE, 21.7% for the MAPE and 43.4% for the RMSE compared to the next best performing classifier. It could therefore be concluded that the proposed RULES-6 algorithm gives the best performance among the six methods tested.

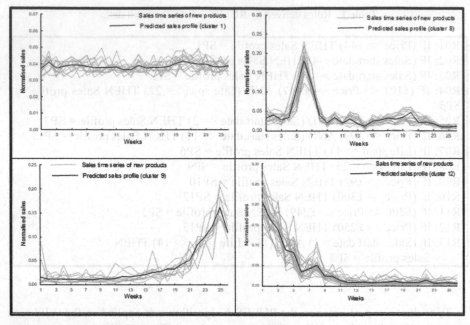

Fig. 5. Comparison between the actual sales of new products and the sales profiles predicted by RULES-6.

Table 2. Forecasting errors of the predicted sales profiles for the tested classification methods.

	MAE	MAPE (%)	RMSE
OneR	70.7	5.2	84.3
Naïve Bayes	63.2	4.5	78.9
k-nearest-neighbours	51.3	3.7	66.8
RIPPER	16.4	2.9	21.1
C4.5	9.7	2.3	12.9
Proposed RULES-6	**5.9**	**1.8**	**7.3**

5 Conclusions and Future Work

The constraints of the retail market make the forecasting of products sales very challenging. This paper has proposed a forecasting approach based on the incremental k-means clustering algorithm and the RULES-6 rule induction classifier to estimate the sales profiles of new products. The performance of the proposed forecasting procedure has been tested on real sales data of a large retail Egyptian company specialised in selling and distribution of high tech products. The incremental k-means algorithm employed in this study has proved to be effective in discovering interesting groupings of historical products, from which general descriptions can be derived by applying the RULES-6 algorithm. From the clusters and their descriptions, sales profiles for new products could be predicted easily. However, some inaccurate profiles could result from the inadequacy of the chosen descriptive attributes to discriminate the sales behaviours of all new products. The RULES-6 algorithm has also been tested against five other state-of-the-art classifiers and attained more accurate forecasts. These results suggest that the combined application of the incremental k-means and RULES-6 techniques is a valid approach for estimating the sales profiles of new products in retail businesses where historical sales data are not available.

Further research could be conducted to apply the proposed forecasting procedure to other short life cycle products and to compare the RULES-6 algorithm with other classifiers such as decision trees, neural networks, or genetic algorithms. Also, considering the various sources of uncertainty that arise in the retail market, it would be interesting to use fuzzy learning techniques which may help to produce more accurate forecasts.

Acknowledgement. The author wishes to thank the University of the West of England for providing a good environment, facilities and financial means to complete this paper.

References

1. Kadam, S., Apte, D.: A survey on short life cycle time series forecasting. Int. J. Appl. Innov. Eng. Manag. **4**(5), 445–449 (2015)
2. Armstrong, J.S.: Principles of Forecasting - A Handbook for Researchers and Practitioners. Kluwer Academic Publishers, Norwell (2001)
3. Xia, M., Wong, W.K.: A seasonal discrete grey forecasting model for fashion retailing. Knowl.-Based Syst. **57**, 119–126 (2014)
4. Beheshti-Kashia, S., Karimic, H.R., Thobenb, K.-D., Lütjenb, M., Teuckeb, M.: A survey on retail sales forecasting and prediction in fashion markets. Syst. Sci. Control Eng. **3**, 154–161 (2015)
5. Giustiniano, L., Nenni, M.E., Pirolo, L.: Demand forecasting in the fashion industry a review. Int. J. Eng. Bus. Manag. **5**(37) (2013). https://doi.org/10.5772/56840
6. Liu, N., Ren, S., Choi, T.-M., Hui, C.-L., Ng, S.-F.: Sales forecasting for fashion retailing service industry: a review. Math. Probl. Eng. **4**, 1–9 (2013). https://doi.org/10.1155/2013/738675. ID 738675
7. Choi, T.M., Hui, C.-L., Liu, N., Ng, S.-F., Yu, Y.: Fast fashion sales forecasting with limited data and time. Decis. Support Syst. **59**, 84–92 (2014)
8. Fajardo-Toro, C.H., Mula, J., Poler, R.: Adaptive and hybrid forecasting models—a review. In: Ortiz, Á., Andrés Romano, C., Poler, R., García-Sabater, J.-P. (eds.) Engineering Digital Transformation. LNMIE, pp. 315–322. Springer, Cham (2019). https://doi.org/10.1007/978-3-319-96005-0_38
9. Choi, T.-M., Yu, Y., Au, K.-F.: A hybrid SARIMA wavelet transform method for sales forecasting. Decis. Support Syst. **51**, 130–140 (2011)
10. Thomassey, S.: Sales forecasts in clothing industry: the key success factor of the supply chain management. Int. J. Prod. Econ. **128**, 470–483 (2012)
11. Choi, T.-M., Hui, C.-L., Yu, Y. (eds.): Intelligent Fashion Forecasting Systems: Models and Applications. Springer, Heidelberg (2014). https://doi.org/10.1007/978-3-642-39869-8

12. Thomassey, S., Fiordaliso, A.: A hybrid sales forecasting system based on clustering and decision trees. Decis. Support Syst. **42**(1), 408–421 (2006)
13. Thomassey, S., Happiette, M.: A neural clustering and classification system for sales forecasting of new apparel items. Appl. Soft Comput. **7**(4), 1177–1187 (2007)
14. Maaß, D., Spruit, M., de Waal, P.: Improving short-term demand forecasting for short-lifecycle consumer products with data mining techniques. Decis. Anal. **1**, 1–17 (2014)
15. Gaku, R.: Demand forecasting procedure for short life-cycle products with an actual food processing enterprise. Int. J. Comput. Intell. Syst. **7**, 85–92 (2014)
16. Basallo, M., Rodríguez-Sarasty, J., Benitez-Restrepo, H.: Analogue-based demand forecasting of short life-cycle products: a regression approach and a comprehensive assessment. Int. J. Prod. Res. **55**, 1–15 (2016)
17. Pham, D.T., Dimov, S.S., Nguyen, C.D.: An incremental k-means algorithm. Proc. Inst. Mech. Eng. Part C: J. Mech. Eng. Sci. **218**(7), 783–795 (2004)
18. Pham, D.T., Afify, A.A.: RULES-6: a simple rule induction algorithm for handling large data sets. Proc. Inst. Mech. Eng. Part C: J. Mech. Eng. Sci. **219**(10), 1119–1137 (2005)
19. Witten, I.H., Frank, E.: Data Mining: Practical Machine Learning Tools and Techniques with Java Implementations. Morgan Kaufmann Publishers, Burlington (2000)
20. Han, J., kamber, M.: Data Mining Concepts and Techniques. Morgan Kaufmann Publishers, San Francisco (2001)
21. Klösgen, W., Żytkow, J.M.: Handbook of Data Mining and Knowledge Discovery. Oxford University Press, New York (2002)
22. Kaufman, L., Rousseeuw, P.J.: Finding Groups in Data: An Introduction to Cluster Analysis, 2nd edn. Wiley, Hoboken (2005)
23. Pham, D.T., Afify, A.A.: Clustering techniques and their applications in engineering. Proc. Inst. Mech. Eng. Part C: J. Mech. Eng. Sci. **221**(11), 1445–1459 (2007)
24. Pham, D.T., Dimov, S.S., Nguyen, C.D.: A two-phase k-means algorithm for large datasets. Proc. Inst. Mech. Eng. Part C: J. Mech. Eng. Sci. **218**(10), 1269–1273 (2004)
25. Pham, D.T., Dimov, S.S., Nguyen, C.D.: Selection of k in k-means clustering. Proc. Inst. Mech. Eng. Part C: J. Mech. Eng. Sci. **215**(1), 103–119 (2005)
26. Pham, D.T., Afify, A.A.: Machine-learning techniques and their applications in manufacturing. Proc. Inst. Mech. Eng. Part B: J. Eng. Manuf. **219**(5), 395–412 (2005)
27. Pham, D.T., Afify, A.A.: A new minimum description length based pruning technique for rule induction algorithms. Proc. Inst. Mech. Eng. Part C: J. Mech. Eng. Sci. **222**(7), 1339–1352 (2008)
28. Pyle, D.: Data Preparation for Data Mining. Morgan Kaufmann Publishers, Burlington (1999)
29. Vesanto, J., Alhoniemi, E.: Clustering of the self-organizing map. IEEE Trans. Neural Netw. **11**(3), 586–600 (2000)
30. Holte, R.C.: Very simple classification rules perform well on most commonly used datasets. Mach. Learn. **11**, 63–91 (1993)
31. Langley, P., Iba, W., Thompson, K.: An analysis of Bayesian classifiers. In: Proceedings of the 10th National Conference on Artificial Intelligence, USA, pp. 223–228 (1992)
32. Aha, D.: Tolerating noisy, irrelevant, and novel attributes in instance-based learning algorithms. Int. J. Man Mach. Stud. **36**(2), 267–287 (1992)
33. Cohen, W.W.: Fast effective rule induction. In: Proceedings of the 12th International Conference on Machine Learning, Lake Tahoe City, California, USA, pp. 115–123 (1995)
34. Quinlan, J.R.: C4.5: Programs for Machine Learning. Morgan Kauffman, Burlington (1993)
35. Armstrong, J.S., Collopy, F.: Error measures for generalizing about forecasting methods: empirical comparisons. Int. J. Forecast. **8**, 69–80 (1992)

Image Processing

Arbitrary Scale Super-Resolution
for Brain MRI Images

Chuan Tan[✉], Jin Zhu, and Pietro Lio'

Department of Computer Science and Technology, University of Cambridge,
Cambridge CB3 0FD, UK
{ct538,jz426,pl219}@cam.ac.uk

Abstract. Recent attempts at Super-Resolution for medical images used deep learning techniques such as Generative Adversarial Networks (GANs) to achieve perceptually realistic single image Super-Resolution. Yet, they are constrained by their inability to generalise to different scale factors. This involves high storage and energy costs as every integer scale factor involves a separate neural network. A recent paper has proposed a novel meta-learning technique that uses a Weight Prediction Network to enable Super-Resolution on arbitrary scale factors using only a single neural network. In this paper, we propose a new network that combines that technique with SRGAN, a state-of-the-art GAN-based architecture, to achieve arbitrary scale, high fidelity Super-Resolution for medical images. By using this network to perform arbitrary scale magnifications on images from the Multimodal Brain Tumor Segmentation Challenge (BraTS) dataset, we demonstrate that it is able to outperform traditional interpolation methods by up to 20% on SSIM scores whilst retaining generalisability on brain MRI images. We show that performance across scales is not compromised, and that it is able to achieve competitive results with other state-of-the-art methods such as EDSR whilst being fifty times smaller than them. Combining efficiency, performance, and generalisability, this can hopefully become a new foundation for tackling Super-Resolution on medical images.

Keywords: Super-Resolution · Medical image analysis ·
Meta-Learning · Image processing

1 Introduction

In this paper, we seek to apply elements of Meta-Learning to Generative Adversarial Networks (GANs) to tackle Super-Resolution in medical images, specifically brain MRI images. We are the first to apply such a scale-free Super-Resolution technique on these images. Super-Resolution is the task of increasing the resolution of images. It helps radiological centres located in rural areas which

Supported by University of Cambridge.

© IFIP International Federation for Information Processing 2020
Published by Springer Nature Switzerland AG 2020
I. Maglogiannis et al. (Eds.): AIAI 2020, IFIP AICT 583, pp. 165–176, 2020.
https://doi.org/10.1007/978-3-030-49161-1_15

do not have high-fidelity instruments achieve comparable diagnostic results as their advanced counterparts in the city. The importance of Super-Resolution is growing as cross modality analysis requires combining different types of information (such as PET and NMR scans) of varying resolution.

Traditionally, Super-Resolution has been done using interpolation, such as bicubic interpolation, but recent attempts have involved the use of deep learning methods to extract high level information from data, which can be used to supply additional information to increase the resolution of the image. Current deep learning techniques for medical images rely heavily on Generative Adversarial Networks (GANs) for they are able to generate realistic and sharper images by using a different loss function that yields high perceptual quality [5]. For example, mDCSRN [5], Lesion-focussed GAN [15], ESRGAN [4], are all GAN-based solutions tackling Super-Resolution for medical images. Most of them are based off SRGAN - a GAN-based network tackling Super-Resolution developed by Ledig et al. [9]. We therefore use SRGAN as a foundation to which we apply our modification.

Despite their better performance, almost all the networks relied on the sub-pixel convolution layer introduced by Shi et al. [13], which tied a particular upscaling factor to a particular network architecture. This incurs high storage and energy costs for medical professionals who may wish to conduct Super-Resolution on different scaling factors. In this paper, we combine the Meta-Upscale Module introduced by Hu et al. [7] with SRGAN to create a novel network lovingly termed Meta-SRGAN. This breaks the constraint imposed by Shi et al's layer and is capable of tackling Super-Resolution on any scale, even non-integer ones, and hence can reduce storage, energy costs and lay the foundations for real-time Super-Resolution. We first show that Meta-SRGAN outperforms the baseline of bicubic interpolation on the BraTS dataset [2,3,11], and also show that Meta-SRGAN is capable of performing similarly to SRGAN, but yet is able to super-resolve images of arbitrary scales. We also compare the memory footprint and show that Meta-SRGAN is ≈98% smaller than EDSR, a state-of-the-art Super-Resolution technique, but yet is able to achieve similar performance.

The rest of the paper is organised as follows. Background introduces Super-Resolution formally and illustrates how Hu et al. reframes the problem. We then introduce the architecture of Meta-SRGAN in Methods. Experiments elaborates on dataset preparation and training details. Finally, we show the outcomes of the experiments in Results.

1.1 Background

Super-Resolution. Deep learning-based Super-Resolution techniques are generally supervised learning algorithms. They analyse relationships between the low-resolution (LR) image and its corresponding high-resolution (HR) image, and use this relationship to obtain the super-resolved (SR) image. This SR image is then evaluated against the HR image to see how well the algorithm is performing. Any algorithm that is capable of extracting these relationships can be

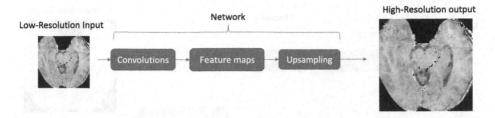

Fig. 1. How a Super-Resolution task is often carried out. A low-resolution input image undergoes some convolutions to generate feature maps. Feature maps are some sort of representation that captures the features of the low-resolution image. Depending on the upscaling factor (it must be a multiple of 2^n), there would be some number n of upscaling modules applied to the representation. This upscaling unfolds the image into the high-resolution output.

used to do Super-Resolution. Dong et al. [6] showed that any such algorithms can be thought of as convolutional neural networks (CNNs). Figure 1 shows a typical high level recipe for a Super-Resolution task.

Meta-Upscale Module. The Upsampling module in the network is often the efficient sub-pixel convolution layer proposed by Shi et al. [13]. Instead of explicitly enlarging feature maps, it expands the channels of the output features to store the extra points to increase resolution. It then rearranges these points to obtain the super-resolved output image. Almost every network tackling Super-Resolution uses this Upsampling module. An example is EDSR [10], a convolutional neural network tackling Super-Resolution. Hu et al. then proposed replacing the Upsampling module with a Meta-Upscale Module [7]. This Meta-Upscale Module consists of a Weight Prediction Network and is an example of Meta-Learning as it generalises a network to tackle more than one upscaling factor. Essentially, given a few training examples (small amount of images from some scales), we train a network to tackle arbitrary scale factors. The Weight Prediction Network is able to predict weights for different upscaling factors. This extra layer of abstraction enables the underlying neural network to generalise across tasks. The framework proposed by Hu et al. for enabling arbitrary scale Super-Resolution is illustrated in Fig. 2.

Hu et al. reframes Super-Resolution as a matrix multiplication problem:

$$Y = Wx + b \tag{1}$$

where Y is the super-resolved image, W is some matrix of weights we learn, x is the input low-resolution image, and b is some constant. This may seem like an oversimplification and in many ways it is (for example it doesn't take account how pixels close to each other tend to exert more influence on each other), but it gives us a rather intuitive understanding for why a separate weight prediction network works for arbitrary scale Super-Resolution. Consider x as a 4×1 vector, and W as a 8×4 matrix. The output Y is therefore a 8×1 vector. In some sense we have doubled the size of x. We can think of W as providing an upscaling of 2.

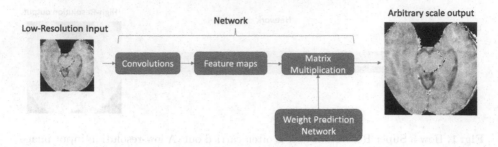

Fig. 2. The framework for tackling Super-Resolution tasks, as proposed by Hu et al. There is no constraint on the upscaling factor. Note that the Upsampling module has been replaced with a matrix multiplication and a Weight Prediction Network. The matrix multiplication and Weight Prediction Network are collectively called the Meta-Upscale Module.

The idea proposed by Hu et al. is to predict W, for every upscaling factor. This is done by passing in the input dimensions to an external function to create the shape of W, before using a weight prediction network to predict the values of W, as in the equation below:

$$W = f_\theta(x) \tag{2}$$

where W is the matrix of weights, f_θ is the weight prediction network with parameters θ, and x is the input low-resolution image.

The architecture of the Meta-Upscale Module is shown in Fig. 3, and includes a weight prediction network that outputs weights which are then matrix multiplied to the input features to obtain the super-resolved image. The Weight Prediction Network consists of three layers, and is trained alongside the main network. Of significant importance is the shape of the input into the Weight Prediction Network. It must contain information relating to the shape of the high-resolution image, as that determines the size of the weight matrix that is multiplied with the input features fed into the Module. Additional details can be found in [7].

2   Methods

2.1   Baselines

We used Bicubic Interpolation as a baseline, and implemented EDSR as a state-of-the-art technique that we compare Meta-SRGAN to. We also wanted to investigate how well using the Meta-Upscale Module on a GAN will affect its performance, so we implemented SRGAN too. EDSR and SRGAN were trained on x2 upsampling tasks, and were trained on DIV2K for 160k updates before transferring their learning onto the BraTS dataset for an additional 160k updates. EDSR was trained with 256 number of features, 32 residual blocks, and a residual scaling of 0.1 (see [10] for more information). EDSR was trained using L1 Loss, whilst SRGAN was trained using the same combination of losses used to train Meta-SRGAN.

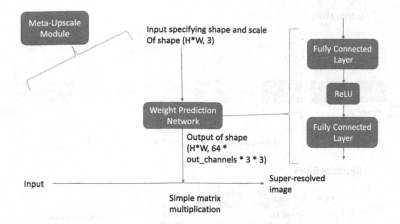

Fig. 3. Architecture of Meta-Upscale Module. H and W refers to the height and width of the *high-resolution* image, and *out_channels* refers to the number of channels in the original image (typically 3 for colour, 1 for grayscale). 64 refers to the number of channels in the input features fed into the Meta-Upscale Module. 3 refers to the kernel size used in the last convolution layer in the network producing the input features fed into the Meta-Upscale Module.

2.2   Meta-SRGAN

To reap the benefits of both GANs and meta-learning, we combine Ledig et al.'s SRGAN with Hu et al.'s Meta-Upscale Module to obtain a new architecture called Meta-SRGAN. The nature of a GAN enables the generator to generate realistic images. The ability to predict weights for different upscaling factors enables the network to tackle arbitrary scales. Combined, they result in a network that is both hugely powerful and highly generalisable.

There are two networks being trained in a Generative Adversarial Network. There is the *generator* which generates an image (in this case it super-resolves an image) which is then fed into the *discriminator* which discriminates between a real and a generated image (in this case it determines whether the image presented is the ground truth high-resolution, or not). This feedback from the discriminator is then used to further train the generator. Both generator and discriminator are playing a game to outbid each other. The generator tries to improve its ability to generate images that can fool the discriminator whilst the discriminator tries to improve its ability to discern real images from generated images. This *adversarial learning* enables us to generate more realistic and detailed images. The architecture of the generator and discriminator are shown in Figs. 4 and 5 respectively.

Generator. The generator is used to generate super-resolved images, which the discriminator will take in and output a classification that says whether it is a generated image or a real image. Residual Blocks are used in the generator to

Generator:

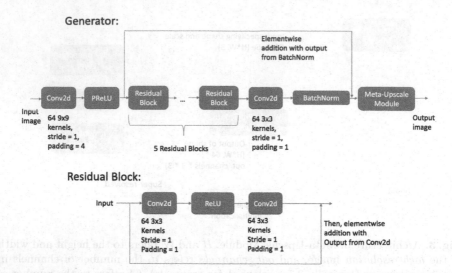

Fig. 4. (Top) Architecture of generator. Generator consists of Convolutions, Residual Blocks, Batch Normalisation, and the Meta-Upscale Module. *PReLU* refers to the Parametric Leaky ReLU activation function. (Bottom) Architecture of Residual Block used in Generator. *ReLU* refers to the ReLU activation layer. The number 64 refers to the number of kernels, and can be modified like any hyperparameter.

stabilise training by incorporating a feedback loop back onto the input. Parametric ReLU was used as it was empirically found to stabilise training [9].

Discriminator. The discriminator consists of several Leaky ReLU layers and an average pooling layer. These was empirically found to stabilise training [9]. A sigmoid layer was used to facilitate binary classification.

L1 Loss. L1 Loss is the typical loss function used by other networks that calculates pixel-wise errors. We incorporate it to train the generator in Meta-SRGAN. We denote an input image by x, and a neural network and its parameters by f_θ. Letting $\hat{y} = f_\theta(x)$, i be just an index, x_i be the low-resolution input image and y_i be the high-resolution target image, $C \times H \times W$ be the dimensions of an image, and $|| \ldots ||$ be the Frobenius norm, L1 Loss is defined as

$$\ell_{l1}(\hat{y}, y) = \frac{1}{CHW} \sum_i ||\hat{y}_i - y_i|| \qquad (3)$$

L1 Loss minimises the mean absolute error between pixels and helps Meta-SRGAN match its output as closely as possible to the high-resolution target.

Adversarial Loss. The generator tries to minimise the following function whilst the discriminator tries to maximise it. $D(x)$ is the discriminator's output that

Discriminator:

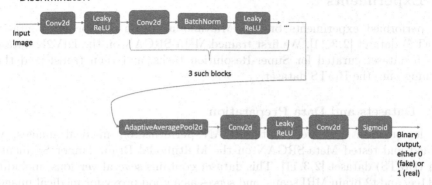

Fig. 5. Architecture of Discriminator. *Leaky ReLU* refers to the Leaky ReLU activation function. *AdaptiveAveragePool2d* refers to an average pooling layer, whilst *Sigmoid* refers to the sigmoid activation function that outputs a binary value.

the high-resolution image is high-resolution, $G(z)$ is the super-resolved image produced by the generator, and $D(G(z))$ is the discriminator's output that the super-resolved image is high-resolution.

$$E_x[log(D(x))] + E_z[log(1 - D(G(z)))] \qquad (4)$$

The above function is termed Adversarial Loss, and it allows Meta-SRGAN to achieve realistic super-resolved images.

Perceptual Loss. Meta-SRGAN also incorporates Perceptual Loss, a type of loss introduced by Johnson et al. to improve performance on Super-Resolution tasks [8]. The idea was to encourage the generated image to have similar feature representations as computed by a separate network. We use a 19-layer VGG network [14] pretrained on the ImageNet dataset [12], and we extract the features before the last layer of the network. Letting $\hat{y}$ and y be the super-resolved and high-resolution images respectively, ϕ be the network that extracts their feature representations, and assuming that the feature representations are of shape $C \times H \times W$, we define the Perceptual Loss as the Euclidean distance between the feature representations:

$$\ell_p(\hat{y}, y) = \frac{1}{CHW} ||\phi(\hat{y}) - \phi(y)||^2 \qquad (5)$$

In the following experiments we combined the loss functions as

$$\ell_{total}(\hat{y}, y) = \ell_{l1}(\hat{y}, y) + 0.001\ell_{adversarial}(\hat{y}, y) + 0.006\ell_p(\hat{y}, y) \qquad (6)$$

and use it to train Meta-SRGAN.

3   Experiments

We performed experiments on the Multimodal Brain Tumor Segmentation (BraTS) dataset [2,3,11]. We first trained Meta-SRGAN on the DIV2K dataset [1] - a dataset curated for Super-Resolution tasks, and then transferred that learning onto the BraTS dataset.

3.1   Datasets and Data Preparation

To demonstrate how well Meta-SRGAN performs on medical images, we trained and tested Meta-SRGAN on the Multimodal Brain Tumor Segmentation (BraTS) dataset [2,3,11]. This dataset contains several versions, including t1, t1ce, and t2 brain MRI scans, and serves as a good proxy for medical images. We picked the 2D t1ce version as it was two dimensional and was simple to work with. There were 15140 training images and 3784 validation images. Training images were normalised using a mean and standard deviation of 0.370 and 0.117 respectively before they were passed to the network to be trained on. These values were calculated using pixel values from the training images, after the maximal informational crop was applied. The BraTS dataset is tricky to deal with because every image has sparse information. Only about 50% of each image contained useful information (i.e. the brain), the rest was black. To tackle this problem, we cropped the image with the most pixel information. This corresponded to crops of the brain. Then, from this crop, we then randomly sampled 16 96 by 96 patches to provide sufficient coverage per image. For an upscaling factor of 2, the low-resolution input image (LR) was of size 48 by 48 whilst the high-resolution target image (HR) and super-resolved output image (SR) were of size 96 by 96. This workflow is summarised in Fig. 6.

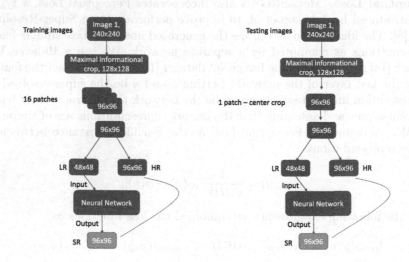

Fig. 6. Workflow for handling the BraTS dataset, for a x2 upsampling task.

3.2 Training Parameters and Experiment Settings

We first trained Meta-SRGAN on the DIV2K dataset for 360k updates, with a mini-batch size of 8. We then did transfer learning and trained Meta-SRGAN on the BraTS dataset for an additional 360k updates. We trained Meta-SRGAN on a range of scales from 1.1 to 4.0 (with 0.1 increments). Higher upscaling factors were not chosen as that would have made the input image too small to be useful. Each minibatch of images was associated with a scale, and each scale was chosen uniformly at random from 1.1 to 4.0. We optimised the network using a combination of L1 Loss, Adversarial Loss, and Perceptual Loss as described in the previous section. We used an Adam optimiser with a learning rate of $1 * 10^{-4}$ on each of the generator and discriminator, and decreased the learning rate by 20% every 60k updates. Pixel values were clamped to between 0 and 255 to give the generator an edge. The network was trained on a NVIDIA TITAN X GPU and took three days. We calculated the PSNR and SSIM scores only on the luminance channel of the maximal information crops.

4 Results

The results of our experiments are tabulated in Table 1. We see that Meta-SRGAN clearly outperformed the baseline of Bicubic Interpolation, which suggests that the inclusion of the Meta-Upscale Module did not hinder the network's performance. This is further reinforced by the fact that Meta-SRGAN was able to achieve similar performance on the x2 upsampling task as SRGAN, the architecture it was based on. It also had the added benefit of a smaller number of parameters compared to SRGAN. Visual quality is indicated by a proxy metric, Structural Similarity Index Metric (SSIM), which measures the perceived image similarity between two images. The high SSIM scores achieved by the GANs bodes well for brain MRI imaging which demands accurate images. We also see that Meta-SRGAN has the lowest number of parameters, which means its memory footprint is really small ($\approx$98% smaller than EDSR), which suggests that it can be deployed easily. Sampled images are shown in Fig. 7. The higher SSIM scores of SRGAN and Meta-SRGAN than the baseline of Bicubic Interpolation highlights the fact that they generate perceptually better images than the baseline. The very fact that Meta-SRGAN is even able to have comparable scores to

Table 1. A comparison of PSNR and SSIM scores between Bicubic Interpolation, SRGAN and Meta-SRGAN on the BraTS dataset. Best results are **bolded**.

x2 Upsampling task, BraTS testing set								
	Bicubic Interpolation		EDSR		SRGAN		Meta-SRGAN	
	PSNR (dB)	SSIM	PSNR (dB)	SSIM	PSNR (dB)	SSIM	PSNR (dB)	SSIM
Mean	17.98	0.6091	**18.93**	0.6885	18.42	**0.7295**	18.72	0.7279
Std	5.674	0.1708	4.2909	0.1814	5.4975	0.1308	5.338	0.1422
#parameters	–		40.7M		0.565M		0.561M	

EDSR despite being trained on a whole range of scales, and with considerably less parameters, attests to not just the power of adversarial training, but also the robustness of the Meta-Upscale Module.

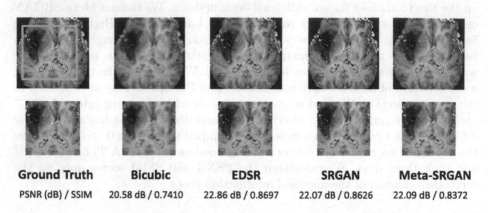

Ground Truth	Bicubic	EDSR	SRGAN	Meta-SRGAN
PSNR (dB) / SSIM	20.58 dB / 0.7410	22.86 dB / 0.8697	22.07 dB / 0.8626	22.09 dB / 0.8372

Fig. 7. Sample images from Bicubic Interpolation, EDSR, SRGAN and Meta-SRGAN. This is from a x2 upsampling task.

Meta-SRGAN was also able to produce a sequence of images corresponding to different upscaling factors. This is shown in Fig. 8. The relatively consistent PSNR and SSIM scores indicate that performance across different scales are not compromised. All these results suggests that we have a memory efficient network capable of generating super-resolved brain MRI images of high visual quality, and can be used to generate images of any scale.

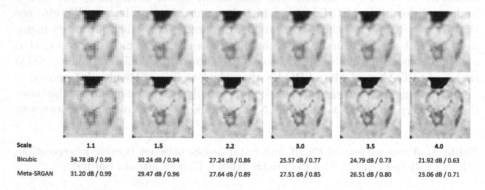

Scale	1.1	1.5	2.2	3.0	3.5	4.0
Bicubic	34.78 dB / 0.99	30.24 dB / 0.94	27.24 dB / 0.86	25.57 dB / 0.77	24.79 dB / 0.73	21.92 dB / 0.63
Meta-SRGAN	31.20 dB / 0.99	29.47 dB / 0.96	27.64 dB / 0.89	27.51 dB / 0.85	26.51 dB / 0.80	23.06 dB / 0.71

Fig. 8. (Top) Image going through various upsampling factors using Bicubic Interpolation. (Bottom) Image going through various upsampling factors using Meta-SRGAN. PSNR and SSIM scores are calculated against a target image downscaled by a factor of 4.

With better visual quality than the baseline, lower memory footprint, and ability to tackle any upscaling factor with negligible loss in performance, Meta-SRGAN clearly can deliver comparable performance to state-of-the-art Super-Resolution techniques.

5   Conclusions

The implications of the results on the BraTS dataset are three-fold. The low memory footprint of Meta-SRGAN enables it to be wrapped as a helper tool to aid medical tasks such as MRI and endoscope videography. The ability to super-resolve images on any arbitrary scale allows one to easily extend this to an application of real-time zooming and enhancing of an image, which will be useful in medical screening and surgical monitoring. The fact that its performance is not hindered across multiple upscaling factors affords it the flexibility to be used in other works such segmentation, de-noising, and registration. Despite only testing it on brain MRI images, Meta-SRGAN can easily be extended to datasets of other modalities such as Ultrasound and CT scans. Its low memory footprint, high visual quality, and generalisability suggests that Meta-SRGAN can become a new foundation on which other architectures can be built to enhance the performance for medical images.

We have built a network that combines the ability to generate images of high visual quality with the ability to tackle arbitrary scales, and are the first to show that this does not compromise performance or memory footprint for brain MRI images. This means that unlike other state-of-the-art methods, our method works on arbitrary scales which means that only a single network is required to perform Super-Resolution on any upscaling factors. In future work we hope to enhance the performance of Meta-SRGAN and apply it to other modalities.

Acknowledgements. Jin Zhu's PhD research is funded by China Scholarship Council (grant No.201708060173), whilst Pietro Lio' is supported by the GO-DS21 EU grant proposal.

References

1. Agustsson, E., Timofte, R.: Ntire 2017 challenge on single image super-resolution: dataset and study. In: The IEEE Conference on Computer Vision and Pattern Recognition (CVPR) Workshops, July 2017
2. Bakas, S., Akbari, H., Sotiras, A., Bilello, M., Rozycki, M., Kirby, J., et al.: Advancing the cancer genome atlas glioma MRI collections with expert segmentation labels and radiomic features. Sci. Data **4** (2017). https://doi.org/10.1038/sdata.2017.117
3. Bakas, S., Reyes, M., Jakab, A., Bauer, S., Rempfler, M., Crimi, A., et al.: Identifying the best machine learning algorithms for brain tumor segmentation, progression assessment, and overall survival prediction in the BRATS challenge. CoRR abs/1811.02629 (2018). http://arxiv.org/abs/1811.02629
4. Bing, X., Zhang, W., Zheng, L., Zhang, Y.: Medical image super resolution using improved generative adversarial networks. IEEE Access **7**, 145030–145038 (2019). https://doi.org/10.1109/ACCESS.2019.2944862

5. Chen, Y., Shi, F., Christodoulou, A.G., Zhou, Z., Xie, Y., Li, D.: Efficient and accurate MRI super-resolution using a generative adversarial network and 3d multi-level densely connected network. CoRR abs/1803.01417 (2018). http://arxiv.org/abs/1803.01417
6. Dong, C., Loy, C.C., He, K., Tang, X.: Image super-resolution using deep convolutional networks. CoRR abs/1501.00092 (2015). http://arxiv.org/abs/1501.00092
7. Hu, X., Mu, H., Zhang, X., Wang, Z., Tan, T., Sun, J.: Meta-sr: A magnification-arbitrary network for super-resolution. CoRR abs/1903.00875 (2019). http://arxiv.org/abs/1903.00875
8. Johnson, J., Alahi, A., Li, F.: Perceptual losses for real-time style transfer and super-resolution. CoRR abs/1603.08155 (2016). http://arxiv.org/abs/1603.08155
9. Ledig, C., et al.: Photo-realistic single image super-resolution using a generative adversarial network. CoRR abs/1609.04802 (2016). http://arxiv.org/abs/1609.04802
10. Lim, B., Son, S., Kim, H., Nah, S., Lee, K.M.: Enhanced deep residual networks for single image super-resolution. CoRR abs/1707.02921 (2017). http://arxiv.org/abs/1707.02921
11. Menze, B.H., Jakab, A., Bauer, S., Kalpathy-Cramer, J., Farahani, K., Kirby, J., et al.: The multimodal brain tumor image segmentation benchmark (BRATS). IEEE Trans. Med. Imaging 34(10), 1993–2024 (2015). https://doi.org/10.1109/TMI.2014.2377694
12. Russakovsky, O., et al.: Imagenet large scale visual recognition challenge. CoRR abs/1409.0575 (2014). http://arxiv.org/abs/1409.0575
13. Shi, W., et al.: Real-time single image and video super-resolution using an efficient sub-pixel convolutional neural network. CoRR abs/1609.05158 (2016). http://arxiv.org/abs/1609.05158
14. Simonyan, K., Zisserman, A.: Very deep convolutional networks for large-scale image recognition (2014)
15. Zhu, J., Yang, G., Lio, P.: How can we make gan perform better in single medical image super-resolution? A lesion focused multi-scale approach. In: 2019 IEEE 16th International Symposium on Biomedical Imaging (ISBI 2019), April 2019. https://doi.org/10.1109/isbi.2019.8759517

Knowledge-Based Fusion for Image Tampering Localization

Chryssanthi Iakovidou, Symeon Papadopoulos$^{(\boxtimes)}$, and Yiannis Kompatsiaris

Information Technologies Institute, Centre for Research and Technology Hellas,
6th km Harilaou - Thermi, 57001 Thessaloniki, Greece
{c.iakovidou,papadop,ikom}@iti.gr
https://mklab.iti.gr/

Abstract. In this paper we introduce a fusion framework for image tampering localization, that moves towards overcoming the limitation of available tools by allowing a synergistic analysis and multiperspective refinement of the final forensic report. The framework is designed to combine multiple state-of-the-art techniques by exploiting their complementarities so as to produce a single refined tampering localization output map. Extensive evaluation experiments of state-of-the-art methods on diverse datasets have resulted in a modular framework design where candidate methods go through a multi-criterion selection process to become part of the framework. Currently, this includes a set of five passive tampering localization methods for splicing localization on JPEG images. Our experimental findings on two different benchmark datasets showcase that the fused output achieves high performance and advanced interpretability by managing to leverage the correctly localized outputs of individual methods, and even detecting cases that were missed by all individual methods.

Keywords: Image tampering localization · Late-decision fusion ·
Passive image forensics

1 Introduction

Image forensics techniques have an important role in determining the authenticity of digital images. This is evident by the plethora of scientific approaches available in the literature that have carefully designed mechanisms to reveal different types of digital manipulations and traces that are expected to be generated during a given tampering process [10]. Producing robust tools for detecting and localizing a specific type of forgery has proven to be challenging, even when testing their effectiveness on benchmark datasets and controlled scenarios [19] and becomes even greater when dealing with real-world scenarios; images are being edited and manipulated in a variety of ways during a single forgery session in order to produce a convincing outcome, and are then forwarded and shared

© IFIP International Federation for Information Processing 2020
Published by Springer Nature Switzerland AG 2020
I. Maglogiannis et al. (Eds.): AIAI 2020, IFIP AICT 583, pp. 177–188, 2020.
https://doi.org/10.1007/978-3-030-49161-1_16

over the Internet, further undergoing transformations (e.g. cropping, re-sizing, re-compression). These uncontrolled factors inevitably force many of the standalone methods to suffer in terms of detection accuracy and localization robustness, presenting noisy outcomes and higher false positive rates when applied to new datasets [3, 9, 19]. Thus, researchers have come to realise that there is a true benefit in acquiring different reports from independent tools and evaluating the multiple clues in conjunction in the context of blind/passive image forensics (i.e. forensic analysis where prior knowledge regarding the original capturing circumstances, the manipulations or other post processing transformations is unknown) [7, 9, 12].

Several fusion approaches have been proposed in the literature, aiming at a synergistic analysis that improves the overall robustness and reliability of the forensic report. The different strategies can be roughly categorized based on the level at which fusion is carried out and the traces that are considered. Frameworks proposed for *feature-level fusion* often suffer from drawbacks related with selecting and handling a large number of features and scalability when adding new tools [3, 4], while approaches based on *measurement level fusion* are best suited for the tampering detection problem as they provide a more high-level response in terms of confidence for particular traces being present or not [7, 9]. On the other hand, *pixel-level fusion* is more effective for tampering localization and techniques proposed in this direction usually involve utilizing probability output maps and a fusion model to refine the final output and improve the localization of the tampered region [11, 12]. However, several important issues have not been comprehensively studied, for example, how to select the appropriate "base" forensics approaches, how to fuse their detection results, and how to refine the fused localization map.

In this paper, we introduce an extensible fusion framework for tampering localization and output refinement. The design strategy focuses on analyzing tampering localization approaches from the literature that are selected and categorized based on a multi-criterion ranking process integrating also expert background knowledge regarding their domain of application (types of images and encoding, supported traces, known limitations, etc.). Next we employ a fusion mechanism based on local and cross-tool statistics to produce a single, refined fused heat-map output for tampering localization. We primarily focus on splicing localization which is a very common and effective type of tampering that occurs when parts of the original image are replaced by alien content, while also prioritizing including methods that base their detection mechanisms on JPEG-related traces as it remains the dominant image codec for digital images in devices and on the Internet.

The main objectives of the delivered framework are: i) to exploit tools that are complementary to each other, such that the robustness and reliability of the overall localization system can be improved, and ii) communicate the tampering detection and localization results to end users in a manner that is easier to interpret compared to existing forensics approaches.

2 The Proposed Tampering Localization Fusion Framework

2.1 Selecting Tampering Localization Base Methods

In our effort to integrate different forensic approaches into a single framework, we begin by investigating the properties of state-of-the-art to assess what background information can benefit the fusion scheme. We first start by grouping candidate methods based on: (i) their known domain of application (type of tampering), (ii) their detection mechanisms (types of trace) and (iii) their reported performance (reliability of localization and readability of outputs). In order to limit the possible choices between methods, we primarily focus on splicing localization, a very common and effective type of tampering, and we prioritize passive methods that base their detection on analysis of the JPEG compression given its dominance on the Internet.

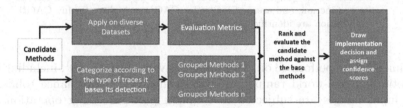

Fig. 1. Diagram of the selection process for localization methods as modules in the image forensics fusion framework.

Figure 1 depicts the general diagram of the selection process for including tampering localization methods as modules in our fusion framework. The various methods are organized in groups depending on the traces they detect, so that, when becoming part of the fusion framework, grouped methods can reinforce each other's results, while results deriving from different groups can be evaluated in a complementary fashion. In parallel, the candidate methods are undergoing a set of evaluation experiments on diverse datasets so as to assess their effectiveness in terms of tampering detection, localization and output readability. For that matter, we utilized the large volume of experiments we conducted in [19] and in [8] as a guide for the selection of the most effective methods. In [19] we evaluated 14 established state-of-the-art methods for image splicing localization that cover the full spectrum of known tampering traces. In [8] we extended the evaluations of seven techniques from [19] (Table 1, rows 1–7) and added a novel algorithm recently developed by us [8] (Table 1, row 8).

The evaluations concern i) the ability of a method to retrieve true positives of tampered images at a low level of false positives (KS@0.05); ii) the ability to achieve good localization of the tampered region within the image (F1); and iii) the readability of the produced heat map, i.e., a high distinction of assigned values for pixels belonging to tampered versus untampered regions, expressed as the range of different binarization thresholds that result in high F1 scores. The

Table 1. Image forensics methods considered as tampering detection modules.

Acronym	Description
DCT [17]	A simple fast detection method for inconsistencies in JPEG DCT coefficient histograms
ADQ1 [14]	Tampering localization by exploiting the characteristics of double DCT quantization
BLK [13]	Detection of disturbances of the JPEG 88 block grid in the spatial domain
NOI1 [16]	Modeling of the local image noise variance using wavelet filtering
NOI2 [15]	Modeling of the local image noise variance utilizing the properties of the Kurtosis of frequency sub-band coefficients in natural images
NOI3 [5]	Computes a local co-occurrence map of the quantized high-frequency component of the image and locates inconsistencies in local statistical properties
CFA1 [6]	Models the Color Filter Array interpolation patterns as a mixture of Gaussian distributions and locates tampering based on detected disturbances
CAGI [8]	Detects JPEG grid abnormalities in the spatial domain, taking into account the contents of the image. Multiple grid positions are evaluated with respect to a fitting function, and areas of lower (for CAGI) or higher (for inv-CAGI) contribution are identified as tampered

experiments were performed on three publicly available datasets[1], including both synthetic and real-world tampering cases while their performance robustness when input images are subjected to common post-processing operations was also investigated. Through these evaluations, summarised in Fig. 2, we were able to assess, rank and correlate their classification ability and overall performance over a wide spectrum of cases and conditions.

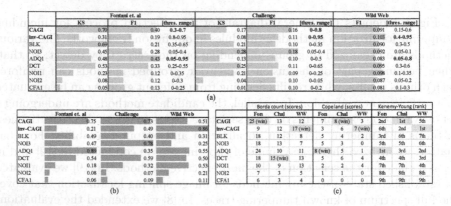

Fig. 2. Summary of results: a) performance: KS score, max F1 score; threshold binarization range, b) average performance of methods based on normalized KS, F1 and threshold range per dataset; c) rank aggregation results based on Borda count, Copeland and Kemeny-Young voting.

[1] The First IFS-TC Image Forensics Challenge training set [1]; the synthetic Fontani et al. dataset [7]; and the real cases of the Wild Web dataset [18].

Based on the evaluation results and taking also into account the selection principles described above, a set of five methods were selected as the "base" building blocks of the framework; these include: i) ADQ1 and DCT that both base their detection on analysis of the JPEG compression, in the transform domain; ii) BLK and CAGI that base their detection on analysis of the JPEG compression in the spatial domain; and iii) NOI3 that is a noise-based detector selected as a complementary tool mainly due to its high reported performance and the good interpretability of its produced outputs. Any new candidate method that will be considered for inclusion in the framework, will go through the same evaluation and grouping steps, being additionally ranked against the base methods on these multiple criteria, so as to decide whether it is expected to contribute to the fusion (include or not), how (in which group/what trace), and by how much (confidence weights based on ranks).

2.2 Fusion and Refinement of Tampering Localization Masks

The objective of designing a fusion framework is to improve the system's robustness and reliability. If one detector produces noisy or erroneous scores, having other detectors at hand makes it possible to complement, correct and refine the final localization. Figure 3 depicts the block diagram of the proposed fusion framework. For each input image I, we calculate a set of different tampering maps obtained according to the selected subset of detection methods M_k. Based on those, we formulate the fusion task as a labeling problem and we work towards denoting forged pixels with label "0" and authentic pixels with label "1".

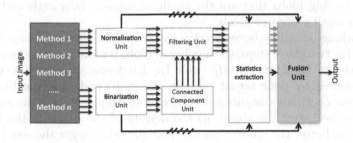

Fig. 3. Block diagram of the proposed fusion framework.

Normalization and Binarization Units: First, output maps are normalized in the $[0, 1]$ range at image level. Next, we are able to cost-effectively automate the binarization of the maps by choosing the appropriate binarization threshold as a value belonging to the respective *safe ranges* per method as these are determined through the analysis that was performed during the selection process (Fig. 2). The binarized maps allow easy analysis of their respective spatial and visual properties. We model as valuable and favor outputs that are easy to interpret visually; we are expecting useful maps to have well-defined tampered pixel areas that are spatially concentrated and form "blob–like" structures of significant size.

Connected Components Unit: For each binarized map M_k^b, we calculate the center of mass (i.e., centroid) for every 8-connected region that is marked as tampered:

$$R_c = 1/N \sum_{i=1}^{N} R_i \ and \ C_c = 1/N \sum_{i=1}^{N} C_i \qquad (1)$$

where (R_C, C_C) are the row and column coordinates of the centre of mass of the region under test, R_i, C_i are the i-th pixel coordinates of the region, i.e., matrix elements with zero value, and, N is the total number of pixels in the region. Next we build a feature vector describing the number of the detected connected regions, the location of their centroids (R_C, C_C), the spatial standard deviation of the pixels belonging to a region from their respective centroid, and the image area of each connected region expressed as the smallest possible rectangle (bounding box) containing the pattern of interest. Additionally, for each method, we produce maps of the connected components M_k^{cc}, where pixels belonging to each region (hereinafter referred to as blobs) are marked with unique labels.

Filtering Unit: The normalized maps, M_k^n, are forwarded to the Filtering Unit together with the outputs of the Connected Component Unit in order to filter the binary maps, M_k^b. Two types of filtering take place. First, we filter based on findings of each method independently from one another:

- Blobs that present bounding boxes of dimensions bigger than 50% or less that 5% of the image's largest dimension are automatically discarded. This contributes towards fast filtering of spurious, noisy and overall falsely detected results i.e. big blobs that are the result of densely, over-activated maps or isolated small groups of pixels.
- Blobs whose bounding boxes overlap by more than 90% are merged.
- If after the two above steps, the number of blobs is more than five, we calculate the Center of Mass for each M_k^n (as in Eq. 1 but now all pixels are considered and weighted by their actual value in the map) and rank the blobs based on i) their centroids distance from the overall Map centroid (the smaller the distance the better the score), ii) the density of the pixels in the blob (the denser the better the score), and iii) their size (the bigger the size the bigger the score). We then keep the top five based on their mean score in all three criteria.

Second, we perform a content-aware filtering step that depends on the particular methods. Utilizing the content annotation process implemented in CAGI [8] that provides information about areas that are expected to present no noise traces at all (i.e., over and under exposed areas) and the fact that DCT also outputs zero pixel map scores for image blocks of 8-by-8 pixels that share the same intensity value, we are able to filter blobs that may occur as false localizations in BLK and NOI3 outputs; BLK areas that lack any kind of grid pattern are considered tampered; for NOI3 complete lack of noise activates false alarms as they are recognized as inconsistencies in noise distributions. ADQ1 is not triggered by content and thus not affected by this filtering step.

Statistics Extraction: Finally, we extract statistics to automate the evaluation of the outputs' usefulness. These constitute an additional layer of confidence in selecting from the various intermediate maps the ones that are appropriate for use in the fusion step. We mainly rely on multilevel measurements of the entropy of the data. Image entropy is a quantity used to express the randomness of an image, computed by:

$$E = -\sum_i p_i \log_2 p_i \tag{2}$$

where p_i is the probability that the difference between two adjacent pixels is equal to i. Measuring the entropy of the visual output maps can give us an immediate rough measure of the interpretability of the result. Low entropy corresponds to clear distinction between foreground to background, while noisy outputs with values ranging over many areas will have high entropy. We calculate the following levels of entropy: i) the overall entropy of the normalized map (M^n), per method; ii) the entropy of its binarized counterpart (M^b), and iii) the entropy of each blob region against the entropy of the remaining image.

Additionally, we calculate the Kolmogorov-Smirnov (KS) statistic to compare the value distribution for the different regions of the maps (tampered and untampered) as follows:

$$KS = \max_u |C_1(u) - C_2(u)| \tag{3}$$

where $C_1(u)$ and $C_2(u)$ are the cumulative probability distributions inside and outside the mask, respectively. High KS values indicate that what we have marked as a blob of tampered pixels presents a very different distribution of values compared to the rest of the image.

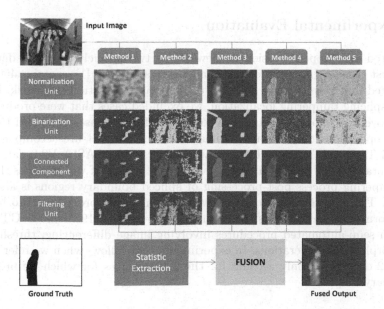

Fig. 4. Intermediate outcomes and final fused output of an example input image (taken from the Challenge dataset [1]).

Fusion Unit: Figure 4 showcases the intermediate steps of the fusion process. This leverages the intermediate calculations to produce a single fused output. To this end, a set of fusion quality properties have been defined:

- *Interpretability of the methods' localization maps:* Maps are ranked and assigned a confidence score, C_i, based on the difference of the entropy before and after the map's binarization.
- *Compatibility between the traces detected by the different methods:* Confidence of a method is reinforced if other methods detecting similar traces also achieve high confidence. Thus, if the C_i is high for more methods from the grouped set of tools (e.g., BLK/CAGI/NOI3 or ADQ1/DCT) the confidence score is boosted.
- *Reliability of the method as measured and assigned after performing extensive evaluations:* The reliability of the tools is also a factor for ranking. All methods are ranked to contribute based on their historical performance (Tables (b) and (c) in Fig. 2) as long as their outcome interpretability score surpasses a given threshold.
- *Confidence in the presence or absence of identified tampered regions:* For labeling regions as tampered or not, we also consider the original values of region pixels in the normalized tampering map. The KS statistic is calculated for regions belonging to blobs and background per method. The blobs with highest KS score of the best ranking method serves as our baseline detected tampered region. The refinement of the localization of the blobs is based on comparing it with the blob masks of the other methods in a ranked weighted order.

3   Experimental Evaluation

We tested our proposed fusion framework on two publicly available datasets. The First IFS-TC Image Forensics Challenge training set [1], contains 450 user-submitted forgeries and was designed to serve as a realistic benchmark. Focusing on splicing tampering localization, we excluded cases that were produced by copy-move operations resulting in a set of 306 forgery cases produced through spicing operations only. Tampered images in this dataset are accompanied by Ground Truth (GT) maps. The second dataset is the CASIA V2.0 dataset [2] that contains 5,123 realistically tampered color images of varying sizes. During the tampering process post-processing of spliced boundary regions is also considered. This dataset does not come with GT maps. In order for us to be able to perform localization tests, we manually produced 2,195 reliable GT maps through semi-automated procedures involving image differencing, thresholding and morphological operations. In experiments that follow, when we refer to the CASIA2 dataset we only account for the 2,195 images for which we produced GT binary maps[2].

---

[2] All produced GT maps are available upon request.

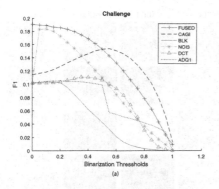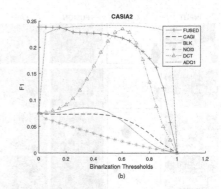

Fig. 5. F1 score curves on the (a) Challenge and (b) CASIA2 datasets for FUSED, CAGI, BLK, NOI3, DCT, ADQ1 of the tampering fusion framework

The overall localization quality and output readability is based on the pixelwise agreement between the reference mask (GT) and the produced tampering localization heat map and it is measured in terms of the achieved F-score (F1). This evaluation methodology requires the output maps to be thresholded prior to any evaluation. To this end, we first normalize all maps in the $[0, 1]$ range and proceed by successively shifting the binarization threshold by 0.05 increments, calculating the achieved F1 score for every step.

Figure 5 presents the mean F1 scores curves per binarizartion step over the Challenge and CASIA2 collection for the outputs of each individual method along with the fused output. The achieved localization is evaluated by the maximum mean F1 score for each method, at its respective best performing binarization threshold (Table 2).

Table 2. Best mean F1 score and binarization range that allows F1 to remain high ($> 70\%$ of respective maximum F1 score) and reported detections for F1 score $>= 0.7$ at each method's best binarization threshold for Challenge and CASIA2 datasets.

	Challenge				CASIA2			
Method	F1 score	Binarization range	Localizations	Unique localizations	F1 score	Binarization range	Localizations	Unique localizations
FUSED	0.191	0.0–0.4	37	7	0.238	0.0–0.6	490	88
CAGI	0.159	0.3–0.8	16	3	0.073	0.0–0.5	29	4
BLK	0.103	0.0–0.35	8	1	0.085	0.0–0.4	48	3
NOI3	0.183	0.05–0.3	38	15	0.075	0.0–0.05	66	29
ADQ1	0.109	0–0.5	4	1	0.241	0.05–0.9	371	22
DCT	0.105	0–0.65	5	0	0.234	0.55–0.65	488	30

As an indicator of a method's output interpretability we consider the range of the binarization threshold values, where the achieved F1 remains high (>0.7 of the best reported score). A wide range suggests that the tampered and untampered image regions are characterized by significantly different values in the

186   C. Iakovidou et al.

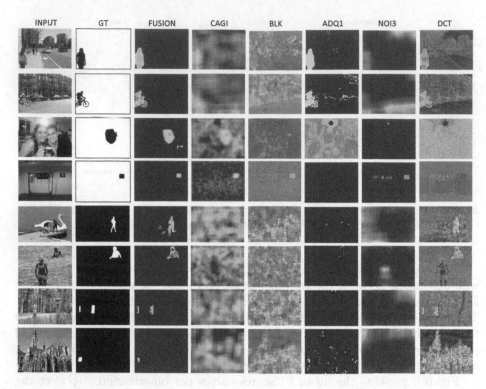

Fig. 6. Various examples of tampering localization outputs from the fusion framework and the individual methods; for the first four rows, images are taken from the Challenge dataset, for the last four rows images are taken from the CASIA2 dataset.

output maps making the respective heat map easy to interpret. Table 2 also reports the best localized detections achieved per method. The detection threshold was set to 0.7 and the search was performed for the best binarization step for each method. Unique Localizations corresponds to the number of detections exclusively achieved by that method.

From the experimental results in both datasets we can see that the fused output heat maps achieve high F1 scores over a wide range of thresholds. This verifies that the method produces outputs that exhibit increased localization ability and interpretability. In the Challenge dataset, the next best method (NOI3) achieves similar localization scores but is somewhat worse in terms of interpretability, while all other methods achieve significantly lower F1 scores. In CASIA2, the fusion method is the second best performing method in terms of F1 scores, while it still presents the best interpretability with F1 scores remaining high for a wider range of binarization steps. DCT, which is the leading method in this dataset, is significantly outperforming the rest of the individual methods, which is probably due to the tampering process followed in this specific dataset. The fusion framework manages to produce outputs that generally localize tampering better than most of the individual methods (Fig. 5(b)) but, in its current state, does not take full advantage of the very good DCT localizations in building its final output. Instead, while trying to construct hybrid outputs with low risk by collectively examining the various outputs and not heavily relying on only one method, the good DCT localizations were undermined by the many unsuccessful localizations of other methods. Motivated by these findings, assigning better weighting factors and ranking criteria will be at the heart of our next efforts.

Finally, in both datasets the fused method reports a high number of absolute localizations, which is indicating that the fusion criteria set in this framework manage to take advantage of the correctly localized outputs of the individual methods, and more importantly the framework contributes additional unique localizations through fusion and refinement, especially so in the CASIA2 database. Various localization outcomes are depicted in the Fig. 6.

Overall, this first set of experimental evaluations verifies the importance of exploiting the available state-of-the-art methods in a manner that improves the robustness and reliability of the system. In our next steps, we will continue to further test and refine the framework, while we also plan to introduce more localization methods in the system.

4   Conclusions

In this paper, we addressed the splicing tampering localization problem focusing on traces and methods that apply to JPEG images. To this end, we proposed an extensible tampering localization fusion and map refinement framework that combines multiple state-of-art techniques by exploiting their complementarities. We performed and took advantage of extensive evaluation experiments with the goal of selecting the most appropriate "base" methods to be fused so as to produce a single refined localization map outcome. Our experimental findings indicate that the fused output achieves high performance and interpretability by managing to exploit the correctly localized outputs of the individual methods while contributing with unique accurate tampering localizations. While we consider the results of our fusion approach promising, we also recognize the fact that the fusion is based on hard-coded expert knowledge that is directly implemented in the fusion criteria and rules. To this end, we plan to also investigate the potential of fusion approaches based on supervised learning.

Acknowledgements. This work was partially funded by the European Commission under contract numbers H2020-825297 WeVerify and H2020-700024 TENSOR.

References

1. Report on the IEEE-IFS Challenge. http://ifc.recod.ic.unicamp.br/. Accessed 20 Mar 2016
2. CASIA TIDE v2 (2009). http://forensics.idealtest.org/casiav2/. Accessed 20 Mar 2017

3. Barni, M., Costanzo, A.: A fuzzy approach to deal with uncertainty in image forensics. Signal Process. Image Commun. **27**(9), 998–1010 (2012)
4. Chetty, G., Singh, M.: Nonintrusive image tamper detection based on fuzzy fusion. Int. J. Comput. Sci. Netw. Secur. **10**(9), 86–90 (2010)
5. Cozzolino, D., Poggi, G., Verdoliva, L.: Splicebuster: a new blind image splicing detector. In: 2015 IEEE International Workshop on Information Forensics and Security (WIFS), pp. 1–6. IEEE (2015)
6. Ferrara, P., Bianchi, T., De Rosa, A., Piva, A.: Image forgery localization via fine-grained analysis of CFA artifacts. IEEE Trans. Inf. Forensics Secur. **7**(5), 1566–1577 (2012)
7. Fontani, M., Bianchi, T., De Rosa, A., Piva, A., Barni, M.: A framework for decision fusion in image forensics based on Dempster-Shafer theory of evidence. IEEE Trans. Inf. Forensics Secur. **8**(4), 593–607 (2013)
8. Iakovidou, C., Zampoglou, M., Papadopoulos, S., Kompatsiaris, Y.: Content-aware detection of JPEG grid inconsistencies for intuitive image forensics. J. Vis. Commun. Image Represent. **54**, 155–170 (2018)
9. Kaur, M., Gupta, S.: A fusion framework based on fuzzy integrals for passive-blind image tamper detection. Cluster Comput. **22**(5), 11363–11378 (2017). https://doi.org/10.1007/s10586-017-1393-3
10. Korus, P.: Digital image integrity-a survey of protection and verification techniques. Digit. Signal Process. **71**, 1–26 (2017)
11. Korus, P., Huang, J.: Multi-scale fusion for improved localization of malicious tampering in digital images. IEEE Trans. Image Proc. **25**(3), 1312–1326 (2016)
12. Li, H., Luo, W., Qiu, X., Huang, J.: Image forgery localization via integrating tampering possibility maps. IEEE Trans. Inf. Forensics Secur. **12**(5), 1240–1252 (2017)
13. Li, W., Yuan, Y., Yu, N.: Passive detection of doctored JPEG image via block artifact grid extraction. Signal Process. **89**(9), 1821–1829 (2009)
14. Lin, Z., He, J., Tang, X., Tang, C.K.: Fast, automatic and fine-grained tampered JPEG image detection via DCT coefficient analysis. Pattern Recogn. **42**(11), 2492–2501 (2009)
15. Lyu, S., Pan, X., Zhang, X.: Exposing region splicing forgeries with blind local noise estimation. Int. J. Comput. Vis. **110**(2), 202–221 (2014). https://doi.org/10.1007/s11263-013-0688-y
16. Mahdian, B., Saic, S.: Using noise inconsistencies for blind image forensics. Image Vis. Comput. **27**(10), 1497–1503 (2009)
17. Ye, S., Sun, Q., Chang, E.C.: Detecting digital image forgeries by measuring inconsistencies of blocking artifact. In: 2007 IEEE International Conference on Multimedia and Expo, pp. 12–15. IEEE (2007)
18. Zampoglou, M., Papadopoulos, S., Kompatsiaris, Y.: Detecting image splicing in the wild (web). In: IEEE International Conference on Multimedia & Expo Workshops (ICMEW), pp. 1–6. IEEE (2015)
19. Zampoglou, M., Papadopoulos, S., Kompatsiaris, Y.: Large-scale evaluation of splicing localization algorithms for web images. Multimed. Tools Appl. **76**(4), 4801–4834 (2016). https://doi.org/10.1007/s11042-016-3795-2

Transfer Learning Using Convolutional Neural Network Architectures for Brain Tumor Classification from MRI Images

Rayene Chelghoum[1]([✉]) [iD], Ameur Ikhlef[1] [iD], Amina Hameurlaine[1] [iD], and Sabir Jacquir[2] [iD]

[1] Frères Mentouri University, LARC, Laboratory of Automatic and Robotic, Constantine, Algeria
rayene.chelghoum@umc.edu.dz, ameikhlef@yahoo.fr, am.hameurlaine@gmail.com
[2] Université Paris-Saclay, CNRS, Institut des Neurosciences Paris Saclay, Gif-sur-Yvette, France
sabir.jacquir@u-psud.fr

Abstract. Brain tumor classification is very important in medical applications to develop an effective treatment. In this paper, we use brain contrast-enhanced magnetic resonance images (CE-MRI) benchmark dataset to classify three types of brain tumor (glioma, meningioma and pituitary). Due to the small number of training dataset, our classification systems evaluate deep transfer learning for feature extraction using nine deep pre-trained convolutional Neural Networks (CNNs) architectures. The objective of this study is to increase the classification accuracy, speed the training time and avoid the overfitting. In this work, we trained our architectures involved minimal pre-processing for three different epoch number in order to study its impact on classification performance and consuming time. In addition, the paper benefits acceptable results with small number of epoch in limited time. Our interpretations confirm that transfer learning provides reliable results in the case of small dataset. The proposed system outperforms the state-of-the-art methods and achieve 98.71% classification accuracy.

Keywords: Convolutional Neural Network · Brain tumor · Classification · Deep learning · Magnetic resonance images · Transfer learning

1 Introduction

Brain tumor diagnosis is very important in order to develop an effective plan of treatment. There are more than 120 types of brain and Central Nervous System (CNS) tumors. Neurologists classify manually the brain MR images using the World Health Organization (WHO) classification [1]. The Automation of the classification procedure, in particular brain MR images classification help radiologist in their diagnosis and reduce enormously their interventions.

© IFIP International Federation for Information Processing 2020
Published by Springer Nature Switzerland AG 2020
I. Maglogiannis et al. (Eds.): AIAI 2020, IFIP AICT 583, pp. 189–200, 2020.
https://doi.org/10.1007/978-3-030-49161-1_17

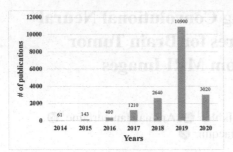 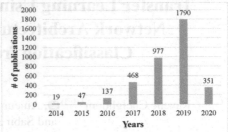

Fig. 1. Number of publications per year in google scholar containing "Convolutional Neural Network" and "medical imaging" keywords from 2014 to 2020. (Queried: February 25th, 2020)

Fig. 2. Number of publications per year in google scholar containing "Convolutional Neural Network" and "brain tumor" keywords from 2014 to 2020. (Queried: February 25th, 2020)

The first automatic classification methods are the machine learning ones. These methods take a long time because they need pre-processing and handcrafted features by experts. The classification accuracy depends on the extracted features which depend on the expert competences. Despite the limitation of machine learning methods, some works [2] achieved between 79% and 85% classification accuracy with their proposed method used tumor extracted features such as shape, rotation invariant texture, intensity characteristics and MR images for brain tumor classification.

To avoid handcrafted features extraction, deep learning (DL) methods involving deep neural networks to classify images in self-learning without the need of handcrafted features extraction are used by Benjio [3] and Litjens et al. [4]. Among several DL methods, CNNs are one of the most useful that have been used to solve complex problems in various applications such as detection [5], localization [6], segmentation [7] and classification [8]. They have also yielded good results in medical image application [9–11].

The first real-word application of CNNs was realized by Yann LeCun in 1998 to recognize hand written digits [12]. Since 2010, ImageNet launched an important visual database project called ImageNet Large Scale Visual Recognition Challenge (ILSVRC) [13]. This challenge runs an annual software contest where research teams evaluate their algorithms on the given dataset and achieved higher accuracy. Moreover, CNNs become more useful when Krizhevsky et al. (2012) proposed their CNN architecture called AlexNet [14]. The later competition allowed creating and improving real deep CNNs architectures that have achieved higher accuracy on several visual recognition tasks.

In recent years, CNNs have achieved good results in medical image applications due to the growth of available labelled training data, the increase of powerful graphics processing GPU, the rise of accuracy to solve complicated applications over time and the appearance of numerous techniques to learn features. According to the statistics that we made from google scholar, the number of publications using CNNs in the field of medical image applications in general and brain tumor in particular is increase since 2014. This trend can be observed in Fig. 1. and Fig. 2.

Today, several public brain MR images datasets for classification are available for researchers. This help medical scientists to develop more automated classification methods [15]. However, the CNNs training become more complicated and can lead to overfitting because of the samples size of medical datasets. Also, applying deep pre-trained CNNs based on transfer learning in medical imaging needs to adjust the hyper-parameters and learning parameters of the models in order to achieve a good result. Training the networks with transfer learning is usually much faster and easier than training the networks with randomly initialized weights [16–18]. In [19], authors used a small CNN architecture and achieved 84.19% classification accuracy. A block-wise fine-tuning has been proposed, this one, based on transfer learning, reached a classification accuracy of 94.82% [20]. Other approach proposed by [21] used pre-trained GoogleNet and transfer learning to classify brain MR images and achieved 97.1% of classification accuracy.

In this paper, we present an automatic classification system designed for three types of brain tumor. We use the brain CE-MRI dataset from figshare [22] which consists of three kind of brain tumors (glioma, meningioma and pituitary tumor) in order to classify only abnormal brain MR images. Based on this dataset, we adopted deep transfer learning for feature extraction from from brain MR images using nine deep CNNs architectures: AlexNet [14], GoogleNet [23], VGG16 [24], VGG19 [24], Residual Networks (ResNet18, ResNet50, ResNet101) [25], Residual Networks and Inception-v2 (ResNet-Inception-v2) [26], Squeeze and Excitation Network (SENet) [27]. This system makes easier the interventions of radiologist, helps them to solve brain tumor classification problem and develop an effective treatment.

We report the overall classification accuracy of the nine pre-trained architectures based on training time and epoch number. We explore the impact of epochs number to minimize the consuming time. We classify the extracted features for three different epochs. We achieve good results compared to related works. Also, with smaller number of epochs, we achieve acceptable results in short time.

The paper is structured as follows. The proposed method and different pre-trained CNNs architectures are given in Sect. 2. The experimental setting, the networks preparation and dataset are shown in Sect. 3. The experimental results with a brief discussion are provided in Sect. 3, the conclusion and an outlook for future work are given in Sect. 4.

2 Method

In this work, we applied nine pre-trained deep networks including AlexNet, GoogleNet, VGG16, VGG19, ResNet18, ResNet50, ResNet101, ResNet-Inception-v2 and SENet for brain tumor classification problem using transfer learning.

2.1 Pre-trained CNNs Architectures for Image Classification

CNNs architectures have been designed to learn spatial hierarchies of features by building multiple blocks: convolution layers with a set of filters, pooling layers, and fully connected layers (FCLs). The real deep architectures created until 2012 through ILSVRC challenge. The classification error of ILSVRC challenge winners is decreased from

15.3% in AlexNet (2012) [14] to 2.251% in SENet (2017) [27]. Also, the number of layers is increased from 8 layers to 152 layers. Table 1 summarizes the differences between those architectures regarding the classification error, the number of layers, the tasks, the execution environment and the training datasets.

ALexNet. AlexNet [14] architecture is deeper and much greater than LeNet architecture [28]. It consists of eight layers, five convolutional layers most of them are followed by max pooling and three fully connected layers. The output is the 1000-way softmax that represents the classes. It is trained on two parallel GTX 580 GPU 3 GB which communicate only in certain layers. This scheme reduces the top-5 error rates. AlexNet is improved with ZFNet architecture [29] which visualizes the AlexNet activities within the layers to debug problems and obtain better results. It allows observing the evolution

Table 1. Comparison between different CNNs architectures for image analysis.

ILSVRC architectures	Number of layers	Top 5 error rate	Tasks	Training dataset	Execution environment
AlexNet (2012) Ranked 1	8	15.3%	Classification	ImageNet	Two GTX 580 GPUs 3Gg (parallel)
ZFNet (2013) Ranked 1	8	14.8%	Classification	ImageNet Caltech-101 Caltech-256	Single GTX 580 GPU
GoogleNet (2014) Ranked 1	22	6.67%	Classification Detection	ImageNet	CPU
VGGNet (2014) Ranked 2	16–19	6.8%	Classification Localization	ImageNet	Four NVIDIA Titan Black GPU
ResNet (2015) Ranked 1	18–34– 50–101 **152**	3.57%	Classification Detection Segmentation Localization	ImageNet COCO	Two GPUs
Inception-v4/ResNet-inception (2016) Ranked 1	50–101 **152**	3.08%	Classification	ImageNet	Twenty replicas with NVIDIA Kepler GPU
SENet (2017) Ranked 1	18–34– 50–101 **152**	**2.251%**	Classification Detection	ImageNet –COCO CIFAR-10- CIFAR-100	Eight GPU NVIDIA Titan

of features during training and maps the activities back to the pixel space in intermediate layers.

GoogleNet. GoogleNet architecture codenamed Inception-v1 is the improved utilization of computing resources inside the network [23]. The network with the inception architecture is faster than the network with non-inception architecture. The GoogleNet architecture including the inception module uses rectified linear activation function, average pooling layer and not fully connected layer and dropout after removing fully connected layer.

Inception-v1 is improved to Inception–v2 by Ioffe and Szegedy [30] who tried to solve the internal covariate shift. They achieved a top-5 error rate of 4.82%. This result is outperformed to 3.5% by Szegedy *et al.* [31] with their new inception architecture called Inception-v3. Table 2 shows a comparison between the three inceptions.

Table 2. Difference between the three inceptions.

Inception-v1	Inception-v2	Inception-v3
Increase the number of units at each stage and shielding the large number of input filters of the last stage to the next layers	Increase the learning rate, remove dropout and local response normalization, shuffle training examples more thoroughly, reduce the L2 weight regularization and the photometric distortions	Trained much faster compared to the other inception and method
Error rate = 6.67%	Error rate = 4.82%	Error rate = 3.5%

VGGNet. Karen Simonyan and Andrew Zisserman [24] investigated the effect of the neural convolutional network depth on its accuracy in image recognition. They pushing depth to 11–19 weight layers of the developed VGGNet using very small (3×3) convolution filters. The configurations that use 16 and 19 weight layers, called VGG16 and VGG19 perform the best. The classification error decreases with the increased depth and saturated when the depth reached 19 layers. Authors confirm the importance of depth in visual representations.

ResNet/ Inception-v4. ResNet [25] used residual learning to ease the training of the deeper networks and reduce the errors from increasing depth. This architecture proposed many structures including: 18-layers, 34-layers, 50-layers, 101-layers and 152-layers structure, where the 152-layers structure is better than the other ones. It is less complex and deeper than VGG, and has similar performances to the Inception-v3 network, this is why Szegedy *et al.* [26] combined the inception architecture with residual connections. They evaluated the three ResNet-Inception and the Inception-v4 architectures: The Inception-ResNet-v1 has similar performances to Inception-v3 while the

ResNet-Inception-v2 performs more than ResNet-Inception-v1. The Inception-v4 is simpler and has more inception modules than Inception-v3 but has similar performances to ResNet-Inception-v2.

SENet. SENet [27] used Squeeze and Excitation (SE) block which improved the representational power of a network by enabling it to perform dynamic channel-wise feature recalibration. It was applied directly in the Residual Network architecture such as SE- Inception-ResNet-v2, SE-ResNet-101, SE-ResNet-50, SE-ResNet-152 and can be applied to the other existing architectures. It has been performed on ImageNet, COCO, CIFAR-10 and CIFAR-100 datasets across multiple tasks.

2.2 Transfer Learning Setting

Transfer learning use the gained knowledge that solve one problem and applied them to solve different related problems by using trained model to learn different set of data. The setting for transfer learning used in this work is explained in the following statements. The pre-trained CNNs architectures: AlexNet, GoogleNet, VGG16, VGG19, ResNet-18, ResNet-50, ResNet-101, ResNet-Inception-v2 and SENet consist of 1000 classes, 1.28 million training images, tested on 100 k test images and evaluated on 50 k validation images. They are challenging the accuracy of human with the best given results. The networks take an image as an input and produce the object label in the image as an output as well as the probabilities of the object categories. In this research, we focus on slice by slice classification of brain tumor using CE-MRI dataset into three types of tumors. First, we modified the last three layers of pre-trained networks in order to adapt them to our classification task. Next, we replaced the fully connected layer in the original pre-trained networks by another fully connected layers, in which the output size represents the three kind of tumor. The transfer learning setting and modification are shown in Fig. 3. Finally, we used transfer learned and fine-tuned deep pre-trained CNNs for experiments using MRI data.

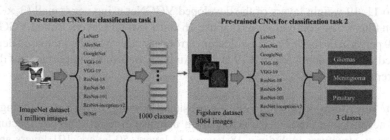

Fig. 3. Transfer learning setting and modification

3 Experiments and Results

The proposed classification model is implemented in MATLAB 2019b on a computer with the specifications of 16 GB RAM and Intel I9 4.50 GHz CPU.

3.1 Experiments

In this section, we describe the dataset used in the experiments, the training parameters and classification accuracy prediction. Figure 4 represents the pre-processing of training dataset and the use of transfer learning networks for brain tumor classification.

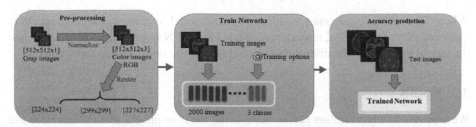

Fig. 4. Preparation and use the transfer learning network for brain tumor classification

Dataset and Pre-processing. The public database used to train and test the networks is available in [22]. It has already used in previous works like in [32, 33]. The dataset is collected from Nanfang Hospital, Guangzhou and General Hospital, Tianjin Medical University, in China during 2005–2010. It contains 3064 abnormal brain CE-MRI from 233 patients with three kinds of brain tumor: meningioma (708 slices), glioma (1426 slices), and pituitary tumor (930 slices). It is based on two dimensional gray images (2D slices). Those data are organized in MATLAB data format (. mat file). The size of images is 512×512 pixels and the pixel size is 49 mm x 49 mm. In our work, we normalize the gray MRI images in the dataset in intensity values and we convert them into RGB images by corresponding color map to RGB format using Matlab function. We specify the slices as an array of data type where the value 1 corresponds to the first color in the color map. RGB images are returned as an m $\times$ n $\times$ 3 numeric array with values in the range of [0, 1]. The value 3 corresponds to red, green and blue colors. Then, we resize them according to the used network: (227×227) in AlexNet and SENet, (224×224) in GoogleNet, VGGNet and ResNet, (299×299) in ResNet-Inception-v2 RGB images. The dataset pre-processing is shown on Fig. 4. We divide the data into training and test datasets, where 60% (1836 slices) of the images are used for training and 40% (1228 slices) used for test. The splitting of data into train and test set is performed on a slice basis.

Training Parameters. For transfer learning, we train the networks by stochastic gradient descent (SGD) with 0.9 momentum. We use a minibatch size of 128 images and a learning rate of 10^{-4}. To speed up the learning in the new layers, we rise the weight learn rate factor and the bias learn rate factor to 10. Even though, the transferred layers are still slower than the new layers. In order to perform the transfer learning, we train for 25, 50 and 90 epochs where an epoch is a full pass during the dataset training. The networks are validated every 50 iterations during training.

Classification Accuracy Prediction. In this part we use the trained networks to classify the test images and calculate the overall classification accuracy. The classification accuracy is the ability to predict correctly and guess the value of predicted attribute for new data. It is defined as the ratio of sum of true positives (TP) and true negatives (TN) to the total number of trials:

$$Accuracy = \frac{TP + TN}{TP + FP + FN + TN} \times 100 \qquad (1)$$

Where TP and TN are outcomes produced when the model correctly classifies the positive class and the negative class, respectively. While FP and FN are outcomes produced when the model incorrectly classifies the positive class and the negative class, respectively.

3.2 Results

We evaluate the classification performance using the nine pre-trained architectures and summarize our results in the form of tables.

In fact, the purpose of this study is to increase the classification accuracy, speed the training time and avoid the overfitting. This can be assessed through the classification accuracy and the training time of our pre-trained networks. The classification accuracy and the training time using different transfer learning architectures trained for different epochs are respectively shown in Table 3 and Table 4. All our pre-trained networks excepting SENet are reached up to 90% classification accuracy for three different epochs. Despite the use of transfer learning, SENet has an overfitting with epoch equal to 25 and 50, but achieves an acceptable result with epoch equal to 90.

Table 3. Classification accuracy using different transfer learning architectures for different epochs.

Architectures	Epoch = 25	Epoch = 50	Epoch = 90
AlexNet	**98.14**	**98.55**	98.22
GoogleNet	95.69	97.16	97.24
VGG-16	**98.06**	98.14	98.71
VGG-19	97.97	**98.55**	98.47
ResNet-18	96.01	97.86	97.81
ResNet-50	96.67	97.65	96.16
ResNet-101	96.67	96.83	95.99
ResNet-inception-v2	93.67	95.03	95.50
SENet	56.66	56.66	95.18

Another characteristic observed during experiments is the impact of epoch number on the classification accuracy. This effect can be seen in Table 3 and Table 4. The training time is increasing gradually with incremental epochs number, which means that we can consume less time using less epoch. However, the classification accuracy is neither influenced by the epochs number, nor the deep architectures. As shown in Table 3 the

Table 4. Training time in minutes using different transfer learning architectures for different epochs.

Architectures	Epoch = 25	Epoch = 50	Epoch = 90
AlexNet	**24**	**48**	**91**
GoogleNet	79	158	281
VGG-16	**495**	**907**	**1953**
VGG-19	532	**1174**	**1979**
ResNet-18	71	148	245
ResNet-50	190	374	678
ResNet-101	409	777	1339
ResNet-inception-v2	766	1481	2643
SENet	42	85	160

majority of the pre-trained networks record well with epoch equal to 50 and achieve acceptable results with epoch equal to 25. Even though, the fewer layers of AlexNet, VGG16 and VGG19 perform more than deeper architectures such as ResNet and ResNet-Inception-v2 for the three chosen epochs. They achieved respectively **98.55%**, **98.71%** and **98.55%** classification accuracy. Also, we observed that with an epoch of 50 AlexNet and VGG16 achieved the same accuracy **98.55%** however VGG16 consume a long training time compared to AlexNet.

Table 5. Samples images classification prediction using different architectures.

	Glioma	Meningioma	Pituitary
Sample Images			
AlexNet	**Glioma 100%**	Meningioma 99.9%	**Pituitary 100%**
GoogleNet	**Glioma 100%**	Meningioma 97%	Pituitary 99.8%
VGG-16	**Glioma 100%**	**Meningioma 100%**	**Pituitary 100%**
VGG-19	**Glioma 100%**	Meningioma 99.9%	**Pituitary 100%**
ResNet-18	**Glioma 100%**	Meningioma 99.9%	Pituitary 98.3%
ResNet-50	**Glioma 100%**	Meningioma 98.3%	Pituitary 98.9%
ResNet-101	**Glioma 100%**	Meningioma 99.2%	Pituitary 99.8%
ResNet-Inception-v2	Glioma 96.9%	Meningioma 93.5%	Meningioma 73.1%
SENet	**Glioma 100%**	Meningioma 96.1%	Pituitary 91.6%

Table 5 shows a classification of three sample instances. We find that all of the pre-rained architectures pertain to the class glioma and meningioma. All of them, excepting ResNet-Inception-v2 pertain to the class pituitary. This confirms that the deeper architectures do not result good with small datasets.

Table 6 provides a broad comparison based on classification accuracy with the existing methods on the same CE-MRI dataset. Abiwinanda *et al.* [19] achieved 84.19% accuracy with their proposed CNN. Swati *et al.* [20] propose a block-wise fine-tuning method based on transfer learning and achieved 94.82% accuracy. Dcepak and Ameer [21] used a pre-trained GoogleNet to extract features from brain MRI images and achieved 97.1% classification accuracy. Our proposed method using the pre-trained VGG16 achieved 98.71% classification accuracy.

Table 6. Related works & classification accuracy comparison using the CE-MRI training dataset.

Methods	Abiwinanda *et al.* (2019) [19]	Swati *et al.* (2019) [20]	Deepak and Ameer (2019) [21]	Proposed
Training Data	–	25–50–75%	56%	**60%**
Classification accuracy (%)	84.19	94.82	97.1	**98.71**

4   Conclusion

This paper presents a fully automatic system for three kind of brain tumor classification using CE-MRI dataset from figshare. The proposed system applied the concept of deep transfer learning using nine pre-trained architectures for brain MRI images classification trained for three epochs. Our system outperforms the classification accuracy compared to related works. It shows a good performance with a small number of training samples and small epochs number, which allows to reduce consuming time. The architectures which have fewer layers perform more than the deeper architectures. In the future work, we will apply our system to classify medical images from different modalities such as X-rays, Positron Emission Tomography (PET) and Computed Tomography (CT) for other body organ. Also, we will address the effect of epochs number to the classification performances.

References

1. Tustison, N.J., et al.: Optimal symmetric multimodal templates and concatenated random forests for supervised brain tumor segmentation (Simplified) with ANTsR. Neuroinform **13**(2), 209–225 (2015). https://doi.org/10.1007/s12021-014-9245-2
2. Zacharaki, E.I., et al.: Classification of brain tumor type and grade using MRI texture and shape in a machine learning scheme. Magn. Reson. Med. **62**(6), 1609–1618 (2009). https://doi.org/10.1002/mrm.22147

3. Bengio, Y.: Learning deep architectures for AI. Found. Trends® Mach. Learn. **2**(1), 1–127 (2009)
4. Litjens, G., et al.: A survey on deep learning in medical image analysis. Med. Image Anal. **42**, 60–88 (2017). https://doi.org/10.1016/j.media.2017.07.005
5. Dung, C.V., Anh, L.D.: Autonomous concrete crack detection using deep fully convolutional neural network. Autom. Constr. **99**, 52–58 (2019). https://doi.org/10.1016/j.autcon.2018.11.028
6. Long, C., Basharat, A., Hoogs, A.: A Coarse-to-fine Deep Convolutional Neural Network Framework for Frame Duplication Detection and Localization in Forged Videos, p. 10 (2018)
7. Nogovitsyn, N., et al.: Testing a deep convolutional neural network for automated hippocampus segmentation in a longitudinal sample of healthy participants. NeuroImage **197**, 589–597 (2019). https://doi.org/10.1016/j.neuroimage.2019.05.017
8. Merdivan, E., et al.: Image-based Text Classification using 2D Convolutional Neural Networks, p. 6 (2019)
9. Dou, Q., et al.: Automatic detection of cerebral microbleeds From MR images via 3D convolutional Neural networks. IEEE Trans. Med. Imaging **35**(5), 1182–1195 (2016). https://doi.org/10.1109/TMI.2016.2528129
10. Suk, H.-I., Wee, C.-Y., Lee, S.-W., Shen, D.: State-space model with deep learning for functional dynamics estimation in resting-state fMRI. NeuroImage **129**, 292–307 (2016). https://doi.org/10.1016/j.neuroimage.2016.01.005
11. Zhang, W., et al.: Deep convolutional neural networks for multi-modality isointense infant brain image segmentation. NeuroImage **108**, 214–224 (2015). https://doi.org/10.1016/j.neuroimage.2014.12.061
12. Lecun, Y.: Gradient-based learning applied to document recognition. Proc. IEEE **86**(11), 47 (1998)
13. Russakovsky, O., et al.: ImageNet Large Scale Visual Recognition Challenge. arXiv:1409.0575 [cs], September 2014
14. Krizhevsky, A., Sutskever, I., Hinton, G.E.: ImageNet classification with deep convolutional neural networks. In: Pereira, F., Burges, C.J.C., Bottou, L., Weinberger, K.Q. (eds.) Advances in Neural Information Processing Systems, vol. 25, pp. 1097–1105. Curran Associates, Inc. (2012)
15. Jui, S., et al.: Brain MRI tumor segmentation with 3D intracranial structure deformation features. IEEE Intell. Syst. **31**(2), 66–76 (2016). https://doi.org/10.1109/MIS.2015.93
16. Toğaçar, M., Ergen, B., Cömert, Z.: BrainMRNet: brain tumor detection using magnetic resonance images with a novel convolutional neural network model. Med. Hypotheses **134**, 109531 (2020). https://doi.org/10.1016/j.mehy.2019.109531
17. Sharif, M.I., Li, J.P., Khan, M.A., Saleem, M.A.: Active deep neural network features selection for segmentation and recognition of brain tumors using MRI images. Pattern Recogn. Lett. **129**, 181–189 (2020). https://doi.org/10.1016/j.patrec.2019.11.019
18. Bernal, J., et al.: Deep convolutional neural networks for brain image analysis on magnetic resonance imaging: a review. Artif. Intell. Med. **95**, 64–81 (2019). https://doi.org/10.1016/j.artmed.2018.08.008
19. Abiwinanda, N., Hanif, M., Hesaputra, S.T., Handayani, A., Mengko, T.R.: Brain tumor classification using convolutional neural network. World Congress Med. Phys. Biomed. Eng. **2019**, 183–189 (2018)
20. Swati, Z.N.K., et al.: Brain tumor classification for MR images using transfer learning and fine-tuning. Comput. Med. Graph. **75**, 34–46 (2019). https://doi.org/10.1016/j.compmedimag.2019.05.001
21. Deepak, S., Ameer, P.M.: Brain tumor classification using deep CNN features via transfer learning. Comput. Biol. Med. **111**, 103345 (2019). https://doi.org/10.1016/j.compbiomed.2019.103345

22. Brain tumor dataset. https://figshare.com/articles/brain_tumor_dataset/1512427. Accessed 17 Feb 2020
23. Szegedy, C., et al.: Going Deeper with Convolutions. arXiv:1409.4842 [cs], September 2014
24. Simonyan, K., Zisserman, A.: Very Deep Convolutional Networks for Large-Scale Image Recognition, arXiv:1409.1556 [cs], September 2014
25. He, K., Zhang, X., Ren, S., Sun, J.: Deep Residual Learning for Image Recognition, arXiv: 1512.03385 [cs], December 2015
26. Szegedy, C., Ioffe, S., Vanhoucke, V., Alemi, A.: Inception-v4, Inception-ResNet and the Impact of Residual Connections on Learning, arXiv:1602.07261 [cs], February 2016
27. Hu, J., Shen, L., Albanie, S., Sun, G., Wu, E.: Squeeze-and-Excitation Networks, arXiv:1709. 01507 [cs], September 2017
28. Handwritten digit recognition with a back-propagation network | Advances in neural information processing systems 2. https://dl.acm.org/doi/10.5555/109230.109279. Accessed 17 Feb 2020
29. Zeiler, M.D., Fergus, R.: Visualizing and Understanding Convolutional Networks, arXiv: 1311.2901 [cs], November 2013
30. Ioffe, S., Szegedy, C.: Batch Normalization: Accelerating Deep Network Training by Reducing Internal Covariate Shift, arXiv:1502.03167 [cs], February 2015
31. Szegedy, C., Vanhoucke, V., Ioffe, S., Shlens, J., Wojna, Z.: Rethinking the Inception Architecture for Computer Vision, arXiv:1512.00567 [cs], December 2015
32. Cheng, J., et al.: Enhanced performance of brain tumor classification via tumor region augmentation and partition. PLoS ONE 10(10), e0140381 (2015). https://doi.org/10.1371/journal. pone.0140381
33. Retrieval of Brain Tumors by Adaptive Spatial Pooling and Fisher Vector Representation. https://journals.plos.org/plosone/article?id=10.1371/journal.pone.0157112. Accessed 17 Feb 2020

Learning Algorithms

A Novel Learning Automata-Based Strategy to Generate Melodies from Chordal Inputs

I. Helmy and B. John Oommen[✉]

School of Computer Science, Carleton University, Ottawa K1S 5B6, Canada
ibyhelmy@gmail.com, oommen@scs.carleton.ca

Abstract. This paper deals with the automated composition of music. Although music within the field of AI has been studied relatively extensively, the arena in which we operate is, to the best of our knowledge, unexplored. When it concerns computer composition, a noteworthy piece of research has involved the automated generation of the chordal notes when the underlying melody is specified. The chordal notes and beats have been generated based on models of well-known composers like Bach, Beethoven, and so on. The problem we study is the converse. We assume that the system is provided with the chords of some unknown melody. Our task is to generate a melody that flows with the given chords and which is also aesthetically and musically fitting. As far as we know, there is no research that has been reported to solve this problem and in that sense, although we have merely provided "baby steps", our work is pioneering.

Keywords: Learning Automata (LA) · Learning systems · Music composition · Markov chain model · Markov property

1 Introduction

From the time-immemorial theory of music, there has been a well-studied pattern for generating fundamental chords for a specific melody. This depends on the key in which the song is played, and the basic set of chords associated with this key are well documented. Thus, for a song in a particular key, for the most part, simplistic chordal arrangements can be generated, using a relatively small subset of chords associated with the key[1].

The problem that we consider is of a much higher dimension. The reader should observe that a sequence of chordal formations could, in turn, be suitable

---

[1] This is, of course, a very simplistic model. Indeed, the "majesty" of the composer is displayed by the wealth and spectrum of the chords that he uses.

B. John Oommen—Chancellor's Professor; Life Fellow: IEEE and Fellow: IAPR. This author is also an *Adjunct Professor* with the University of Agder in Grimstad, Norway.

© IFIP International Federation for Information Processing 2020
Published by Springer Nature Switzerland AG 2020
I. Maglogiannis et al. (Eds.): AIAI 2020, IFIP AICT 583, pp. 203–215, 2020.
https://doi.org/10.1007/978-3-030-49161-1_18

for multiple melodies. Thus, even though the key of a song is known, and the chordal arrangements are specified, these chordal arrangements could fit to blend well for a myriad of underlying melodies. In other words, unlike the work that has been reported in the literature, our goal, which is the converse of the prior art, is to automatically generate melodies for chordal formations and sequences.

As human beings, we typically associate a musical piece with its primary melody. For example, while Mozart's clarinet concerto is accompanied by the orchestra, the melody is primarily played by the clarinet. Clearly, the problem that we study is fascinating, because we are attempting to generate a primary melodic tune that can be associated with the accompanying chordal music. This is by no means trivial, and even if we do accomplish such a goal, it will be the task of the musically-minded ear to decide whether the melody that has been generated is aesthetically beautiful. Although determining this is beyond the scope of our research, we have developed a AI-based system to achieve this.

AI has numerous tools such as problem solving, searching, Neural Networks (NNs), classification, and many more. The tool that we will use to achieve our goal forms the skeleton of reinforcement learning and is known as Learning Automata (LA). The beauty of using LA to solve this problem is that the action that the LA takes is chosen stochastically. The goal, of course, is to minimize a certain penalty function. If the actions of the LA are used to represent the musical notes, the output can be the melody sought for. Further, if the "Teacher" to the LA is, in some way, governed by the input chordal pattern, we can envision a scheme by which the output (the composed melody) fits in line with the provided chords. We believe that we have achieved this, our goal, to some extent[2].

1.1   Prior Art for Automated Music Composition

The two earliest notable works done on automated music composition appeared in 1988. The first, due to Lewis, [5], created an multi-layer Perceptron NN trained to form a certain desired mapping using a gradient-descent scheme. The second approach by Todd [4] used an auto-regressive NN to generate music sequentially. The computations of both these systems were limited by the hardware of that generation. Consequently, neither of these works could really succeed in taking their research any further. A common problem characterized by NN-composed music is that it often lack structure. The music generated in this way is limited in scope, and it thus does not possess a method by which it can represent the musical sequences created a few iterations before the current note.

The next notable work was done by Eck and Schmidhuber [6]. Their work aimed to address the lack of structure that the previous works suffered from. They proposed the Long Short-Term Memory (LSTM) algorithm, which was successful in the composition of well-structured and timed music that did not sway from its desired skeleton. Further, Eck who leads an AI team at Google under its *Google Brain*, also worked on the Magenta Project [7], which aims to

---

[2] A version of this system can be currently demonstrated and can be made available to the reader.

analyze the roles of machine learning in different aspects of the artistic world. Among the developments that emerged from the Magenta Project are Magenta Studio, which works with plugins and tools to be utilized with other music composition programs to generate and modify music. The Magenta project should also be credited for MusicVAE [8], which is a tool that takes two musical melodies and blends them together.

1.2 Learning Automata (LA)

We now briefly review[3] the field that we shall work in, namely, that of LA. The concept of LA was first introduced in the early 1960's in the pioneering by Tsetlin [22]. He proposed a computational learning scheme that can be used to learn from a random (or stochastic) *Environment* which offers a set of actions for the automaton to choose from. The automaton's goal is to pick the best action that maximizes the average *Reward* received from the *Environment* and minimizes the average *Penalty*. The evaluation is based on a function that permits the Environment to stochastically measure how good an action is, and to thereafter, send an appropriate feedback signal to the automaton.

After the introduction of LA, different structures of LA, such as the deterministic and the stochastic schemes, were introduced and studied by the famous researchers Tsetlin, Krinsky and Krylov in [22], and Varshavskii *et al.* in [23].

LA, like many of the Reinforcement Learning techniques, has been used in a variety of problems, mainly optimization problems, and in many fields of AI. Among other applications, it has been used in NN adaptation [11], solving communication and networking problems [12,14,16] and [17], in distributed scheduling problems [19], and in the training of Hidden Markov Models [10].

Figure 1 displays the general stochastic learning model associated with LA. The components of the model are the *Random Environment* (RE), the *Automaton*, the *Feedback* received from the RE and the *Actions* chosen by the LA.

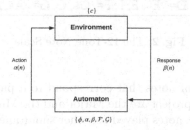

Fig. 1. The Automaton-Environment Feedback Loop.

The Environment: At every time instant, the LA picks an action from the set of actions available to it and communicates it to the RE. The RE evaluates that action as per a random function, and responds with a *Reward* or a *Penalty*.

---

[3] We will not go through irrelevant details and/or the proofs of the LA-related claims. They can be found in [24] and the reference book by the pioneers [13].

The Automaton: The LA achieves its learning process as an iterative operation that is based on the RE and their mutual interaction. The RE first *evaluates* the policy, which is how it appraises the selected action. It then undertakes *policy improvement*, by which it modifies the probability of selecting actions so as to maximize the *Reward* received from the RE.

Estimator and Pursuit Algorithms: The concept of Estimator Algorithms was introduced by Thathachar and Sastry in [20,21] when they realized that the family of Absolutely Expedient algorithms would be absorbing, and that they possessed a small probability of not converging to the best action. Estimator algorithms were initially based on Maximum Likelihood estimates (and later on Bayesian Estimates). By doing this, the LA converged faster to the actions that possessed the higher reward estimates. The original Pursuit Algorithms are the simplest versions of those using the Estimator paradigm introduced by Thathachar and Sastry in [20], where they *pursued* the current best-known action based on the corresponding reward estimates. The concept of designing ϵ-optimal discretized Pursuit LA was introduced by Oommen and Lanctot in [15].

2   Fundamental Concepts of Music Theory

Before delving into the details and structure of the design and creation of our AI module, in the interest of completeness, we briefly state some of the fundamental concepts[4] of the Theory of Music [1]. The most elemental term is that of a **note**, which is the most primitive unit that music is composed of. More specifically, a note is a symbol to represent sounds or pitches. Music is composed of notes, and in music, there are a collection of 12 notes in what is known as the **12-Tone Note Scale** depicted in Fig. 2 below.

| C, C# | Db, D, D# | Eb, E, F, F# | Gb, G, G# | Ab, A, A# | Bb, B |

Fig. 2. The 12-Tone Note Scale

A **scale** is a subset of notes that correspond to a particular pitch. The two main scales used in this project are the Major and the Minor scales. A **chord** is a collection of two or more notes played together simultaneously. In the interest of simplicity and consistency, within the scope of this project, all referenced chords will be assumed to be triads, where a **triad** chord is one made up of exactly three notes. Thus, an "A Major" triad is composed of the notes: A, C Sharp, E.

The main difference between the two scales is the pitch, and this marginally affects the notes that constitute the chords being used. Thus, in the context of the above example for the "A Major chord", one observes that it is made up

---

[4] The terms and concepts which concern music are very fundamental, and are really included only in the interest of completeness.

of the notes: A, C Sharp and E. If one starts from the root note 'A', a Major chord moves 4 notes (3 to the right and then back again at the start to C) down the scale to the C Sharp note, and then 4 more notes to 'E' that completes the chord's formation. The "A Minor" chord, however, is composed of the notes: A, C and E. Here the middle note is increased by 3 notes rather than 4, which is essentially the difference between the Major and Minor chords. Finally, a **bar** defines a segment of music based on the tempo (number of beats per bar). Throughout the project, we assume a tempo of 120 bpm (beats per minute). Hence the bars will be 8 beats long, denoted as (4, 4) musically.

2.1   The Circle of Fifths

Our goal is to have an AI program learn to create melodies for chord progressions that are musically coherent. This means that through studying the input music, it will be able to determine which chords typically come after others, and hence form satisfying sequences. The tool that aids musicians in this task is known as the "Circle of Fifths", where as displayed in the figure below, the chords are all placed side-by-side in a circular fashion [18] (Fig. 3).

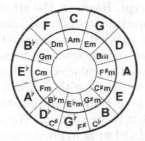

Fig. 3. A diagram displaying the Circle of Fifths

The circle creates a mapping for 12 chords ordered in such a way that when deciding what chord should follow the current one, one can moving down the circle in either direction to infer which chord will provide the best sounding match. The further away that a chord is in the Circle, the less likely the chord switch will satisfy the melody, which is central to creating cohesion in musical pieces. However, it is sometimes used as an artistic tool to generate a sense of "discomfort", and the use of such a device is left to the discretion of the artist.

2.2   Keys

Keys are a collection of chords that all fit within the same tonal basis as each other. Throughout this project, we will only consider melodies and chords within the Key of C Major. The triad chords in the key are as follows [2]:

- i – C major, (C)
- ii – D minor, (Dm)
- iii – E minor, (Em)
- iv – F major, (F)

- v – G major, (G)
- vi – A minor, (Am)
- vii – B diminished, (B)

2.3 Assumptions

Since we are attempting a *prima facie* solution, we make a few simplifying assumptions and restrictions namely, that the key in which the music will be handled and created is the Key of C Major, and that all the chords in the progressions are exactly *one* bar long. We also assume that each bar will be limited to exactly *four* notes, and that are all of even duration as well.

3   The Markov Chain

Fundamentally, music is based on patterns, and more specifically, patterns of sounds that mimic emotions and particular feelings. This is what renders music to seem to be uniquely human. To formalize how we could teach this phenomenon to an AI program, our hypothesis is that we have to relay the non-randomness aspects of "music" as a concept. However, the art of *writing* music is, in itself, a connected and partly dependant journey, implying that any single part of a song could be heavily reliant on what was composed earlier. However, one must still retain an important sense of independence between the notes and sections. Our model is that structure can be enforced or achieved in a formulaic manner, but the flow of music has to be taught with a certain sense of spontaneity. Thus, we consider stepping away from traditional progressions every now and then, thus adding to the "spice" that music contains. We achieve this by modeling the process using a Markovian Model so as to provide a mix of randomness which, in turn, is based on the probability aspects of the music that has been established.

A Markov chain is a mathematical system built to model statistical probabilities of random processes. These processes move from one state to another while satisfying a particular property, which states that the probability of a particular transition from any state to another is completely independent of any past sequence of states that preceded it. This is particularly useful in Computer Science, since there would then be no need to store large amount of information.

A Markov model follows the following formula:

$$Pr(X_{n+1} = a | X_1 = x_1, X_2 = x_2, \ldots, X_n = x_n) = Pr(X_{n+1} = a | X_n = x_n).$$

This specifically deals with a sequence of previous states $X_1, X_2, \ldots, X_n$, where the current state is X_{n+1}. The formula demonstrates that moving from one of these states to another state is solely dependent on the most recent state, since $Pr(X_{n+1})$ is only based on of the probability of moving from the state X_n.

Figure 4 depicts a state diagram for three chords; C major, D minor, E minor.

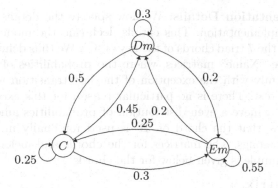

Fig. 4. An example of a Markov Chain for a set of three chords

Here, the transition matrix is a 3×3 matrix, where the chords on the left are the current chords, and the chords on the top are those to be transitioned to:

	C	Dm	Em
C	0.25	0.45	0.3
Dm	0.5	0.3	0.2
Em	0.25	0.2	0.55

3.1 The Model for Chords and Melodies

The probabilities of the state transitions are maintained in a stochastic matrix, the *transition matrix*, whose row sums are unity. Indeed, one can observe that from any given chord the possible state transitions sum up to unity.

Because there are 7 chords within the key of C, the matrix is a 7×7 matrix, and the $[i, j]$ entry is the probability of switching from chord 'i' to 'j'.

To help with the book-keeping, we keep track of which changes are mapped, and to the corresponding chords. To do this, we keep an identical matrix, the "Name" matrix, which contains the two chord symbols rather than the probabilities at the same entry. For example, in the transition matrix, the entry [0,1] has probability 0.4. In the "Name" matrix, entry [0, 1] contains the string "CDm" implying that chord changes C to Dm occurred with probability 0.4.

Analogously, we maintain a similar matrix for the melodies. However, the novelty of our strategy is that instead of mapping the notes to each other, like the chords, the notes are mapped to the *chords* being played. Modeling the matrix in this way allows the AI module to possess a greater control over the notes being played. This matrix is, therefore, also an 7×7 matrix where each chord contains the probabilities that a note is played alongside it in a song. Observe that as per the model, a random *note* is chosen based on that particular probability distribution depending on the *chord* selected.

Code Implementation Details: We now specify the details of the Markov Model's code implementation. This entails declaring the model's entire state-space, which are the 7 triad chords of the key of 'C'. We thus define the transition matrices and the "Name" matrices, where the probabilities of the former are initialized uniformly with the exception of the last transition which is slightly higher than the rest. There is no particular reason for this except for the fact that one cannot achieve a even distribution of probabilities among 7 numbers. Besides, we know that the chord of 'B' is used minimally in the key of 'C', especially in pop songs. The matrices for the chords and melodies are similar, and a typical example is given below for the chords.

```
transitionMatrix =
[[0.14, 0.14, 0.14, 0.14, 0.14, 0.14, 0.16],
 [0.14, 0.14, 0.14, 0.14, 0.14, 0.14, 0.16],
 ...
 [0.14, 0.14, 0.14, 0.14, 0.14, 0.14, 0.16],
 [0.14, 0.14, 0.14, 0.14, 0.14, 0.14, 0.16]].
transitionName =
[['CC', 'CDm', 'CEm', 'CF', 'CG', 'CAm', 'CB'],
 ['DmC', 'DmDm', 'DmEm', 'DmF', 'DmG', 'DmAm', 'DmB'],
 ...
 ['AmC', 'AmDm', 'AmEm', 'AmF', 'AmG', 'AmAm', 'AmB'],
 ['BC', 'BDm', 'BEm', 'BF', 'BG', 'BAm', 'BB']].
```

We also need the imports relevant to the Markov Model, namely, numpy, the package for scientific computing in Python, as "np", and random as "rm".

The helper functions (and their tasks) that we define are summarized below:

- load(*str*): This function takes a song file name and loads it into the program;
- getTransIndex(*str, matrix*): Gets the index of a certain transition from the transition matrix passed in;
- getShorthand(*chord*): Returns a list of the notes contained within a *chord*;
- adjustMatrix(*change, degree*): Adjusts the transition matrix probability for a certain chord *change* by a certain *degree* (percentage);
- adjustMelodyMatrix(*melodyDict*): For every chord in the dictionary, this function adjusts the probabilities of the notes played over it;
- convertNotes(*notes*): Takes the data from the MIDI files and converts them into notes which are to be used in the learning process;
- convertChords(*chordz*): Takes the data from the MIDI files and converts them into chords to be used in the learning process;
- normalize(): Normalizes the values in the transition matrix to ensure that it retains its stochastic property after any adjustments;
- formVerse(*bars*): This function yields a verse structure to be filled with chords, where a sequence of *bars* is passed in;
- formChorus(*bars*): The function yields a chorus structure to be filled with chords with a sequence of *bars* passed in.

The Chord Creator: The final part of the code to be discussed is the function chordCreator(*bars, sChord*). This is the module where the Markov Model itself is implemented for the chord progressions. To initialize the module, we specify the starting chord selected based on the input passed into the function. This, is, in turn, displayed to the user. This initializes an empty list containing only the starting chord. Thereafter, we create a counter used to track how many bars have been created so far. The largest chunk of the function deals with the selection of the next chord. Using the numpy and random libraries, the next chord is selected based on the Markov Chain, and added to the list of chords. The function used to construct the chords is:

np.random.choice(a, size=None, replace=True, p=None)

The above function takes 4 parameters. The first, 'a', is a list of possible elements from which we can generate a random sample. The second parameter defines the shape of the output, if needed. The next is a boolean variable, labeled 'replace', which accounts for the fact whether the sample is generated "with" or "without" replacement. The final parameter is the quantity 'p', which specifies the probabilities associated with each entry in the chain. In our case these parameters are as follows: We pass the list of chord transitions for that particular starting chord as the parameter 'a', we place no size restriction, we set the variable 'replace' to be *True*, and we finally set 'p' to be the chain's transition matrix.

The code is designed to handle 7 individual cases, one for each possible current starting chord. For each of the possibilities, the random choice function (discussed above) decides on what chord the system transitions into. Once this is done, the new chord is set as the starting chord and appended to the list of chords contained within the chord progression so far (i.e., chordList).

After the next chord in the progression is decided on, it is set as the new starting chord, and the counter is incremented. Once the number of bars is completed and the while loop terminates, the program yields as its output a chord progression along with a probability value of that specific chord sequence.

The code for the "Melody Creator", melodyCreator(), is quite similar. However, for each chord, the program selects four notes to be played. This is due to the previously-stated assumption that each note lasts for a quarter of a bar, and that a chord is played for a whole bar. Again, the np.random.choice function is used to determine the note that is played. Once verified, the note is added to the list of notes, and once the code terminates, the function returns the list of notes played, and the associated probability of the created melody.

4   The Learning Process

We execute reinforcement learning by means of a LA specifically constructed for this purpose. The actions that are chosen as based as per a specific probability distribution, which in our case is the Markov Model previously described. In our case, however, we deal with two parallel LA, the first of which deals with the learning involved with the chords, the second with the melodies. The interactions between the LA and the RE (the music-related data) is explained below.

4.1 Learning the Chords

To learn to generate chord progressions, the AI program is fed with the appropriate form of data. To achieve this, we went through a list of songs in the key of C major. After selecting the songs, we invoke the website ChordsWorld.com, to retrieve the chords used for each song. Once the chords for each song are collected, they are placed into .txt files, where the chords are separated by tabs. For the parsing of the chords data, we use a "load" function which opens the file and parses it based on the tab spacing. Once the file is parsed, the list of chords is returned by the function to be used in the learning process.

Before proceeding, the code examines the two cases where the current chord or the next chord are both not those commonly used for the key of C major. If this occurs, the code skips these chords and resumes the learning. With every chord change made, we resort to a call to the adjustMatrix function, after which the chord is added to the list of chords. After all the adjustments are made, we normalize the rows so as to maintain the stochastic property of the matrix.

4.2 Learning the Melodies

It is far more difficult to find a large public source of sheet music for the melodies of the songs presented and "taught" to the AI program. Consequently, we created short song clips written by the First Author using a Digital Audio Workstation (DAW) known as FL Studio. Each mini-song utilizes the chords from the key of C and has a very different set of melodies. By connecting it to a MIDI keyboard, one could play the music directly.

As is well known, a .mid file (MIDI file) is a form of an audio file that contains many channels. When using FL Studio, we write the chords onto one channel and the melodies onto another. We then combine the two into a single .mid file for learning. The .mid files were read using the Python MIDI library, readily available on Github [3], using which we are able to open and read the different channels and to extract the pitch data. Thus, we are able to analyze the pitches to determine the chords/notes being played using the convertNotes and convertChords functions (Fig. 5).

The above snippet of code shows how the pitch data from the chords passed in, is translated into the actual chord names. The list is traversed in increments of three, since the chords are all triads, and only the first note is needed to be able to identify the chord. This process is again repeated for the convertNotes function, but this time every note is converted and stored.

In order to adequately judge the interaction with the Environment when it comes to the melodies, we use a different approach than what we is done with the chords. For each chord played, we set a key in a python dictionary. From there, we add a value under that key for every note played alongside that chord. This is always the next four notes, as per the assumption that for every bar of music there is a single chord and four notes. Once this is accomplished, we iterate through the dictionary for the matrix of note probabilities and adjust the matrix entries as needed using the adjustMelodyMatrix function. Again,

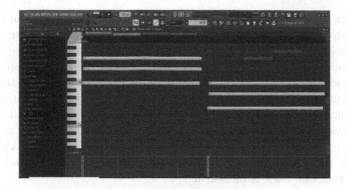

Fig. 5. Example of an FL Studio Session

after all the adjustments are made, we invoke a call to the `normalize` function to ensure that stochastic property of the matrix is maintained.

5   Results and Discussions

Understandably, it is not possible to adequately discuss the audio results that we have obtained in a paper document. Indeed, a prototype solution for the system that we have proposed is available, and a version of this system can be currently demonstrated. This prototype can be made available for interested listeners.

To produce the final output we used another library called `midiutil`. Using this, we were able to write the music that the AI system created into MIDI files. The chords and notes generated were passed into the `writeMusicOutput` function where the chords were first written on a channel and the corresponding notes on another, after which both were written to the same file and saved so as to be played by any basic media player that could play .mid files. The entire project was written and executed on Windows 10, and in it we used the native Windows Media Player application to listen to all the music that was produced. As mentioned earlier, while most of the other works in the field of music composition using AI were focused strictly on making music, our approach focused more on the generation of melodies based on the chords that were also generated. Thus, given a sequence of chords, the AI was capable of improvising a coherent melodic structure as an accompaniment. Although our solution is a *prima facie* prototype, we have succeeded in the task that we undertook.

6   Conclusions

In this paper, we have dealt with the automated composition of melodic (as opposed to chordal) music. Although music within the field of AI has been studied relatively extensively, the problem that we have studied is, to the best of our knowledge, unexplored. The existing research when it concerns computer composition, has involved the composition of the chordal notes when the underlying melody is specified, where these chords and beats have been generated based on models of well-known composers. The problem studied here is the converse, where we assume that the system is provided with the chords of some unknown melody. Using the foundations of Markov Chains and the tools of Learning Automata, we have been able to generate a melody that flows with the given chords. The melody generated is aesthetically and musically fitting. A prototype system that achieves this is currently available, and as far as we know, there is no research that has been reported to solve this problem. Although we have merely provided a primary solution for this rather complicated endeavor, we believe that our work is pioneering.

References

1. Khan Academy: Glossary of Musical Terms (2018). https://www.khanacademy.org/humanities/music/music-basics2/notes-rhythm/a/glossary-of-musical-terms
2. Cazaubon, M.: Chords in the Key of C Major (2018). http://www.piano-keyboard-guide.com/key-of-c.html
3. vishnubob: pyhton-midi (2013). https://github.com/vishnubob/python-midi
4. Todd, P.: A sequential network design for musical applications. In: Proceedings of the Connectionist Models Summer Schooli (1988)
5. Lewis, J.: Creation by refinement: a creativity paradigm for gradient descent learning networks. In: International Conference on Neural Networks (1988)
6. Eck, D., Schmidhuber, J.: Finding temporal structure in music: blues improvisation with LSTM recurrent networks. In: IEEE Workshop on Neural Networks for Signal Processing (2002)
7. Google Brain: Magenta (2018). https://ai.google/research/teams/brain/magenta/
8. Google Brain: MusicVAE: creating a palette for musical scores with machine learning (2018). https://magenta.tensorflow.org/music-vae
9. Ghaleb, O., Oommen, B.J.: Learning automata-based solutions to the multi-elevator problem (2019, to be Submitted)
10. Kabudian, J., Meybodi, M.R., Homayounpour, M.M.: Applying continuous action reinforcement learning automata (CARLA) to global training of hidden Markov models. In: 2004 International Conference on Information Technology: Coding and Computing. Proceedings, ITCC 2004, vol. 2, pp. 638–642. IEEE (2004)
11. Meybodi, M.R., Beigy, H.: New learning automata based algorithms for adaptation of backpropagation algorithm parameters. Int. J. Neural Syst. **12**(01), 45–67 (2002)
12. Misra, S., Oommen, B.J.: GPSPA: a new adaptive algorithm for maintaining shortest path routing trees in stochastic networks. Int. J. Commun. Syst. **17**(10), 963–984 (2004)
13. Narendra, K.S., Thathachar, M.A.L.T.: Learning Automata: An Introduction Prentice-Hall, New Jersey (1989)
14. Obaidat, M.S., Papadimitriou, G.I., Pomportsis, A.S., Laskaridis, H.S.: Learning automata-based bus arbitration for shared-medium ATM switches. IEEE Trans. Syst. Man Cybern. Part B (Cybern.) **32**(6), 815–820 (2002)
15. Oommen, B.J., Lanctot, J.K.: Discretized pursuit learning automata. IEEE Trans. Syst. Man Cybern. Part B (Cybern.) **20**(4), 931–938 (1990)
16. Oommen, B.J., Roberts, T.D.: Continuous learning automata solutions to the capacity assignment problem. IEEE Trans. Comput. **49**(6), 608–620 (2000)

17. Papadimitriou, G.I., Pomportsis, P.A.S.: Learning-automata-based TDMA protocols for broadcast communication systems with bursty traffic. IEEE Commun. Lett. **4**(3), 107–109 (2000)
18. Paszakowski, S.: The Circle of Fifths: The Clock of Key Signatures (2019). https://www.libertyparkmusic.com/the-circle-of-fifths/
19. Seredyński, F.: Distributed scheduling using simple learning machines. Eur. J. Oper. Res. **107**(2), 401–413 (1998)
20. Thathachar, M.A.L., Sastry, P.S.: A new approach to the design of reinforcement schemes for learning automata. IEEE Trans. Syst. Man Cybern. **SCM-15**(1), pp. 168–175 (1985)
21. Thathachar, M.A.L., Sastry, P.S.: A class of rapidly converging algorithms for learning automata. In: IEEE International Conference on Systems, Man and Cybernatics. IEEE (1984)
22. Tsetlin, M.: On behaviour of finite automata in random medium. Avtomatika i Telemekh **22**(10), 1345–1354 (1961)
23. Varshavskii, V., Vorontsova, I.P.: On the behavior of stochastic automata with a variable structure. Avtomatika i Telemekhanika **24**(3), 353–360 (1963)
24. Zhang, X.: Advances in the theory and applications of estimator-based learning automata. Ph.D. thesis, University of Agder, Norway (2015)

Graph Neural Networks to Advance Anticancer Drug Design

Asmaa Rassil[1]([✉]), Hiba Chougrad[2], and Hamid Zouaki[1]

[1] Faculty of Science, Laboratory of Computer Science and Mathematics
and their Applications, University Chouaib Doukkali, El Jadida, Morocco
asmaarassil@gmail.com, hamid_zouaki@yahoo.fr
[2] Laboratory of Intelligent Systems, Georesources and Renewable Energies,
National School of Applied Sciences, University Sidi Mohamed Ben Abdellah,
Fez, Morocco
chougrad.hiba@gmail.com

Abstract. Predicting the activity of chemical compounds against cancer is a crucial task. Active chemical compounds against cancer help pharmaceutical drugs producers in the conception of anticancer medicines. Still the innate way of representing chemical compounds is by graphs, the machine learning algorithms can not handle directly the anticancer activity prediction problems. Dealing with data defined on a non-Euclidean domain gave rise to a new field of research on graphs. There has been many proposals over the years, that tried to tackle the problem of representation learning on graphs. In this work, we investigate the representation power of Node2vec for embedding learning over graphs, by comparing it to the theoretical framework Graph Isomorphism Network (GIN). We prove that GIN is a deep generalization of Node2vec. We then exert the two models Node2vec and GIN to extract regular representations from chemical compounds and make predictions about their activity against lung and ovarian cancer.

Keywords: Graph neural networks · Transductive learning · Inductive learning · Representation vectors · Anticancer activity prediction

1 Introduction

Application domains of Artificial Intelligence (AI) became concerned in the last few years with treating complex problems defined on irregular data structure as graphs. The conception of pharmaceutical drugs [9,18,29] through the prediction of chemical compounds activities against cancer [22,23] need more than tabular representations to be solved. The natural way of modeling molecular data for an anticancer activity prediction problem is by graphs. Representing chemical compounds in the shape of graphs keep the structure of the data unchanged, and so no information from the initial collected data will get lost. Graphs are

© IFIP International Federation for Information Processing 2020
Published by Springer Nature Switzerland AG 2020
I. Maglogiannis et al. (Eds.): AIAI 2020, IFIP AICT 583, pp. 216–226, 2020.
https://doi.org/10.1007/978-3-030-49161-1_19

irregular types of data which are defined on non-Euclidean domains, therefore, all the classic machine learning algorithms cannot be directly applied on this type of structure [4]. To deal with such complex data, a number of works on representation learning on graphs have been proposed and proved to be effective in resolving these tasks. These representation approaches aim to map nodes or the entirely given graph to an embedding space of low-dimension (Fig. 1). The embedding vectors are supposed to summarize the information on top of nodes and their surrounding neighbors [13]. To do representation learning on graphs two types of methods are proposed transductive and inductive learning. Inductive approaches [2,7,11,13,28] for representation learning on graphs do not learn the embedding vectors directly from the graph but use instead the training graphs to learn a parametric encoding function to generate embeddings for unseen graphs. Transductive approaches [1,3,6,19,21,26] associate input graphs to embedding matrices of N rows and d columns. The parameter $\mathbf{N}$ refers to the number of nodes in the given graph and $\mathbf{d}$ refers to the dimensionality of the embedding space. Therefore, by transforming the irregular structure of graphs into Nxd matrices, we can straightforwardly apply Machine Learning algorithms to the data. In this work, we investigate the representation power of Node2vec for embedding learning over graphs, by comparing it to the inductive theoretical framework Graph Isomorphism Network (GIN) that proved to be equivalent to the Weisfeiler-Lehman (WL) test of isomorphism. We demonstrate theoretically that GIN is a deep generalization of Node2vec. We employ Node2vec and GIN models to define regular representations of chemical compounds represented by graphs and use these regular representations to predict the activities of the corresponding chemical compounds against lung and ovarian cancer in NCI1 and NCI109 [22,23] datasets respectively.

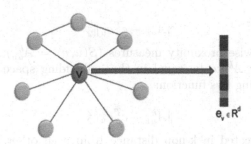

Fig. 1. Embedding learning. Each node v in the graph is mapped to a d-dimensional embedding vector e_v which summarizes the local information of the nodes around v.

2 Problem Formulation

A graph $\mathbf{G} = (\mathbf{V}, \mathbf{E})$ is defined by the vertex set $\mathbf{V} = \{v_i : i = 1, 2, ..., N\}$ and the edges $e_{ij} \in \mathbf{E}$ connecting pairs of nodes v_i and v_j in the graph. For each node v in the graph , we note its neighbors set by $\mathcal{N}(v) = \{u : (u, v) \in E\}$. The adjacency matrix $\mathbf{A} = (a_{ij})_{1 \leq i \leq N, 1 \leq j \leq N}$ outlines the graph structure by setting $a_{ij} = 1$ for connected nodes $e_{ij} \in E$ and $a_{ij} = 0$ if $e_{ij} \notin E$.

2.1   Graph Embedding

Graph embedding frameworks [1,3,6,19,21,26] are unsupervised approaches for learning on graph structured data. The main purpose of network embedding is to map input graphs to a regular low dimensional space where the use of standard machine learning algorithms become feasible [5,8,14]. To preserve the graph topology and the structural information on graphs in the embedding space, the network embedding variants propose pairwise similarity measures upon the graph structure $S_G(u, v)$ and learn the node embeddings by preserving the defined similarity measure in the embedding space $S_E(e_u, e_v)$ through the following loss function:

$$L[S_G(u, v), S_E(e_u, e_v)]$$

where $e_v \in \mathbf{R}^d$ is the corresponding embedding vector of the node $v \in \mathbf{V}$ in the embedding space $E = \mathbf{R}^d$ and $L(.)$ is a loss function.

Two principal branches of graph embedding are proposed in the literature [14] matrix factorization [1,3,6,19] and random walk [12,20]. Matrix factorization approaches define the proximity measure over the graph $S_G(u, v)$ either by an adjacency-based pairwise proximity measure ($S_G(u, v) = A_{uv}$) [1], or by multi-hop pairwise proximity measure captured by the k-th power of the adjacency matrix of the graph ($S_G(u, v) = A_{uv}^k$) [6] (see Fig. 2). The adjacency-based approaches push adjacent nodes $(u, v) \in E$ in the graph $S_G(u, v) = A_{uv} = 1$ to be equal in the embedding space $S_E(u, v) = e_v^T e_u = 1$ through the following loss function:

$$\|A_{uv} - e_v^T e_u\|_2^2 \tag{1}$$

Multi-hop pairwise proximity measures ($S(u, v) = A_{uv}^k$) [6] push adjacent nodes captured by A^k to be equal in the embedding space $S_E(u, v) = e_v^T e_u$ through the following loss function:

$$\|A_{uv}^k - e_v^T e_u\|_2^2 \tag{2}$$

Thus, nodes located in k-hop distance from each other, will have similar embedding vectors. Random walk [12,20] approaches for graph embedding are heuristic methods for embedding learning. To learn embeddings, authors in [12, 20] optimize the probability of visiting nodes v that tend to co-occur in random walks on the graph starting from a node u

$$p(v \mid u) = \frac{exp(e_v^T e_u)}{\sum_{v_i \in \mathbf{V}} exp(e_v^T e_{v_i})} \tag{3}$$

Matrix factorization gives a deterministic definition of similarity upon the graph. Otherwise, random walk methods adopt a stochastic proximity measure

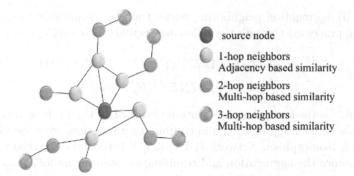

source node

1-hop neighbors
Adjacency based similarity

2-hop neighbors
Multi-hop based similarity

3-hop neighbors
Multi-hop based similarity

Fig. 2. For every node (i.e. source node), matrix factorization approaches learn embedding vectors of source nodes based on information from local neighbors (i.e. 1-hope neighbors) or distant neighbors.

based on randomly generated walks on the graph. Through two searching parameters p and q, Node2vec [12] controls the propagation of the generated walk. For $p > \max(q, 1)$ we reduce the probability of sampling an already visited node in the next two steps, therefore distant nodes from a source node u are pushed to have similar embeddings [12]. However, for $p < \min(q, 1)$, the walk stays local, then closest nodes to the source are pushed to have similar embeddings (Fig. 3).

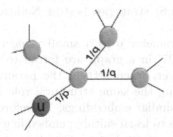

Fig. 3. The parameters p and q control the propagation of the generated walk on the graph. For small values of p, local neighbors are likely to be visited. However, when setting the parameter q to a small value, distant nodes tend to co-occur in the random walk

2.2   Graph Neural Networks

Contrary to graph embedding, graph neural networks (GNNs) [2,7,11,13,28] are deep and inductive approaches for representation learning on graphs. Through an end-to-end network, GNNs learn jointly the embeddings or representation vectors of the nodes and solve the defined problem on the graph structure. Thus, the representation vectors are learned based on the intrinsic topology of the graph and the task we aim to solve. The representation learning process involves two steps (i) aggregating neighboring nodes then (ii) combining them with the node being processed through the following recursive schema:

$$a_v^{(k)} = AGGREGATE^{(k)}(\{h_u^{(k-1)} : u \in \mathcal{N}(v)\}) \tag{4}$$

$$h_v^{(k)} = COMBINE^{(k)}(h_v^{(k-1)}, a_v^{(k)}) \tag{5}$$

where $h_v^{(k)}$ is the learned representation vector of node v in the kth layer of the network. Variants of aggregation and combination functions give the GNN variants. Graph Isomorphism Network (GIN) [25] is a theoretical variant of GNNs which combines the aggregation and combination step in the following formula:

$$h_v^{(k)} = MLP^{(k)}[(1 + \epsilon^{(k)})h_v^{(k-1)} + \sum_{u \in \mathcal{N}(v)} h_u^{(k-1)}] \tag{6}$$

where $\epsilon^{(k)}$ is a model parameter. Graph Isomorphism Network (GIN) [25] is as powerful as the Weisfeiler-Lehman (WL) test of isomorphism [24]. Thus, the learned graph representation vectors are as powerful as WL in capturing differences of graph topologies.

3   Deep Generalization of Node2vec

Learning embedding vectors by maximizing the likelihood of visiting the neighbors generated through a combination of Breadth-first Sampling (BFS) and Depth-first Sampling (DFS) strategies bestow Node2vec [12] with an impressive representation power.

When setting the parameter p to a small value (Fig. 3), nodes belonging to similar communities [27] in a graph are forced to learn approximately the same embeddings. However, when setting the parameter q to a small value (Fig. 3), nodes representing the same structural role in different communities [15] are pushed to learn similar embeddings. Therefore, according to the given task Node2vec [12] is able to learn suitable embeddings.

In this work, we demonstrate the analogy between Node2vec [12] and the powerful theoretical framework GIN [25]. And, we prove that Node2vec is closely equivalent to the Weisfeiler-Lehman(WL) test of isomorphism [24].

We consider a shallow neural network with one hidden layer **h** as described in Fig. 4. Given a node u, we maximize the probability of visiting a node $v \in \mathcal{N}(v)$ by generating a random walk starting from u. We initialize the embedding of u by its corresponding column $a_u \in \mathbf{R}^N$ in the adjacency matrix $\hat{A} = A + \mathbb{I}_N$, then we feed it as input to the network (Fig. 4). We denote the $N \times d$ matrices corresponding to the model parameters in the input-to-hidden and the hidden-to-output sections of the network by W and W' respectively. We assume that W and W' are embedding matrices of the input graph. Each row of W and W' is a d-dimensional embedding vector e_v of a node $v \in \mathbf{V}$.

In the forward pass,

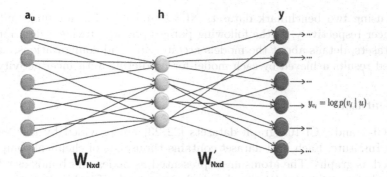

Fig. 4. Unsupervised representation Learning on graphs. Starting from a source node u defined by its corresponding column in the adjacency matrix $\hat{A} = A + \mathbb{I}_N$, we update the embedding of u by maximizing the probability of visiting the neighbors $v \in \mathcal{N}(u)$ generated by a random walk. The embedding we are looking for are defined by the parameters W and W' of the model.

$$W^T a_u = \sum_{v \in \mathcal{N}(u) \cup u} W_{\cdot,k}^T = e_u + \sum_{v \in \mathcal{N}(u)} e_v^T = h_u = \mathbf{h}$$

the updated embedding vector $h_u \in \mathbf{R}^d$ of the input node u is a summation of the embedding vectors e_u of u and its surrounding neighbors $\{e_v : v \in \mathcal{N}(u)\}$ given by the embedding matrix W. Node2vec adopts a sum-aggregator to aggregate neighboring nodes and utilizes the collected information to update the initial node embedding a_u. Thus, we may proclaim that node2vec is a shallow approximation of the deep and inductive Graph Isomorphism Network GIN (Eq. 6).

The output of the network before applying activation is,

$$z_i = W_{\cdot,i}^{'T} \mathbf{h} = W_{\cdot,i}^{'T} h_u = h_{v_i}^T h_u \qquad \forall i = 1, ..., N$$

using a log-softmax activation function $\Phi(.)$, we obtain (Eq. 3) corresponding to Node2vec:

$$\Phi(h_{v_i}^T h_u) = \log p(v_i \mid u) = \log \frac{exp(h_{v_i}^T h_u)}{\sum_{v \in \mathbf{V}} exp(h_v^T h_v)} \qquad (7)$$

We can conclude that, Node2vec learns embedding vectors of the graph nodes using approximately the same process of one layer GIN. Now, considering that GIN can capture and distinguish between different graph topologies as would the WL test of isomorphism, we may assert that embeddings learned by Node2vec will also be powerful in capturing differences of graph topologies.

4   Experiments and Results

To predict anticancer activity of chemical compounds against lung and ovarian cancer, we investigate the representation power of Node2vec for embedding learning on graphs and two variants of the inductive Graph Isomorphism Network

(GIN) using two benchmark datasets NCI-1 and NCI-109 for lung and ovarian cancer respectively. In the following paragraphs we provide a description of the datasets, details about the models architecture and configurations, and the obtained results achieved by each model for predicting anticancer activity.

4.1 Datasets

The NCI-1 and NCI-109 graph datasets [22,23] are provided by the National Cancer Institute. Each NCI dataset contains thousands of chemical compounds modelled as graphs. The atoms are represented as nodes and bonds as edges in the graph representation of the chemical compounds (Fig. 5). Statistics about NCI-1 and NCI-109 graph datasets are reported in Table 1.

NCI-1 dataset contains two categories of chemical compounds active and non-active against the non-small cell lung cancer. Active compounds are labeled by +1 and non-active by −1. NCI-109 dataset contains two categories of chemical compounds active and non-active against ovarian cancer. Active compounds are labeled by +1 and non-active by −1.

Table 1. Datasets statistics. The table summarizes the number of graphs, the average number of nodes, the average number of edges and the number of node labels in NCI-1 and NCI-109 datasets.

Datasets	Dataset statistics			
	# Graphs	# Nodes(avg.)	# Edges(avg.)	# Node Labels
NCI-1	4110	29.8	32.3	37
NCI-109	4127	29.6	32.1	38

4.2 Implementation Details and Results

To predict the anticancer activity of chemical compounds against lung and ovarian cancer, we train the embedding model Node2vec [12] on NCI-1 and NCI-109 datasets with the following hyper-parameters: embedding dimensionality $\in \{7, 12\}$, random walk length $\in \{3, 4\}$, size of neighbors set (i.e the context size) $\in \{5, 7\}$ and for each node we generate 10 random walks.

To train the deep Graph Isomorphism Network (GIN) [25] on NCI-1 and NCI-109 datasets, we use two variants of the model using (1) multilayer perceptron and (2) single-layer perceptron as combination functions of neighboring node representations. To build the anticancer activity prediction model, we use five GIN layers each followed by batch normalization [16]. We learn the model parameters by optimizing the cross-entropy loss between the predicted labels and the true labels using the Adam optimizer [17]. We tune the following hyper-parameters: batch size $\in \{32, 42\}$, hidden units $\in \{16, 32\}$, learning rate $\in \{0.01, 0.001\}$ and dropout coefficient $\in \{0, 0.2\}$. All implementations were carried out using PyTorch Geometric [10] and the Cuda enabled GPU 12GB NVIDIA Tesla K80.

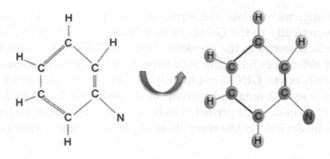

Fig. 5. Graph representation of a chemical compound. The left figure represents a chemical compound and the right figure represents its corresponding graph representation where C refers to carbon atom, H to Hydrogen atom and N to Nitrogen atom.

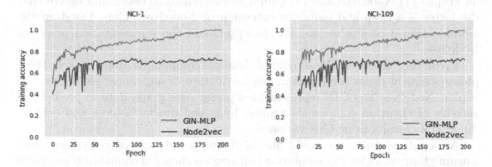

Fig. 6. Training set performance of GIN-MLP compared to Node2vec on NCI-1 and NCI-109 datasets over 100 epochs

Table 2. Binary-classification results (%). The table reports the average and standard deviations of accuracies on binary-graph classification

Datasets	Representation learning models		
	Node2vec [12]	*GIN-1-Layer* [25]	*GIN-MLP* [25]
NCI-1	54.9 ± 1.2	83.0 ± 3.5	83.4 ± 2.5
NCI-109	51.4 ± 2.3	86.8 ± 6.0	85.0 ± 6.7

Results reported in Table 2 and Fig. 6 show that GIN models perfectly fit the data and report approximately the same results for predicting the anticancer activity of chemical compounds on both NCI-1 (GIN-1-layer gives an average accuracy of 83.0 ± 3.5% and GIN-MLP gives 83.4 ± 2.5%) and NCI-109 (GIN-1-layer gives an average accuracy of 86.8 ± 6.0% and GIN-MLP gives 85.0 ± 6.7% accuracy) datasets, while Node2vec has a low and unstable performance on the two datasets (average accuracy of 54.9 ± 1.2% on NCI-1 for lung cancer and 51.4 ± 2.3% on NCI-109 for ovarian cancer). Those results are maybe due to the fact that GIN models are deep while Node2vec is a shallow model.

Subsequently, we validate the representation power of Node2vec by comparing it theoretically to the Graph Isomorphism Network (GIN) [25] which is proved to be as powerful as the Weisfeiler-Lehman(WL) test of isomorphism [24] in capturing differences on graph structures and topologies. We use Node2vec and two variants of the GIN model to auspiciously predict the activity of chemical compounds against lung and ovarian cancer over NCI-1 and NCI-109 respectively. Chemical compounds predicted to be active against lung and ovarian cancer can then be decisive in the conception of pharmaceutical anticancer drugs.

5 Conclusion

In this work, we investigate the representation learning power of Graph Neural Networks on graph data structure. We explore two powerful models that operate on graphs (1) Node2vec and (2) Graph Isomorphism Network and derive that the latter is a deep and inductive extension of Node2vec. Thus, based on the aforementioned results we are able to apprehend the representation power of Node2vec.

Predicting effectively the activity of chemical compounds against cancer helps pharmaceutical drugs producers in the fabrication of anticancer medicines. We exploit Node2vec and GIN representation power over graph data to predict the activity of chemical compounds against lung and ovarian cancer and achieve remarkable results using GIN variants.

As a future work, we would like to define a new specially-designed deep learning framework for representation learning on chemical compounds modelled as graphs. We also plan to go beyond neighbor aggregation and adopt attention mechanisms to learn the connective strength between nodes.

References

1. Ahmed, A., Shervashidze, N., Narayanamurthy, S., Josifovski, V., Smola, A.J.: Distributed large-scale natural graph factorization. In: Proceedings of the 22nd International Conference on World Wide Web, pp. 37–48 (2013)
2. Atwood, J., Towsley, D.: Diffusion-convolutional neural networks. In: Advances in Neural Information Processing Systems, pp. 1993–2001 (2016)
3. Belkin, M., Niyogi, P.: Laplacian eigenmaps and spectral techniques for embedding and clustering. In: Advances in Neural Information Processing Systems, pp. 585–591 (2002)
4. Bronstein, M.M., Bruna, J., LeCun, Y., Szlam, A., Vandergheynst, P.: Geometric deep learning: going beyond Euclidean data. IEEE Signal Process. Mag. **34**(4), 18–42 (2017)
5. Cai, H., Zheng, V.W., Chang, K.C.C.: A comprehensive survey of graph embedding: problems, techniques, and applications. IEEE Trans. Knowl. Data Eng. **30**(9), 1616–1637 (2018)
6. Cao, S., Lu, W., Xu, Q.: GraRep: learning graph representations with global structural information. In: Proceedings of the 24th ACM International on Conference on Information and Knowledge Management, pp. 891–900 (2015)

7. Chen, J., Ma, T., Xiao, C.: FastGCN: fast learning with graph convolutional networks via importance sampling. arXiv preprint arXiv:1801.10247 (2018)
8. Cui, P., Wang, X., Pei, J., Zhu, W.: A survey on network embedding. IEEE Trans. Knowl. Data Eng. **31**(5), 833–852 (2018)
9. De Cao, N., Kipf, T.: MolGAN: an implicit generative model for small molecular graphs. arXiv preprint arXiv:1805.11973 (2018)
10. Fey, M., Lenssen, J.E.: Fast graph representation learning with PyTorch geometric. In: ICLR Workshop on Representation Learning on Graphs and Manifolds (2019)
11. Gilmer, J., Schoenholz, S.S., Riley, P.F., Vinyals, O., Dahl, G.E.: Neural message passing for quantum chemistry. In: Proceedings of the 34th International Conference on Machine Learning, vol. 70, pp. 1263–1272. JMLR. org (2017)
12. Grover, A., Leskovec, J.: node2vec: scalable feature learning for networks. In: Proceedings of the 22nd ACM SIGKDD International Conference on Knowledge Discovery and Data Mining, pp. 855–864 (2016)
13. Hamilton, W., Ying, Z., Leskovec, J.: Inductive representation learning on large graphs. In: Advances in Neural Information Processing Systems, pp. 1024–1034 (2017)
14. Hamilton, W.L., Ying, R., Leskovec, J.: Representation learning on graphs: methods and applications. arXiv preprint arXiv:1709.05584 (2017)
15. Henderson, K., et al.: RolX: structural role extraction & mining in large graphs. In: Proceedings of the 18th ACM SIGKDD International Conference on Knowledge Discovery and Data Mining, pp. 1231–1239 (2012)
16. Ioffe, S., Szegedy, C.: Batch normalization: accelerating deep network training by reducing internal covariate shift. arXiv preprint arXiv:1502.03167 (2015)
17. Kingma, D.P., Ba, J.: Adam: a method for stochastic optimization. arXiv preprint arXiv:1412.6980 (2014)
18. Li, Y., Vinyals, O., Dyer, C., Pascanu, R., Battaglia, P.: Learning deep generative models of graphs. arXiv preprint arXiv:1803.03324 (2018)
19. Ou, M., Cui, P., Pei, J., Zhang, Z., Zhu, W.: Asymmetric transitivity preserving graph embedding. In: Proccedings of the 22nd ACM SIGKDD International Conference on Knowledge Discovery and Data Mining, pp. 1105–1114 (2016)
20. Perozzi, B., Al-Rfou, R., Skiena, S.: DeepWalk: online learning of social representations. In: Proceedings of the 20th ACM SIGKDD International Conference on Knowledge Discovery and Data Mining, pp. 701–710 (2014)
21. Shen, X., Pan, S., Liu, W., Ong, Y.S., Sun, Q.S.: Discrete network embedding. In: Proceedings of the 27th International Joint Conference on Artificial Intelligence, pp. 3549–3555 (2018)
22. Shervashidze, N., Schweitzer, P., Leeuwen, E.J.V., Mehlhorn, K., Borgwardt, K.M.: Weisfeiler-Lehman graph kernels. J. Mach. Learn. Res. **12**, 2539–2561 (2011)
23. Wale, N., Watson, I.A., Karypis, G.: Comparison of descriptor spaces for chemical compound retrieval and classification. Knowl. Inf. Syst. **14**(3), 347–375 (2008)
24. Weisfeiler, B., Leman, A.: A reduction of a graph to a canonical form and an algebra arising during this reduction. Nauchno-Technicheskaya Informatsia **9**, 12–16 (1968)
25. Xu, K., Hu, W., Leskovec, J., Jegelka, S.: How powerful are graph neural networks? In: International Conference on Learning Representations (ICLR) (2019)
26. Yang, H., Pan, S., Zhang, P., Chen, L., Lian, D., Zhang, C.: Binarized attributed network embedding. In: 2018 IEEE International Conference on Data Mining (ICDM), pp. 1476–1481. IEEE (2018)

226   A. Rassil et al.

27. Yang, J., Leskovec, J.: Overlapping communities explain core-periphery organization of networks. Proc. IEEE **102**(12), 1892–1902 (2014)
28. Ying, Z., You, J., Morris, C., Ren, X., Hamilton, W., Leskovec, J.: Hierarchical graph representation learning with differentiable pooling. In: Advances in Neural Information Processing Systems, pp. 4800–4810 (2018)
29. You, J., Liu, B., Ying, Z., Pande, V., Leskovec, J.: Graph convolutional policy network for goal-directed molecular graph generation. In: Advances in Neural Information Processing Systems, pp. 6410–6421 (2018)

Optimizing Self-organizing Lists-on-Lists Using Transitivity and Pursuit-Enhanced Object Partitioning

O. Ekaba Bisong and B. John Oommen$^{(\boxtimes)}$

School of Computer Science, Carleton University, Ottawa K1S 5B6, Canada
ekaba.bisong@carleton.ca, oommen@scs.carleton.ca

Abstract. The study of Self-organizing lists deals with the problem of lowering the average-case asymptotic cost of a list data structure receiving query accesses in Non-stationary Environments (NSEs) with the so-called *"locality of reference"* property. The de facto schemes for *Adaptive* lists in such Environments are the Move To Front (MTF) and Transposition (TR) rules. However, significant drawbacks exist in the asymptotic accuracy and speed of list re-organization for the MTF and TR rules. This paper improves on these schemes using the design of an Adaptive list data structure as a hierarchical data *"sub"*-structure. In this framework, we employ a hierarchical Singly-Linked-Lists on Singly-Linked-Lists (SLLs-on-SLLs) design, which divides the list data structure into an outer and inner list context. The inner-list context is itself a SLLs containing sub-elements of the list, while the outer-list context contains these sublist partitions as its primitive elements. The elements belonging to a particular sublist partition are determined using reinforcement learning schemes from the theory of Learning Automata. In this paper, we show that the Transitivity Pursuit-Enhanced Object Migration Automata (TPEOMA) can be used in conjunction with the hierarchical SLLs-on-SLLs as the dependence capturing mechanism to learn the probabilistic distribution of the elements in the Environment. The idea of *Transitivity* builds on the Pursuit concept that injects a noise filter into the EOMA to filter divergent queries from the Environment, thereby increasing the likelihood of training the Automaton to approximate the "true" distribution of the Environment. By taking advantage of the Transitivity phenomenon based on the statistical distribution of the queried elements, we can infer "dependent" query pairs from non-accessed elements in the transitivity relation. The TPEOMA-enhanced hierarchical SLLs-on-SLLs schemes results in superior performances to the MTF and TR schemes as well as to the EOMA-enhanced hierarchical SLLs-on-SLLs schemes in NSEs. However, the results are observed to have superior performances to the PEOMA-enhanced hierarchical schemes in Environments with a Periodic non-stationary distribution but were inferior in Markovian Switching Environments.

B. John Oommen—This author is also an *Adjunct Professor* with the University of Agder in Grimstad, Norway.

© IFIP International Federation for Information Processing 2020
Published by Springer Nature Switzerland AG 2020
I. Maglogiannis et al. (Eds.): AIAI 2020, IFIP AICT 583, pp. 227–240, 2020.
https://doi.org/10.1007/978-3-030-49161-1_20

Keywords: Learning Automata (LA) · Transitivity and Pursuit-Enhanced Object Migration Automaton (TPEOMA) · "Adaptive" Data Structure (ADS) · Singly-Linked-Lists on Singly-Linked-Lists (SLLs-on-SLLs)

1 Introduction

Research in self-organizing lists is aimed at mitigating the worst-case linear cost of list retrieval by adaptively rearranging the list nodes in response to query accesses from a Non-Stationary Environment (NSE). This paper optimizes the singly linked-list data structure. The models of NSEs are covered in Sect. 3.1. The goal of the rearrangement is to move towards the head of the list, elements that are more frequently accessed. This has the result of improving the asymptotic average cost of retrievals. However, this action requires information of the *"unknown"* probability distribution of the Environment. The Environment has a dependency property where element O_i is not conditionally independent of O_j, $O_i \not\perp O_j$. This property is called "locality of reference". Queries are requests coming from the Environment to retrieve elements from the data structure.

The asymptotic cost is an empirical measure for assessing the algorithmic performance of the list re-organization strategies [1]. This is performed by implementing the corresponding strategy, and taking an ensemble average of the averages of the cost as the algorithm converges. Since the schemes are ergodic (meaning that they have a final solution that is independent of the starting states of a Markov chain), they can be seen to provide an accurate estimation of the asymptotic cost.

The formal expression for the asymptotic cost of an adaptive strategy, A, is given as:

$$E[A] = \sum_{1 \leq j \leq J!} P\{\pi_j\}_A \, C(\pi_j) \tag{1}$$

$$= \sum_{1 \leq j \leq J!} [\, P\{\pi_j\}_A \sum_{1 \leq i \leq J} s_i \pi_j(i) \,]. \tag{2}$$

In Eq. (1) and Eq. (2), the expression $P\{\pi_j\}_A$ represents the steady-state (or stationary) probability of choosing the list permutation π_j in the Markov chain that involves the adaptive strategy, A. Further, $C(\pi_j)$ is the ordering cost or the average-access cost of the list permutation π_j. For more mathematical details on the asymptotic cost and Markov chains, the reader is directed to [1,5].

The de facto schemes for list self-organization in NSEs are the Move To Front (MTF) and Transposition (TR) rules. In the MTF update scheme, the queried element is moved to the list head, except when it is the first element, because, in that case, it is already at the head (Fig. 1). As opposed to this, in the TR adaptive scheme, if the queried element is not already at the list head, it is moved one position towards the front of the list (Fig. 2).

Another deterministic scheme for self-organization in NSEs is the Frequency Count (FC) scheme. The FC rule maintains an accumulator for recording the

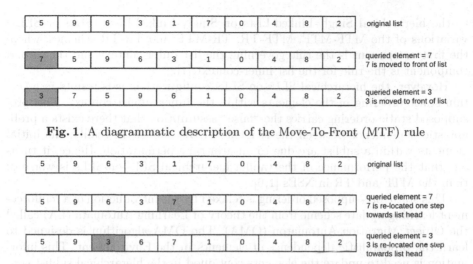

Fig. 1. A diagrammatic description of the Move-To-Front (MTF) rule

Fig. 2. A diagrammatic description of the Transposition (TR) rule

access frequencies of the list elements. The resulting list is re-arranged according to the descending order of the counters. This relatively simple scheme yields rather impressive results with regard to its asymptotic cost being close to optimal, and its amortized cost being about two times the optimal cost [4]. Notwithstanding, the FC scheme has some obvious drawbacks, the first being that the memory cost scales poorly for large lists [12]. Secondly, in environments exhibiting *"locality of reference"*, the FC scheme yields an unacceptable performance [11,15].

The MTF and TR rules are of greater interest to this work for two reasons. The first being that they have been shown to empirically out-perform the other schemes [2], and the second, that the time and space complexities involved in implementing other surveyed schemes render most of them impractical for real-world settings. It is for these sort of Environments with dependent query accesses that this present work seeks to provide novel solutions.

Indeed, while these schemes (MTF, TR, FC) show superior performances in minimizing the asymptotic average-cost of Singly-Linked-Lists (SLLs) in NSEs [13], they, however, suffer peculiar drawbacks in NSEs exhibiting *"locality of reference"*. An empirical analysis of TR responding to query accesses from different probability distributions shows that TR outperforms MTF for the Zipf's distribution [17], and the results of [9] showed that the asymptotic cost of TR outperforms MTF for the Lotka, exponential, linear, and 80-20 probability distributions. However, the MTF has a faster adaptive rate and quickly converges early-on in the algorithm's execution [8].

This work combines the MTF and TR rules to take advantage of the quick updates of the MTF rule and the asymptotically stable convergence of the TR rule in designing our improved hierarchical adaptive strategies. This design led to the hierarchical Singly-Linked-Lists on Singly-Linked-Lists (SLLs-on-SLLs) variations of the MTF-MTF, MTF-TR, TR-MTF and TR-TR schemes, where the first component is the rule governing the list outer-context and the second component is the rule for the list inner-context [1].

However, the hierarchical SLLs-on-SLLs configuration as-is results in a certain static ordering of the elements within the inner sublist context. This presupposed static ordering carries the "false" assumption that there exists a probabilistic dependence between the elements in the sublist. In reality, the initial elements within a sublist are due to an arbitrary permutation. Hence, it turns out that the performance of the vanilla hierarchical SLLs-on-SLLs is worse of than the MTF and TR in NSEs [1,6].

This *static* pre-supposed ordering is relaxed by the introduction of a reinforcement learning update scheme from the theory of Learning Automata (LA) called the Object Migration Automaton (OMA). The OMA algorithm is designed to learn the probabilistic dependence of elements in the Environment. This information is used to update the elements contained in the hierarchical sublist context represents their dependence ordering in the Environment. This formulation led to the OMA-hierarchical SLLs-on-SLLs consisting of the MTF-MTF-OMA, MTF-TR-OMA, TR-MTF-OMA and TR-TR-OMA schemes [1].

However, the OMA suffers from the "deadlock" problem[1] which occurs when an accessed element is swapped from one action to another and then back to the original action and thus prevented from converging to their optimal ordering (this concept is further explained in Sect. 2.2). To mitigate this, the Enhanced Object Migration Automaton (EOMA) was introduced by [10] that imposes conditions to restrict unnecessary swaps on action-state boundaries as well as to redefine the convergence criteria to when the automaton is in the two-innermost states as the "final" states instead of when the automaton is in the innermost state.

Bisong and Oommen [6] employed the EOMA reinforcement scheme in designing the EOMA-augmented hierarchical SLLs-on-SLLs which resulted in superior performances to the de facto MTF and TR schemes and the OMA-augmented hierarchical schemes in NSEs. Further, the work by [7] further improved the performance of the hierarchical SLLs-on-SLLs by incorporating the PEOMA reinforcement scheme. The PEOMA algorithm by [18] employed the Pursuit concept to filter divergent query pairs from the Environment. To the best of our knowledge, the work by [7] is currently the state-of-the-art for self-organizing lists in NSEs.

This paper further explores the state of the art by incorporating the concept of Transitivity in the PEOMA algorithm, that is the Transitivity Pursuit-Enhanced Object Migration Automaton (TPEOMA) to design a TPEOMA-augmented hierarchical SLLs-on-SLLs. This proposed hierarchical formulation consists of the MTF-MTF-TPEOMA, MTF-TR-TPEOMA, TR-MTF-TPEOMA and TR-TR-TPEOMA schemes. The Transitivity concept is used to

---

[1] Albeit referred to as a "deadlock" in the literature, it could more appropriately be described as a "livelock".

infer "dependent" queries from the Environment to further train the automaton to approximate its "unknown" distribution. The Transitivity concept is discussed in more detail in Sect. 2.3.

The novel contributions of this paper include:

- The design and implementation of the TPEOMA-enhanced SLLs-on-SLLs;
- Demonstrating the superiority of the TPEOMA-augmented hierarchical schemes to the MTF and TR rules and to the original OMA-augmented schemes that pioneered the idea of a hierarchical LOL approach;
- Demonstrating the superiority of the TPEOMA-augmented hierarchical schemes to the EOMA-augmented hierarchical schemes;
- Highlighting the performances of the TPEOMA-augmented hierarchical schemes to the PEOMA variants;
- Showing that the "Periodic" and "UnPeriodic" versions of the TPEOMA-augmented hierarchical schemes yielded superior and comparable performances respectively in PSEs to those without such additions.

Section 1 introduces the idea of a hierarchical SLLs-on-SLLs for minimizing the asymptotic average case for list retrieval in NSEs. It also lays out the case for incorporating the OMA-family of reinforcement schemes for learning the probability dependence distribution of elements in the Environment. Section 2 reviews the theory of LA^2 as the foundational theory for the TPEOMA reinforcement scheme. In addition, this section accesses the idea of the Transitivity concept that builds on the Pursuit concept in the PEOMA algorithm. Section 3 explains the design of the TPEOMA-augmented hierarchical SLLs-on-SLLs for NSEs. Section 4 presents the Results and Discussions, and Sect. 5 concludes the paper.

2 Theoretical Background

2.1 Learning Automata

Learning automata (LA) arose in the Soviet Union in the 1960s by Tsetlin as a computational adaptive scheme for learning. [20] The LA task can be modeled by means of a feedback loop between the Environment and the automaton. The automaton interacts with the Environment by choosing from a set of actions based on the feedback it receives from the Environment. The LA model is set up as an adaptive process [3] in which little or no information is known *a priori* about the Environment. The goal of the learner (the LA) is then to learn the optimal action that maximizes a utility function or improves a performance index [1,5,18].

---

² Due to space limitations, the background material is only briefly surveyed. The seminal work by [14] and the theses by Shirvani [18] and the first author of this paper [5] contain exhaustive details of the theory and applications of LA.

2.2 The OMA and EOMA

The OMA improves on the Tsetlin/ Krinsky strategies for partitioning objects into groups in the Equal-Partitioning Problem (EPP) [16]. In the OMA, the number of actions represents the number of groups, or partitions, R, where each action contains a set number of states, N. The OMA partitions the object group W into R partitions by moving the abstract objects $\mathcal{O}$ around the action-states of the automaton. To learn more about the OMA algorithm the reader is referred to [5–7,18].

The EOMA algorithm improves on the OMA to mitigate the inability of the OMA algorithm to converge due to a "deadlock" scenario. The deadlock scenario occurs when there is a query pair $\langle O_i, O_j \rangle$ in a stream of query pairs belonging to different actions, α_h and α_g. If one object is in the boundary state of its action, and the other is not, the query pairs are prevented from converging to their optimal ordering, and this can lead to an "infinite" loop scenario. The boundary state of the OMA is the outermost memory state of an action α (see Fig. 3). To learn more about the EOMA algorithm the reader is referred to [5–7,18].

To resolve the deadlock scenario for the query pair $\langle O_i, O_j \rangle$, let us say that there exists an object in the boundary state of the action-group, α_g containing O_j, given that the other element O_i is not in the boundary state of its action-group α_h. The EOMA moves moves O_j from action-group α_g to be in the boundary state of action-group, α_h. By taking this step, the partitions become unequal. To regain the equi-partitioning, the object that is closest to O_i in α_h, which we will call O_l, is moved to the boundary state of α_g.

Hence, given a set of query elements, the transitions of the EOMA for the abstract objects $\langle O_i, O_j \rangle$ on reward and on penalty are illustrated in Fig. 3. In addition, the EOMA also modifies the convergence criteria to reduce its vulnerability to divergent queries by setting the two-innermost states as the "final" states, as opposed to just the innermost state in the vanilla OMA.

To learn more about how the EOMA algorithm mitigates the "deadlock" scenario in the OMA, the reader is referred to [5–7,18].

2.3 The Case for Transitivity

The Pursuit concept is incorporated into the EOMA design to filter *divergent* queries from the Environment using Maximum Likelihood Estimates (MLEs). It works by updating the joint query probabilities using ranked estimates of the reward probabilities to asymptotically choose or "pursue" better actions. For a detailed discussion of the pursuit concept, the reader is referred to [18]. The Pursuit-EOMA scheme was the best-known object-partitioning algorithm until the authors of [19] introduced the TPEOMA.

The TPEOMA algorithm is based on the observation that the Pursuit matrix can also be used to infer underlying relations in the Environment [18]. It then invokes a policy that spins off reward/penalty operations by incorporating the statistics obtained between objects that have been previously accessed. The Pursuit matrix M is defined as a $\frac{W}{R} \times \frac{W}{R}$ matrix whose element $[i,j]$ are the probabilities that $\langle O_i, O_j \rangle$ are simultaneously accessed where $1 \leq i, j \leq W$. The Pursuit

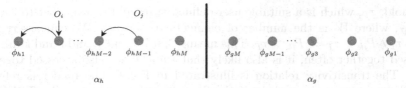

(a) On reward: Move the accessed abstract objects $\langle O_i, O_j \rangle$ towards the extreme states.

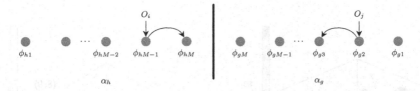

(b) On penalty: Move the accessed abstract objects $\langle O_i, O_j \rangle$ towards their boundary states.

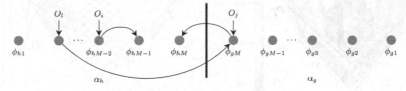

(c) On penalty: Move the accessed abstract objects $\langle O_i, O_j \rangle$ to be in the same group. An extra object O_l in the old group of O_i is moved to the old group of O_j.

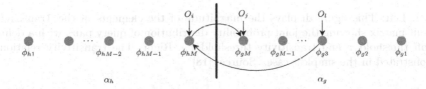

(d) On penalty: If both abstract objects $\langle O_i, O_j \rangle$ are in the boundary states, move one of them, say O_i, to the boundary state of the other group. An extra object O_l in the group of O_j is moved to the old group of O_i.

Fig. 3. The EOMA algorithm

matrix M is computed using Maximum Likelihood estimates, where each entry m_i in the matrix is the ratio of the frequency of occurrence of m_i to the total number of query accesses. The probabilities in M sum to unity. The reader is referred to [19] for a details analysis of the Pursuit matrix.

When an estimate of the Pursuit matrix is obtained, the transitivity property can be used to *infer* queries to further train the automaton to learn the model of dependence of the Environment. In other words, if the current query received from the Environment is $\langle O_i, O_j \rangle$ and O_j is in relation with $O_k, k \in \{1, \cdots, W_j\}$, $k \neq i$, we can infer that O_i is also in a relation with O_k, where i, j, and k are the indices of the entries in the Pursuit matrix. Thus, given the transitivity threshold, τ_T, which is a suitable user-defined threshold which is set to a value $\frac{1}{W^2 - W}$, where W is the number of elements in the list. We can assert that $P_{ij} > \tau_T \wedge P_{jk} > \tau_T \Rightarrow P_{ik} > \tau_T$. This means that if i and j, and i and k are also accessed together often, it is also likely that i and k are also accessed together often. The transitivity relation is illustrated in Fig. 4. The reader is referred to [19] for a thorough analysis of the Transitivity property in the TPEOMA algorithm.

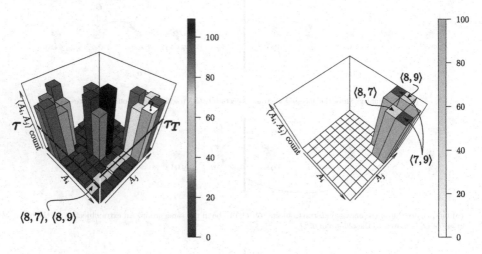

Fig. 4. Left: This figure displays the magnitude of the elements in the Transitivity-Pursuit matrix showing the joint probability distribution of query pairs with a defined cut-off threshold τ and transitivity threshold τ_T. Right: The transitivity relation is demonstrated in the simplest case. Source: [18]

3   TPEOMA-Augmented Hierarchical SLLs-on-SLLs

The concept of a hierarchical data "sub"-structure involves dividing a list of size W into k sub-lists. The re-organization strategy is then hierarchically applied to the list by first considering the elements within the sub-list (also called the sub-context) and then operating over the sublists (or sub-contexts) themselves.

The re-organization strategies involved are the MTF and TR rules. When used in a hierarchical scheme, this yields MTF-preceding-MTF, (MTF-MTF), MTF-preceding-TR, (MTF-TR), TR-preceding-MTF, (TR-MTF), and TR-preceding-TR, (TR-TR) schemes. For example, in the case of MTF-TR, the element within a sub-context is first moved to the front of the list, and then the sub-context is moved to the front of the list context. Again, the fundamental idea of combining the MTF and TR schemes in this hierarchical formulation is principally to take advantage of the fast convergence properties of the MTF rule and the more accurate asymptotic convergence of the TR rule.

In NSEs with *"locality of reference"*, let us assume that query accesses m are made to a sub-context (or local context) Q_a. For a given query access m_i

from Q_a, the probability that the next query access, m_j will come from the same local context, Q_a is high. Hence, it is useful to take advantage of this dependency relationship by moving the entire sublist of elements of the sub-context *en masse* towards the head of the list to cut-down the access-time cost when an element within the sub-context Q_a is requested. The hierarchical schemes mentioned above are preferred in Environments characterized by such a *"locality of reference"*. Observe that the stand-alone MTF will require at least $\frac{J}{k}$ distinct requests to promote the entire sub-context to the head of the list.

Further, if a record m_u is accessed that is not in the re-organized sub-context, the hierarchical schemes will promote *en masse* all records that are part of m_u's sub-context towards the head of the list thereby reducing the subsequent access costs. As opposed to this in the MTF and TR schemes, for example, the entire context is promoted towards the list head one record at a time.

In the TPEOMA-Augmented Hierarchical SLLs-on-SLLs, the Transitivity phenomenon, included in the TPEOMA, takes advantage of the statistical distribution of the queried elements to infer good query pairs from non-accessed elements in the transitivity relation. Both of these are used to improve the dependence-capturing aspect to be included in the sub-lists when it concerns the Lists-on-Lists hierarchy. The augmentation of hierarchical SLLs with the TPEOMA reinforcement scheme gives rise to the MTF-MTF-TPEOMA, MTF-TR-TPEOMA, TR- MTF-TPEOMA and the TR-TR-TPEOMA.

3.1   Models of NSEs

In NSEs, the penalty probabilities for each action vary with time. In the context of adaptive data structures, this variation affects the expected query cost because the Environment exhibits the so-called *"locality of reference"*, or is characterized by dependent accesses. The *"locality of reference"* occurs when there exists a probabilistic dependence between the consecutive queries [2]. In other words, there is a considerably small number of distinct or unrelated queries within a segment of the access sequence.

To initiate the design of adaptive data structures in NSEs, we introduce and examine two dependent query generators for simulating an Environment producing queries with dependent accesses. They are the Markovian and Periodic query generators. Given a set of n distinct elements, if we split it into k disjoint and equal partitions with m elements where $n = k.m$, the k subsets can be considered to be local or "sub"-contexts. The elements within a sub-context k_i exhibit *"locality of reference"*. This implies that if an element from set k_i is queried at time t, there exists a high likelihood that the next queried element at time $t+1$ will also arrive from the same set k_i. In other words, the Environment itself can be seen to have a finite set of states $\{Q_i | 1 \leq i \leq k\}$, and the dependent model defines the transition from one Environmental state to another.

3.2   NSEs and Their Distributions

The Environment generates queries according to a probability distribution. In recording the behaviour of the hierarchical list schemes proposed in this work, we considered five different types of query distributions, namely, the Zipf, Eighty-Twenty, Lokta, Exponential and Linear distributions. For a given list of size W, divided into k sub-lists, with each sub-list containing $\frac{W}{k}$ elements, the probability distribution $\{s_i\}$ where $1 \leq i \leq m$ describes the query accesses for the elements in the subset k. Notice that in this way, the total probability mass for the query accesses in each group is the same, and the distribution within each group has the respective distribution.

A rationale for conducting the simulations with these query distributions is that, for the most part, they result in "L-shaped" graphs. This is true in particular, for the Exponential and Lotka distribution, and to an extent for the Zipf distributions. Such "L-shaped" distributions assign high probabilities to a small number of the sub-list elements. By working in this manner, we can compare our hierarchical variants against the MTF and TR schemes, which were the *de facto* schemes for adaptive lists in NSEs.

4   Results and Discussions

The experimental setup for the simulations involving the TPEOMA-augmented hierarchical schemes in MSEs involved splitting a list of size 128 into k sublists with $k \in 2, 4, 8, 16, 32, 64$. The degree of dependence of the MSE, α, was set to 0.9 and the period for the PSE, $T = 30$. For all the results discussed in this section, the simulation involved an ensemble of 10 experiments, each evaluating $300,000$ query accesses, and for the various aforementioned query generators. For conciseness sake, we present results for $k = 8$.

From Table 1, when $k = 8$, we observed that the TPEOMA-augmented hierarchical schemes were superior to the MTF and TR standalone schemes for all query distributions in the MSE. As an example in the Lotka distribution, the asymptotic and amortized costs for the MTF-MTF-TPEOMA, MTF-TR-TPEOMA, TR-MTF-TPEOMA and TR-TR-TPEOMA were (6.60, 6.11, 4.39, 4.37) and (8.09, 7.84, 6.57, 6.54) respectively. As opposed to this, the corresponding asymptotic and amortized costs for the MTF and TR were significantly higher at (39.30, 48.25) and (39.17, 48.66) respectively. Further, in the Exponential distribution for the MSE, the MTF-MTF-TPEOMA, MTF-TR-TPEOMA, TR-MTF-TPEOMA and TR-TR-TPEOMA had asymptotic and amortized costs of (7.01, 6.99, 7.50, 7.64) and (8.98, 8.83, 9.66, 9.87), while the MTF and TR rule had asymptotic and amortized costs of (8.72, 10.52) and (8.71, 10.93) respectively. Hence, showing the superiority of the TPEOMA-augmented schemes to the MTF and TR rules respectively in the MSE. Also, from Table 1, observe that while the TPEOMA-augmented hierarchical schemes are superior to the EOMA-augmented hierarchical schemes, the PEOMA-augmen-ted hierarchical still boasts superior performances in MSEs.

Figure 5 accesses the asymptotic cost ratio of the MTF-MTF-TPEOMA scheme to the MTF for varying number of sublist partitions ranging from $k = \{2, 4, 8, 10, 16, 32, 64\}$. For all the query distributions under consideration, the MTF-MTF-TPEOMA scheme possesses a superior asymptotic cost to the MTF except for the Exponential scheme where the MTF has a better asymptotic cost when $k = \{2, 4\}$.

Table 1. Asymptotic (top) and Amortized (bottom) costs in **MSE** with $\alpha = 0.9$ and $k = 8$.

Scheme	Zipf	80-20	Lotka	Exp.	Linear
MTF	43.35	43.76	39.30	8.72	43.60
TR	55.44	56.74	48.25	10.52	56.79
MTF-MTF-EOMA	19.14	19.23	18.70	12.34	19.31
MTF-TR-EOMA	27.80	27.77	27.17	16.89	28.04
TR-MTF-EOMA	18.84	18.99	18.37	12.87	18.96
TR-TR-EOMA	27.55	27.62	26.96	17.17	27.70
MTF-MTF-PEOMA	5.80	6.73	1.25	2.45	6.76
MTF-TR-PEOMA	5.35	6.21	1.25	2.97	6.77
TR-MTF-PEOMA	4.67	6.00	0.96	2.88	5.61
TR-TR-PEOMA	5.07	6.49	0.98	2.99	6.34
MTF-MTF-TPEOMA	13.45	10.73	6.60	7.01	9.62
MTF-TR-TPEOMA	11.51	10.97	6.11	6.99	8.74
TR-MTF-TPEOMA	12.50	13.42	4.39	7.50	8.09
TR-TR-TPEOMA	14.80	12.38	4.37	7.64	8.30
MTF	43.25	43.82	39.17	8.71	43.64
TR	55.85	56.96	48.66	10.93	57.26
MTF-MTF-EOMA	19.35	19.40	19.26	12.90	19.45
MTF-TR-EOMA	27.93	28.02	27.54	16.57	28.08
TR-MTF-EOMA	10.09	19.18	19.07	13.35	19.18
TR-TR-EOMA	27.72	27.80	27.25	17.10	27.87
MTF-MTF-PEOMA	6.97	7.77	2.32	4.00	7.79
MTF-TR-PEOMA	7.14	7.87	2.63	4.23	8.31
TR-MTF-PEOMA	5.95	7.13	2.04	4.65	6.71
TR-TR-PEOMA	6.84	8.05	2.29	4.77	7.91
MTF-MTF-TPEOMA	15.44	13.04	8.09	8.98	11.56
MTF-TR-TPEOMA	14.08	13.41	7.84	8.83	11.00
TR-MTF-TPEOMA	14.81	15.62	6.57	9.66	10.15
TR-TR-TPEOMA	16.88	15.07	6.54	9.87	10.67

Table 2. Asymptotic (top) and Amortized (bottom) costs in **PSE** with $T = 30$ and $k = 8$.

Scheme	Zipf	80-20	Lotka	Exp.	Linear
MTF	49.64	50.24	44.52	8.46	50.08
TR	55.65	56.91	48.51	11.18	57.19
MTF-MTF-EOMA	14.63	14.70	14.12	8.59	14.72
MTF-TR-EOMA	25.82	25.90	25.32	13.88	25.92
TR-MTF-EOMA	14.39	14.49	13.76	8.92	14.50
TR-TR-EOMA	25.58	25.69	24.97	13.70	25.70
MTF-MTF-EOMA-P	7.16	7.24	6.66	6.14	7.26
MTF-MTF-EOMA-UP	7.69	7.78	7.19	8.90	7.79
MTF-MTF-PEOMA	11.80	11.29	10.57	4.10	10.30
MTF-TR-PEOMA	23.31	23.22	21.64	5.56	21.42
TR-MTF-PEOMA	12.00	11.04	10.21	4.32	10.04
TR-TR-PEOMA	20.04	20.11	19.28	5.56	21.17
MTF-MTF-PEOMA-P	7.13	7.21	6.53	6.10	7.21
MTF-MTF-PEOMA-UP	7.40	7.57	5.45	6.31	7.49
MTF-MTF-TPEOMA	12.30	12.16	12.40	11.57	11.50
MTF-TR-TPEOMA	12.79	12.62	12.01	10.37	12.03
TR-MTF-TPEOMA	12.40	12.26	12.48	10.60	12.26
TR-TR-TPEOMA	12.56	12.41	12.37	11.00	12.18
MTF-MTF-TPEOMA-P	8.95	9.19	9.56	12.56	10.49
MTF-MTF-TPEOMA-UP	12.25	11.91	13.65	12.50	14.50
MTF	49.62	50.23	44.53	8.48	50.06
TR	56.09	57.28	48.91	11.58	57.60
MTF-MTF-EOMA	14.76	14.84	14.31	8.68	14.84
MTF-TR-EOMA	25.93	26.01	25.44	12.49	26.02
TR-MTF-EOMA	14.54	14.62	14.03	9.69	14.63
TR-TR-EOMA	25.71	25.80	25.12	13.11	25.80
MTF-MTF-EOMA-P	7.28	7.38	6.82	7.53	7.40
MTF-MTF-EOMA-UP	7.86	7.95	7.48	10.57	7.92
MTF-MTF-PEOMA	12.44	11.35	10.81	4.88	10.37
MTF-TR-PEOMA	23.84	23.78	21.64	5.90	21.63
TR-MTF-PEOMA	12.59	12.13	10.41	5.25	10.13
TR-TR-PEOMA	17.33	17.05	19.32	6.32	21.22
MTF-MTF-PEOMA-P	7.27	7.34	6.75	7.07	7.36
MTF-MTF-PEOMA-UP	7.44	7.68	5.51	6.90	7.59
MTF-MTF-TPEOMA	12.63	12.50	12.49	12.34	11.92
MTF-TR-TPEOMA	12.72	12.57	12.51	12.13	11.96
TR-MTF-TPEOMA	12.69	12.54	12.51	12.00	11.97
TR-TR-TPEOMA	12.65	12.50	12.50	11.93	11.98
MTF-MTF-TPEOMA-P	9.55	9.47	9.50	10.12	10.35
MTF-MTF-TPEOMA-UP	13.33	13.08	13.09	14.86	15.22

In Table 2, the performance of the TPEOMA-augmented hierarchical schemes in the PSEs were superior to the standalone MTF and TR rules for all distributions under consideration except for the Exponential scheme which boasted slightly comparable results. As an example, consider the 80-20 distribution where the MTF-MTF-TPEOMA, MTF-TR-TPEOMA, TR-MTF-TPEOMA and TR-TR-TPEOMA had asymptotic and amortized costs of (12.16, 12.62, 12.26, 12.41) and (12.50, 12.57, 12.54, 12.50) respectively. Whereas the MTF and TR had asymptotic and amortized costs of (50.24, 56.91) and (50.23, 57.28) respectively showing the superiority of the TPEOMA-augmented schemes in such Environments. Moreover, the results from Table 2 also indicate that the TPEOMA-augmented schemes have superior performances to EOMA-augmented and PEOMA-augmented hierarchical schemes in PSEs for the query distributions under consideration.

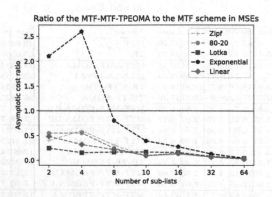

Fig. 5. The asymptotic cost ratio of the MTF-MTF-TPEOMA to the MTF scheme for different values of the sub-list partitions $k = \{2, 4, 8, 10, 16, 32, 64\}$ for MSE-dependent query Environments in which $\alpha = 0.9$.

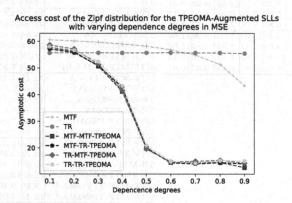

Fig. 6. Changes in the asymptotic cost of the stand-alone and hierarchical schemes with TPEOMA in the MSE. In this experiment, a list of 128 elements is partitioned into 8 sub-lists.

Also, when the knowledge of the Periodic state change is passed to the TPEOMA-augmented hierarchical schemes, we observed that the *"periodic"* variations yielded superior performances to their vanilla versions, whereas the *unperiodic* variations had comparable performances to the vanilla versions.

From Fig. 6, we see that the TPEOMA-augmented hierarchical schemes were superior to the MTF and TR schemes in noisy Environments when the dependence degree $\alpha > 0.2$. Actually, when $\alpha = 0.2$, the TPEOMA enhanced schemes already displayed comparable (and in some cases better) performances to the MTF and TR schemes for the Zipf distribution.

5   Conclusion

In this paper, we designed a TPEOMA-augmented hierarchical Singly-Linked-Lists on Singly-Linked-Lists (SLLs-on-SLLs) using reinforcement learning schemes from the theory of Learning Automata, which led to the MTF-MTF-TPE-OMA, MTF-TR-TPEOMA, TR-MTF-TPEOMA and TR-TR-TPEOMA schemes. The TPEOMA-augmented hierarchical schemes showed superior performances to the standalone MTF and TR schemes in MSEs. Also, they are superior to the EOMA-augmented hierarchical schemes. However, the PEOMA-augmented hierarchical schemes are still superior to the TPEOMA-augmented hierarchical schemes in the MSE. In the PSE, the TPEOMA-augmented hierarchical were for the most part superior to the MTF and TR schemes for the query distributions under consideration. Further, the TPEOMA-augmented schemes were also superior to the EOMA-augmented and PEOMA-augmented hierarchical schemes in the PSE. However, when the knowledge of the periodic state change were incorporated to the hierarchical schemes, the "periodic" case had superior performances to their vanilla versions while the "unperiodic" case had comparable performances.

References

1. Amer, A.: Adaptive list organizing strategies for non-stationary distributions (2004)
2. Bachrach, R., El-Yaniv, R., Reinstadtler, M.: On the competitive theory and practice of online list accessing algorithms. Algorithmica **32**(2), 201–245 (2002)
3. Bellman, R.: Adaptive Control Processes: A Guided Tour. Princeton University Press, Princeton (1961)
4. Bentley, J.L., McGeoch, C.C.: Amortized analyses of self-organizing sequential search heuristics. Commun. ACM **28**(4), 404–411 (1985)
5. Bisong, E.O.: On designing adaptive data structures with adaptive data "sub"-structures. Master's thesis, Carleton University (2018)
6. Bisong, O.E., Oommen, B.J.: Optimizing self-organizing lists-on-lists using enhanced object partitioning. In: MacIntyre, J., Maglogiannis, I., Iliadis, L., Pimenidis, E. (eds.) AIAI 2019. IAICT, vol. 559, pp. 451–463. Springer, Cham (2019). https://doi.org/10.1007/978-3-030-19823-7_38
7. Bisong, O.E., Oommen, B.J.: Optimizing self-organizing lists-on-lists using pursuit-oriented enhanced object partitioning. In: Huang, D.-S., Huang, Z.-K., Hussain, A. (eds.) ICIC 2019. LNCS (LNAI), vol. 11645, pp. 201–212. Springer, Cham (2019). https://doi.org/10.1007/978-3-030-26766-7_19

8. Chassaing, P.: Optimality of move-to-front for self-organizing data structures with locality of references. Ann. Appl. Probab. 1219–1240 (1993)
9. Dong, J.: Time reversible self-organizing sequential search algorithms (1998)
10. Gale, W., Das, S., Yu, C.T.: Improvements to an algorithm for equipartitioning. IEEE Trans. Comput. **39**(5), 706–710 (1990)
11. Hester, J.H., Hirschberg, D.S.: Self-organizing linear search. ACM Comput. Surv. (CSUR) **17**(3), 295–311 (1985)
12. Knuth, D.E.: The Art of Computer Programming, vol. 3. Pearson Education (1997)
13. McCabe, J.: On serial files with relocatable records. Oper. Res. **13**(4), 609–618 (1965)
14. Narendra, K.S., Thathachar, M.A.L.: Learning Automata: An Introduction. Courier Corporation (2012)
15. Oommen, J., Dong, J.: Generalized swap-with-parent schemes for self-organizing sequential linear lists. In: Leong, H.W., Imai, H., Jain, S. (eds.) ISAAC 1997. LNCS, vol. 1350, pp. 414–423. Springer, Heidelberg (1997). https://doi.org/10.1007/3-540-63890-3_44
16. Oommen, B.J., Ma, D.C.Y.: Deterministic learning automata solutions to the equipartitioning problem. IEEE Trans. Comput. **37**(1), 2–13 (1988)
17. Rivest, R.: On self-organizing sequential search heuristics. Commun. ACM **19**(2), 63–67 (1976)
18. Shirvani, A.: Novel solutions and applications of the object partitioning problem. Ph.D. thesis, Carleton University, Ottawa (2018)
19. Shirvani, A., Oommen, B.J.: The advantages of invoking transitivity in enhancing pursuit-oriented object migration automata (2017)
20. Tsetlin, M.L.: Finite automata and models of simple forms of behaviour. Russ. Math. Surv. **18**, 1–27 (1963)

Task-Projected Hyperdimensional Computing for Multi-task Learning

Cheng-Yang Chang[✉], Yu-Chuan Chuang, and An-Yeu (Andy) Wu

Graduate Institute of Electronics Engineering, National Taiwan University, Taipei, Taiwan
{kevin,frankchuang}@access.ee.ntu.edu.tw, andywu@ntu.edu.tw

Abstract. Brain-inspired Hyperdimensional (HD) computing is an emerging technique for cognitive tasks in the field of low-power design. As an energy-efficient and fast learning computational paradigm, HD computing has shown great success in many real-world applications. However, an HD model incrementally trained on multiple tasks suffers from the negative impacts of catastrophic forgetting. The model forgets the knowledge learned from previous tasks and only focuses on the current one. To the best of our knowledge, no study has been conducted to investigate the feasibility of applying multi-task learning to HD computing. In this paper, we propose Task-Projected Hyperdimensional Computing (TP-HDC) to make the HD model simultaneously support multiple tasks by exploiting the redundant dimensionality in the hyperspace. To mitigate the interferences between different tasks, we project each task into a separate subspace for learning. Compared with the baseline method, our approach efficiently utilizes the unused capacity in the hyperspace and shows a 12.8% improvement in averaged accuracy with negligible memory overhead.

Keywords: Hyperdimensional Computing · Multi-task learning · Redundant dimensionality

1 Introduction

In the era of IoT, edge computing with energy-efficient machine learning models keeps data processing close to end-users. This brings out numerous advantages, including lower latency, user security, and cost savings [1]. Meanwhile, multi-task learning (MTL) is grabbing attention recently since a single model can accommodate multiple cognitive tasks is more desirable for the future of IoT [2].

Brain-inspired Hyperdimensional (HD) computing emulates the operations of brains and handles cognitive tasks in a hyperdimensional space with well-defined vector space operations [3]. As an energy-efficient and fast-learning computational paradigm, HD computing has shown successful progress in many real-world applications such as gesture recognition [4], language recognition [5], and general bio-signal processing [6, 7]. Moreover, HD computing can operate at an ultra low-power condition with lower latency through massively parallel bitwise operation [8]. These advantages make HD computing suitable for efficient signal processing, e.g., 2× lower energy at iso-accuracy when

© IFIP International Federation for Information Processing 2020
Published by Springer Nature Switzerland AG 2020
I. Maglogiannis et al. (Eds.): AIAI 2020, IFIP AICT 583, pp. 241–251, 2020.
https://doi.org/10.1007/978-3-030-49161-1_21

compared to a highly-optimized SVM on an ARM Cortex M4 [9]. However, an HD model incrementally trained on multiple tasks forgets the knowledge learned from previous tasks and only focuses on the current one. The phenomenon is called catastrophic forgetting [10]. To the best of our knowledge, no study has been conducted to overcome this problem and investigate the feasibility of applying MTL to HD computing.

This paper aims to establish a reliable MTL framework based on HD computing to minimize the negative impact of catastrophic forgetting. Over-parameterization in DNN implies that only a small subspace spanned by the optimal parameters is occupied by a given task [11]. Based on this phenomenon, [12] exploits the redundant subspace in DNN to superimpose multiple models into one. We are inspired by this concept and propose to exploit the unused capacity in the hyperspace to project each task into a separate subspace for learning. Our approach efficiently mitigates the interferences between different tasks and keeps the knowledge learned from numerous tasks stored in one HD model with minimal accuracy degradation.

The rest of the paper is organized as follows. Section 2 provides a review of HD computing. Section 3 describes the proposed Task-projected Hyperdimensional Computing (TP-HDC) for multi-task learning. Section 4 shows our experiment setting and simulation results. Finally, we conclude this paper in Sect. 5.

2 Review of Hyperdimensional Computing

HD computing is based on high-dimensional and dense binary vectors, called HD vectors. The components of HD vectors are binary with equally probable (-1)s and 1s. The processing flow chart of a general HD computing is shown in Fig. 1 and can be divided into the following four stages:

Nonlinear Mapping to Hyperspace: The main goal of mapping is to project a feature vector x to HD vectors with dimensionality (d), where $x \in R^m$ with m components. Feature identifier (ID) is regarded as a basic field, and the actual value of the feature is the filler of the field. HD computing starts by constructing Item Memory (IM) and Continuous item Memory (CiM). $IM = \{ID_1, ID_2, \ldots, ID_m\}$, where $ID_k \in (-1, 1)^d$, $k \in \{1, 2, \ldots m\}$ corresponds to the ID of the kth feature component. When d is large enough, any two different HD vectors in IM are nearly orthogonal, implying that $Cos(ID_i, ID_j) \cong 0$, $Ham(ID_i, ID_j) \cong 0.5$, if $i \neq j$ [13]. $Cos(\cdot)$ and $Ham(\cdot)$ are cosine similarity metric and normalized Hamming distance between the two vectors, respectively.

Continuous item memory (CiM) serves as the look-up table for the actual value of a feature. The procedure of establishing CiM first finds the maximum value and minimum value of each feature denoted as V_{max} and V_{min}. The range between V_{max} and V_{min} is quantized to ℓ levels, and then an HD vector $L_1 \in (-1, 1)^d$ is assigned to V_{max}, and $L_\ell \in (-1, 1)^d$ is assigned to V_{min}. The HD model determines L_1 and L_ℓ at random, making them approximately orthogonal. $CiM = \{L_1, L_2, \ldots, L_\ell\}$, where $L_k \in (-1, 1)^d$, $k \in \{1, 2, \ldots \ell\}$, and every vector in CiM corresponds to a range of actual value. The spatial relation of levels is preserved through adjusting the Hamming distance between L_i and L_j according to the difference of value to which the two HD vectors correspond. In other words, each value of the specific feature component will be associated with a vector proportionate to L_1 and L_ℓ. Mapping of each feature value to hyperspace comprises quantizing and looking up the corresponding vectors $\{S_1, S_2, \ldots, S_m\}$ in CiM. After mapping each feature component of data to HD vectors, a set of two-vector pairs $I = \{(ID_1, S_1), (ID_2, S_2), \ldots, (ID_m, S_m)\}$ can readily be used in the next stage with the vector space operations.

Encoding: The HD model conducts the binding operation, bitwise XOR operation ($\oplus$) between two HD vectors, for each two-vector pair in I. After that, the resulting m HD vectors in the set I is accumulated by the bundling operation, bitwise addition ($+$) between HD vectors. Followed by binarization with sign function denoted as $[\cdot]$, data can be encoded as (1) and represented by the resulting binary HD vector $T \in (-1, 1)^d$.

$$T = \sum_{i=1}^{m} ID_i \oplus S_i = [ID_1 \oplus S_1 + ID_2 \oplus S_2 + \ldots + ID_m \oplus S_m], \tag{1}$$

Training: All training samples go through the previous two stages and the resulting vector T is sent to the associative memory (AM) for training. Training samples of the same class denoted as T_i for the ith class are bundled together to form a class HD vector, as shown in (2). n_i means the number of training samples of the ith class. For a k-class classification task, AM comprises k class HD vectors, denoted as $\{C_1, C_2, \ldots, C_k\}$.

$$C_i = \sum_j T_i^j = \left[T_i^1 + T_i^2 + \ldots + T_i^{n_i}\right], \tag{2}$$

Classification: In the inference phase, an unseen testing data would go through the same processing flow of mapping and encoding in the training phase and be encoded as a query vector $Q \in (-1, 1)^d$. To perform classification, the HD model checks the similarity between Q and all class HD vectors stored in AM by the Hamming distance metric. Finally, the HD model outputs the class with the minimum distance as the prediction.

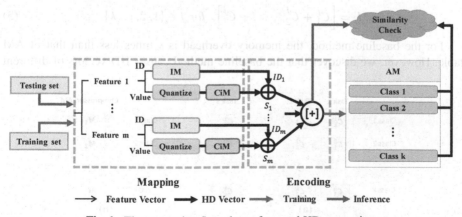

Fig. 1. The processing flow chart of general HD computing.

3 Proposed Task-Projected Hyperdimensional Computing

In this section, we propose Task-projected Hyperdimensional Computing (TP-HDC) to realize multi-task learning in an HD model. We are inspired by [12], which exploits the over-parameterization in DNN to superimpose multiple models into one. This implies that only a small subspace spanned by the optimal parameters is occupied by a given task. We observe that HD computing shows a similar phenomenon, where only a small subspace spanned by class HD vectors in AM is relevant to a given task. Based on this observation, MTL can be feasible if the massive hyperspace is partitioned appropriately for each task.

3.1 AM Table for Multi-task Learning

Before diving into the illustration of the proposed scheme, we first introduce the definition of AM table supporting multiple tasks and its notation. Following the training flow described in Sect. 2, each task in task sequence $\{T_1, T_2, \ldots, T_s\}$ generates its own AM. A total of s AM are present and form a 2-dimensional AM table, as shown in Fig. 2(a). Each column of the table comprises k class HD vectors. We notate the jth class vector of the ith task as C_i^j. If an original HD model needs to support multiple tasks, the memory requirement of storing the AM table grows linearly with the number of tasks. For resource-constrained edge devices, the memory overhead could hinder HD computing from MTL. As a result, it is more desirable to store a compressed AM with a size that is independent of the number of tasks, as shown in Fig. 2(b).

Baseline Method: Considering the jth class in Fig. 2(a), the baseline method bundles the class HD vectors of the same class from all involved tasks. As shown in (3), s HD vectors in the jth class $\left\{C_1^j, C_2^j, \ldots, C_s^j\right\}$ are bundled together and form the vector M_j, which is shared across all tasks. Compressed AM comprises $\{M_1, M_2, \ldots, M_k\}$, where M_j represents the jth class HD vector used by all tasks. That is, the baseline method naïvely finds the most representative vector in the hyperspace regardless of the spatial relation between tasks.

$$M_j = \left[C_1^j + C_2^j + \ldots + C_s^j\right], \; for \; j \in \{1, 2, \ldots k\}, \tag{3}$$

For the baseline method, the memory overhead is s times less than that of AM table. However, we discover that the baseline method causes HD vectors of different

	Task 1	Task 2	Task s		Compressed AM
Class 1	C_1^1	C_2^1	C_s^1	Class 1	M_1
Class 2	C_1^2	C_2^2 ...	C_s^2	Class 2	M_2
	⋮				⋮
Class k	C_1^k	C_2^k	C_s^k	Class k	M_k
	(a)				(b)

Fig. 2. (a) AM table for original HD computing to support multiple tasks. (b) Compressed AM.

tasks to occupy overlapping subspace. This induces interference between tasks and significant accuracy degradation. In the next section, we concretize the proposed TP-HDC to efficiently realize MTL in an HD model with a lower accuracy drop.

3.2 Orthogonalization with Task-Oriented Keys

The TP-HDC consists of the following three parts, including generation of task-oriented keys, composition with task-oriented projection, and decomposition:

Generation of Task-Oriented Keys: We propose to leverage the peculiar property in the Hamming space, the normalized Hamming distance from any given point in the hyperspace to a randomly drawn point highly concentrates at 0.5 [3]. Namely, two random HD vectors are approximately orthogonal (unrelated) due to hyper-dimensionality. Based on this fact, each task is assigned a task-oriented key generated at random, denoted as $\{P_1, P_2, \ldots, P_s\}$. These keys can be used for projection to achieve a division of the hyperspace in the following step of TP-HDC.

Composition with Task-Oriented Projection: To utilize the unused capacity of the HD model more efficiently, orthogonalization of class HD vectors of the same class, e.g., $\left\{C_1^j, C_2^j, \ldots, C_s^j\right\}$ for the jth class is required. With task-oriented keys generated in the previous step, we bind the keys and the class HD vectors for each task. The effect of binding projects originally close HD vectors $\left\{C_1^j, C_2^j, \ldots, C_s^j\right\}$ to different zones of the hyperspace since pseudo-randomly generated keys are approximately orthogonal. The new class HD vector (M_j) is formed by bundling the generated s HD vectors, as shown in (4). By projecting the class HD vectors of different tasks into near-orthogonal hyperspaces, TP-HDC can mitigate the information loss caused by directly bundling class HD vectors, as implemented by the baseline method.

$$M_j = \sum_{i=1}^{s} C_i^j \oplus P_i = \left[C_1^j \oplus P_1 + C_2^j \oplus P_2 + \ldots + C_s^j \oplus P_s \right], \, for \, j \in \{1, 2, \ldots k\},$$

(4)

Decomposition: Retrieval of class HD vector of the jth class in m^{th} task C_m^j, is ensured by binding M_j with P_m, as shown in (5). The resulting vector consists of C_m^j and noise $\in$ because the vectors are stored in superposition. Despite the presence of $\in$, TP-HDC can still be reliable because HD computing is robust against noise [3].

$$\hat{C}_j = M_j \oplus P_m = \left[C_1^j \oplus P_1 + C_2^j \oplus P_2 + \ldots + C_s^j \oplus P_s \right] \oplus P_m$$

$$= \left[C_1^j \oplus P_1 \oplus P_m + \ldots + C_m^j \oplus P_m \oplus P_m + \ldots + C_s^j \oplus P_s \oplus P_m \right]$$

$$= \left[C_m^j + \in \right], for \, j \in \{1, 2, \ldots k\}$$

(5)

3.3 Training and Inference in TP-HDC

The framework of TP-HDC is depicted in Fig. 3, and the procedure of training and inference is summarized in Algorithm 1:

Training: Given a task sequence, $T = \{T_1, T_2, \ldots, T_s\}$ each with k classes, s different AMs are updated using the general HD computing training flow described in Sect. 2. $C \in \mathbb{R}^{k \times s \times d}$ is a three-dimensional matrix. The first two axes of C represent the number of classes and tasks, respectively, and the last axis of C represents the dimensionality of HD vectors. We first generate task-oriented projection keys and denoted them as $P = \{P_1, P_2, \ldots, P_s\}$. P helps achieve a division of space and project class HD vectors of different tasks to separate subspaces with Eq. (4). The compressed AM is $M \in \mathbb{R}^{k \times d}$, whose size is independent of the number of tasks.

Inference: Given a task, the HD model produces the query HD vector Q in the inference phase by processing a testing sample with the same mapping and encoding modules used in the training phase. After retrieving all class HD vectors of the specific task $\hat{C} = \left\{\hat{C}_1, \hat{C}_2, \ldots, \hat{C}_k\right\}$ with Eq. (5), the classification result is the class in which the corresponding class HD vector has the smallest Hamming distance with Q, see Eq. (6).

$$Prediction = \underset{j}{argmin}\, Ham\left(\hat{C}_j, Q\right), \tag{6}$$

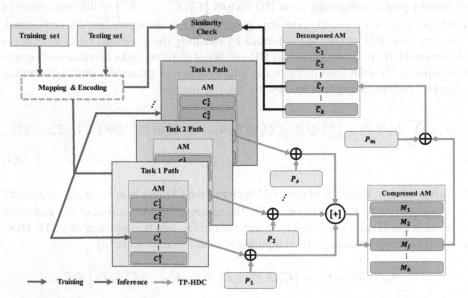

Fig. 3. The framework of the proposed task-projected HD computing (TP-HDC).

Algorithm 1 Task-projected HD Computing

    Input: $T = \{T_1, T_2, T_3, \ldots, T_s\}$ – task sequence, each with k classes

              $mode$ – training phase or inference phase of the mth task

              d – dimensionality of the HD model

    Output: $P = \{P_1, P_2, P_3, \ldots, P_s\}$ – task-oriented projection keys

              M – compressed AM

              Y – prediction

1:    **Initialize** $C \leftarrow 0_{k,s,d}$; $M \leftarrow 0_{k,d}$; $\hat{C} \leftarrow 0_{k,d}$

2:    **if** $mode$ = training **do**

3:       **for** each task in T **do**

4:          **for** each training data **do**

5:             update C

6:          generate projection keys P

7:          **for** $(j = 1:k)$ **do**

8:             $M[j] = \sum_{i=1}^{s} C[j][i] \oplus P_i$

9:    **if** $mode$ = inference **do**

10:      **for** $(j = 1:k)$ **do**

11:         $\hat{C}[j] = M[j] \oplus P_m$

12:     make prediction Y based on $\hat{C}$

4 Experimental Settings and Simulation Results

4.1 Comparisons

We compare the proposed TP-HDC with two different approaches tackling the multi-task learning problem in HD computing, including the baseline method and the ideal method.

Baseline Method: As mentioned in Sect. 3.1, the baseline method naïvely finds the most representative vector among all tasks with bundling operation, causing severe interference between different tasks. Therefore, the baseline method can be regarded as the model telling us what happens if we do nothing to explicitly retain information from the previous tasks.

Ideal Method: The ideal benchmark considers the case where computing resources are unconstrained so that all tasks can have their own AM. Therefore, the performance of the ideal method can be viewed as the upper bound of our evaluation since the class HD vectors are stored without any information loss.

4.2 Dataset and Experimental Setup

We evaluate the effectiveness of our proposed TP-HDC on Split MNIST, a standard benchmark for multi-task learning [14]. Following the experiment setting of [14] with minor modifications, we split ten digits into disjoint sets. Each set corresponds to a specific task in $T = \{T_1, T_2, \ldots, T_s\}$, where T_i aims at discriminating between k_i digits $\left\{D_i^1, D_i^2, \ldots, D_i^{k_i}\right\}$. We fix the dimensionality of HD computing at $d = 5000$, where the performances of all HD computing models saturate. Mapping and encoding modules are shared across all tasks, meeting the expectations of MTL. Moreover, we vectorize the gray images of digits in MNIST to form 784-dimensional feature vectors and pre-process pixel values using min-max normalization. All experiments are conducted on 100 independent runs to get the final averaged simulation results.

4.3 Performance Analysis

First, we evaluate our proposed TP-HDC with a three-task MTL configuration. Each of the tasks, namely task A, task B, and task C contains three digits different from those of the other two tasks. HD models are sequentially trained on task A, task B, and task C. We observe that 100 training samples are enough for the convergence of all HD models in each task. Therefore, we train each task for 100 steps, and a training sample is randomly drawn to update AM in each step.

Figure 4 illustrates the learning curve of the different methods on split MNIST. Compared with the ideal method, the baseline method suffers from catastrophic forgetting, resulting in around 20% accuracy drop on task A and task B. Furthermore, task A has occupied the subspace that task B and task C need to learn for classification, causing information loss for task B and task C and bringing about a 15% accuracy drop. On the other hand, the accuracy of the proposed TP-HDC only drops by 3.6%, 3.9%, 2.9% on task A, task B, and task C, respectively.

To validate the generalization ability of our model, we also evaluate TP-HDC on the five-task case. The experimental setup is almost the same as the three-task case except that each task contained two digits. For the baseline method, Fig. 5 shows that the five tasks tend to interfere with each other severely like that in the three-task case, leading to a 16.5% accuracy drop. In comparison, TP-HDC provides around 12.8% improvement in averaged accuracy compared with the baseline method and performs closely to the ideal benchmark consistently, with a slight 3.7% accuracy drop on average. By efficiently separating the subspaces, TP-HDC mitigates the effect of interference between tasks and improves the performance of sequential training on multiple tasks.

Table 1 shows the performance of TP-HDC on split MNIST in different cases. The high standard deviation of the accuracy of the baseline method implies instability. By contrast, the results demonstrate both effectiveness (<4% accuracy drop compared with the ideal benchmark) and stability (lower variance compared with the baseline method) of TP-HDC.

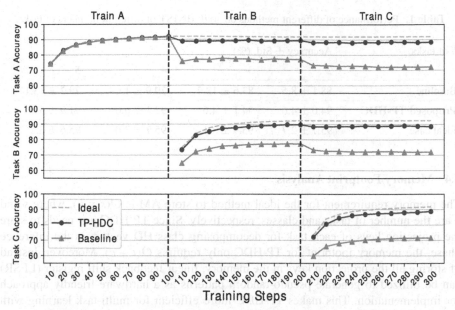

Fig. 4. The learning curves of different methods trained on split MNIST. Tasks A, B, and C are assigned with disjoint sets of MNIST digits. The vertical dashed lines imply the transitions of the training procedure of different tasks. The top plot shows the accuracy of task A. The middle plot and bottom plot indicate the accuracy of tasks B and C, respectively.

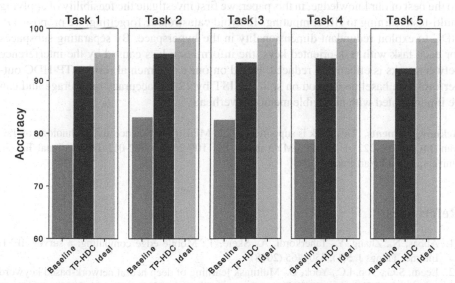

Fig. 5. Classification accuracy of the five-task case of split MNIST. TP-HDC provides around 12.8% improvement in averaged accuracy compared with the baseline method.

Table 1. Performance of different methods on split MNIST of different numbers of tasks.

# of tasks	Accuracy ± Std. (%)			
	2	3	4	5
Baseline	88.1 ± 8.5	83.6 ± 12.2	79.9 ± 13.7	79.2 ± 13.1
Proposed TP-HDC	94.0 + 4.6	94.1 ± 4.6	92.1 ± 6.0	91.9 ± 5.9
Ideal case	95.7 ± 2.7	95.9 ± 3.1	95.7 ± 3.0	95.6 ± 3.2

4.4 Memory Footprint Analysis

The memory requirement for the ideal method to store AM is $\mathcal{O}(s \times k)$, where s and k are the number of tasks and classes, respectively. Since TP-HDC just needs to store the projection keys of each task for decomposing class HD vectors in the inference phase, the memory footprint for TP-HDC only requires $\mathcal{O}(t + c)$. Moreover, instead of storing all the projection keys in the memory, linear feedback shift register (LFSR) can be utilized to generate pseudo-random patterns as a hardware-friendly approach for implementation. This makes TP-HDC more efficient for multi-task learning with negligible memory overhead.

5   Conclusions

To the best of our knowledge, in this paper, we first investigate the feasibility of applying multi-task learning to HD computing. To avoid catastrophic forgetting, we propose TP-HDC to exploit redundant dimensionality in the hyperspace. By separating subspaces for each task with task-oriented keys, the information loss caused by the interference between tasks is effectively reduced. Based on our experimental results, TP-HDC outperforms the baseline method on split MNIST by 12.8% accuracy on average and can be implemented with negligible memory overhead.

Acknowledgments. This work is supported by the Ministry of Science and Technology of Taiwan (MOST 106-2221-E-002-205-MY3 and MOST 109-2622-8-002-012-TA), National Taiwan University, and Pixart Imaging Inc.

References

1. Abbas, N., Zhang, Y., Taherkordi, A., Skeie, T.: Mobile edge computing: a survey. IEEE Internet Things J. **5**(1), 450–465 (2018)
2. Leem, S.G., Yoo, I.C., Yook, D.: Multitask learning of deep neural network-based keyword spotting for IoT devices. IEEE Trans. Consum. Electron. **65**(2), 188–194 (2019)
3. Kanerva, P.: Hyperdimensional computing: an introduction to computing in distributed representation with high-dimensional random vectors. Cogn. Comput. **1**(2), 139–159 (2009)
4. Rahimi, A., Benatti, S., Kanerva, P., Benini, L., Rabaey, J.M.: Hyperdimensional biosignal processing: a case study for EMG-based hand gesture recognition. In: IEEE International Conference on Rebooting Computing (ICRC), San Diego, CA, pp. 1–8 (2016)

5. Imani, M., Hwang, J., Rosing, T., Rahimi, A., Rabaey, J.M.: Low-power sparse hyperdimensional encoder for language recognition. IEEE Design Test **34**(6), 94–101 (2017)
6. Rahimi, A., Kanerva, P., Benini, L., Rabaey, J.M.: Efficient biosignal processing using hyperdimensional computing: network templates for combined learning and classification of ExG signals. Proc. IEEE **107**(1), 123–143 (2018)
7. Chang, E.J., Rahimi, A., Benini, L., Wu, A.Y.: Hyperdimensional computing-based multi-modality emotion recognition with physiological signals. In: IEEE International Conference on Artificial Intelligence Circuits and Systems (AICAS), pp. 137–141 (2019)
8. Salamat, S., Imani, M., Khaleghi, B., Rosing, T.: F5-HD: fast flexible FPGA-based framework for refreshing hyperdimensional computing. In: Proceedings of the 2019 ACM/SIGDA International Symposium on Field-Programmable Gate Arrays (2019)
9. Montagna, F., Rahimi, A., Benatti, S., Rossi, D., Benini, L.: PULP-HD: accelerating brain-inspired high-dimensional computing on a parallel ultra-low power platform. In: 55th ACM/ESDA/IEEE Design Automation Conference (DAC), pp. 1–6 (2018)
10. French, R.M.: Catastrophic forgetting in connectionist networks. Trends Cogn. Sci. **3**(4), 128–135 (1999)
11. Allen-Zhu, Z., Li, Y., Song, Z.: A convergence theory for deep learning via over-parameterization. arXiv preprint arXiv:1811.03962 (2018)
12. Cheung, B., Terekhov, A., Chen, Y., Agrawal, P., Olshausen, B.: Superposition of many models into one. In: Advances in Neural Information Processing Systems, pp. 10867–10876 (2019)
13. Plate, T.A.: Holographic reduced representations. IEEE Trans. Neural Netw. **6**(3), 623–641 (1995)
14. Zenke, F., Poole, B., Ganguli, S.: Continual learning through synaptic intelligence. In: Proceedings of the 34th International Conference on Machine Learning, vol. 70, pp. 3987–3995. JMLR.org (2017)

Neural Network Modeling

Cross-Domain Authorship Attribution
Using Pre-trained Language Models

Georgios Barlas$^{(\boxtimes)}$ and Efstathios Stamatatos

University of the Aegean, 83200 Karlovassi, Greece
barlasgeorgios@gmail.com, stamatatos@aegean.gr

Abstract. Authorship attribution attempts to identify the authors
behind texts and has important applications mainly in cyber-security,
digital humanities and social media analytics. An especially challenging but very realistic scenario is cross-domain attribution where texts
of known authorship (training set) differ from texts of disputed authorship (test set) in topic or genre. In this paper, we modify a successful
authorship verification approach based on a multi-headed neural network
language model and combine it with pre-trained language models. Based
on experiments on a controlled corpus covering several text genres where
topic and genre is specifically controlled, we demonstrate that the proposed approach achieves very promising results. We also demonstrate the
crucial effect of the normalization corpus in cross-domain attribution.

Keywords: Authorship Attribution · Neural network language
models · Pre-trained language models

1 Introduction

Authorship Attribution (AA) is a very active area of research dealing with the
identification of persons who wrote specific texts [12,20]. Typically, there is a list
of suspects and a number of texts of known authorship by each suspect and the
task is to assign texts of disputed authorship to one of the suspects. The basic
forms of AA are closed-set attribution (where the list of suspects necessarily
includes the true author), open-set attribution (where the true author could be
excluded from the list of suspects), and author verification (where there is only
one candidate author). The main applications of this technology are in digital
forensics, cyber-security, digital humanities, and social media analytics [8,15].

In real life scenarios the known and the unknown texts may not share the
same properties. The topic of the texts may differ but also the genre (e.g., essay,
email, chat). Cross-domain AA examines those cases where the texts of known
authorship (training set) differ with respect to the texts of unknown authorship
(test set) in topic (cross-topic AA) or in genre (cross-genre AA) [19,22]. The
main challenge here is to avoid the use of information related to topic or genre of

Published by Springer Nature Switzerland AG 2020
I. Maglogiannis et al. (Eds.): AIAI 2020, IFIP AICT 583, pp. 255–266, 2020.
https://doi.org/10.1007/978-3-030-49161-1_22

documents and focus only on stylistic properties of texts related to the personal style of authors.

Recently, the use of pre-trained language models (e.g., BERT, ELMo, ULM-FiT, GPT-2) has been demonstrated to obtain significant gains in several text classification tasks including sentiment analysis, emotion classification, and topic classification [2,7,13,14]. However, it is not yet clear whether they can be equally useful for style-based text categorization tasks. Especially, in cross-topic AA, information about the topic of texts can be misleading.

An approach based on neural network language models achieved top performance in recent shared tasks on authorship verification and authorship clustering (i.e., grouping documents by authorship) [16,23]. This method is based on a character-level recurrent (RNN) neural network language model and a multi-headed classifier (MHC) [1]. So far, this model has not been tested in closed-set attribution which is the most popular scenario in relevant literature. In this paper, we adopt this approach for the task of closed-set AA and more specifically the challenging cases of cross-topic and cross-genre AA.

We examine the use of pre-trained language models (e.g., BERT, ELMo, ULMFiT, GPT-2) in AA and the potentials of MHC. We also demonstrate that in cross-domain AA conditions, the effect of an appropriate normalization corpus is crucial.

2   Previous Work

The vast majority of previous work in AA focus on the closed-set attribution scenario. The main issues is to define appropriate stylometric measures to quantify the personal style of authors and the use of effective classification methods [12,20].

A relatively small number of previous studies examine the case of cross-topic AA. In early approaches, features like function words or part-of-speech n-grams have been suggested as less likely to correlate with topic of documents [10,11]. However, one main finding of several studies is that low-level features, like character n-grams, can be quite effective in this challenging task [19,21]. Typed character n-grams provide a means for focusing on specific aspects of texts [17]. Interestingly, character n-grams associated with word affixes and punctuation marks seem to be the most useful ones in cross-topic AA. Another interesting idea is to apply structural correspondence learning using punctuation-based character n-gram as pivot features [18]. Recently, a text distortion method has been proposed as a pre-processing step to mask topic-related information in documents while keeping the text structure (i.e., use of function words and punctuation marks) intact [22].

There have been attempts to use language modeling for AA including traditional n-gram based models as well as neural network-based models [1,4,5]. The latter is closely related to representation learning approaches that use deep learning methods to generate distributed text representations [3,9]. In all these cases, the language models are extracted from the texts of known authorship. As a result, they heavily depend on the size of the training set per candidate author.

3   The Proposed Method

An AA task can be expressed as a tuple (A, K, U) where A is the set of candidate authors (suspects), K is the set of known authorship documents (for each $a \in A$ there is a $K_a \subset K$) and U is the set of unknown authorship documents. In closed-set AA, each $d \in U$ should be attributed to exactly one $a \in A$. In cross-topic AA, the topic of documents in U is distinct with respect to the topics found in K, while in cross-genre AA, the genre of documents in U is distinct with respect to the genres found in K.

Bagnall introduced an AA method[1] [1] and obtained top positions in shared tasks in authorship verification and authorship clustering [16,23]. The main idea is that a character-level RNN is produced using all available texts by the candidate authors while a separate output is built for each author (MHC). Thus, the recurrent layer models the language as a whole while each output of MHC focuses on the texts of a particular candidate author. To reduce the vocabulary size, a simple pre-processing step is performed (i.e., uppercase letters are transformed to lowercase plus a symbol, punctuation marks and digits are replaced by specific symbols) [1].

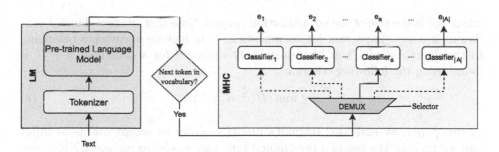

Fig. 1. The proposed model consists of two parts, the language model (LM) and the multi-headed classifier (MHC). The DEMUX layer in MHC part functions as a demultiplexer, its state is defined by the selector. During training phase the selector is defined by the author of the input text and during calculation of normalization vector or test phase the input of DEMUX is connected to all its outputs.

The model, as shown in Fig. 1, consists of two parts, LM and MHC. LM consists of a tokenization layer and the pre-trained language model. MHC comprises a demultiplexer which helps to select the desirable classifier and a set of $|A|$ classifiers, where $|A|$ is the number of candidate authors. Each classifier

---

[1] https://github.com/pan-webis-de/caravel.

has N inputs, where N is the dimensionality of the LM's representation, and V outputs, where V is the size of the vocabulary. The vocabulary is created using the most frequent tokens. The output of LM is a representation of each token in text. If the token exists in vocabulary its representation propagates to MHC, otherwise is ignored (despite the fact that the representation is not further useful, the calculations that took place in LM to produce the representation are mandatory to update the hidden states of the pre-trained language model). If the sequence of input tokens is modified, the representation is also affected.

The function of LM remains the same during training, calculation of normalization vector n and test phase. The MHC layer during training propagates the LM's representations only to the classifier of the author a which is the author of the given text. Then the cross-entropy error is back-propagated to train MHC. During the test phase (as well as the calculation of normalization vector n explained below) the LM's representation is propagated to all classifiers.

The MHC calculates the cross-entropy $H(d, K_a)$ for each input text d and the training texts of each candidate author K_a. The lower cross-entropy is, the more likely for author a to write document d. However, the scores obtained for different candidate authors are not directly comparable due to different bias at each head of MHC. To handle this problem, a normalization vector n is used which is equal to zero-centered relative entropies produced by using an unlabeled normalization corpus C [1]:

$$n = \frac{1}{|C|} \sum_{d_i \in C} H(d_i, K_a) \tag{1}$$

where $|C|$ is the size of the normalization corpus. Note that in cross-domain conditions it is very important for documents in C to include documents belonging to the domain of d. Then, the most likely author a for a document $d \in U$ is found using the following criterion:

$$\arg \min_a (H(d, K_a) - n) \tag{2}$$

In this paper, we extended Bagnall's model in order to accept tokens as input and we propose the use of a pre-trained language model to replace RNN in the aforementioned AA method. The RNN proposed by Bagnall [1] is trained using a small set of documents (K for closed-set AA). In contrast, pre-trained language models have been trained using millions of documents in the same language. Moreover, RNN is a character-level model while the pre-trained models used in this study are token-level approaches. More, specifically, the following models are considered:

- *Universal Language Model Fine-Tuning* (ULMFiT): It provides a contextual token representation obtained from a general domain corpus of millions of unlabeled documents [7]. It adopts left-to-right and right-to left language modeling in separate networks and follows auto-encoder objectives.
- *Embeddings from Language Models* (ELMo): It extracts context-sensitive features using a left-to-right and a right-to-left language modeling [13]. Then,

the representation of each token is a linear combination of the representation of each layer.

- *Generative Pretrained Transformer 2* (GPT-2): It is based on a multi-layer unidirectional *Transformer* decoder [24]. It applies a multi-headed self-attention operation over the input tokens followed by position-wise feed-forward layers [14].
- *Bidirectional Encoder Representations from Transformer* (BERT): It is based on a bidirectional *Transformer* architecture that can better exploit contextual information [2]. It masks a percentage of randomly-selected tokens which the language model is trained to predict.

4 Experiments

4.1 Corpus

We use the CMCC corpus introduced in [6] and also used in previous cross-domain AA works [19,22]. CMCC is a controlled corpus in terms of genre, topic and demographics of subjects. It includes samples by 21 undergraduate students as candidates authors (A), covering six genres (blog, email, essay, chat, discussion, and interview) and six topics (catholic church, gay marriage, privacy rights, legalization of marijuana, war in Iraq, gender discrimination) in English. To ensure that the same specific aspect of the topic is followed, a short question was given to subjects (e.g., Do you think the Catholic Church needs to change its ways to adapt to life in the 21th Century?). In two genres (discussion and interview) the samples were audio recordings and they have been transcribed into text as accurately as possible maintaining information about pauses, laughs etc. For each subject, there is exactly one sample for each combination of genre and topic. More details about the construction of this corpus are provided in [6].

4.2 Experimental Setup

In this study, our focus is on cross-topic and cross-genre AA. In cross-topic, we assume that the topic of training texts (K) is different from the topic of test texts (U) while all texts (both K and U) belong in the same genre. Similar to [22] and [19], we perform leave-one-topic-out cross-validation where all texts on a specific topic (within a certain genre) are included in the test corpus and all remaining texts on the remaining topics (in that genre) are included in the training corpus. This is repeated six times so that all available topics to serve exactly once as the test topic. Mean classification accuracy over all topics is reported.

Similar to cross-topic, in cross-genre we perform leave-one-genre-out cross-validation as in [22], where all texts on a specific genre (within a certain topic) are included in the test corpus and all remaining texts on the remaining genres (in that topic) are included in the training corpus. The number of available genres is also six like topics, and though we repeat the leave-one-genre-out cross-validation

Table 1. Dimensionality of representation (N) for each language model in this study.

Model	RNN	BERT	ELMo	GPT-2	ULMFiT
N	149	768	1024	768	400

six times and report the mean classification accuracy. In both scenarios, cross-topic and cross-genre, the candidates authors set A consists of 21 undergraduate students as mentioned in Sect. 4.1.

All the examined models use a MHC on top of a language modeling method. First, we study the original Bagnall's approach where a character-level RNN is trained over K. Then, each one of the pre-trained language models described in previous section. In our experiments, all of the pre-trained LMs was fine-tuned for the specific AA task with MHC as classifier without further training the language model , since our goal is to explore the potential of pre-trained models obtained from general domain corpora.

In MHC, each author corresponds to a separate classifier with N inputs and M outputs, where N is the dimensionality of text representation, Table 1, and M is equal to vocabulary size V. During training, each classification layer is trained only with the documents of the corresponding author. The vocabulary is defined as the most frequent tokens in the corpus. These are less likely to be affected by topic shifts and the reduced input size increases the efficiency of our approach. The selected values of V are 100, 500, $1k$, $2k$ and $5k$. Each model used its own tokenization stage except from ELMo (where ULMFiT's tokenization was used). Note that RNN is a character-level model while all pre-trained models are token-based.

Since RNN is trained from scratch for a corpus of small size, it is considerably affected by initialization. As a result, there is significant variance when it is applied several times to the same corpus. To compensate this, we report average performance results for 10 repetitions. Regarding the training phase of each method, we use 100 epochs for RNN and examine four cases for the pre-trained models: the minimal training of 1 epoch and the cases of 5, 10 and 20 epochs of training.

4.3 Results on Cross-Topic AA

Table 2 presents the leave-one-topic-out cross-validation accuracy results for each one of the six available genres as well as the average performance over all genres for each method. Two cases are examined: one using the (unlabeled) training texts as normalization corpus $(C = K)$ and another where the (unlabeled) test texts are used as normalization corpus $(C = U)$. The former means that C includes documents with distinct topics with respect to the document of unknown authorship while the latter ensures that there is perfect thematic similarity. As can be seen, the use of a suitable normalization corpus is crucial to enhance the performance of the examined methods.

Table 2. Accuracy results (%) on Cross-Topic AA. The reported performance of baseline models is taken from the corresponding publications.

‡	LM	V	epochs	Blog	Email	Essay	Chat	Disc.	Interv.	Avg.
$C = K$	RNN	–	100	47.94	44.37	41.00	73.41	75.71	72.54	59.16
	BERT	2k	1	57.14	49.21	60.32	84.92	79.37	80.16	68.52
	BERT	5k	1	53.97	52.38	58.73	86.51	77.78	78.57	67.99
	ELMo	2k	1	56.35	55.56	56.35	80.95	72.22	76.98	66.40
	ELMo	5k	1	55.56	53.17	57.14	82.54	70.64	76.19	65.87
	GPT-2	2k	20	60.32	57.94	54.76	76.98	63.49	79.37	65.48
	GPT-2	5k	20	58.73	59.52	61.11	84.13	63.49	76.98	67.33
	ULMFiT	2k	10	50.00	43.65	52.38	79.37	72.22	71.43	61.51
	ULMFiT	5k	20	46.83	40.48	50.79	80.16	69.84	70.64	59.79
$C = U$	RNN	–	100	61.67	56.43	68.36	81.27	**86.90**	84.52	73.19
	BERT	2k	5	72.22	64.29	76.98	90.48	84.13	90.48	79.76
	BERT	5k	5	**73.81**	61.11	77.78	**92.86**	84.13	90.48	**80.03**
	ELMo	2k	10	72.22	65.08	75.40	89.68	76.19	**91.27**	78.31
	ELMo	5k	10	72.22	**67.46**	**77.78**	88.10	76.98	89.68	78.57
	GPT-2	2k	20	72.22	64.29	73.02	80.16	67.46	82.54	73.28
	GPT-2	5k	20	69.84	65.87	69.84	84.13	73.81	85.71	74.87
	ULMFiT	1k	10	64.29	57.94	73.02	87.30	80.16	88.89	75.26
	ULMFiT	2k	10	64.29	54.76	73.81	88.89	78.57	88.10	74.74
	ULMFiT	5k	20	58.73	54.76	75.40	88.89	75.40	84.13	72.88
	C3G-SVM [19]	–	–	33.41	36.53	36.66	57.46	49.91	56.35	45.05
	PPM5 [22]	–	–	52.38	39.68	50.00	57.94	36.51	47.62	47.35
	DV-MA [22]	–	–	43.65	65.87	60.32	71.43	80.16	67.46	64.81

As concerns individual pre-trained language models, BERT and ELMo are better able to surpass the RNN baseline while ULMFiT and GPT-2 are not that competitive. In addition, BERT and ELMo methods need small number of training epochs while ULMFiT and GPT-2 improve with increased number of epochs.

Table 2 also shows the corresponding results from previous studies on cross-topic AA using exactly the same experimental setup. These baselines are based on character 3-grams features and a SVM classifier (C3G-SVM) [19], a compression-based method (PPM5) [22], and a method using text distortion to mask thematic information (DV-MA) [22]. As can be seen, when $C = U$ all of the examined methods surpass the best baseline in average performance and the improvement is high in all genres. It is remarkable that all models except ULMFiT achieve to surpass the baselines (in average performance) even when $C = K$.

Figure 2 presents the mean classification accuracy with respect to vocabulary size on cross-topic AA. Each sub-figure correspond to different LM. The type of the line indicates the normalization corpus, dashed line indicates the use of training texts $(C = K)$ as normalization corpus, while the continues line indicates the use of test texts $(C = U)$, as noted in Sect. 3 in both cases the texts are unlabeled. The shape of each point correspond to epochs of training, 1, 5, 10 and 20 for circle, triangle, square and x-mark respectively.

From the aspect of vocabulary size, in contradiction to the state of the art [22], where the best results achieved for vocabularies that consisted of less than $1k$ words (most frequent), in our set up the most appropriate value seems to be above $2k$. Despite the gap between $2k$ and $5k$ words in vocabulary size BERT and ELMo have minor difference in accuracy indicating that above $2k$ words the affect of vocabulary size is minor. GPT-2 continues to increment the accuracy and ULMFiT started to decrement for values above $1k$ words, Table 2. Experiments with values over $5k$ were prohibitive due to runtime of training, with $5k$ words the runtime was approximately 4 days for each model running on GPU.

From the aspect of training epochs, BERT and ELMo achieved their best performance in $C = K$ case with minimal training. In $C = U$ case their performance is slightly affected by the number of training epochs. This behavior raises the question of over-fitting. As mentioned in Sect. 3 the selection criterion Eq. 2, is based on the cross-entropy of each text. MHC is trained on predicting the text flow and thus the cross-entropy decreases after each epoch of training. Having in mind the cross-entropy, if we have a second look on Fig. 2, the case of over-fitting is rejected since the behavior of accuracy in relevance with the number of training epochs (indicated by the shape of point) do not have the characteristics of over-fitting (increment of training epochs decrements the accuracy).

4.4   Results on Cross-Genre AA

The experiments on cross-genre performed on the same set up as in cross-topic. Table 3 presents the accuracy results on leave-one-genre-out cross-validation for each one of the six available topics and the average performance over all topics, similar to Table 2. Based on the results of Sect. 4.3 the most reasonable value of V in order to check the performance of each method is $V = 2k$. The case of $V = 5k$ is very time consuming without offering valuable gain and below $1k$ the performance is not remarkable. For the experiments on cross-genre the values of $1k$ and $2k$ were selected for V. Comparing the two cases, the results with $V = 2k$ surpass in all experiments the results with $V = 1k$ and thus we selected to present only the case of $V = 2k$ on Table 3.

Table 3. Accuracy results (%) on cross-genre AA for vocabulary size $2k$ $(V = 2k)$ and each topic (Church (C), Gay Marriage (G), War in Iraq (I), Legalization of Marijuana (M), Privacy Rights (P), Gender Discrimination (S)). The reported performance of the baseline models (only available in average across all topics) is taken from the corresponding publications.

‡	LM	epochs	C	G	I	M	P	S	Avg.
$C = K$	RNN	100	58.89	68.33	71.59	60.24	50.40	62.22	61.94
	BERT	10	70.63	77.78	83.33	73.81	62.70	76.98	74.21
	ELMo	10	68.25	78.57	78.57	71.43	55.56	65.08	69.58
	GPT-2	20	52.38	67.46	61.11	57.94	50.79	53.17	57.14
	ULMFiT	20	72.22	77.78	79.37	70.63	61.11	68.25	71.56
$C = U$	RNN	100	75.32	75.95	86.11	79.52	69.37	74.21	76.75
	BERT	5	84.13	87.30	88.10	82.54	**77.78**	78.57	83.07
	ELMo	20	87.30	88.89	**88.89**	**83.33**	76.98	**81.75**	**84.52**
	GPT-2	20	69.84	76.98	74.60	67.46	61.11	72.22	70.37
	ULMFiT	10	**88.10**	**89.68**	85.71	82.54	**77.78**	79.37	83.86
	C3G-SVM [19]	–							
	PPM5 [22]	–							60.00
	DV-MA [22]	–							33.00

BERT and ELMo achieved high results as expected from their performance on cross-topic, with ELMo achieving the highest accuracy result. Unexpectedly, ULMFiT which had the worst performance in cross-topic achieved the second best performance. GPT-2 performed lower than RNN baseline in both cases of $C = K$ and $C = U$. Comparing Table 2 and Table 3 is noticeable that ELMo and BERT are more stable in performance than GPT-2 and ULMFiT. The main difference between the former and the latter is the directionality, the former two are bidirectional while the latter are unidirectional, we suspect that this is the main reason that affects the stability in performance.

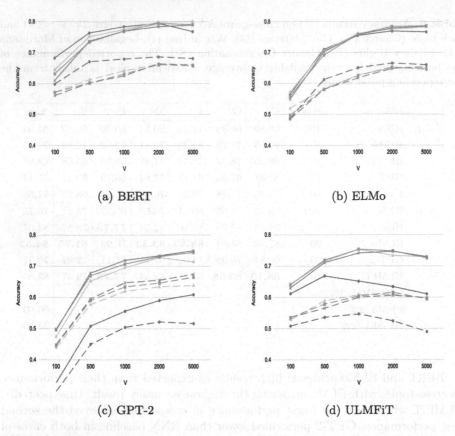

(a) BERT   (b) ELMo

(c) GPT-2   (d) ULMFiT

Fig. 2. Accuracy results for vocabulary sizes (V) 100, 500, $1k$, $2k$ and $5k$ for each pre-trained model. Colored symbols blue circle, red triangle, yellow square and green x-mark correspond to 1, 5, 10 and 20 training epochs, respectively. The type of the line indicates the normalization corpus, a dashed line indicates the use of training texts ($C = K$) as normalization corpus, while a solid line indicates the use of test texts ($C = U$). (Color figure online)

5   Conclusions

In this paper, we explore the usefulness of pre-trained language models in cross-domain AA. Based on Bagnall's model [1], originally proposed for authorship verification, we compare the performance when we use either the original character-level RNN trained from scratch in the small-size AA corpus or pre-trained token-based language models obtained from general-domain corpora. We demonstrate that BERT and ELMo pre-trained models achieve the best results while being the most stable approaches with respect to the results in both scenarios.

A crucial factor to enhance performance is the normalization corpus used in the MHC. In cross-domain AA, it is very important for the normalization corpus to have exactly the same properties with the documents of unknown authorship.

In our experiments, using a controlled corpus, it is possible to ensure a perfect match in both genre and topic. In practice, this is not always feasible. A future work direction is to explore how one can build an appropriate normalization corpus for a given document of unknown authorship. Other interesting extensions of this work is to study the effect of extending fine-tuning to language model layers and focus on the different layers of the language modeling representation.

References

1. Bagnall, D.: Author identification using multi-headed recurrent neural networks. In: Working Notes of CLEF 2015 - Conference and Labs of the Evaluation forum (2015)
2. Devlin, J., Chang, M.W., Lee, K., Toutanova, K.: BERT: pre-training of deep bidirectional transformers for language understanding. In: Proceedings of the 2019 Conference of the North American Chapter of the Association for Computational Linguistics: Human Language Technologies, Volume 1 (Long and Short Papers), pp. 4171–4186 (2019)
3. Ding, S., Fung, B., Iqbal, F., Cheung, W.: Learning stylometric representations for authorship analysis. IEEE Trans. Cybern. **49**(1), 107–121 (2019)
4. Fourkioti, O., Symeonidis, S., Arampatzis, A.: Language models and fusion for authorship attribution. Inf. Process. Manag. **56**(6), 102061 (2019)
5. Ge, Z., Sun, Y., Smith, M.J.T.: Authorship attribution using a neural network language model. In: Proceedings of the Thirtieth AAAI Conference on Artificial Intelligence, AAAI 2016, pp. 4212–4213. AAAI Press (2016)
6. Goldstein-Stewart, J., Winder, R., Sabin, R.E.: Person identification from text and speech genre samples. In: Proceedings of the 12th Conference of the European Chapter of the Association for Computational Linguistics, pp. 336–344. Association for Computational Linguistics (2009)
7. Howard, J., Ruder, S.: Universal language model fine-tuning for text classification. In: Proceedings of the 56th Annual Meeting of the Association for Computational Linguistics (Volume 1: Long Papers), pp. 328–339 (2018)
8. Kestemont, M., Stover, J., Koppel, M., Karsdorp, F., Daelemans, W.: Authenticating the writings of Julius Caesar. Expert Syst. Appl. **63**, 86–96 (2016)
9. Kocher, M., Savoy, J.: Distributed language representation for authorship attribution. Digital Sch. Humanit. **33**(2), 425–441 (2018)
10. Madigan, D., Genkin, A., Lewis, D.D., Argamon, S., Fradkin, D., Ye, L.: Author identification on the large scale. In: Proceedings of the Meeting of the Classification Society of North America (2005)
11. Menon, R., Choi, Y.: Domain independent authorship attribution without domain adaptation. In: Proceedings of the International Conference Recent Advances in Natural Language Processing, pp. 309–315 (2011)
12. Neal, T., Sundararajan, K., Fatima, A., Yan, Y., Xiang, Y., Woodard, D.: Surveying stylometry techniques and applications. ACM Comput. Surv. **50**(6), 1–36 (2018)
13. Peters, M., et al.: Deep contextualized word representations. In: Proceedings of the 2018 Conference of the North American Chapter of the Association for Computational Linguistics: Human Language Technologies, Volume 1 (Long Papers), pp. 2227–2237 (2018)

14. Radford, A., Wu, J., Child, R., Luan, D., Amodei, D., Sutskever, I.: Language models are unsupervised multitask learners. OpenAI Blog **1**(8), 9 (2019)
15. Rocha, A., et al.: Authorship attribution for social media forensics. IEEE Trans. Inf. Forensics Secur. **12**(1), 5–33 (2017)
16. Rosso, P., Rangel, F., Potthast, M., Stamatatos, E., Tschuggnall, M., Stein, B.: Overview of PAN'16. In: Fuhr, N., et al. (eds.) CLEF 2016. LNCS, vol. 9822, pp. 332–350. Springer, Cham (2016). https://doi.org/10.1007/978-3-319-44564-9_28
17. Sapkota, U., Bethard, S., Montes, M., Solorio, T.: Not all character n-grams are created equal: a study in authorship attribution. In: Proceedings of the 2015 Conference of the North American Chapter of the Association for Computational Linguistics: Human Language Technologies, pp. 93–102 (2015)
18. Sapkota, U., Solorio, T., Montes, M., Bethard, S.: Domain adaptation for authorship attribution: improved structural correspondence learning. In: Proceedings of the 54th Annual Meeting of the Association for Computational Linguistics (Volume 1: Long Papers), pp. 2226–2235 (2016)
19. Sapkota, U., Solorio, T., Montes, M., Bethard, S., Rosso, P.: Cross-topic authorship attribution: will out-of-topic data help? In: Proceedings of COLING 2014, the 25th International Conference on Computational Linguistics: Technical Papers, pp. 1228–1237 (2014)
20. Stamatatos, E.: A survey of modern authorship attribution methods. J. Am. Soc. Inform. Sci. Technol. **60**(3), 538–556 (2009)
21. Stamatatos, E.: On the robustness of authorship attribution based on character n-gram features. J. Law Policy **21**, 421–439 (2013)
22. Stamatatos, E.: Masking topic-related information to enhance authorship attribution. J. Assoc. Inf. Sci. Technol. **69**(3), 461–473 (2018)
23. Stamatatos, E., Potthast, M., Rangel, F., Rosso, P., Stein, B.: Overview of the PAN/CLEF 2015 evaluation lab. In: Mothe, J., et al. (eds.) CLEF 2015. LNCS, vol. 9283, pp. 518–538. Springer, Cham (2015). https://doi.org/10.1007/978-3-319-24027-5_49
24. Vaswani, A., et al.: Attention is all you need. In: Advances in Neural Information Processing Systems, pp. 5998–6008 (2017)

Indoor Localization with Multi-objective Selection of Radiomap Models

Rafael Alexandrou[(✉)], Harris Papadopoulos[(✉)],
and Andreas Konstantinidis[(✉)]

Frederick University, 7 Y. Frederickou Street Pallouriotisa, 1036 Nicosia, Cyprus
{res.ar,com.ph,com.ca}@frederick.ac.cy

Abstract. Over the last years, Indoor Localization Systems (ILS) evolved, due to the inability of Global Positioning Systems (GPS) to localize in indoor environments. A variety of studies tackle indoor localization with technologies such as Bluetooth Beacons and RFID that require costly installation, or techniques such as Google Wi-Fi/Cell DB and fingerprinting that leverage from the already existing Wi-FI and telecommunication infrastructure. Additionally, recent studies attempt to solve the same problem using Bio-Inspired techniques, such as Artificial Neural Networks (ANNs) and Deep Neural Networks (DNN). In this paper, we introduce a Multi-Objective Optimization Radiomap Modelling (MOO-RM) based ILS. The MOO-RM ILS divides the dataset into clusters using a K-Means algorithm and trains ANN models on the data of each cluster. The resulting models are fed into a Multi-Objective Evolutionary Algorithm based on Decomposition (MOEA/D), which minimizes the required storage space and the localization error, simultaneously. Our experimental studies demonstrate the superiority of the proposed approach on real datasets of Wi-Fi traces with respect to various existing techniques.

Keywords: Indoor Localization · Smartphones · Fingerprinting · Artificial Neural Networks · Multi-Objective Optimization

1 Introduction

Currently, Global Navigation Satellite Systems (GNSS), such as GPS, are unable to localize accurately in indoor environments due to the satellite signal attenuation while passing through solid objects. This has led to an increased interest for alternative localization techniques by the scientific community. In particular, the indoor localization community has proposed a variety of solutions that include technologies such as Bluetooth Beacons (BLE), Infrared, Li-Fi technologies, RFID, Sensor Networks and their combinations, for localizing a device in an indoor environment with a fine-grain accuracy [1]. These techniques, however, require the deployment of specialized equipment such as antennas, beacons, and custom transmitters a priori [2], which is time consuming and costly.

© IFIP International Federation for Information Processing 2020
Published by Springer Nature Switzerland AG 2020
I. Maglogiannis et al. (Eds.): AIAI 2020, IFIP AICT 583, pp. 267–278, 2020.
https://doi.org/10.1007/978-3-030-49161-1_23

In order to alleviate the aforementioned challenges, a variety of Indoor Localization Systems (ILS) that rely on geolocation data retrieved from the existing infrastructure of a building, such as wireless signals, have been implemented. These ILS, such as *Google, Indoo.rs, Navizon, IndoorAtlas, ByteLight* and *Anyplace*[1] [3] have managed to provide accurate indoor localization of a user without the need of any additional hardware.

Geolocation database entries act as reference points for the localization tasks. A comparison between those reference points and a sensed point from a smartphone device, either on the service (Server-Side) or the smartphone itself (Client-Side), can help determine a user's location within an area. As [4] suggests, Server-Side violates user's privacy, since calculations, that are processed on the server, reveal the actual location of a user to the service. Contrary, such concern does not apply for Client-Side, since the service only knows the area that surrounds the user and not the user's actual location. However, since the main processing of the ILS is done on the smartphone, this leads to performance concerns, mainly, due to the vast amount of data being downloaded and processed locally.

In this paper, we examine the possibility of providing the user with a variety of solutions based on his preferences, by introducing a novel Multi-Objective Optimization Radiomap Modelling (MOO-RM) based ILS. The solutions include Artificial Neural Network (ANN) models, associated to clusters of location data, which are generated using K-Means and can be used to localize users. Then, a Multi-Objective Evolutionary Algorithm Based on Decomposition (MOEA/D) is used to minimize the storage space requirements on the smartphone device and the localization error, by selecting a near-optimal set of models from the set of all models that represent the whole location-based dataset.

The rest of this paper is organized as follows. Section 2 covers the background of indoor localization and provides solutions in the literature that are similar to our proposed method. Section 4 presents the proposed ILS method, which is evaluated in Sect. 5. Finally, Sect. 6 discusses future work.

2   Related Work

In this section, background on indoor localization that lies at the foundations of the proposed approach is introduced.

2.1   Indoor Localization with Smartphones

A wide range of technologies dealing with localization in outdoor and indoor environments is provided in the literature. For outdoor environments, Global Navigation Systems (GNSS), such as GPS and Galileo, are considered the leading technologies for localization. However, they require high energy consumption and their signal attenuation while passing through solid objects, such as concrete walls, is negatively affecting navigation in indoor environments. A variety of indoor localization solutions [1] have been proposed including technologies such

---

[1] Available at: http://anyplace.cs.ucy.ac.cy/.

as BLE sensors, visual or acoustic analysis, RFID, Wireless Sensor Networks, laser and LiFi, IMUs and their combinations into hybrid systems. Even though the performance of the aforementioned suggest positive votes, most of these techniques require the costly deployment of additional equipment a priori to localization. Contrary, the indoor localization community introduced localization approaches, such as Cell/Wi-Fi Database and Wi-Fi Fingerprinting, that rely on existing Wi-Fi infrastructure already deployed in most buildings.

Wi-Fi Fingerprints construct a database of radio signals from Wi-Fi Access Points (Wi-Fi APs) within the area. In the literature, Anyplace [5–7] achieved the second highest known accuracy [8], with an average error of 1.96 meters. In particular, Anyplace is divided into two phases: The offline phase, or "logging" phase, records the Wi-Fi Fingerprints, which are Received Signal Strength indications (RSSi) of Wi-Fi APs at certain locations (x, y) of a building, into an $N \times M$ matrix known as a radiomap. The radiomap itself consists of N unique fingerprints and M Wi-Fi APs. In the online phase, or "localization" phase, a smartphone user observes the RSSi from the surrounding Wi-Fi APs and compares it against the radiomap using either K-Nearest Neighbors (KNN) or Weighted K-Nearest Neighbors (WKNN) to find the best match. Finally, Anyplace provides two different approaches: In *Server-Side Fingerprinting (SS)*, the main localization process happens on a server that has unlimited energy, storage and processing budget. Therefore, the smartphone user requires little network messaging and minimal energy consumption to localize. However, this approach violates the user's privacy, since the server knows the actual location of the user. Contrary, *Client-Side Fingerprinting (CS)* focuses on processing locally on the smartphone to eliminate the privacy concerns. However, this leads to high resource consumption on the client device.

In our previous work [9], we proposed a CS fingerprinting approach that utilizes Artificial Neural Networks (ANNs) to train a model representing the whole radiomap data. Even though this approach reduces the energy consumption and storage space required on the smartphone, while preserving user's privacy and maintaining an acceptable localization error compared to conventional fingerprinting techniques, we realized that storage space requirements are still high for smartphone devices. In this work, we investigate the trade-off between storage space requirements and localization error by providing multiple solutions, that consist of a subset of ANN models corresponding to clusters within the radiomap. A variety of solutions will allow users to localize in indoor environments based on their preferences (e.g. preserve battery/storage levels on the device, or minimize localization error during localization).

2.2 Smartphone Indoor Localization Applications

In [10], authors present a deep learning-based fingerprinting localization schema. Similarly,[11] presents a deep-learning approach that utilizes deep architectures and Channel State Information (CSI) on fingerprinting for localization. Additionally, authors of [12] present the WiDeep approach that uses a probabilistic denoising auto-encoder. All three aforementioned research studies provide solutions for improving the fingerprinting technique. However, the use of deep neural networks requires high volume of resources during localization, due to networks' structure complexity.

Furthermore, the indoor localization problem is, also, tackled as an optimization problem, in the literature. An indoor localization problem is formulated in [13] and optimized with a Particle Swarm Optimization (PSO) approach, named JADE. Likewise, [14] proposes a fingerprinting localization and tracking system with PSO and Kalman Filter (KF). Finally, the authors of [15] present MILos, a multi-objective indoor localization service that utilizes a Multi-Objective Evolutionary Algorithm Based on Decomposition (MOEA/D) to maximize the coverage and minimize the energy consumption at the same time. The latter optimizes coverage and energy consumption objectives on conventional fingerprinting techniques, whereas our proposed method optimizes the storage space required on the device and the localization error using ANN models, which are clustered using K-Means.

3   System Overview

3.1   System Model

We assume an area A divided into several building areas $A_1, .., A_n$, which consist of several floors $F_1, .., F_m$ (see Fig. 1). Each A_i and its corresponding F_j contain a finite set of (x, y) points and is covered by a set of Wi-Fi Access Points $\{ap_1, ap_2, \cdots, ap_M\}$, each covering a planar points. Area A_i is not necessarily continuous and can be considered as the joint area of all $ap_i \in AP$ (i.e., global coverage). Each ap_i has a unique ID (i.e., MAC address) that is publicly broadcasted and passively received by anyone moving in the a points of ap_i. The signal intensity at which the ID of ap_i is received at location (x, y), is termed the *Received Signal Strength (RSS)* of ap_i at (x, y), having a value in the range $[-30... - 110]dB$.

Let a static (cloud-based) localization service s hosting an $N \times M$ table, coined *RadioMap (RM)*, which records the RSS of the $ap_i \in AP$ broadcasts at specified $(x, y) \in A$ locations. When an ap_i is not seen at a certain (x, y) the *RM* records "-110" in its respective cell. A user u localizes through the indoor positioning service s, using the ID and RSS broadcasts of surrounding $ap_i \in AP$ while moving. This information is termed, hereafter, *RSS Vector* or *Fingerprint* (V_u) of user u, which changes from location to location and over time. Contrary to *RM* rows having M attributes, V_u has only $M' \ll M$ attributes.

3.2   Research Goal and Metrics

Research Goal. *Allow a smartphone user to localize with fine accuracy and minimum storage resources by selecting a subset of ANN models representing clusters of Radiomap data.*

The efficiency of the proposed technique to achieve the above research goal is measured by the *Localization Error*, and the *Storage Space Requirements* on the client device u.

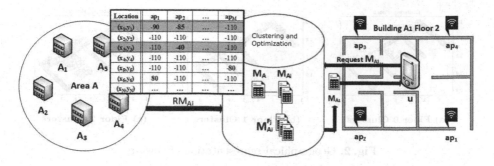

Fig. 1. System model

The **Localization Error** *is defined as the Haversine distance (in m), between the predicted location* (x_2, y_2) *and the actual location* (x_1, y_1) *of u,*

$$d = 2r \times \arcsin(\sqrt{\sin^2(\frac{x_2 - x_1}{2}) + \cos(x_1)\cos(x_2)\sin^2(\frac{y_2 - y_1}{2})})/1000 \qquad (1)$$

where $r = 6371$ km is the radius of earth.

Storage Space $(Size_s)$ *is the storage space required on the client device in order to perform indoor localization.*

4   Proposed Methodology

Our proposed Indoor Localization Method is divided into four phases, namely, Data Clustering, Modeling, Optimization, and Decision Making. Each phase is explained next.

4.1   Data Clustering

Data Clustering utilizes a K-Means algorithm to divide the dataset into clusters. In particular, K-Means is a vector quantization method that is widely used in data mining and aims to partition n observations into k clusters based on their distances.

In our proposed method, K-Means is applied to the multi-dimensional finger-printing dataset for forming clusters with respect to the whole vector, considering both the sensed Received Signal Strength Intensity (RSSI) and the mapped geo-location. Figure 2 provides an example of the resulting clusters for the three floors of our dataset.

4.2   Modeling

After the clustering process, each cluster is modelled with an Artificial Neural Network (ANN). In particular, the trained models are able to predict the user's

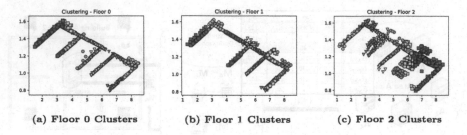

(a) **Floor 0 Clusters**   (b) **Floor 1 Clusters**   (c) **Floor 2 Clusters**

Fig. 2. Geographical representation of clusters

current location based on a vector V_u that contains the RSSIs of the user's surrounding Wi-Fi APs, arranged in the same order for all patterns. The models used were fully connected feed-forward ANNs with sigmoid hidden and linear output activation functions and were trained with a stochastic gradient descent optimizer called Adam. Implementation was performed using the Scikit-Learn library in a Python environment with the ANN regression algorithm MLPRegressor. Following a trial and error approach for the network's structure led to the average optimum scenario of 100 hidden units on a single hidden layer.

The trained models were extracted and stored on the server in order to be readily available for being sent to a smartphone device. The model extraction was performed manually since Scikit-learn uses a python tool to extract models commonly known as *pickle* (or its updated version called *joblib*) which extracts unnecessary information along with the model. This lead to a larger model file compared to the original *RMs*, which conflicts with one of the major goals of the proposed method i.e. to minimize the storage space required on the smartphone device. Therefore, the model's coefficients and intercepts were extracted and saved into two small files. On the smartphone device, both files were loaded and used to reconstruct the model locally, before localization. Note that this reconstruction does not imply re-training the model, since the coefficients and intercepts are known from the training phase.

4.3   Multi-Objective Optimization

A Multi-objective Optimization Problem (MOP) can be mathematically formulated as

$$\text{minimize } F(X) = (f_1(X), ..., f_k(X)), \text{ subject to } X \in \Omega \tag{2}$$

where Ω is the decision space and $X \in \Omega$ is a decision vector. $F(X)$ consists of k objective functions $f_i : \Omega \to R, i = 1, ..., k$, where R^k is the objective space. The objectives often conflict with each other and improving on one objective may lead to deterioration of another. Thus, no single solution exists that can optimize all objectives simultaneously. In that case, the best trade-off solutions, called the set of Pareto optimal (or non-dominated) solutions, is often required by a decision maker. The image of the Pareto Set (PS) in the objective space is called the Pareto Front (PF).

In this paper the MOP aims at minimizing the storage space required on the smartphone device and the localization error, and it is tacked by the Multi-Objective Evolutionary Algorithm Based on Decomposition (MOEA/D). The initial population is generated by a random generator that provides binary vectors, of length equal to number of clusters, where 1 and 0 denote that the corresponding model is included to the set or not. Finally, MOEA/D follows a single point crossover and the termination condition is set to n maximum generations.

4.4 Decision Making

Decision Making is the last phase and has access to a set of solutions instead of a single one. It relies on the user's input and the smartphone's available storage. Considering a device that has limited storage availability, the user is asked to decide whether he/she wants to save storage by sacrificing localization accuracy. The user's choice will dictate the need for a more accurate or less storage-dependent model subset. Finally, note that the less storage-dependent model subset implies reduced accuracy within the whole area but not within the part of the area that is covered by the subset. (e.g. model subsets that include only 1 model will be able to predict user's location accurately in 14% of the whole area).

5 Experimental Evaluation

5.1 Dataset

For our experimental evaluation, we used a real dataset consisting of $\approx 45,000$ reference fingerprints taken from ≈ 120 Wi-Fi APs installed in three floors of a building in Cyprus. Therefore, it contains three radiomaps that correspond to the three floors of the building. Firstly, each floor's radiomap was divided into smaller clusters, which further divided into a training and a test set corresponding to 80% and 20% of the data, respectively. The training sets were used to model each cluster's data and the models were evaluated in the corresponding test sets.

The proposed indoor localization method is evaluated in terms of *Localization Error* and *Storage Space* required on the device, as defined in Sect. 3.2.

5.2 Evaluation Metrics

The metrics utilized during the optimization phase are defined as follows:

Hypervolume (I_H) *indicates the area dominated by at least one solution in an obtained non-dominated set A. Therefore, high I_H suggests better diversity. The metric is formally defined as*

$$I_H(A) = \int_{x \in U_{x \in A}} \cdots \int_{HV(f(x),f^*)} 1.dz \tag{3}$$

where $HV(f(x), f*) = [f_1(x), f_1^*] \times ... \times [f_m(x), f_m^*]$ is the Cartesian product of the closed intervals $[f_i(x), f_i^*], i = 1, .., m$. Since we consider minimization objectives the reference point $f^* = (f_1^*...f_m^*)$ is the ideal worst point.

Number of Non-dominated Solutions (*NDS*) *is the number of non-dominated solutions in set A.*

$$NDS(A) = |A| \tag{4}$$

C-Metric *is defined by the percentage of non-dominated solutions of set A with respect to solutions of a non-dominated set B and can be annotated as C(A,B).*

5.3 Experimental Study 1: Number of Clusters

In the first experimental study, we evaluated the K-means algorithm for various cluster sizes in terms of localization error. Note that various cluster sizes were examined but only three are presented, which are representative for all cases. Mainly, the focus of this experimental study is to provide information regarding the expected output of the whole set of models.

Figure 3 shows the comparison of three cluster sizes, in terms of localization error and storage space requirements. Figure 3a shows a similar average localization error of clusters for the three different clusters sizes at each floor. Figure 3b examines the storage space requirements on the smartphone device and shows that for floor 0 and floor 2 are similar in all three cases. Contrary, floor 1 results show that the lowest storage space required is achieved using seven clusters. In general, seven clusters seem to provide promising results, for both objectives, and therefore it is adopted in all experimental studies that follow.

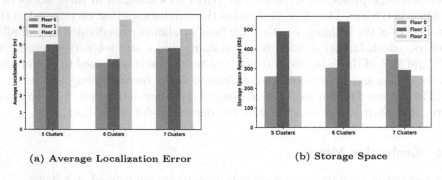

(a) Average Localization Error   (b) Storage Space

Fig. 3. Clusters performance

5.4 Experimental Study 2: Optimization Integrity

The second experimental study focuses on ensuring the integrity of MOEA/D while minimizing the two conflicting objectives with respect to the state-of-the-art in MOO, Non-Dominated Sorting Genetic Algorithm II (NSGA-II). Table 1 shows a comparison between NSGA-II and MOEA/D in terms of the metrics defined in Sect. 5.2. The results show similar performance with respect to I_H and C, but it seems that MOEA/D provides a higher number of NDS to the decision maker.

Table 1. NSGA-II(A) vs MOEA/D(B)

Floor	$I_H(A)$	$NDS(A)$	$C(A,B)$	$I_H(B)$	$NDS(B)$	$C(B,A)$
0	0.846	7	0.571	0.819	9	0.666
1	0.778	14	0.714	0.834	22	0.545
2	0.73	14	0.8	0.73	20	0.8

5.5 Experimental Study 3: Localization Error and Storage Space

In order to ensure the usability of the proposed indoor localization method, we evaluate our models with the floor level test set obtained in Sect. 4.2 and over three randomly generated paths within the area, namely, a visitor, a professor, and a student path. Finally, our proposed method was evaluated over an error-dependent scenario (solution with minimum localization error), a storage-dependent scenario (solution with minimum storage space required), a conventional fingerprinting technique such as Weighted K-Nearest Neighbors (WKNN), and an Artificial Neural Network (ANN) model generated from the whole training set (without clustering).

Figure 4 provides the Pareto Front (PF) solutions obtained from the optimization phase. Additionally, the solutions of WKNN and ANN single model are provided within the same figure for each floor, as baselines for minimum localization error and minimum storage requirements, respectively. As shown in Fig. 4a, for floor 0, all MOO-RM PF solutions require less storage space than WKNN, and 71% of the MOO-RM PF solutions require less storage space than the ANN model, as well. As expected, since the WKNN solution is one of our baselines, it has the minimum localization error, whereas the majority of MOO-RM PF solutions are similar to the localization error of the ANN. Figure 4b shows that all MOO-RM PF solutions require less storage than WKNN and 64% of MOO-RM PF need less storage space than the ANN, for floor 1. Additionally, various MOO-RM PF solutions are close to the localization error provided by the ANN. Lastly, Fig. 4c represents the MOO-RM PF solutions for floor 2. The results demonstrate the superiority of the proposed approach in real life large scale datasets. In particular, all MOO-RM PF solutions require less storage space compared to the WKNN and ANN baselines. Additionally, there are solutions that are similar to WKNN's localization error with a significant decrease in storage space requirements.

Figure 5 shows the results on the user defined paths using the two extreme solutions in the PF obtained by MOO-RM that is: *i) Error-Dependent solution*

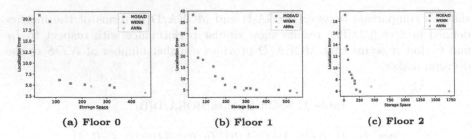

(a) Floor 0   (b) Floor 1   (c) Floor 2

Fig. 4. Floor level evaluation

provides the least possible localization error without any storage space concerns, *ii) Storage-Dependent solution* provides the least possible storage space on the smartphone device without considering the localization error levels. The results demonstrate that in almost all cases, MOO-RM maintains an acceptable localization error and less storage space requirements, compared to WKNN and ANN when considering the Error-Dependent solution. Additionally, the Storage-Dependent solution provides a localization error, which is higher compared to WKNN and ANN, offering however a storage space decrease of 99%.

5.6 Experimental Study 4: Building Level

In this experimental study, the proposed MOO-RM was evaluated over a building level dataset. In particular, we utilized all 21 models of the three floors, and we examined the performance of MOO-RM on a building level (larger dataset) than on a floor level.

The comparison between MOO-RM, the conventional fingerprinting (WKNN), and the ANN model is shown in Table 2. In particular, the proposed method has an increase of 6% in terms of localization error and minimizes the storage space by 56%, compared to WKNN. Additionally, MOO-RM decreases the localization error by 54% and the storage space requirements by 67%, compared to the ANN single model. The ANN model attempts to predict the radiomap of the whole building area A, while MOO-RM utilizes a MOO selection between the 21 clusters converging to better localization error.

Table 2. Building level comparison

	WKNN	ANNs	MOO-RM
Error(m)	4.274	9.96	4.55
Storage(Kb)	2729	3697	1203

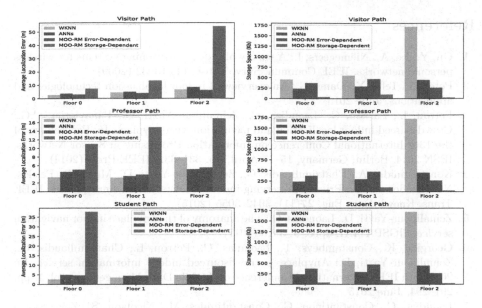

Fig. 5. Paths comparison

6 Conclusions and Future Work

In this paper, a novel MOO-RM-based ILS is introduced. MOO-RM provides the user with a variety of solutions in different granularities, each providing different levels of localization error and storage space requirements. This allows the user to localize in an indoor environment based on his preferences e.g. limited storage availability levels might require a smaller model with high localization error trade-off. The experimental results demonstrate the superiority of the proposed approach when compared with conventional client-side and ANN fingerprinting approaches, in terms of resource consumption, while preserving user privacy and maintaining localization accuracy.

Finally, this research study provides solutions based on a fixed ANN structure. As a future challenge, we would like to examine the use of variable structures between the cluster models provided to the MOEA/D. In particular, different ANN model structures result in different levels of storage space requirements and localization error. This will allow us to provide better solutions to smartphone users while navigating based on their preferences.

Acknowledgements. This work is part of the "EnterCY" Integrated Project with project number INTEGRATED/0609/0020, which is co-funded by the European Regional Development Fund and the Cyprus Government through the Research & Innovation Foundation (RIF) program RESTART 2016–2020.

References

1. Gu, Y., Lo, A., Niemegeers, I.: A survey of indoor positioning systems for wireless personal networks. IEEE Commun. Surv. Tutor. **11**, 13–32 (2009)
2. Peter, A., Tella, Y., Dams, G.: An overview of indoor localization technologies and applications, June 2015
3. Petrou, L., Larkou, G., Laoudias, C., Zeinalipour-Yazti, D., Panayiotou, C.G.: Crowdsourced indoor localization and navigation with anyplace. In: Proceedings of the 13th International Conference on Information Processing in Sensor Networks, IPSN 2014, Berlin, Germany, 15–17 April, pp. 331–332. IEEE Press (2014)
4. Konstantinidis, A., Chatzimilioudis, G., Zeinalipour-Yazti, D., Mpeis, P., Pelekis, N., Theodoridis, Y.: Privacy-preserving indoor localization on smartphones. IEEE Trans. Knowl. Data Eng. **27**(11), 3042–3055 (2015)
5. Zeinalipour-Yazti, D., Laoudias, C.: The anatomy of the anyplace indoor navigation service. SIGSPATIAL Spec. **9**, 3–10 (2017)
6. Georgiou, K., Constambeys, T., Laoudias, C., Petrou, L., Chatzimilioudis, G., Zeinalipour-Yazti, D.: Anyplace: a crowdsourced indoor information service. In: 2015 16th IEEE International Conference on Mobile Data Management, vol. 1, pp. 291–294, June 2015
7. Laoudias, C., Constantinou, G., Constantinides, M., Nicolaou, S., Zeinalipour-Yazti, D., Panayiotou, C.: The airplace indoor positioning platform for android smartphones, July 2012
8. Lymberopoulos, D., et al.: A realistic evaluation and comparison of indoor location technologies: experiences and lessons learned, April 2015
9. Alexandrou, R., Papadopoulos, H., Konstantinidis, A.: Smartphone indoor localization using bio-inspired modeling. In: Yang, X.-S., Zhao, Y.-X. (eds.) Nature-Inspired Computation in Navigation and Routing Problems. STNC, pp. 149–167. Springer, Singapore (2020). https://doi.org/10.1007/978-981-15-1842-3_7
10. Félix, G., Siller, M., Álvarez, E.N.: A fingerprinting indoor localization algorithm based deep learning. In: 2016 Eighth International Conference on Ubiquitous and Future Networks (ICUFN), pp. 1006–1011, July 2016
11. Wang, X., Gao, L., Mao, S., Pandey, S.: CSI-based fingerprinting for indoor localization: a deep learning approach. IEEE Trans. Veh. Technol. **66**, 763–776 (2017)
12. Abbas, M., Elhamshary, M., Rizk, H., Torki, M., Youssef, M.: WiDeep: WiFi-based accurate and robust indoor localization system using deep learning, January 2019
13. Bergenti, F., Monica, S.: A bio-inspired approach to WiFi-based indoor localization. In: Cagnoni, S., Mordonini, M., Pecori, R., Roli, A., Villani, M. (eds.) WIVACE 2018. CCIS, vol. 900, pp. 101–112. Springer, Cham (2019). https://doi.org/10.1007/978-3-030-21733-4_8
14. Ding, G., Tan, Z., Wu, J., Zeng, J., Zhang, L.: Indoor fingerprinting localization and tracking system using particle swarm optimization and Kalman filter. IEICE Trans. Commun. **E98.B**, 502–514 (2015)
15. Pericleous, S., Konstantinidis, A., Demetriades, A.: A multi-objective indoor localization service for smartphones, April 2019

STDP Plasticity in TRN Within Hierarchical Spike Timing Model of Visual Information Processing

Petia Koprinkova-Hristova[1]([✉])(iD), Nadejda Bocheva[2](iD), Simona Nedelcheva[1],
Miroslava Stefanova[2], Bilyana Genova[2], Radoslava Kraleva[3](iD),
and Velin Kralev[3](iD)

[1] Institute of Information and Communication Technologies,
Bulgarian Academy of Sciences, Sofia, Bulgaria
pkoprinkova@bas.bg, croft883@gmail.com
[2] Institute of Neurobiology, Bulgarian Academy of Sciences, Sofia, Bulgaria
nadya@percept.bas.bg, {mirad_st,b.genova}@abv.bg
[3] Department of Informatics, South West University, Blagoevgrad, Bulgaria
{rady_kraleva,velin_kralev}@swu.bg

Abstract. We investigated age related synaptic plasticity in thalamic
reticular nucleus (TRN) as a part of visual information processing system
in the brain. Simulation experiments were performed using a hierarchical spike timing neural network model in NEST simulator. The model
consists of multiple layers starting with retinal photoreceptors through
thalamic relay, primary visual cortex layers up to the lateral intraparietal
cortex (LIP) responsible for decision making and preparation of motor
response. All synaptic inter-' and intra-layer connections of our model
are structured according to the literature information. The present work
extends the model with spike timing dependent plastic (STDP) synapses
within TRN as well as from visual cortex to LIP area. Synaptic strength
changes were forced by teaching signal typical for three different age
groups (young, middle and elderly) determined experimentally from eye
movement data collected by eye tracking device from human subjects
preforming a simplified simulated visual navigation task.

Keywords: Spike timing neural model · Spike timing dependent
plasticity · Visual system · Decision making · Saccade generation

1 Introduction

The visual information coming through our eyes is processed by a hierarchy
of multiple consecutive brain areas having different functionality. The sensory

This work was financially supported by the Bulgarian Science Fund, grant No DN02-
3-2016 "Modeling of voluntary saccadic eye movements during decision making".

© IFIP International Federation for Information Processing 2020
Published by Springer Nature Switzerland AG 2020
I. Maglogiannis et al. (Eds.): AIAI 2020, IFIP AICT 583, pp. 279–290, 2020.
https://doi.org/10.1007/978-3-030-49161-1_24

layer (retina) consists of photo-receptive cells. It transforms the incoming light into electrical signals fed into the brain via retina ganglion cells (RGC). Next a relay structure (lateral geniculate nucleus (LGN) and thalamic reticular nucleus (TRN)) transmits the signals to the primary visual cortex (V1). Higher brain areas (middle temporal area (MT) and medial superior temporal area (MST)) are responsible for motion information processing. Based on perceived sensory information our brain makes decisions and initiates motor responses. The decisions based on processed visual information are taken in the lateral intraparietal area (LIP) that is also responsible for preparation of the eyes' motor response (change of gaze direction) called saccade. Most of the existing motion information processing models are restricted to the interactions between some of the mentioned areas like: V1 and MT in [1,2,5,25], V1, MT and MST in [21]; MT and MST in [9,19]. Many models consider only the feedforward interactions (e.g. [25,26]) disregarding the feedback connectivity; others employ rate-based equations (e.g. [10,20]) considering an average number of spikes in a population of neurons. In our preliminary research [11] we have developed and implemented in NEST 2.12.0 simulator ([16]) a spike-timing neural network model having static inter- and intra-layer synaptic connections structured according to the literature information that includes all mentioned above structures. Determination of the parameters' values for such kind of models is usually done using electrophysiological recordings directly from the brain that are rarely available for all modelled brain areas for human subjects. Our preliminary attempt to tune the synaptic connections between the last two layers (MST and LIP) using final outcome from experiments on visual information perception and decision making, i.e. recorded motor reaction (saccade generation) revealed that spike timing dependent plasticity (STDP) led to age-related changes in these synaptic weights [12]. Training data was collected from the human decisions during experiment with visual stimuli simulating optic flow patterns of forward self-motion on a linear trajectory to the left or to the right of the center of the visual field with a gaze in the direction of heading. The subjects had to indicate the perceived direction of heading by saccade movement. The mean latency of the eye movements for each one of the three age groups (young, middle age and elderly) was used as training signal fed into the output layer of the model structures.

In the present work we extended our model [11] with feedback connectivity between each pair of consecutive layers thus allowing for complete feedback propagation of training signals. We also allow STDP plasticity in the feedback/feedforward connections within thalamic relay structure in order to test whether training signal propagates deeper in the model. The thalamus provides sensory input to the cortex, but it also receives feedback from the cortex that is considered to be modulatory [24]. It receives also indirect inhibitory feedback from TRN, a structure that is considered to be related to attention (e.g., [6]) and thus, it is also expected to have a modulatory effect on the activity in the thalamus. Hence, a propagation of the effects of training to the thalamus would support the biological relevance of the proposed model. The paper is organized as follows: Sect. 2 describes briefly the overall model structure and parameters; next we describe briefly the experimental set-up and data processing; Sect. 4 presents results from STDP training of the model and obtained parameters typical for mean behaviour of each one of the three tested age groups; the concluding section comments obtained results and determines directions for our future work.

2   Model Structure

The hierarchical model organization is shown on Fig. 1. It is based on structure developed first in [11] based on literature information about each layer neurons' functionality, structure and connectivity according to [7,8,15,17,18,23,27]. The difference is in additional feedback connections from MST to MT as well as STDP connections within thalamic relay (to and from LGN to TRN and interneurons IN) and from MST to LIP area.

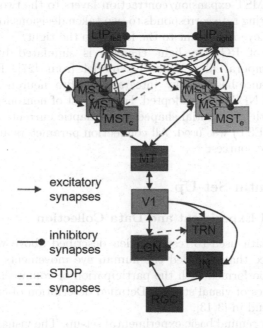

Fig. 1. Model structure. (Color figure online)

Detailed description of model structure and connectivity design can be found in [11]. Here we briefly explain it. Each coloured rectangle on Fig. 1 represents a layer of neurons positioned on a regular two-dimensional grid. Connections between layers are denoted by arrows having color corresponding to the sign of their weights (red for positive, called excitatory and blue for negative, called inhibitory connections respectively). Connections denoted by solid lines have constant weights while those denoted by dashed lines are able to change their weights in dependence on activity of pre- and postsynaptic neurons, i.e. they have spike timing dependent plasticity (STDP) [22]. Sensory layer (RGC) as well as thalamic relay (LGN) consist of two sub-layers of neurons reacting positively (ON1 and ON2) and negatively (OFF1 and OFF2) to the increase of luminosity. Each neuron within LGN has its own interneuron (IN) and thalamic reticular nucleus (TRN) that processes feedback from the next (V1) layer. Layers V1 and MT have identical structure and connectivity adopted from [15] that makes them sensitive to orientation and direction of movement of visual objects. The MST consists of two layers, sensitive to expansion (MSTe) and contraction (MSTc) movement patterns respectively, like in [17]. These sub-structures are represented as three groups on Fig. 1 that are able to detect expansion or contraction of moving objects from imaginary centers positioned left, right or at the center of the visual scene denoted by l, r and c on Fig. 1. Since our model aims to decide whether the expansion center of a moving dot stimulus is left or right from the stimulus center, we proposed a task-dependent design of excitatory/inhibitory connections from MST expansion/contraction layers to the two LIP sub-regions whose increased firing rate corresponds to two taken decisions for two alternative motor responses - eye movement to the left or to the right.

The reaction of RGC to light changes is simulated by a convolution with a spatio-temporal filter following model from [27]. For the neurons in LGN conductance-based leaky integrate-and-fire neuron model as in [4] (iaf_chxk_2008 in NEST) was adopted. For the rest of neurons, leaky integrate-and-fire model with exponential shaped postsynaptic currents according to [28] (iaf_psc_exp in NEST) was used. All connection parameters are the same as in the cited literature sources.

3   Experimental Set-Up

3.1   Behavioral Experiment and Data Collection

The time series data used to test the idea described above were collected by eye tracking device that recorded the human eye movements during a behavioral experiment performed with the participation of volunteer human subjects responding to series of visual stimuli. Detailed description of experimental conditions can be found in [3,13].

Here we briefly remind basic experimental set-up. The visual stimulation was performed by projection on a gray screen of different patterns of 50 white moving dots in a circular aperture with radius of 7.5 cm positioned in the middle of a computer screen. The patterns of dot movements were designed to mimic dots expansion from an imaginary center positioned left or right from the screen center respectively. The subject sat at 57 cm from the monitor screen. Each stimulus presentation was preceded by a warning sound signal. A red fixation point with size of 0.8 cm appeared in the center of the screen for 500 ms. The stimuli were presented immediately after the disappearance of the fixation point. The Subject's task was to continue looking at the position where the fixation point was presented until he/she made a decision where the center of the pattern was and to indicate this position by a saccade (fast eye movement). The subjects also had to press the left or the right mouse button depending on the perceived position of the center - to the left or to the right from the middle of the screen. If the subject could not make a decision during the stimulus presentation (3.3 s for 100 consecutive frames), the stimulus disappeared and the screen remained gray until the subject made a response. Each experimental session consisted of consecutive presentation in random order of 10 patterns for each of the 14 possible stimulus types from chosen experimental condition, i.e. totally 140 stimuli were observed by test subjects during every session. The eye movements of the participants in the experiment were recorded by a specialized hardware – Jazz novo eye tracking system [29]. All recordings from all the sensors of the device for one session per person were collected with 1 KHz frequency and the information is stored in files. These include: the calibration information; records of horizontal and vertical eye positions in degrees of visual angle eye_x and eye_y; screen sensor signal for presence/absence of a stimulus on the monitor; microphone signal recording sounds during the experiment; information about tested subjects (code) and type of the experimental trial for each particular record.

Three age groups took part in the experiment: young (between 20 and 34 years old), elderly (from 57 to 84 years old) and middle aged group (between 36 and 52 years old). All participants have given a written informed consent for participating in the study after explanation of the experimental procedure. The experiments were approved by the ethical committee of the Institute of Neurobiology, Bulgarian Academy of Sciences and are in accordance with the Declaration of Helsinki.

3.2 Data Processing

The raw data was collected in a relational database [14]. It allowed us to process all sensors data in order to extract only the records from the presentation of a stimulus on the screen to the mouse button press. The data between the stimulus presentations were excluded since it is not relevant to the eye movements during task performance. The processed eye movement data were refined by removing the outliers and a drift diffusion model of mean response time of each one of the age groups for all four experimental conditions was derived [3].

Based on the identified mean reaction times from [3] we've created training signals as generating currents I_{left} and I_{right} for the left and right LIP neurons respectively as follows:

$$I_{left/right} = A_{left/right}/(1 + \exp(k_{left/right}t))$$ (1)

Amplitude $A_{left/right}$ defines maximal input current (in pA) while $k_{left/right}$ determines settling time of the exponent that corresponds to the mean reaction time determined from experiments for each age group and experimental condition. For all three age groups amplitude values were the same: $A_{left} = 200$ and $A_{right} = 100$. In order to achieve approximately the settling time determined from experimental data, parameter $k_{left/right}$ has different values for three age

284   P. Koprinkova-Hristova et al.

groups (Y - young, M - middle, O - old) with opposite signs for left and right case of stimulus respectively as follows: $k^Y_{left/right} = -/+0.02$; $k^M_{left/right} = -/+0.01$; $k^O_{left/right} = -/+0.005$.

4 Simulations

The overall model was tested using visual stimulation simulating an observer's motion on a linear trajectory with eyes fixed in the heading direction. Example from the stimuli used in the behavioural study with a position of the imaginary center to the left was selected so our aim is to teach the model to react correctly with increased spiking activity in the left LIP area after a time interval typical for group mean reaction time for each age. Training was performed in iterations, each one consisting of presentation of visual stimulation to the model input and the teaching signal to its output layer respectively.

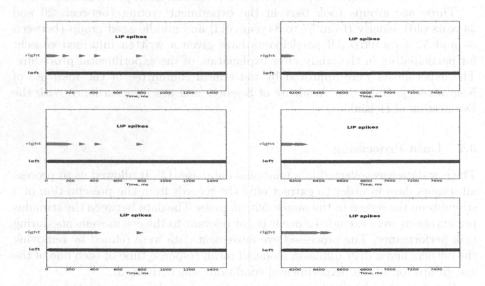

Fig. 2. LIP reactions for left (blue arrows) and right (red arrows) during first (left column) and fifth (right column) iteration for three age groups (top - young, second row - middle age and bottom row - elderly people). (Color figure online)

Figure 2 presents spikes in both left (blue) and right (red) LIP layers obtained during first and last (fifth) iteration. It shows that the frequency of spikes induced in the right LIP layer decrease with time and after a varying delay for the three age groups no more spikes occur. The frequency of spikes for the left LIP region increases with time (though this change is not evident from the figure). Figure 3 compares the weights of connections from MST to LIP area obtained after five iterations for the three age groups. For clarity only weights that are different

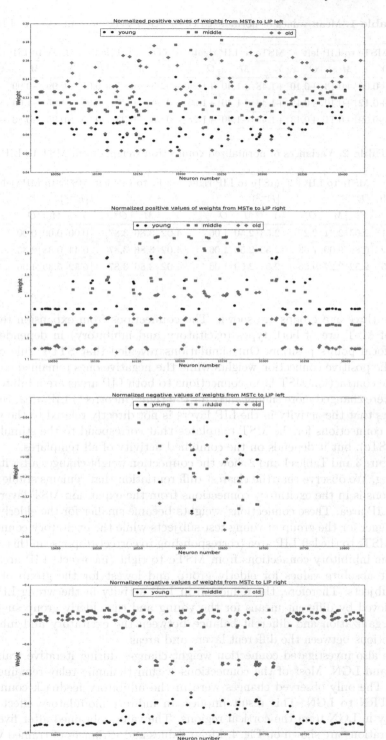

Fig. 3. Connections from MST to LIP.

Table 1. Mean values of normalized connection weights from MST to LIP.

It.	MSTe to LIP left			MSTe to LIP right			MSTc to LIP left			MSTc to LIP right		
	Y	M	O	Y	M	O	Y	M	O	Y	M	O
1	+0.08	+0.13	+0.10	+1.48	+1.40	+1.25	−0.95	−0.95	−0.95	−0.99	−0.99	−0.99
2	+0.12	+0.10	+0.12	+1.48	+1.40	+1.25	−0.95	−0.95	−0.95	−1.05	−1.04	−0.89
5	+0.09	+0.11	+0.13	+1.48	+1.40	+1.25	−0.95	−0.95	−0.95	−0.95	−0.83	−0.84

Table 2. Variances of normalized connection weights from MST to LIP.

It.	MSTe to LIP left $[E^{-4}]$			MSTe to LIP right $[E^{-2}]$			MSTc to LIP left $[E^{-2}]$			MSTc to LIP right $[E^{-2}]$		
	Y	M	O	Y	M	O	Y	M	O	Y	M	O
1	2.03	2.21	2.27	2.53	2.50	1.06	4.02	3.86	3.87	0.00	0.00	0.00
2	6.89	9.09	7.63	2.53	2.50	1.06	4.02	3.86	3.87	0.44	0.38	2.35
5	2.52	2.27	9.55	2.53	2.50	1.06	4.02	3.86	3.87	0.52	5.35	4.96

for the three age groups are shown. The connections from expansion template area of MST are of both types (excitatory and inhibitory) in dependence on their focal points position. Our simulations revealed that STDP rule changes only the positive connection weights while the negative ones remained constant. For the contraction MST layer connections to both LIP areas are inhibitory and they were changed more significantly for the right (incorrect) LIP area. Hence, it appears that the activity in the LIP layers is not directly related to the weights of the connections for the MST templates that correspond to the stimulus (i.e. the MSTe), but it depends on the combined activity of all templates.

Figure 3 and Tables 1 and 2 show the connection weight changes after iterative training. We observe that the clearest differentiation that remains stable during iterations is in the excitatory connections from the expansion MST layer to the right LIP area. These connections' weights became smaller for the elderly group and bigger for the group of young test subjects while the excitatory connections from MSTe to the left LIP area (corresponding to correct response in this case) as well the inhibitory connections from MSTc to right (incorrect) LIP area reach highest absolute values for elderly group and lowest for the group of young test subjects. Therefore, the reduction in LIP activity in the wrong LIP layer is achieved by different means for the young and the elderly group suggesting a re-organization and different balance between the excitatory and inhibitory connections between the different layers and areas.

We also investigated connection weight changes during iterative training in TRN and LGN. Most of the connections within thalamic relay remained constant. The only observed changes were in the inhibitory feedback connections from TRN to LGN. This result implies an indirect modulatory effect of the activity in LGN from the cortical regions. The values obtained after five training iterations are shown on Fig. 4. Mean values and variances of trained weights

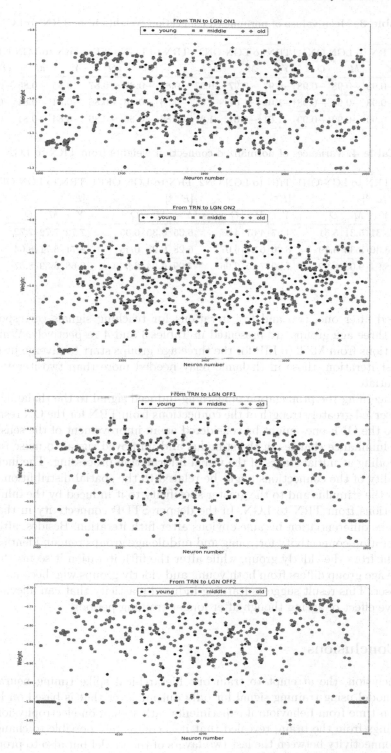

Fig. 4. Connections from TRN to LGN.

Table 3. Mean values of normalized connection weights from TRN to LGN.

It.	TRN to LGN ON1			TRN to LGN ON2			TRN to LGN OFF1			TRN to LGN OFF2		
	Y	M	O	Y	M	O	Y	M	O	Y	M	O
1	−0.98	−0.98	−0.98	−0.97	−0.97	−0.97	−0.87	−0.87	−0.87	−0.85	−0.85	−0.85
2	−0.98	−0.98	−0.98	−1.01	−1.01	−1.01	−0.86	−0.86	−0.86	−0.84	−0.84	−0.84
5	−0.99	−0.99	−0.99	−1.01	−1.01	−1.01	−0.86	−0.86	−0.86	−0.84	−0.84	−0.84

Table 4. Variances of normalized connection weights from TRN to LGN.

It.	TRN to LGN ON1 $[E^{-3}]$			TRN to LGN ON2 $[E^{-3}]$			TRN to LGN OFF1 $[E^{-3}]$			TRN to LGN OFF2 $[E^{-3}]$		
	Y	M	O	Y	M	O	Y	M	O	Y	M	O
1	5.31	5.31	5.31	4.67	4.67	4.67	6.35	6.35	6.35	7.72	7.72	7.72
2	3.30	3.30	3.39	6.26	6.26	6.07	6.28	6.28	6.45	8.71	8.71	8.64
5	8.94	9.03	8.92	7.00	7.45	6.96	7.09	6.94	6.96	8.68	8.59	8.57

obtained after one, two and five iterations for teaching signals corresponding to the three age groups are presented in Tables 3 and 4 respectively. While the connections from MST to LIP for the three age groups start to diverge just after the first iteration, those in thalamic relay needed more than two iterations to differentiate.

Concerning the propagation of learning-induced signal to the thalamic relay, the observed greater strength of the connections from TRN for the ON responses than to the OFF ones might be considered as an improvement of the sensitivity to the luminance changes in the stimulus and improved signal to noise ratio in its encoding at initial stages of the visual information processing. The increased variability of the connections might be related to the spatial distribution of the dots in the stimulus and to the greater specificity to it induced by the inhibitory connections from TRN to LGN. In the deeper STDP connectivity in thalamic relay, age differentiation became obvious after fifth iteration. Besides, after second iteration connectivity for young and middle age groups remain identical and different from the elderly group, while after the fifth iteration it seems that the middle age group differs from both young and elderly groups who became a little bit closer. This result suggests learning-induced plasticity that can alleviate the negative effect of ageing though after longer training periods.

5   Conclusions

In conclusion, the attempt to train our hierarchical spike timing neural network model using training signal for only output layer that is based on human reaction time from behavioural experiments rather than on electrophysiological recordings from the brain, revealed that it is completely possible to change not only connectivity between the last two layers of the model but also to propagate the teaching signal much deeper in the hierarchical structure. Such feedback modulatory propagation from the cortical areas to the thalamus is in accordance with the known physiological data and gives credit to our modeling efforts. Our preliminary results also revealed typical for aging differentiation of connection weights especially for the connections between MST and LIP areas. While for all age groups the activity in the LIP layer corresponding to the incorrect response terminates with time, this effect is achieved by different recombination of the connection weights for the different age groups. Concerning deep thalamic relay, although age-related changes propagate much slowly, they were also observed after more training iterations indicating that learning-induced activity may reduce the age-related changes induced by the imbalance of inhibitory and excitatory activity in the cortical regions.

Further direction of our investigations will be enriching our hierarchical model with more STDP connections in order to complete training of its connection weights in dependence on typical reactions of different age groups.

References

1. Bayerl, P., Neumann, H.: Disambiguating visual motion through contextual feedback modulation. Neural Comput. **16**(10), 2041–2066 (2004)
2. Bayerl, P.: A Model of Visual Perception. Ulm University, Germany (2005). PhD Thesis
3. Bocheva, N., Genova, B., Stefanova, M.: Drift diffusion modeling of response time in heading estimation based on motion and form cues. Int. J. Biol. Biomed. Eng. **12**, 75–83 (2018)
4. Casti, A., Hayot, F., Xiao, Y., Kaplan, E.: A simple model of retina-LGN transmission. J. Comput. Neurosci. **24**, 235–252 (2008)
5. Chessa, M., Sabatini, S., Solari, F.: A systematic analysis of a V1-MT neural model for motion estimation. Neurocomputing **173**, 1811–1823 (2016)
6. Crick, F.: Function of the thalamic reticular complex: the searchlight hypothesis. Proc. Natl. Acad. Sci. USA **81**, 4586–4590 (1984)
7. Escobar, M.-J., Masson, G.S., Vieville, T., Kornprobst, P.: Action recognition using a bio-inspired feedforward spiking network. Int. J. Comput. Vis. **82**, 284–301 (2009)
8. Ghodratia, M., Khaligh-Razavic, S.-M., Lehky, S.R.: Towards building a more complex view of the lateral geniculate nucleus: recent advances in understanding its role. Prog. Neurobiol. **156**, 214–255 (2017)
9. Grossberg, S., Mingolla, E., Pack, C.: A neural model of motion processing and visual navigation by cortical area MST. Cereb. Cortex **9**(8), 878–895 (1999)
10. Grossberg, S., Mingolla, E., Viswanathan, L.: Neural dynamics of motion integration and segmentation within and across apertures. Vis. Res. **41**(19), 2521–2553 (2001)
11. Koprinkova-Hristova, P., Bocheva, N., Nedelcheva, S., Stefanova, M.: Spike timing neural model of motion perception and decision making. Front. Comput. Neurosci. **13**, 20 (2019). https://doi.org/10.3389/fncom.2019.00020
12. Koprinkova-Hristova, P., et al.: STDP training of hierarchical spike timing model of visual information processing. In: World Congress on Computational Intelligence (2020). (accepted)
13. Koprinkova-Hristova, P., Stefanova, M., Genova, B., et al.: Features extraction from human eye movements via echo state network. Neural Comput. Appl., 1–14 (2019). https://doi.org/10.1007/s00521-019-04329-z

14. Kraleva, R., Kralev, V., Sinyagina, N., Koprinkova-Hristova, P., Bocheva, N.: Design and analysis of a relational database for behavioral experiments data processing. Int. J. Online Eng. **14**(2), 117–132 (2018)

15. Kremkow, J., et al.: Push-pull receptive field organization and synaptic depression: mechanisms for reliably encoding naturalistic stimuli in V1. Front. Neural Circ. **10**, 37 (2016). https://doi.org/10.3389/fncir.2016.00037

16. Kunkel, S., et al.: NEST 2.12.0. Zenodo (2017). https://doi.org/10.5281/zenodo.259534

17. Layton, O.W., Fajen, B.R.: Possible role for recurrent interactions between expansion and contraction cells in MSTd during self-motion perception in dynamic environments. J. Vis. **17**(5), 1–21 (2017)

18. Nedelcheva, S., Koprinkova-Hristova, P.: Orientation selectivity tuning of a spike timing neural network model of the first layer of the human visual cortex. In: Georgiev, K., Todorov, M., Georgiev, I. (eds.) BGSIAM 2017. SCI, vol. 793, pp. 291–303. Springer, Cham (2019). https://doi.org/10.1007/978-3-319-97277-0_24

19. Perrone, J.: A neural-based code for computing image velocity from small sets of middle temporal (MT/V5) neuron inputs. J. Vis. **12**(8), 1 (2012). https://doi.org/10.1167/12.8.1

20. Raudies, F., Neumann, H.: A neural model of the temporal dynamics of figure-ground segregation in motion perception. Neural Netw. **23**(2), 160–176 (2010)

21. Raudies, F., Mingolla, E., Neumann, H.: Active gaze control improves optic flow-based segmentation and steering. PLoS ONE **7**(6), 1–19 (2012)

22. Rubin, J., Lee, D., Sompolinsky, H.: Equilibrium properties of temporally asymmetric Hebbian plasticity. Phys. Rev. Lett. **86**, 364–367 (2001)

23. Sadeh, S., Rotter, S.: Statistics and geometry of orientation selectivity in primary visual cortex. Biol. Cybern. **108**(5), 631–653 (2013). https://doi.org/10.1007/s00422-013-0576-0

24. Sherman, S.M., Guillery, R.W.: Exploring the Thalamus and Its Role in Cortical Function, 2nd edn. MIT Press, Cambridge (2009)

25. Simoncelli, E., Heeger, D.: A model of neuronal responses in visual area MT. Vis. Res. **38**, 743–761 (1998)

26. Solari, F., Chessa, M., Medathati, K., Kornprobst, P.: What can we expect from a V1-MT feedforward architecture for optical flow estimation? Sign. Process. Image Commun. **39**, 342–354 (2015)

27. Troyer, T.W., Krukowski, A.E., Priebe, N.J., Miller, K.D.: Contrast invariant orientation tuning in cat visual cortex: thalamocortical input tuning and correlation-based intracortical connectivity. J. Neurosci. **18**, 5908–5927 (1998)

28. Tsodyks, M., Uziel, A., Markram, H.: Synchrony generation in recurrent networks with frequency-dependent synapse. J. Neurosci. **20**(RC50), 1–5 (2000)

29. http://www.ober-consulting.com/product/jazz/

Tensor-Based CUDA Optimization for ANN Inferencing Using Parallel Acceleration on Embedded GPU

Ahmed Khamis Abdullah Al Ghadani ⓘ, Waleeja Mateen(✉) ⓘ,
and Rameshkumar G. Ramaswamy

National University of Science and Technology, Muscat 111, Oman
{ahmed150309,waleeja160359,rameshkumar}@nu.edu.om

Abstract. With image processing, robots acquired visual perception skills; enabling them to become autonomous. Since the emergence of Artificial Intelligence (AI), sophisticated tasks such as object identification have become possible through inferencing Artificial Neural Networks (ANN). Be that as it may, Autonomous Mobile Robots (AMR) are Embedded Systems (ESs) with limited on-board resources. Thus, efficient techniques in ANN inferencing are required for real-time performance. This paper presents the process of optimizing ANNs inferencing using tensor-based optimization on embedded Graphical Processing Unit (GPU) with Computer Unified Device Architecture (CUDA) platform for parallel acceleration on ES. This research evaluates renowned network, namely, You-Only-Look-Once (YOLO), on NVIDIA Jetson TX2 System-On-Module (SOM). The findings of this paper display a significant improvement in inferencing speed in terms of Frames-Per-Second (FPS) up to 3.5 times the non-optimized inferencing speed. Furthermore, the current CUDA model and TensorRT optimization techniques are studied, comments are made on its implementation for inferencing, and improvements are proposed based on the results acquired. These findings will contribute to ES developers and industries will benefit from real-time performance inferencing for AMR automation solutions.

Keywords: Artificial Neural Networks · Embedded GPU · TensorRT · Real-time · NVIDIA Jetson · Image processing · YOLO · CUDA

1 Introduction

Artificial Intelligence (AI) is being increasingly adopted into ES due to its synergistic relation in the virtue of efficient data analysis, capability to rationalize tasks and performing system optimization. Over the past few years, AI has been rapidly developed to be utilized in applications such as safe autonomous driving which involves object avoidance and collision mitigation [1] and in visual perception tasks like scene understanding, object detection, and localization [2, 3]. Machine Learning, a prime constituent of AI comprises of two main processes: training and inferencing. These are power-intensive and computationally demanding tasks [4].

© IFIP International Federation for Information Processing 2020
Published by Springer Nature Switzerland AG 2020
I. Maglogiannis et al. (Eds.): AIAI 2020, IFIP AICT 583, pp. 291–302, 2020.
https://doi.org/10.1007/978-3-030-49161-1_25

As observed in examples [5, 6], inferencing is a critical step in implementing and deploying an effective ES. It encompasses the procedures involved in determining a viable conclusion from the evidence collected during the training phase. Observing from the perspective of ESs, inferencing which is a power-intensive utility, is challenging to run with limited available resources [7]. In such a scenario, optimization is the key to achieving best performance with lower power consumption [8].

System optimization is a challenging step in achieving higher performance and therefore, the relatively more popular alternative is usually opted for: scaling the system over several GPUs [9]. This solution incurs high cost and decreases projects' economic feasibility. The research [10] supports this observation through a study conducted on ImageNet competition which aims to achieve maximum accuracy through complex Deep Neural Networks (DNNs). To maximize performance, the participants neglect system optimization and invest in hardware upgrades, which shows inefficient resources utilization as the existing system was not executed up to its full potential.

Optimization encompasses processes ranging from data-handling, parameters fine-tuning, architecture selection and process pipeline remodeling [11]. This can be seen in examples like [12] open source GPU-Accelerated feature extraction tool and Bonnet framework for semantic segmentation in robotics [13]. This is also evident in the emergence of GPU acceleration frameworks like CUDA and OpenGL [8]. Another key technology in optimization is TensorRT [14]. Yamane [15] stressed on the importance of inferencing on ESs. Also, highlighting the fact that ESs are still lacking in real-time performance, rendering edge devices unreliable for critical missions.

In this paper, You-Only-Look-Once (YOLOv3) object detection system is inferred on Jetson TX2. YOLO model predicts objects by running a single network evaluation [16]. Making it lighter than most networks with similar accuracy; hence, ideal candidate for ESs. Running on the GPU for parallelization, the inference performance is evaluated with and without optimization using TensorRT inference accelerator. The acquired results present an eye-catching increase in performance up to 3.5x times. This provides an insight into optimization trends and their effectiveness. Furthermore, with the acquired insights, CUDA framework pipeline is studied under limited scope and areas of potential improvements are suggested.

2   Methods

This research is a part of a bigger research that focuses on applying visual Simultaneous Localization And Mapping (vSLAM) for AMRs in industrial environments. With focus on material handling in warehouses; this work has been presented in the Logistic Research Competition organized by Oman Logistic Center. Additionally, for ore mining industry, a dataset with one class "Power Screens" has been initiated [17]. Both of which targeted real-time inferencing of YOLOv3 ANN. This research continues the previous work by optimizing the inferencing speed to meet real-time needs with a pretrained YOLOv3 ANN with COCO dataset with 80 classes.

2.1 Hardware Setup

The hardware setup follows certain specifics for AMRs: limited on-board compute capability, limited power source, System-On-Module (SOM). The physical set-up of the robot adopts the open source Turtlebot3, with adjustments for Jetson TX2, ZED Stereo Camera, 2D RP-LIDAR, and a Logitech C920 HD Pro webcam. On the execution layer, as can be seen on the AMR system block diagram in Fig. 1, the robot uses differential drive for movements with magnetic encoders for odometry. Be that as it may, in this AMR context, this research focuses solely on the optimization process of ANN inferencing.

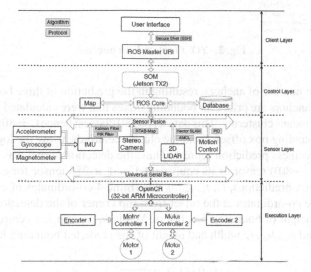

Fig. 1. AMR system block diagram

2.2 YOLO ANN for Object Detection

YOLO ANN implements a smart approach for object detection. Achieving real-time performance using a single Convolutional Neural Network (CNN) structure. Quoting its developers from their research paper [18] "It achieves 57.9 AP50 in 51 ms on a Titan X, compared to 57.5 AP50 in 198 ms by RetinaNet, similar performance but 3.8× faster".

YOLO ANN architecture is based on an open-source CNN called Darknet-53 for image classification. Darknet-53 serves as the network's backbone with an additional 53 layers for detection. Resulting in a total of 106 layers, out of which 75 are convolutional, 23 shortcuts, 4 routes, 2 upsamples, 3 detections [19].

With reference to the architecture illustration in Fig. 2, the network performs detection at 3 different scales at 82nd, 94th, and 106th layers. At each one of the detection layers, the image is downsampled by a factor called stride of the network. Respective to the detection layers, the input image is downsampled by a factor of 32, 16, and 8.

In this architecture, the detection kernel shape is defined by $N \times N \times (B \times (5 + C))$. $N \times N$ being the alternating 1×1 convolutional layers for feature space reduction.

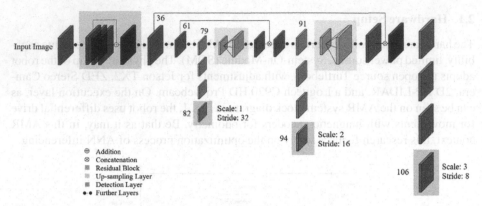

Fig. 2. YOLO network architecture

$B = 3$ being the number of anchors, resulting in the prediction of three bounding boxes per cell. These anchors are priors for bounding boxes that were calculated on the COCO dataset using k-means clustering. This dataset has 80 classes; hence $C = 80$. The '5' is the sum of the 4 bounding box offsets from the cluster centroids, using the logistic sigmoid Eq. (1) and objectness prediction score 1. Thus, the detection kernel was designed as 1 $\times$ 1 $\times$ (3 $\times$ (5 + 80)) tensor or its equivalent 1 $\times$ 1 $\times$ 255 tensor for each scale. For the bounding box predictions, t_x, t_y, t_w, t_h determine 4 co-ordinates of each bounding box, (c_x, c_y) are co-ordinates at the offset from top corner of the detection grid. p_w, p_h are width and height of bounding box priors. While b_x, b_y are x, y centre co-ordinates of prediction and b_w, b_h are width and height of the predicted bounding box.

$$\sigma(x) = \frac{1}{1 + e^{-x}} \tag{1}$$

$$b_x = \sigma(t_x) + c_x \tag{2}$$

$$b_y = \sigma(t_y) + c_y \tag{3}$$

$$b_w = p_w e^{t_w} \tag{4}$$

$$b_h = p_h e^{t_h} \tag{5}$$

Considering the above, the output value is always between 0 and 1 due to the logistic sigmoid function. YOLO ANN predicts the relative offsets of the bounding box center rather than the absolute coordinates. Normalized by the feature map dimensions which is 1 [20].

2.3   Complexity Class

To verify the complexity class in which the adopted YOLO ANN falls in; it is necessary to examine the decision-making procedure in determining a prediction. Bearing that in mind, it is worth noting that "object recognition is not a formally defined problem, so is not in itself either polynomial time solvable or NP-complete" [21].

YOLO ANN recognizes an object based on class-specific confidence score for each bounding box. This is achieved by first computing the confidence as follows:

$$\text{Confidence} = \text{Pr(Object)} \times \text{IOU}_{\text{pre}}^{\text{truth}} \tag{6}$$

The value of Pr(Object) is 1 if a ground truth box is present within the grid cell, otherwise it is 0. While the intersection over union value between the predicted bounding box and the actual ground truth is denoted by $\text{IOU}_{\text{pre}}^{\text{truth}}$ [22].

With reference to the detection kernel in Sect. 2.2, each grid generates conditional class probabilities represented as follow: $\text{Pr(Class}_i \mid \text{Object})$. YOLO ANN output is class-specific confidence, computed by multiplying the conditional class probabilities and the individual box confidence predictions as represented by [22]:

$$\text{Pr(Class}_i|\text{Object}) \times \text{Pr(Object)} \times \text{IOU}_{\text{pre}}^{\text{truth}} = \text{Pr(Class}_i) \times \text{IOU}_{\text{pre}}^{\text{truth}} \tag{7}$$

2.4 Tensor-Based Optimization

YOLO ANN was originally developed using Darknet framework written in C and CUDA [16]. Optimization using TensorRT adds an extra step between training a model and inferencing it. That step requires the trained ANN model to be converted into a format that is optimizable by TensorRT. TensorRT is a programmable inference accelerator built on CUDA for parallel programming. In this research, the conversion of the YOLO ANN model is done through an open-source format defined by the Open Neural Network Exchange (ONNX) ecosystem. This format ensures interoperability and easy hardware access optimization.

When approaching an optimization problem on an AMR system, specific considerations pertaining to ESs must be put into account. Memory management, hardware architecture utilization, inference precision, and power efficiency. TensorRT memory allocation is done using TensorFlow allocators by setting argument `config.gpu_options.per_process_gpu_memory_fraction` [23]. For hardware architecture utilization, ANN model is converted on target Jetson TX2.

Inference in INT8 and mixed precision reduces memory footprint; which is important on an ES. Using symmetric linear quantization, models running in 32-bit floating-point precision are scaled down to 8 bits with preserved symmetry. FP32 represents billions of numbers while the INT8 represents 256 possible values only [24]. Equation below formulates this process of getting quantized INT8 value, where input, floating point range, and scaling factor are denoted by x, r and s respectively:

$$\text{Quantize}\,(x, r) \;=\; \text{round}\,\big(s^{*}\,\text{clip}\,(x,\; -r, r)\big) \tag{8}$$

Using calibration and quantization aware training, accuracy is preserved when model is scaled to INT8. Taking a specific range where most of the activation values fall.

The network's performance is measured through latency and throughput. The former is determined by the time elapsed between input presence until output is acquired. The latter is determined by the number of inferences performed in a set amount of time. "Both can be measured using high precision timers present in C++ std::chrono::high_resolution_clock, and monitored by profiling CUDA and memory utilization during runtime [25]."

3 Experimental Procedure

The experiment was conducted on the following setup with the webcam mounted on an AMR. The robot was set to move at a constant speed of 10 cm/s detecting objects set in its field of view. The objects detected are apparent on the screen attached (Fig. 3).

Fig. 3. Objects detected displayed in real-time

The process of optimizing an ANN, especially for inferencing on an ES; requires a deep understanding of the software and the hardware architecture. This research is carried on Jetson TX2 SOM running Linux for Tegra (L4T). The key technical specification of this system can be seen in the Table 1 taken from [26].

Table 1. Jetson TX2 module specifications

GPU	256-core NVIDIA Pascal™ GPU architecture with 256 NVIDIA CUDA cores		
CPU	Dual-Core NVIDIA Denver 2 64-Bit CPU Quad-Core ARM® Cortex®-A57 MPCore		
Memory	8 GB 128-bit LPDDR4 Memory 1866 MHx - 59.7 GB/s		
Storage	32 GB eMMC 5.1		Power	7.5 W/15 W

It is noted that the Jetson board is running a 256-core Nvidia GPU which is based on Pascal architecture [26]. An efficient optimisation technique for GPU parallelization should be able to optimally distribute the workload on the 256 cores available.

Since YOLO was originally developed using Darknet framework, optimizing this ANN will require adjustments starting from the framework. Hence, the framework was built from source on the Jetson TX2 with CUDA, cuDNN, and OpenCV enabled. This is done by setting their respective flags to 1 in the Makefile. Adding to that, it is important to specify the CUDA architecture, which is 62, for Jetson TX2.

On the time of conducting this experiment, Jetson TX2 was running L4T, CUDA, and other relevant libraries as seen in Table 2 below:

Table 2. Software components and their versions on Jetson TX2

Component	Version	Component	Version
JetPack	4.2.1	cuDNN	7.5.0.66
OS	L4T R32.2 (K4.9)	TensorRT	5.1.6.1
CUDA	10.0.326	OpenCV	3.4.0

At first YOLOv3-416 was run using a direct implementation of Darknet detector as documented by Redmon & Farhadi [27]. Even though Darknet was built with CUDA and cuDNN enabled, the GPU utilization was low. This resulted in poor performance; maxing at 2 FPS. For efficient GPU utilization, YOLO ANN is formatted according to ONNX open format. Facilitating the conversion to TensorRT using PyCUDA API to access parallel computation of the GPU.

4 Result and Discussion

4.1 System Performance

Obtained results were analyzed for comparison between FPS with and without TensorRT optimization at different power modes as plotted in Fig. 4:

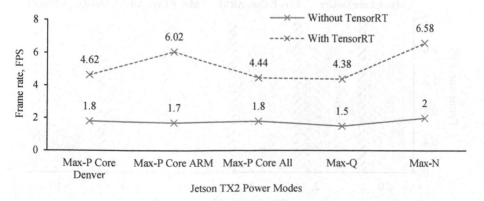

Fig. 4. FPS performance analysis with and without TensorRT optimization

As inferred from Fig. 4, when YOLOv3 is executed on CUDA along with OpenCV, the maximum achieved FPS was 2.0 at Max-N. This power mode offers maximum GPU frequency and utilizes both, ARM A57 and Denver cores. The lowest FPS was obtained from the Max-Q power mode at 1.5 FPS and was also observed to drop to 1.3 FPS when input traffic for object detection was increased. Max-Q and Max-P Core All modes had fluctuations in output. However, Max-P Core ARM was observed to be the most stable as the FPS stayed constant at 1.7. Max-N and Max-P Core Denver were observed to have minimum drop of −0.1 FPS.

In contrast to the above, with TensorRT optimization, the performance showed a drastic improvement from 147% in Max-P Core All to 254% improvement in Max-P Core ARM. The highest FPS achieved was on the Max-N at 6.58 units; an increase of 3.5× in performance. This is important for object detection applications on ESs from a live stream. Table 3 summarizes the performance gain with and without TensorRT.

Table 3. Percentage increase in FPS obtained without versus with TensorRT

Power modes	Without TensorRT	With TensorRT	Percentage increase, %
Max-P Core Denver	1.8	4.62	157
Max-P Core ARM	1.7	6.02	254
Max-P Core All	1.8	4.44	147
Max-Q	1.5	4.38	192
Max-N	2.0	6.58	229

4.2 Precise Object Detection

Adding one more dimension to the comparison, the effect of FPS on the number of objects detected is observed. Performance in FPS and detection rate were monitored on a test scene including 15 objects belonging to the ANN classes. As observed in Fig. 5, without optimization, maximum 10 objects were detected out of 15.

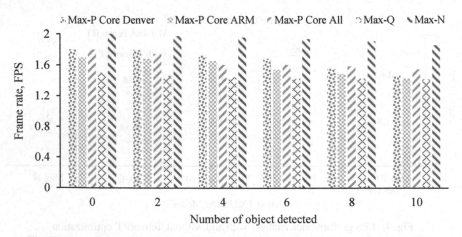

Fig. 5. Frame rate versus number of objects detected without TensorRT optimization

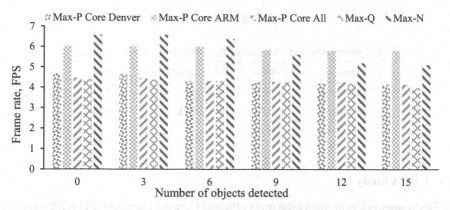

Fig. 6. Frame rate versus number of objects detected with TensorRT optimization

When tested with TensorRT optimization, all 15 objects in the scene were detected while the FPS was relatively higher than without optimization as shown in Fig. 6.

Both Fig. 5 and Fig. 6 illustrate the relation between FPS, number of objects detected, and power modes. It can be deduced that the increase in FPS results in more objects detected. However, more processing power is required as more objects must be processed per frame. The system allocates more resources towards recognition of objects present in the frame. This results in drop in FPS, consequently lowering the accuracy as Table 4 and Table 5 clarify. Additionally, 5 objects failed to be detected without optimization as highlighted in Table 4.

Table 4. Observation of objects detected without TensorRT and their percentage accuracy

#	Detected	Accuracy	#	Detected	Accuracy	#	Failed to detect
1	Bottle	100%	6	Keyboard	91%	11	Mouse
2	Bottle	93%	7	Backpack	88%	12	Laptop
3	Person	100%	8	Cell phone	87%	13	Keyboard
4	Person	99%	9	Cell phone	84%	14	Clock
5	Keyboard	99%	10	Cup	99%	15	Table

Table 5. Observation of objects detected with TensorRT and their percentage accuracy

#	Objects detected	Accuracy	#	Objects detected	Accuracy
1	Bottle	100%	9	Cell phone	96%
2	Bottle	92%	10	Cup	100%
3	Person	97%	11	Backpack	91%
4	Person	89%	12	Mouse	37%

(continued)

Table 5. (*continued*)

#	Objects detected	Accuracy	#	Objects detected	Accuracy
5	Keyboard	99%	13	Laptop	99%
6	Keyboard	67%	14	Table	73%
7	Keyboard	67%	15	Clock	99%
8	Cell phone	96%			

4.3 CUDA Study

CUDA framework is arguably the most efficient General-Purpose GPU (GPGPU) platform. This is mainly due to its maturity, continuous update and support from NVIDIA [28]. However, to accommodate the ever-increasing range of GPUs and their architectures, NVIDIA moved into generalization and multi-layering of CUDA 5.0. This caused a noticeable reduction in performance when compared to CUDA 4.0 [29]. Adding to that, it was noted in [30] that CUDA GPU acceleration was performing best on certain workloads and for full utilisation, GPU kernels must be optimised as well.

5   Conclusion and Future Work

Optimization is the key in achieving maximum efficiency of any given ES running on AI. This paper identifies the causes of low accuracy and decreased performance of inferencing ANNs on ESs. Using Jetson TX2 SOM, a comparison was made between YOLOv3 ANN running on CUDA with and without tensor-based optimization using TensorRT inference accelerator. Making use of open-source format (ONNX) for conversion from YOLO native format to TensorRT.

Inferencing performance and accuracy are seen to be linked. As seen in Sect. 4.2, high FPS allow for accurate perception of the current scene. Enabling AMRs to detect more objects faster, resulting in prompt response. Nevertheless, complex scenes pose computational challenges that may affect the resource allocation, lowering the FPS.

The optimization technique focuses on reducing the memory footprint and computational demands. Accomplished through downscaling the ANN model from FP32 to INT8 while targeting intermediate activation layers. Although this increases the performance in FPS significantly by 3.5× and number of objects detected but reduces the detection accuracy. This analysis conjoins the aforementioned parameters with power efficiency, pertaining to AMR specifics.

This comprehensive study gives a thorough insight towards optimization of ESs for effective resources utilization. Future works include in-depth analysis of the processes in terms of mean Average Precision (mAP) and studying the relationships between the several elements of the existing frameworks and APIs using a visualization software such as NVIDIA Nsight Systems.

Acknowledgment. This research has been funded conjointly by Oman Logistics Center in Asyad Group and the National University under NU Grant Call 2019. Special appreciation to Mr. D. Ragavesh for his help with image processing. Additionally, sincere gratitude to Mr. M. J. Varghese for his help with Artificial Neural Networks and guidance in performance evaluation. Lastly, special thanks to J. K. Jung for sharing his experience on tensor optimization for Embedded Systems; which was a prominent guide in the making of this research.

References

1. Okuyama, T., Gonslaves, T., Upadhay, J.: Autonomous driving system based on deep Q learning. In: 2018 International Conference on Intelligent Autonomous Systems (ICoIAS), pp. 201–205. IEEE; Singapore (2018). https://doi.org/10.1109/icoias.2018.8494053
2. Feng, X., Jiang, Y., Yang, X., Du, M., Li, X.: Computer vision algorithms and hardware implementations: a survey. Integration **69**, 309–320 (2019). https://doi.org/10.1016/j.vlsi.2019.07.005
3. Ray, J., Thompson, B., Shen, W.: Comparing a high and low-level deep neural network implementation for automatic speech recognition. In: Kellenberger, P. (ed.) First Workshop for High Performance Technical Computing in Dynamic Languages, pp. 41–46. IEEE Service Center, Piscataway (2016). https://doi.org/10.1109/HPTCDL.2014.12
4. Rungsuptaweekoon, K., Visoottiviseeth, V., Takana, R.: Evaluating the power efficiency of deep learning inference on embedded GPU systems. In: 2017 2nd International Conference on Information Technology (INCIT), pp. 1–5. IEEE, Nakhon Pathom (2017). https://doi.org/10.1109/incit.2017.8257866
5. Marco, V.S., Taylor, B., Wang, Z., Elkhatib, Y.: Optimizing deep learning inference on embedded systems through adaptive model selection. ACM Trans. Embed. Comput. Syst. **19**(1), 1–28 (2020). https://doi.org/10.1145/3371154
6. Yoo, S., et al.: Structure of deep learning inference engines for embedded systems. In: The 10th International Conference on Information and Communication Technology Convergence (ICTC), Jeju Island, South Korea, pp. 920–922 (2019). https://doi.org/10.1109/ictc46691.2019.8939843
7. Zhou, X., Wu, P., Zhang, H., Gou, W., Liu, Y.: Learn to navigate: cooperative path planning for unmanned surface vehicles using deep reinforcement learning. IEEE Access **7**, 165262–165278 (2019). https://doi.org/10.1109/access.2019.2953326
8. Aldegheri, S., Manzato, S., Bombieri, N.: Enhancing performance of computer vision applications on low-power embedded systems through heterogeneous parallel programming, In: Proceedings of the IFIP/IEEE International Conference on Very Large-Scale Integration (VLSI-SoC), Verona, Italy, pp. 119–124 (2018). https://doi.org/10.1109/vlsi-soc.2018.8644993
9. Saxerud, A.L., Ferrell, J.P., Dunn, E.A.: Application of the CUDA® toolkit multi-GPU libraries to an out-of-core MoM solver. In: Proceedings of the 2016 IEEE Antennas and Propagation Society International Symposium (APSURSI), Fajardo, Puerto Rico, pp. 2013–2014 (2016). https://doi.org/10.1109/aps.2016.7696713
10. Canziani, A., Culurciello, E., Paszke, A.: Evaluation of neural network architectures for embedded systems. In: 2017 IEEE International Symposium on Circuits and Systems (ISCAS), Baltimore, MD, pp. 1–4 (2017). https://doi.org/10.1109/iscas.2017.8050276
11. Kodra, A.: Machine Learning Model Optimization (2019). https://heartbeat.fritz.ai/machine-learning-model-optimization-9fbcfd9f6990. Accessed 20 Feb 2020
12. Michálek, J., Vaněk, J.: An open-source GPU-accelerated feature extraction tool. In: Yuan, B., Ruan, Q., Tang, X. (eds,) 2014 IEEE 12th International Conference on Signal Processing Proceedings, HangZhou, China, pp. 450–454 (2019). https://doi.org/10.1109/iscas.2017.8050276

13. Milioto, A., Stachniss, C.: BonNet: an open-source training and deployment framework for semantic segmentation in robotics using CNNs. In: 2019 International Conference on Robotics and Automation (ICRA), Montreal, QC, Canada, pp. 7094–7100 (2019). https://doi.org/10.1109/icra.2019.8793510

14. Xu, R., Han, F., Ta, Q.: Deep learning at Scale on NVIDIA V100 accelerators. In: Proceedings of the 2018 IEEE/ACM Performance Modeling, Benchmarking and Simulation of High-Performance Computer Systems (PBMS), Dallas, TX, USA, pp. 23–32 (2018). https://doi.org/10.1109/pmbs.2018.8641600

15. Yamane, S.: Deductively verifying embedded software in the era of artificial intelligence = machine learning + software science., In: 2017 IEEE 6th Global Conference on Consumer Electronics (GCCE), Nagoya, Japan, pp. 1–4 (2017). https://doi.org/10.1109/gcce.2017.8229475

16. Redmon, J., Farhadi, A.: YOLO: real-time object detection (2010). https://pjreddie.com/darknet/yolo/. Accessed: 19 Jan 2020

17. Al Ghadani, A.K.A., Ramaswamy, R.G.: Collecting datasets on ore mining industry and training using YOLO ANNs for object identification in mining sites. J. Big Data Smart City 1, 1–6 (2019). MEC, Muscat, Oman

18. Redmon, J., Divvala, Girshick, Farhadi A.: YOLO9000: better, faster, stronger. In 2017 IEEE Conference on Computer Vision and Pattern Recognition (CVPR), Honolulu, HI, USA, pp. 6517–6525 (2016). https://doi.org/10.1109/cvpr.2017.690

19. GitHub: Darknet/YOLOv3.config at master – PJReddie/Darknet (2018). https://gist.github.com/fabito/a49bb6a5593594f26275bc90baba6e32. Accessed 20 January 2020

20. Smith, S.W.: Neural network architecture (1999). https://www.dspguide.com/ch26/2.htm. Accessed 18 Jan 2020

21. Impagliazzo, R.: A personal view of average case complexity. In: Proceedings of Structure and Complexity Theory, Minneapoilis, MN, USA, pp. 134–147 (1995). https://doi.org/10.1109/sct.1995.514853

22. Wu, J.: Complexity and accuracy analysis of common artificial neural networks on pedestrian detection. In: 2018 2nd International Conference on Electronic Information Technology and Computer Engineering (EITCE), Shanghai, China (2018). https://doi.org/10.1051/matecconf/201823201003

23. NVIDIA: Deep learning frameworks documentation (2020). https://docs.nvidia.com/deeplearning/frameworks/tf-trt-user-guide/index.html. Accessed 1 Feb 2020

24. Davoodi, P., Lai, G., Morris, T., Sharma, S.: High performance inference with TensorRT Integration (2019). https://blog.tensorflow.org/2019/06/high-performance-inference-with-TensorRT.html. Accessed 2 Feb 2020

25. NVIDIA: TensorRT documentation (2020). https://docs.nvidia.com/deeplearning/tensorrt/best-practices/index.html. Accessed 2 Feb 2020

26. NVIDIA: Jetson TX2 module (2020). https://developer.nvidia.com/embedded/jetson-tx2. Accessed 19 Jan 2020

27. Redmon, J., Farhadi, A.: YOLOv3: an incremental improvement (2018). https://arxiv.org/pdf/1804.02767.pdf. Accessed 3 Feb 2020

28. Cook, S.: CUDA Programming: A Developer's Guide to Parallel Computing with GPUs, pp. 13–19. Elsevier Science & Technology, Saint Louis (2012)

29. Wezowicz, M., Taufer, M.: On the cost of a general GPU framework – The strange case of CUDA 4.0 vs. CUDA 5.0. In: SC Companion: High Performance Computing, Networking Storage and Analysis, Salt Lake City, UT, USA, pp. 1535–1536 (2012). https://doi.org/10.1109/sc.companion.2012.310

30. Hwu, W.M., Rodrigues, C., Ryoo, S., Stratton, J.: Compute unified device architecture application suitability. Comput. Sci. Eng. 11(3), 16–26 (2009). https://doi.org/10.1109/MCSE.2009.48

The Random Neural Network in Price Predictions

Will Serrano[(⊠)]

Alumni Imperial College London, London, UK
g.serrano11@alumni.imperial.ac.uk

Abstract. Everybody likes to make a good prediction, in particular, when some sort of personal investment is involved in terms of finance, energy or time. The difficulty is to make a prediction that optimises the reward obtained from the original contribution; this is even more important when investments are the core service offered by a business or pension fund generated by monthly contributions. The complexity of finance is that the human predictor may have other interests or bias than the human investor, the trust between predictor and investor will never be completely established as the investor will never know if the predictor has generated, intentionally or unintentionally, the optimum possible reward. This paper presents the Random Neural Network in recurrent configuration that makes predictions on time series data, specifically, prices. The biological model inspired by the brain structure and neural interconnections makes predictions entirely on previous data from the time series rather than predictions based on several uncorrelated inputs. The model is validated against the property, stock and Fintech market: 1) UK property prices, 2) stock markets indice prices, 3) cryptocurrency prices. Experimental results show that the proposed method makes accurate predictions on different investment portfolios.

Keywords: Artificial Intelligence · Price predictions · Stock market · Fintech · Random Neural Network

1 Introduction

Everybody likes to make a good prediction, this skill is not only a sign of superior wisdom but also exceptional common sense, both acquired from an extensive experience in a mature life through plenty of decisions, rewards and punishments. The traditional prediction model has always been that the older a person is, the larger experience the predictor should have acquired therefore the best prediction would be made by the greater wisdom or common sense. Accurate predictions are more relevant when some sort of personal investment is involved in terms of finance, energy or time. The challenge is to make a prediction that optimises the reward from the original contribution; the reason is simple. This traditional prediction model has three flaws: 1) the predictor may have a different economic

© IFIP International Federation for Information Processing 2020
Published by Springer Nature Switzerland AG 2020
I. Maglogiannis et al. (Eds.): AIAI 2020, IFIP AICT 583, pp. 303–314, 2020.
https://doi.org/10.1007/978-3-030-49161-1_26

interest or bias than the investor therefore not providing intentionally the optimum reward. At the end, the reason the investor uses the predictor is due to its limited knowledge or time to make the prediction itself. 2) Artificial Intelligence and Machine Learning increase the learning experience as they are capable to memorise and store larger and more data feeds for longer time than a human could do during its life spam. Artificial Intelligence therefore makes faster predictions than humans and also removes the predictor bias and service fees. In addition, Machine Learning accepts and learns from its own mistakes. 3) Finally, the traditional model does not efficiently work when there are only a few predictors for all the investors as the reward form investments can be standardised or the market can be manipulated. In order to reduce prediction risks, the predictor approach has always been the "investment portfolio diversification" concept that splits the initial investment into numerous smaller portions and invest them into all possible options therefore averaging the returns. As the predictor charges a fee for its services, then the investor may as well diversify its investments itself or use Artificial Intelligence to automate the process.

Artificial Neural Networks are the biological model of the human brain that emulate neurons and their connections developed in numerous configurations such a Deep Learning [1] and Reservoir Computing [2] with a large number of different learning algorithms such as Gradient Descent and Reinforcement Learning. The key concept is that Artificial Neural Networks must achieve stability in their firing patterns during the learning process, similar to humans, Artificial Intelligence requires time to learn from experiences. This paper presents the Random Neural Network in feedforward configuration that makes predictions on time series data, in particular, prices. The biological model makes predictions entirely on previous data from the same variable series rather than predictions based on several uncorrelated parameters. This proposed model could also be applied to make predictions based on predictions rather than real and factual data as the inputs could also be predicted. The Random Neural Network model is validated against the prices of three key markets:

- Property market: General, House and Flat Prices;
- Stock market: FTSE100, NASDAQ, NIKKEI225 and SZSE;
- Fintech market: Bitcoin, Ethereum and Ripple.

The metric used is the Root Mean Square Error (RMSE) as it penalises large errors. This research is based on java software in an Eclipse platform due its data integration with third party APIs. This paper explores and analyses the number of learning iterations and the optimum memory to make the best prediction that balances memory with recency. Section 2 on this paper presents the related work for this research based on neural networks for price predictions. Section 3 presents the mathematical model of the Random Neural Network for price predictions including the its learning algorithm based on a sling window. Section 4 provides the validation and experimental results. Finally, conclusions are shared on Sect. 5.

2   Related Work

Artificial Intelligence and Neural Networks have already been applied as predictions models. The novelty and innovation presented on this paper is the use of the Random Neural Network as a price prediction algorithm. An Artificial Intelligence decision support system guides investors to buy and sell stocks based on financial data listed on the BSE/NSE markets [3]. The mathematical model uses a three multilayer perceptron to predict stock price based on the fundamental value and intrinsic stock value such as investment valuation, cash flow indicator, liquidity measurement, profitability indicator and debt. Five neural network models based on back propagation neural network, radial basis function neural network, general regression neural network, support vector machine regression and least squares support vector machine regression make price prediction on three individual stocks: Bank of China, Vanke A and Kweichou Moutai [4]. The back propagation neural network outperforms in terms of mean square error and average absolute percentage error criteria for different activation functions (linear, polynomial, sigmoid and radial basis). A two-layer feed forward neural network with sigmoid hidden neurons and linear output neurons predicts next day closing in Dow Jones stock market [5]. Three algorithms Levenberg-Marquardt algorithm, bayesian regularization and the scaled conjugate gradient algorithm are trained using several stock parameters such as open, high, low, close price and volume. The bayesian regularization has the best performance whereas the Levenberg- Marquardt algorithm requires less iterations to train.

Any time series function can be modelled based on four systematic components: base level, trend, seasonality and noise. A Long Short-Term Memory (LSTM) network predict the price of Bitcoin based on a regression model [6] with an Adam optimization and the tanh activation function. The LSTM uses 35 days of previous data in case of monthly prediction and 65 days for the two month prediction. A model that forecasts daily closing price series of Bitcoin uses data on prices and volumes of prior days [7]. The model compares single linear regression for univariate series forecast based on only closing prices, multi linear regression model for multivariate series based on both price and volume data, a multilayer perceptron and a LSTM Network against data from Intel, National Bank and Microsoft daily NASDAQ closing prices. A pricing predict model for options in bitcoins and cryptocurrencies is based on a backpropagation neural network where classical pricing models (trinomial tree, Monte Carlo simulation and explicit finite difference method) are used as input layers [8]. The neural network outperforms in terms of prediction accuracy for financial data based on the 15-days historical volatility for the Bitcoin index and the 2-months Libor interest rate due to its adaptation to the complexity and non-linearity of the option and cryptocurrency market. The relationship between the next day binary change in the price of Bitcoin and its features such as transaction volume, cost per transaction is analysed by a neural Network with a Genetic Algorithm [9]. The network is formed of five multi-layered perceptrons trained with Levenberg-Marquardt supervised learning algorithm and hyperbolic tangent activation function that predicts the next day direction given a set of approximately 200 features.

Deep neural networks combine the advantages of deep learning and neural networks to predict and analyse data within the nonlinear and time-dependent financial model. A deep learning method based on Convolutional Neural Network predicts the stock price movement of Chinese stock market based on the opening price, high price, low price, closing price and volume of the stock [10]. Financial data is considered as one-dimensional time series generated by the projection of a chaotic system composed of multiple factors into the time dimension following a phase-space reconstruction method [11] whereas the deep neural network model is based on a LSTM network with three hidden layers, each of which contains 32 memory cells. The model is validated against several prediction models such as the conventional Autoregressive Integrated Moving Average linear analytical method, the conventional Support Vector Regression machine learning method, a deep Multi Layer Perceptron model and deep LSTM model data for six stock indices for various market environments: the S&P 500, the Dow Jones industrial average, the Nikkei 225, the Hang Seng index, the China Securities index 300 and the ChiNext index. A study of three neural networks with deep learning algorithms: convolutional neural networks, LSTM networks and a combination of both that uses 1D convolutional layers as preprocessing step to extract and reduce the key features are analysed for a stock price prediction application [12]. The deep learning models are based on 40 and 32 memory cells respectively with a RELU activation function and trained with the last five days of stock price data which includes Open, High, Low, and Close price values. A deep recurrent neural network model with multiple inputs and multiple outputs based on LSTM network [13]. The deep learning structure simultaneously predicts the opening price, the lowest price and the highest price of the Shanghai composite index, PetroChina and ZTE with an Adam optimization algorithm. Deep learning models based on multi-layer perceptron, dynamic artificial neural network and the hybrid neural networks which use generalized autoregressive conditional heteroscedasticity are analysed in the NASDAQ Stock Exchange [14].

An adversarial training that improves the generalisation of a neural network prediction model by adding intentional perturbations to simulate the stochasticity of the price variable [15] is validated on 88 high trade volume stocks in NASDAQ and NYSE markets. The neural network is an attentive LSTM with four components: feature mapping, LSTM, temporal attention and prediction. A three-stage hybrid forecasting stock market index model is composed of a self-organizing map that reduces the size of the original data, a Genetic Algorithm that selects relevant features and a backpropagation neural network that predicts future stock price based on features of the previous day [16]. The model uses ten technical indicators as inputs where the model is applied to three indices (S&P 500, IBM and NASDAQ) and validated against support vector regression and random forest algorithms. A discrete wavelet transform that removes noise within the data based on frequency components of financial time series versus the Fourier transform is combined with a recurrent neural network trained via the Back Propagation Through Time (BPTT) and the Adam optimisation method to predict Saudi stock market [17].

3 The Random Neural Network in Price Predictions

The Random Neural Network was introduced in [18] and many of its properties have been developed in [19]. It is a spiking stochastic model which can be used as either a feedforward or recurrent network (Fig. 1).

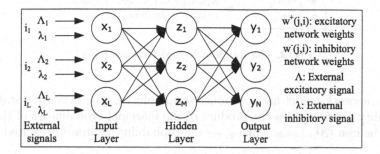

Fig. 1. The Random Neural Network structure.

The Random Neural Network consists on n-neurons. The state of the n neuron network at time t is represented by the vector of non-negative integers $K(t) = [K_1(t), \ldots K_i(t)]$ where $K_i(t)$ is the potential of neuron i at time t. Neurons interact with each other by interchanging signals in the form of spikes of unit amplitude:

- A positive spike is interpreted as an excitation signal because it increases by one unit the potential of the receiving neuron m, $K_m(t^+) = K_m(t) + 1$;
- A negative spike is interpreted as inhibition signal decreasing by one unit the potential of the receiving neuron m, $K_m(t^+) = K_m(t) - 1$, or has no effect if the potential is already zero, $K_m(t) = 0$.

Each neuron accumulates signals and it will fire if its potential is positive. Firing will occur at random and spikes will be sent out at rate r(i) with independent, identically and exponentially distributed inter-spike intervals.

3.1 The Random Neural Network Model

The Random Neural Network excitatory and inhibitory weight parameters $w^+(j, i)$ and $w^-(j, i)$, respectively are directly related to the r(i) and the $p^+(i, j)$, $p^-(i, j)$, and are expressed as:

$$w^+(j, i) = r(i)p^+(i, j) \geq 0 \tag{1}$$

$$w^-(j, i) = r(i)p^-(i, j) \geq 0 \tag{2}$$

Thus information in this model is transmitted by the rate or frequency at which spikes travel. Each neuron i, if it is excited, behaves as a frequency modulator emitting spikes at rate $w(i, j) = w^+(i, j) + w^-(i, j)$ to neuron j (Fig. 2).

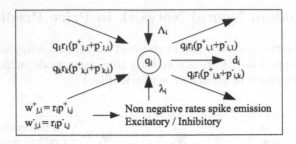

Fig. 2. The Random Neural Network model.

This network model has a product form solution; the network's stationary probability distribution is the product of the marginal probabilities of the state of each neuron [20]. Let's define q_i as the probability neuron i is excited:

$$q_i = \frac{\lambda^+(i)}{r(i) + \lambda^-(i)} \tag{3}$$

$$r(i) + \sum_{j=1}^{n} \left[w^+(i,j) + w^-(i,j) \right] \quad for \quad 1 \le i \le n \tag{4}$$

where the $\lambda^+(i)$, $\lambda^-(i)$ for i = 1, ..., n satisfy the system of nonlinear simultaneous equations:

$$\lambda^+(i) = \sum_{j=1}^{n} \left[q_j r(j) p^+(j,i) \right] + \Lambda(i) \tag{5}$$

$$\lambda^-(i) = \sum_{j=1}^{n} \left[q_j r(j) p^-(j,i) \right] + \lambda(i) \tag{6}$$

3.2 Learning Algorithm

The Random Neural Network learning algorithm [21] is based on gradient descent of a quadratic error function. The backpropagation model requires the solution of n linear and n nonlinear equations each time the n neuron network learns a new input and output pair. Gradient Descent learning algorithm optimizes the network weight parameters w in order to learn a set of k input-output pairs (i,y) where successive inputs are denoted $i = i_1, ..., i_k$ and the successive desired outputs are represented $y = y_1, ..., y_k$. The desired output vectors are approximated by minimizing the cost function E_k:

$$E = \frac{1}{2} \sum_{i=1}^{n} (q_i - y_i)^2 \tag{7}$$

where y_i represents the desired output. Each of the n neurons of the network is considered as an output neuron.

3.3 The Random Neural Network in Price Predictions

The input i_k extracts the S window values of the time series $x(t - 1 - s)$ at time t for s = 1, ... , S. The output y_k corresponds to x(t). The network learns i_k and y_k. As time increases, t = t + 1 the model includes the newest value, removes its oldest value, learns the new input and finally predicts the next value at t + 2. This method iterates for the entire time series data (Fig. 3):

1. time t, the Random Neural Network learns i_k and y_k based on $x(t - 1 - s)$ and x(t) respectively.
2. time t, the model window slides one unit to include (x(t) and predict x(t + 1)
3. time t + 1, the model learns x(t + 1)
4. time t + n, the model window slides n-1 units to include x(t + n) and predict x(t + n + 1)
5. time t + n + 1, the model learns x(t + n + 1).

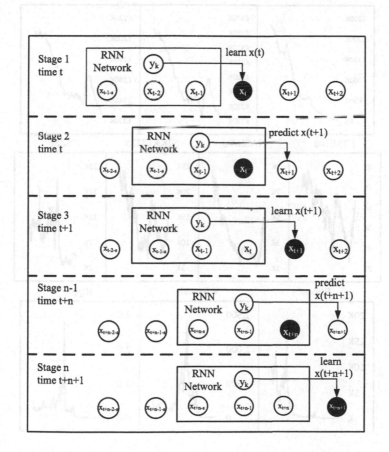

Fig. 3. The Random Neural Network in price predictions.

4   Validation and Experimental Results

The Random Neural Network in price predictions model is validated against three uncorrelated markets where real market values are shown on Fig. 4:

- Property market: General, House and Flat Prices;
- Stock market: FTSE100, NASDAQ, NIKKEI225 and SZSE;
- Fintech market: Bitcoin, Ethereum and Ripple.

The validation exercise covers value of S for the optimum Window that makes the best prediction and balance between memory with recency. The error between the predicted value and the current value is calculated following the Root Mean Square Error (RMSE) for a window of five, ten and twenty input neurons that represent low, medium and high memory respectively.

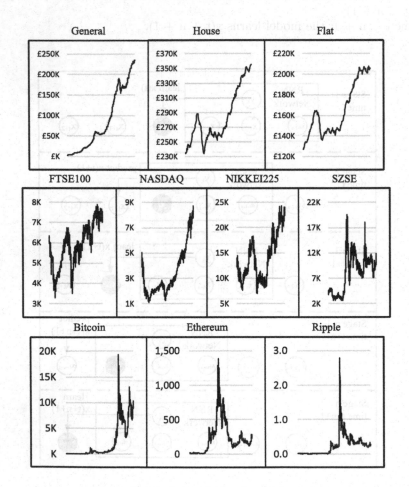

Fig. 4. Validation and experimental results.

4.1   Property Market Validation

The property market validation covers general house prices, detached houses and flats based on the datasets obtained from Land Registry where key parameters, including statistical information, are shown on Table. 1. The period of the data points is monthly rather than daily therefore the standard deviation is reasonable high for the general price values.

Table 1. Property market dataset.

Property	Period	Data points	Max	Min	Average	σ
General	Apr 1968 - Dec 2019	620	234,742.00	3,595.00	85,324.25	73,110.62
House	Jan 2005 - Dec 2019	180	359,542.00	234,509.00	284,096.09	36,016.62
Flats	Jan 2005 - Dec 2019	180	207,847.00	127,018.00	164,425.71	24,857.47

The validation results presented on Table 2 show that additional memory does not reduce the error. The RMSE assessment covers two current values: the real and the normalised value [min 0.0, max 1.0]. The experimental results show RMSE figures differ significantly from real to normalised values.

Table 2. Property market validation.

Property	5-Window		10-Window		20-Window	
Type	Real	Normalised	Real	Normalised	Real	Normalised
General	1166.77	1.1581E−02	1194.90	7.6847E−03	1222.54	5.2890E−04
House	2620.72	2.0960E−03	2667.90	2.1338E−03	2669.57	2.1351E−03
Flats	1457.86	2.0579E−03	1506.79	2.1270E−03	1505.53	2.1252E−03

4.2   Stock Market Validation

The stock market validation covers British FSTE 100, American NASDAQ, Chinese SZSE and Japanese Nikkei 225 based on the datasets obtained from www. uk.investing.com where key parameters are shown on Table 3. The standard deviation for this market is lower than the previous validation.

Table 3. Stock market dataset.

Stock	Period	Data Points	Max	Min	Average	σ
FSTE 100	03 Jan 2001 - 26 Feb 2020	4,838	7,877.45	3,287.00	5,869.38	1,041.52
NASDAQ	26 Jan 2000 - 09 Dec 2019	5,000	8,705.17	1,114.11	3,451.45	1,882.49
Nikkei 225	05 Jan 2001 - 26 Feb 2020	4,717	24,270.62	7,054.98	14,188.07	4,520.57
SZSE 225	26 Jan 2000 - 26 Feb 2020	4,867	19,531.15	2,622.03	8,166.77	3,803.31

312   W. Serrano

The validation results shown on Table 4 represent similar trend as the previous validation; additional memory does not reduce the error and RMSE figures differ significantly from real to normalised values. The real error of the stock market validation is lesser than the Property market validation due to its reduced standard deviation.

Table 4. Stock market validation.

Stock	5-Window		10-Window		20-Window	
Type	Real	Normalised	Real	Normalised	Real	Normalised
FTSE100	61.42	1.3380E−03	61.24	1.3340E−03	61.26	1.3344E−03
NASDAQ	49.22	6.4844E−04	49.06	6.4628E−04	48.89	6.4405E−04
Nikkei 225	195.29	1.1344E−03	194.97	1.1332E−03	195.01	1.1328E−03
SZSE	175.66	1.0388E−03	175.81	1.0398E−03	175.96	1.0406E−03

4.3   Fintech Market Validation

The Fintech market validation covers Bitcoin, Ethereum and Ripple cryptocurrency prices based on the datasets from www.uk.investing.com where key parameters are shown on Table 5. The standard deviation for the Fintech market is quite high as the Bitcoin values have a large range whereas Ethereum and Ripple data points are reduced.

Table 5. Fintech market dataset.

Crypto	Period	Data Points	Max	Min	Average	σ
Bitcoin	18 Jul 2010 - 26 Feb 2020	3,511	19,345.5	0.1	2275.53	3,528.87
Ethereum	10 Mar 2016 - 26 Feb 2020	1,449	1,380	6.7	232.29	239.87
Ripple	22 Jan 2015 - 26 Feb 2020	1,862	2.78	0.0036	0.2375	0.3215

Validation results are shown on Table 6. The Fintech market validation also confirms the trends of the previous validations. The real error is dependent of the maximum and minimum data point range Ripple predictions have the lowest error as the real values have very limited range.

Table 6. Fintech market validation.

Property	5-Window		10-Window		20-Window	
Type	Real	Normalised	Real	Normalised	Real	Normalised
Bitcoin	205.51	1.0623E−03	205.61	1.0628E−03	206.02	1.0649E−03
Ethereum	21.47	1.5631E−03	21.51	1.5662E−03	21.57	1.5704E−03
Ripple	0.0378	1.3614E−03	0.0380	1.3676E−03	0.0381	1.3733E−03

5   Conclusions

This paper has presented the Random Neural Network in back propagation configuration that makes predictions on time series data, in particular, prices. The biological model inspired by the brain structure and neural interconnections makes predictions entirely on previous data from the same variable rather than predictions based on several uncorrelated parameters. The proposed model has been validated against the property, stock and Fintech market: 1) UK property prices, 2) Stock markets indice prices, 3) Cryptocurrency prices. Experimental results show that the proposed method makes accurate predictions on different investment portfolios and the effect of the memory, or input neurons, does not have a major impact on the quality of the prediction.

Future work will expand this research in two aspects: the addition of uncorrelated variables into input layer and the increment of the number of output neurons. The effect from these variations to the model will be analysed based on the enhancement of the quality prediction.

References

1. Serrano, W.: Genetic and deep learning clusters based on neural networks for management decision structures. Neural Comput. Appl. **32**, 1–25 (2019). https://doi.org/10.1007/s00521-019-04231-8
2. Schrauwen, B., Verstraeten, D., Campenhout, J.: An overview of reservoir computing. theory, applications, and implementations. In: Proceedings of the European Symposium on Artificial Neural Networks, pp. 471–482 (2007)
3. Patalay, S., MadhusudhanRao, B.: Design of a financial decision support system based on artificial neural networks for stock price prediction. J. Mech. Continua Math. Sci. **14**(5), 757–766 (2019)
4. Songa, Y.-G., Zhoub, Y.-L., Han, R.-J.: Neural networks for stock price prediction. J. Differ. Equn. Appl. 1–13 (2018, to be published)
5. Al-Shayea, Q.-K.: Neural networks to predict stock market price. World Congr. Eng. Comput. Sci. **1**, 371–377 (2017)
6. Struga, K., Qirici, O.: Bitcoin price prediction with neural networks. In: International Conference on Recent Trends and Applications in Computer Science and Information Technology, pp. 1–9 (2018)
7. Uras, N., Marchesi, L., Marchesi, M., Tonelli, R.: Forecasting Bitcoin closing price series using linear regression and neural networks models, pp. 1–25 (2020, to be published)
8. Pagnottoni, P.: Neural network models for bitcoin option pricing front. Artif. Intell. Financ. J. Front. Artif. Intell. **2**(5), 1–9 (2019)
9. Sin, E., Wang, L.: Bitcoin price prediction using ensembles of neural networks. In: International Conference on Natural Computation, Fuzzy Systems and Knowledge Discovery, pp. 666–671 (2017). https://doi.org/10.1109/FSKD.2017.8393351
10. Chen, S., He, H.: Stock prediction using convolutional neural network. In: International Conference Artificial Intelligence Applications and Technologies IOP Conference Series: Materials Science and Engineering, vol. 435, pp. 1–9 (2018). https://doi.org/10.1088/1757-899X/435/1/012026

11. Yu, P., Yan, X.: Stock price prediction based on deep neural networks. Neural Comput. Appl. **32**(6), 1609–1628 (2019). https://doi.org/10.1007/s00521-019-04212-x
12. Jain, S., Gupta, R., Moghe, A.: Stock prediction on daily stock data using deep neural networks. In: 2018 International Conference on Advanced Computation and Telecommunication (ICACAT), Bhopal, India, pp. 1–13 (2018). https://doi.org/10.1109/ICACAT.2018.8933791
13. Ding, G., Qin, L.: Study on the prediction of stock price based on the associated network model of LSTM. Int. J. Mach. Learn. Cybern. **11**, 1307–1317 (2020). https://doi.org/10.1007/s13042-019-01041-1
14. Guresen, E., Kayakutlu, G., Daim, T.: Using artificial neural network models in stock market index prediction. Expert Syst. Appl. **38**(8), 10389–10397 (2011). https://doi.org/10.1016/j.eswa.2011.02.068
15. Feng, F., Chen, H., He, X., Ding, J., Sun, Ma., Chua, T.-S.: Enhancing stock movement prediction with adversarial training. In: International Joint Conference on Artificial Intelligence, pp. 5843–5849 (2019). arXiv:1810.09936v2
16. Jawad, N., Kurdy, M.: Stock market price prediction system using neural networks and genetic algorithm. J. Theor. Appl. Inf. Technol. **97**(152005), 4175–4187 (2019)
17. Jarrah, M., Salim, N.: A recurrent neural network and a discrete wavelet transform to predict the saudi stock price trends. Int. J. Adv. Comput. Sci. Appl. **10**(4), 155–162 (2019)
18. Gelenbe, E.: Random neural networks with negative and positive signals and product form solution. Neural Comput. **1**, 502–510 (1989)
19. Gelenbe, E.: G-networks with triggered customer movement. J. Appl. Probab. **30**, 742–748 (1993)
20. Gelenbe, E.: Stability of the random neural network model. Neural Comput. **2**(2), 239–247 (1990)
21. Gelenbe, E.: Learning in the recurrent random neural network. Neural Comput. **5**, 154–164 (1993)

Object Tracking/Object Detection Systems

Joint Multi-object Detection and Segmentation from an Untrimmed Video

Xinling Liu[1], Le Wang[1(✉)], Qilin Zhang[2], Nanning Zheng[1], and Gang Hua[3]

[1] Xi'an Jiaotong University, Xi'an 710049, Shannxi, People's Republic of China
lewang@xjtu.edu.cn
[2] HERE Technologies, Chicago, IL 60606, USA
[3] Wormpex AI Research, Beijing 100028, People's Republic of China

Abstract. In this paper, we present a novel method for jointly detecting and segmenting multiple objects from an untrimmed video. Unlike most existing video object segmentation methods that can only handle a trimmed video in which all video frames contain the target objects, we address a more practical and difficult problem, *i.e.*, joint multi-object detection and segmentation from an untrimmed video where the target objects do not always appear per frame. In particular, our method consists of two modules, *i.e.*, object decision module and object segmentation module. The object decision module is used to detect the objects and decide which target objects need to be separated out from video. As there are usually two or more target objects and they do not always appear in the whole video, we introduce the data association into object decision module to identify their correspondences among frames. The object segmentation module aims to separate the target objects identified by object decision module. In order to extensively evaluate the proposed method, we introduce a new dataset named UNVOSeg dataset, in which 7.2% of the video frames do not contain objects. Experimental results on four datasets demonstrate that our method outperforms most of the state-of-the-art approaches.

Keywords: Video object segmentation · Data association · Object detection

1 Introduction

Video object segmentation aims at segmenting the primary objects from the background across all frames. It is a fundamental yet important task in computer vision, which can be used in many applications, such as autonomous driving, video surveillance, action recognition. In this paper, our objective is to separate two or more target objects and link the object instances for each target object across the whole video, simultaneously.

Despite the success of image object segmentation with the development of convolutional neural networks in recent years, it is still challenging when it comes

© IFIP International Federation for Information Processing 2020
Published by Springer Nature Switzerland AG 2020
I. Maglogiannis et al. (Eds.): AIAI 2020, IFIP AICT 583, pp. 317–329, 2020.
https://doi.org/10.1007/978-3-030-49161-1_27

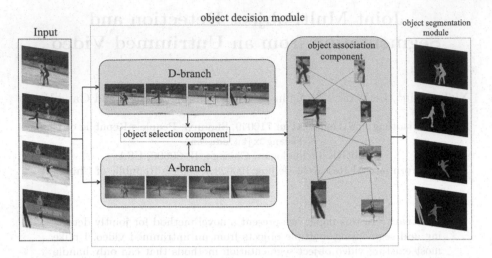

Fig. 1. Illustration of the proposed multiple video object detection and segmentation method.

to video object segmentation. Training a video object segmentation network requires a large amount of annotated video frames, which is very expensive and time-consuming. This invisibly increases the difficulty of video object segmentation. The video object segmentation methods can be roughly divided into two categories, i.e., semi-supervised methods [1–7] and unsupervised methods [8–10]. Semi-supervised methods need some extra user annotations (e.g., segmentation mask for the first frame) to indicate which object(s) need to be separated; while unsupervised methods automatically extract the primary object(s).

Although many methods have achieved significant performances on standard benchmarks [11–15], there are still several problems that need to be further addressed. For example, previous methods [16–20] are more suitable for trimmed videos that the target objects exist in (almost) all frames; while for untrimmed videos, the objects disappear intermittently throughout the whole video. Previous video object segmentation methods for trimmed videos cannot be directly leveraged to handle the untrimmed videos with noisy frames not containing the target objects. Taking the above issues into account, we propose a multiple video object detection and segmentation method to separate the target objects from an untrimmed video, which consists of two modules, i.e., object decision module and object segmentation module. Figure 1 illustrates the proposed multiple video object detection and segmentation method.

The object decision module is designed to detect the objects and decide which target objects need to be separated out from the video, and then identify their correspondence across adjacent frames. The object segmentation module is built to segment out the target objects identified by the object decision module. Specifically, we first feed an untrimmed video into the object decision module, which contains two components, i.e., object selection and object association. In object selection, we combine the object detector and an attention mechanism through a gating mechanism to make the network focus on the main objects denoted by bounding boxes in the video. In object association, we take the bounding boxes between adjacent frames as input to train a Siamese network, and then obtain the correspondences of these objects throughout the whole video. As a result, the object decision module can output the bounding boxes of multiple objects and their associations across the whole video. Then, we propose a weakly supervised segmentation method as our object segmentation module, and the objective is to segment the target objects from these bounding boxes generated by the object decision module. In detail, we first use multi-scale combinatorial grouping (MCG) [21] to generate a set of region proposals as pseudo-ground truth, as as to train a segmentation network which is employed to produce the final segmentation results.

We conduct extensive experiments on four datasets, including SegTrack [22], DAVIS2016 [23], DAVIS2017 [24], and our UNVOSeg. Experimental results show that our method significantly outperforms most of the state-of-the-art approaches, which validates that it can jointly detect and segment multiple objects from an untrimmed video.

The key contributions of this paper are summarized as follows:

- We present a joint multi-object detection and segmentation method, which is able to segment the primary objects in an untrimmed video in which multiple objects co-exist but disappear intermittently throughout the video.
- We propose a object decision module to identify the target objects and their correspondence across the untrimmed video.
- We establish a noisy video object segmentation dataset, named UNVOSeg dataset, including 63 untrimmed long videos and per-frame ground truth segmentation annotations, to verify the proposed method.

2   Related Work

Since our work addresses the problem of automatically segmenting multiple objects from an untrimmed video, we briefly review recent work on semi-supervised and unsupervised video object segmentation.

2.1   Semi-supervised Video Object Segmentation

Given some user annotations in the first frame or several frames, semi-supervised video object segmentation aims at segmenting out the target objects from a video. The research work in this line can benefit from the appearance model initialized by the segmentation mask annotations. Varun et al. [2] proposed a bilateral network followed by a standard spatial network to propagate information across frames. Sergi et al. [3] cast video object segmentation as a per-frame segmentation problem, given the object model from one or more manually segmented frames. Federico et al. [4] first trained a propagating network with static images, then took the same technology to fine-tune online learning strategies. Jingchun et al. [5] proposed a framework which is trained offline to learn a generic notion, and fine-tuned online for specific objects iteratively. The framework is capable of simultaneously predicting pixel-wise object segmentation and optical flow in videos. Hu et al. [6] classified each pixel in succeeding frames, using pixel-wise embeddings learned from supervision in the first frame. Xu et al. [7] modified traditional method Multiple Hypotheses Tracking (MHT) to semi-supervised video object segmentation, adapting a combination of mask propagation score and motion score to determine the affinity between hypotheses.

2.2 Unsupervised Video Object Segmentation

Unsupervised video object segmentation aims to discover and segment the most primary objects from the background with only the video as input. Some previous methods tackled this task based on clustering of point trajectories [9,10], motion characteristics [11], appearance [12,13], or saliency [14–16]. Li et al. [17] proposed a motion-based bilateral network to filter false positive region, and combined the background with instance embeddings. Liu et al. [18] coupled two dynamic Markov Networks to make video object discovery and video object segmentation tasks mutually beneficial. CDTS [19] developed a collaborative detection, tracking, and segmentation technique to extract multiple segment tracks accurately. RVOS [20] proposed a fully end-to-end trainable recurrent network and was the first one to report quantitative results for multiple object video object segmentation on DAVIS2017 [24] and YouTube-VOS [25] benchmarks.

3　Problem Formulation

In this section, we present the details of the proposed method, including two modules, i.e., object decision module and object segmentation module. Given an input video $V = \{v_t\}_{t=1}^T$ of T frames, our objective is to identify the target objects $O = \{O_1, \cdots, O_N\}$ that need to be separated out from V, where N is the total number of video objects. Meanwhile produce their segmentation masks $S = \{s_O^t\}_{t=1}^T$ along with the correspondences throughout the whole video (i.e., giving an object identity $n \in \{1, \cdots, N\}$ for each segmented object instance). The video frames are first put into the object decision module, which includes an object selection component and an object association component. The object selection component combines object detection, attention mechanism and a gating mechanism together to focus on the target objects that need to be segmented out. Specifically, the object detection is to detect the objects per frame, and then generate bounding boxes $B_t = \{b_k^t\}_{k=1}^K$ for each frame v_t, where K is the bounding box number in frame v_t. The attention mechanism is to make the location cues more prominent, and thus we can filter out irrelevant objects using a gating strategy. The bounding boxes of the target objects need to be segmented are then determined as $Q_t = \{q_n^t\}_{n=1}^N$. Subsequently, we feed them to the object association component, where we adopt the Siamese network [26] to identify

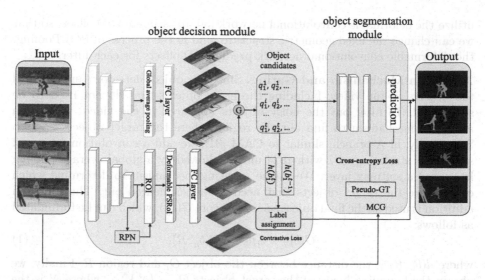

Fig. 2. The framework of our proposed method, which consists of two modules, *i.e.*, object decision module (shown in green) and object segmentation module (shown in blue). The object decision module is to detect the objects and decide which target objects need to be separated out in the video, in which an object association component is leveraged to identify the correspondences of object instances of a same object among frames. The object segmentation module is to segmented out the target objects identified by the object decision module. (Color figure online)

their correspondences between adjacent frames. The object segmentation module proceeds to perform a binary segmentation on bounding boxes of each frame v_t along with the object identities produced by object association component as the final segmentation result. Figure 2 presents the framework of our proposed method.

3.1 Object Decision Module

The purpose of object decision module is to generate reliable bounding boxes that will aid the object segmentation module to segment the primary objects in the video. Taking a video sequence $V = \{v_t\}_{t=1}^{T}$ of T frames as input, we first apply a backbone network ResNet-101 [27] to extract their features F.

Object Detection. Then, we feed F into two branches, *i.e.*, detection-branch (D-branch) and attention-branch (A-branch). In D-branch, we use a RPN network [28] to generate region proposals, and then use a position-sensitive ROI-Pooling [29] followed by a classification and a bounding box regression subnetwork to obtain the bounding boxes of object proposals $B_t = \{b_k^t\}_{k=1}^{K}$ and their detection sores $P_t = \{p_k^t\}_{k=1}^{K}$, where K is the total number of detected objects in frame t. Because video object segmentation is at instance level, we set the category number $C = 1$ in this paper. To enhance the detection capability, we

322   X. Liu et al.

utilize the deformable convolutional network [30] to estimated 2D offsets so that we can change the fixed geometric structure, and in the process of ROI-Pooling, the deformable convolutional network predicts 2D offsets for each filter.

Attention Mechanism and Gating. Since the object detector can detect all the objects in the video, including the noisy objects that do not need to be segmented. To filter out the noisy objects appeared in the video, we use an attention mechanism to find out the region where the target objects appear. Specifically, in A-branch, similar to CAM [31], we add a convolutional layer of size 3×3, stride 1, pad 1 with 1024 units, followed by a global average pooling layer and a softmax layer. We utilize a gating mechanism to determine which objects are the target objects that need to be segmented out. The weighted addition is used as the final object selection score $G(b_k^t)$, which can be formulated as follows:

$$G(b_k^t) = \alpha p_k^t + \beta d(b_k^t, R) \tag{1}$$

where $d(b_k^t, R)$ is the distance between the object O_i and region R. Finally, we obtain the bounding boxes of the target objects $Q_t = \{q_n^t\}_{n=1}^N$, where N is the total number of target objects.

Object Association. Since there are more than one object in a video, it is necessary to design an object association method to correlate the object instances of the same object among different frames, *i.e.*, the temporal consistency of each video object. We employ the Siamese network in object association module. In particular, we collect a dataset containing diverse object pairs to improve the robustness of network for discriminating different object pairs, which contains three cases: 1) positive pairs of the same object; 2) negative pairs in the same object category; and 3) negative pairs in different object categories. We take the contrastive loss [32] to supervise the training process. With the filtered objects in every frame, we first rescale them to the same size, then send the objects in adjacent frames into the trained Siamese network to get the similarity score $A_b(box_i^t, box_j^{t+1})$. When an object in the current frame did not find a corresponding object in the previous frame, we need to calculate another similarity score A_v as follows:

$$A_v = \frac{1}{N} \sum_{q=1}^N A_b(box_i^t, Z_q) \tag{2}$$

where Z_i denotes the bounding box collection of i-th object. Because A_v use more frame information than A_b, it can achieve better results in object association.

3.2 Object Segmentation Module

After getting the bounding boxes of the target objects need to be segmented out, we try to train a segmentation network by feeding the regions in these bounding boxes as input. In particular, we implement the whole training process in the following two steps:

Fig. 3. Some examples of the proposed UNVOSeg dataset: (a) case1: all targets appear in each video frames; (b) case2: one of targets is absent in some video frames; (c) case3: all targets are absent from certain video frames.

Generation of Pseudo Ground Truth. In this step, we generate all the object segmentation candidates from bounding boxes. We adopt a proposal generator, named multi-scale combinatorial grouping (MCG) trained on BSD500 [33] to generate segmentation proposals. Firstly, we use a multiresolution image pyramid to generate edge probability map, which can independently generate the ultrametric contour map (UCM) at each scale, from which we can obtain the connected regions, then a hierarchical segmentation can be acquired by merging these connected regions. In summary, we can get four lists of proposals in each UCM, a total of 16 lists of proposals. After obtaining a complete proposal set, we extract 2D basic features, such as area, perimeter, and boundary strength of each proposal. Finally, we use these features to form a vector to represent these proposals, and then train a random forest regression to rank these proposals.

Object Segmentation. In the second step, we train a standard fully convolutional networks (FCN) [34] by using the segmentation candidates that have the highest overlap rate with bounding boxes as groundtruth. Specifically, for a frame with m bounding boxes, we train a segmentation network with m corresponding pseudo-groundtruth, and the pixels outside the bounding boxes are not calculated.

4 Experiments and Discussions

4.1 Evaluation Datasets

We conduct extensive experiments on four datasets, including SegTrack, DAVIS 2016, DAVIS2017 and our UNVOSeg, to evaluate our method. These datasets are introduced as follows:

SegTrack is a widely used dataset in video object segmentation task. It consists of 14 short videos of 1,080 frames, including 8 single-object videos and 6 multi-object videos with pixel-wise annotations.

Table 1. The statistics of SegTrack, DAVIS2016, DAVIS2017, and UNVOSeg datasets.

Video	SegTrack	DAVIS2016	DAVIS2017	UNVOSeg
Number of videos	14	50	150	63
Number of frames	1080	3455	10474	13129
Noise rate	0.0%	0.90%	0.93%	7.2%

DAVIS2016 is a frequently used dataset, including 50 videos with pixel-wise labels. The goal is to reconstruct real video scenes, such as camera shake, background blur, occlusion, and other complex conditions.

DAVIS2017 is a recently proposed dataset consists of 150 short videos of 10459 frames with multiple objects. The noisy frames only accounts for a small portion of the total video.

UNVOSeg is a newly introduced dataset which can handle the task of multi-object segmentation from an untrimmed video. Because the existing benchmark datasets do not fit our requirement, we establish a new video object segmentation dataset, named UNVOSeg dataset, which meets the following characteristics: 1) The background of the video is more complicated; 2) There are frames that not containing the objects, which is referred as noisy frames; 3) There are cases where the video objects disappear intermittently throughout the video. Figure 3 shows some examples of the proposed UNVOSeg dataset. We collect 63 videos of 13,129 frames from YouTube that satisfies the above criteria, and provide pixel-level annotations for each frame. Considering that most of the videos in existing benchmark datasets contain dozens of frames, the videos in our new dataset contain about 200 to 300 frames. Note that the noisy rate of our new dataset is significantly higher than SegTrack, DAVIS2016, and DAVIS2017 datasets. The statistics of the four datasets are summarized in Table 1.

4.2 Evaluation on Single-Object Datasets

Although the proposed method is designed to solve multi-object detection and segmentation, it can still be applied to single-object video segmentation task. Therefore, we compare our method with other unsupervised single video object segmentation approaches, including FST [11], KEY [12], FSEG [15], SFM [35], UOVOS [36], SAL [37], JOS [18], CVOS [38], TRC [10], and LMP [39]. On Seg-Track and DAVIS2016 datasets. The results are evaluated by using J-measure, which are are presented in Table 2 and Table 3, respectively.

Table 2. The comparison results of our method with other competing methods on SegTrack by using J-measure.

Method	FST	KEY	FSEG	UOVOS	JOS	Ours
J-measure	53.5	57.3	61.4	64.3	80.6	51.1

Table 3. The comparison results of our method with other competing methods on DAVIS2016 by using J-measure.

Method	SAL	CVOS	TRC	SFM	LMP	FST	Ours
J-measure	42.6	51.4	50.1	53.2	69.7	57.5	48.3

From the results, we can see that our method does not obtain the best results, which is most likely because we use pseudo-ground truth in the object segmentation module. Nevertheless, our method still outperforms SAL by a margin of 5.7%, and the comparison with other methods is also competitive. Besides, we notice that JOS and LMP achieve best results on SegTrack and DAVIS2016, respectively. In JOS, two coupled dynamic Markov networks are utilized so that video object discovery and video object segmentation tasks can facilitate each other, while LMP exploit optical flow to assist in segmenting moving objects, yet these information are not used in this paper. It is worth mentioning that our method outperforms all other state-of-the-art methods on video Mallard-Fly with a value of 66.1%.

4.3 Evaluation on Multi-object Datasets

We compare our method with RVOS on both DAVIS2017 and UNVOSeg datasets, so as to evaluate its effectiveness in multi-object segmentation. Table 4 shows the results on DAVIS2017 and UNVOSeg dataset with J-measure and F-measure. The results show that the performance of our method is slightly lower than that of RVOS. The reason is that the video classification standards are not uniform. For example, people and backpacks are seen as one target in one video and two targets in another video. Therefore, this dataset is more suitable for semi-supervised video object segmentation. It should be noticed that our method still surpasses RVOS on several video sequences. In Table 5, we give some evaluation results of several videos on the DAVIS2017 and UNVOSeg datasets. Note that the J-measure and F-measure in video gold-fish is extremely low, because it is very challenging to distinguish so many similar objects. However, our method outperforms RVOS by a significant margin, from 11.2% to 35.8% with J-measure, in video people-dog. The reason is that our method is able to reconnect the same video object, even after a long time before re-entering the field of vision. Figure 4 show some visual example results of both our method and RVOS on DAVIS2017 and the UNVOSeg dataset, respectively.

Table 4. The results on the DAVIS2017 and UNVOSeg dataset with J-measure and F-measure.

Dataset	RVOS		Ours	
	J-Mean	F-Mean	J-Mean	F-Mean
DAVIS2017	39.0	48.3	27.9	29.9
UNVOSeg	16.4	18.2	31.7	34.3

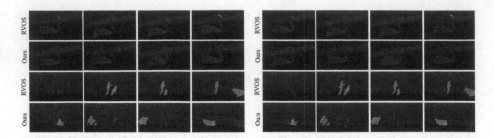

Fig. 4. Some visual example results of our method and RVOS on DAVIS2017 and UNVOSeg datasets.

Table 5. Some of the results on DAVIS2017 and UNVOSeg dataset with J-measure and F-measure.

DAVIS2017	RVOS		Ours		UNVOSeg	RVOS		Ours	
	J	F	J	F		J	F	J	F
Dogs-jump	40.7	51.3	**60.4**	**65.8**	Cats	**26.5**	**22.8**	21.6	19.4
India	22.8	25.5	**45.6**	**44.5**	People-dog	11.2	13.6	**35.8**	**37.8**
Lab-coat	10.0	**22.3**	**20.9**	19.1	Bullfight	22.8	25.5	**45.6**	**44.5**
Soapbox	27.0	**42.3**	**30.3**	39.9	Birds	15.7	16.9	**28.6**	**31.3**
Gold-fish	**25.9**	**38.9**	8.9	8.6	Monkey-dog	5.3	**10.2**	**11.7**	9.6
Kite-surf	**29.7**	**43.8**	26.2	32.3	Elephants	40.2	35.1	**61.4**	**60.4**
Judo	**25.1**	**52.3**	24.4	28.9	Bears	11.4	8.6	**34.8**	**38.6**

To summarize, the results on the above four datasets clearly reveal that: Although our method produces inferior results on single-object segmentation evaluation on video datasets without noisy frames (*i.e.*, SegTrack, DAVIS2016, and DAVIS2017 datasets), we are able to obtain better results on multi-object segmentation evaluation on video dataset with noisy frames (*i.e.*, UNVOSeg dataset), when compared with state-of-the-art methods. This strongly demonstrates that our method is capable of jointly detecting and segmenting the multiple objects from an untrimmed video, where object detection and object segmentation work in a joint way.

5   Conclusion

We proposed a multi-object detection and segmentation method to separate the target objects from an untrimmed video, which benefits from an object decision module and an object segmentation module. The object decision module can identify the target objects and their correspondences across the whole video, and the object segmentation module can segment the target objects out from the background. What's more, we present a new UNVOSeg dataset to fully evaluate the multi-object segmentation performance of our method. Extensive comparisons with the state-of-the-art methods on four datasets validated the efficacy of our method.

Acknowledgement. This work was supported partly by National Key R&D Program of China Grant 2018AAA0101400, NSFC Grants 61629301, 61773312, and 61976171, China Postdoctoral Science Foundation Grant 2019M653642, and Young Elite Scientists Sponsorship Program by CAST Grant 2018QNRC001.

References

1. Shin Yoon, J., Rameau, F., Kim, J., Lee, S., Shin, S., So Kweon, I.: Pixel-level matching for video object segmentation using convolutional neural networks. In: Proceedings of the IEEE International Conference on Computer Vision, pp. 2167–2176 (2017)
2. Jampani, V., Gadde, R., Gehler, P.V.: Video propagation networks. In: Proceedings of the IEEE Conference on Computer Vision and Pattern Recognition, pp. 451–461 (2017)
3. Caelles, S., Maninis, K.K., Pont-Tuset, J., Leal-Taixé, L., Cremers, D., Van Gool, L.: One-shot video object segmentation. In: Proceedings of the IEEE Conference on Computer Vision and Pattern Recognition, pp. 221–230 (2017)
4. Perazzi, F., Khoreva, A., Benenson, R., Schiele, B., Sorkine-Hornung, A.: Learning video object segmentation from static images. In: Proceedings of the IEEE Conference on Computer Vision and Pattern Recognition, pp. 2663–2672 (2017)
5. Cheng, J., Tsai, Y.H., Wang, S., Yang, M.H.: Segflow: joint learning for video object segmentation and optical flow. In: Proceedings of the IEEE International Conference on Computer Vision, pp. 686–695 (2017)
6. Hu, Y.T., Huang, J.B., Schwing, A.G.: Videomatch: matching based video object segmentation. In: Proceedings of the European Conference on Computer Vision, pp. 54–70 (2018)
7. Xu, S., Liu, D., Bao, L., Liu, W., Zhou, P.: MHP-VOS: multiple hypotheses propagation for video object segmentation. In: Proceedings of the IEEE Conference on Computer Vision and Pattern Recognition, pp. 314–323 (2019)
8. Haller, E., Leordeanu, M.: Unsupervised object segmentation in video by efficient selection of highly probable positive features. In: Proceedings of the IEEE International Conference on Computer Vision, pp. 5085–5093 (2017)
9. Brox, T., Malik, J.: Object segmentation by long term analysis of point trajectories. In: Daniilidis, K., Maragos, P., Paragios, N. (eds.) ECCV 2010. LNCS, vol. 6315, pp. 282–295. Springer, Heidelberg (2010). https://doi.org/10.1007/978-3-642-15555-0_21
10. Fragkiadaki, K., Zhang, G., Shi, J.: Video segmentation by tracing discontinuities in a trajectory embedding. In: Proceedings of the IEEE Conference on Computer Vision and Pattern Recognition, pp. 1846–1853 (2012)
11. Papazoglou, A., Ferrari, V.: Fast object segmentation in unconstrained video. In: Proceedings of the IEEE International Conference on Computer Vision, pp. 1777–1784 (2013)
12. Lee, Y.J., Kim, J., Grauman, K.: Key-segments for video object segmentation. In: Proceedings of the IEEE International Conference on Computer Vision, pp. 1995–2002 (2011)

13. Khoreva, A., Galasso, F., Hein, M., Schiele, B.: Classifier based graph construction for video segmentation. In: Proceedings of the IEEE Conference on Computer Vision and Pattern Recognition, pp. 951–960 (2015)
14. Jang, W.D., Lee, C., Kim, C.S.: Primary object segmentation in videos via alternate convex optimization of foreground and background distributions. In: Proceedings of the IEEE Conference on Computer Vision and Pattern Recognition, pp. 696–704 (2016)
15. Jain, S.D., Xiong, B., Grauman, K.: Fusionseg: learning to combine motion and appearance for fully automatic segmentation of generic objects in videos. In: Proceedings of the IEEE Conference on Computer Vision and Pattern Recognition, pp. 2117–2126 (2017)
16. Song, H., Wang, W., Zhao, S., Shen, J., Lam, K.M.: Pyramid dilated deeper convlstm for video salient object detection. In: Proceedings of the European Conference on Computer Vision, pp. 715–731 (2018)
17. Li, S., Seybold, B., Vorobyov, A., Lei, X., Jay Kuo, C.C.: Unsupervised video object segmentation with motion-based bilateral networks. In: Proceedings of the European Conference on Computer Vision, pp. 207–223 (2018)
18. Liu, Z., et al.: Joint video object discovery and segmentation by coupled dynamic markov networks. IEEE Trans. Image Process. 27(12), 5840–5853 (2018)
19. Jun Koh, Y., Kim, C.S.: CDTS: collaborative detection, tracking, and segmentation for online multiple object segmentation in videos. In: Proceedings of the IEEE International Conference on Computer Vision (2017)
20. Ventura, C., Bellver, M., Girbau, A., Salvador, A., Marques, F., Giro-i Nieto, X.: RVOS: end-to-end recurrent network for video object segmentation. In: The IEEE Conference on Computer Vision and Pattern Recognition (2019)
21. Arbeláez, P., Pont-Tuset, J., Barron, J.T., Marques, F., Malik, J.: Multiscale combinatorial grouping. In: Proceedings of the IEEE Conference on Computer Vision and Pattern Recognition, pp. 328–335 (2014)
22. Li, F., Kim, T., Humayun, A., Tsai, D., Rehg, J.M.: Video segmentation by tracking many figure-ground segments. In: Proceedings of the IEEE International Conference on Computer Vision, pp. 2192–2199 (2013)
23. Perazzi, F., Pont-Tuset, J., McWilliams, B., Van Gool, L., Gross, M., Sorkine-Hornung, A.: A benchmark dataset and evaluation methodology for video object segmentation. In: Proceedings of the IEEE Conference on Computer Vision and Pattern Recognition, pp. 724–732 (2016)
24. Pont-Tuset, J., Perazzi, F., Caelles, S., Arbeláez, P., Sorkine-Hornung, A., Van Gool, L.: The 2017 davis challenge on video object segmentation. arXiv:1704.00675 (2017)
25. Xu, N., et al.: Youtube-vos: A large-scale video object segmentation benchmark. arXiv preprint arXiv:1809.03327 (2018)
26. Bromley, J., Guyon, I., LeCun, Y., Säckinger, E., Shah, R.: Signature verification using a "siamese" time delay neural network. In: Advances in neural information processing systems, pp. 737–744 (1994)
27. He, K., Zhang, X., Ren, S., Sun, J.: Deep residual learning for image recognition. In: Proceedings of the IEEE Conference on Computer Vision and Pattern Recognition, pp. 770–778 (2016)
28. Ren, S., He, K., Girshick, R., Sun, J.: Faster R-CNN: towards real-time object detection with region proposal networks. In: Advances in neural information processing systems, pp. 91–99 (2015)

29. Dai, J., Li, Y., He, K., Sun, J.: R-FCN: object detection via region-based fully convolutional networks. In: Advances in neural information processing systems, pp. 379–387 (2016)
30. Dai, J., et al.: Deformable convolutional networks. In: Proceedings of the IEEE International Conference on Computer Vision (2017)
31. Zhou, B., Khosla, A., Lapedriza, A., Oliva, A., Torralba, A.: Learning deep features for discriminative localization. In: Proceedings of the IEEE Conference on Computer Vision and Pattern Recognition, pp. 2921–2929 (2016)
32. Hadsell, R., Chopra, S., LeCun, Y.: Dimensionality reduction by learning an invariant mapping. In: Proceedings of the IEEE Conference on Computer Vision and Pattern Recognition, pp. 1735–1742 (2006)
33. Arbelaez, P., Maire, M., Fowlkes, C., Malik, J.: Contour detection and hierarchical image segmentation. IEEE Trans. Pattern Anal. Mach. Intell. **33**(5), 898–916 (2010)
34. Long, J., Shelhamer, E., Darrell, T.: Fully convolutional networks for semantic segmentation. In: Proceedings of the IEEE conference on computer vision and pattern recognition, pp. 3431–3440 (2015)
35. Yang, C., Zhang, L., Lu, H., Ruan, X., Yang, M.H.: Saliency detection via graph-based manifold ranking. In: Proceedings of the IEEE Conference on Computer Vision and Pattern Recognition, pp. 3166–3173 (2013)
36. Zhuo, T., Cheng, Z., Zhang, P., Wong, Y., Kankanhalli, M.: Unsupervised online video object segmentation with motion property understanding. IEEE Trans. Image Process. **29**, 237–249 (2019)
37. Wang, W., Shen, J., Porikli, F.: Saliency-aware geodesic video object segmentation. In: Proceedings of the IEEE Conference on Computer Vision and Pattern Recognition, pp. 3395–3402 (2015)
38. Taylor, B., Karasev, V., Soatto, S.: Causal video object segmentation from persistence of occlusions. In: Proceedings of the IEEE Conference on Computer Vision and Pattern Recognition, pp. 4268–4276 (2015)
39. Tokmakov, P., Alahari, K., Schmid, C.: Learning motion patterns in videos. In: Proceedings of the IEEE Conference on Computer Vision and Pattern Recognition, pp. 3386–3394 (2017)

Robust 3D Detection in Traffic Scenario with Tracking-Based Coupling System

Zhuoli Zhou[1,2], Shitao Chen[1,2], Rongyao Huang[1,2], and Nanning Zheng[1,2(✉)]

[1] Institute of Artificial Intelligence and Robotics, Xi'an Jiaotong University,
Xi'an, Shaanxi, People's Republic of China
{zjsx5408,chenshitao,hryglory}@stu.xjtu.edu.cn,
nnzheng@mail.xjtu.edu.cn
[2] National Engineering Laboratory for Visual Information
Processing and Applications, Xi'an Jiaotong University,
Xi'an, Shaanxi, People's Republic of China

Abstract. Autonomous driving is conducted in complex scenarios, which requires to detect 3D objects in real time scenarios as well as accurately track these 3D objects in order to get such information as location, size, trajectory, velocity. MOT (Multi-Object Tracking) performance is heavily dependent on object detection. Once object detection gives false alarms or missing alarms, the multi-object tracking would be automatically influenced. In this paper, we propose a coupling system which combines 3D object detection and multi-object tracking into one framework. We use the tracked objects as a reference in 3D object detection, in order to locate objects, reduce false or missing alarms in a single frame, and weaken the impact of false and missing alarms on the tracking quality. Our method is evaluated on kitti dataset and is proved effective.

Keywords: Multi-object tracking · LiDAR 3D detection ·
Autonomous vehicles

1 Introduction

In recent years, autonomous driving has gradually attracted people's eyes and entered a rapid development period. Object detection and multi-object tracking technology are important components of autonomous driving technology, with which autonomous vehicles can understand the surrounding environment and make decisions. Autonomous driving usually integrates with multiple sensors, and the rich sensor information fused by multiple sensors can enhance robustness. For example, the camera can obtain the RGB texture information of the object but cannot accurately obtain the depth and 3D position information of the target. LiDAR sensor can obtain the position information of the object in the

This work was supported by the National Natural Science Foundation of China (NO. 61773312,61790563).

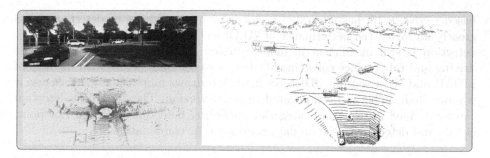

Fig. 1. Proposed 3D detection and tracking coupling system result. Left top image and left bottom image shows the 2D detection results and the raw points cloud as input. Right image shows the 3D detection and multi-object tracking result. Objects' trajectories are represented as deep green curve, green numbers are IDs of each object. (Color figure online)

3D space, not the texture information. By combining the information of camera and LiDAR, the object's RGB texture information and position information in 3D space can be obtained at the same time and the accuracy of object detection can be enhanced (Fig. 1).

As mentioned above, the effect of the object detection algorithm has been greatly improved, but the false detection or miss detection are still urgent problems to be solved. The current mainstream object detection algorithms only consider a single frame, ignoring the connection between the upper and lower frames. Actually, the object detection of autonomous driving usually consumes a period of time. Therefore, the upper and lower frame information is beneficial to object detection. It can not only reduce the false alarms and missing alarms in a single frame, but also can locate objects in the current frame through the historical positions. To a large extent, multi-object tracking depends on the result of object detection. Namely, an efficient object detection can improve multi-object tracking. Combining these two concepts is an interesting research direction.

In this work, we propose a 3D detection and tracking coupling system to complete 3D object detection and multi-object detection tasks. We take the advantage of mature 2D object detectors and project the 2D boxes onto 3D phase to filter the frustum range of point clouds. Then we use the prediction of objects' 3D boxes which have been tracked to locate and segment the objects points in frustum point clouds. We associate the objects in this frame with the tracked objects, and determine whether false alarms or missing alarms would occur according to the tracked objects and handle if it occurs.

2 Related Work

2.1 Object Detection

In this section, we will briefly review the object detection and multi object tracking. Recent years, The emergence and development of region of interests (RoIs)-

based CNNs [6,16] has generated high-confident candidates to detect, and has greatly improved the performance of 2D Object Detection. However, 2D object detection is still insufficient in more complex scenarios, such as autonomous driving and robot, since collecting 3D data is easier than before with the help of LiDAR and other sensors. 3D object detection draws more attention, because it is more challenging and complicated than 2D version. The 3D object detection can be divided in two main categories, including detection based on raw point clouds and detection based on data conversion or combination.

Methods Working on Raw Point Clouds. The VoxelNet-style methods [10,21,22] try to solve the issue about instance segmentation and T-Net alignment part before predicting, but they have a drawback of object unawareness in 3D point clouds. PointNet [13], PointNet++ [14] propose a novel type of network architecture that predicts and segments instance directly based on 3D raw point clouds. PointPillars [10] explores pillar shape instead of the mainstream voxel design to aggregate features. PointRCNN [18] generates a 3D solution directly from the point cloud in a bottom-up manner, which has a higher recall rate than the previous method.

Methods Working on Data Conversion or Other Data Combination. In MV3D [3], the LiDAR point clouds were projected to bird eye view (BEV), and then processed by a Faster-RCNN [16]. To generate more reliable 3D object proposals in MV3D, AVOD [9] fuses the multi-modal features. Some existing methods also use RGB-D or RGB data to improve the performance. ComplexerYOLO [19] is the first method introducing semantic segmentation into 3D object detection, which generates a better voxelized sematic point cloud used in 3D predictions afterward. F-PointNet [12] segments point cloud based on the 2D image detection result.

2.2   Multi-object Tracking

3D MOT systems are frameworks trying to detect and associate multiple identical objects in different frames. Most MOT systems follow tracking-by-detection paradigm [2,5,17], which has two steps. One is 3D object detection and the other is data association. The latter problem could be tackled from various perspectives like min-cost flow [5,11], Markov decision processes (MDP) [20], partial filtering [2]. However, most of these methods are not trained in an end-to-end manner thus many parameters are heuristic (e.g., weights of costs). Therefore, they are susceptible to local optima. DSM [5] proposes an end-to-end tracking and matching method by accurately solving linear programming. It is rarely considered to optimize the detection part through tracking part. [8] boosts the bottom-up object detection with information integrating the top down knowledge about tracking. This method is experimentally validated in inner-city traffic scenes. Inspired by that, we consider the object detection and association as a whole, which means that we use object detection to support association and improve detection after tracking.

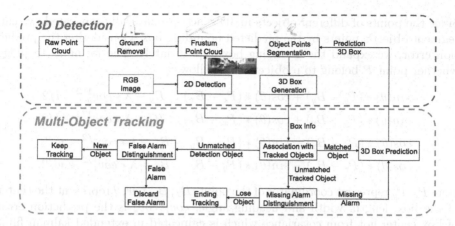

Fig. 2. Method overview. Our input data is processed in three steps. We remove ground of the raw point cloud and detect 2D boxes on image, and then generate frustum point clouds of 2D boxes. We reference the estimate 3D boxes of tracked objects and segment object points. After segmentation, we estimate the 3D boxes and push 3D boxes and 2D box info together into the tracking management. In the MOT part, detection objects are associate with tracked objects. Then we handle the unmatched detection and tracked objects and predict matched objects' box.

3 3D Detection and Tracking Coupling System

3.1 Tracking-Based 3D Detection

Frustum Point Cloud Generation. The framework of our system is shown in Fig. 2. Like the Frustum PointNets network, we first obtain the amodal 2D boxes and categories of objects on the RGB image through the proposed 2D object detector. Based on the known camera projection matrix R_{cam} and camera-to-lidar transformation matrix, we project 2D bounding boxes into the LiDAR coordinate frame and get frustums of each box. Before generating the frustum point cloud, We first preprocess the point cloud to remove the ground [7]. The purpose of ground points removal is to avoid the ground points that are considered to belong to objects located by the tracked objects. Then we filter out the non-ground points in each frustum generated by the 2D boxes.

Point Cloud Segmentation and 3D Box Estimation. Let $c_i^t \in C^t$ represents the 2D detection result in t frame, $f_i^t \in F^t$ represents the point cloud of frustums in t frame. According to the vehicle positioning and heading information and the transformation of the IMU coordinate system to LiDAR coordinate system, the point cloud of the frustums is transformed into the world coordinated system. Meanwhile we predict status of the stably tracked objects $x_j \in X$ and get their positions, orientations and sizes in the world coordinated system. If the predicted object bounding box $\tilde{x}_j^t \in \tilde{X}^t$ intersects with a frustum point cloud f_i^t, the tracked objects x_i are associated with the 2D detection objects c_i^t.

Since the points of different objects in 3D space are naturally separated, we can segment object's points by the predicted bounding box. Considering the prediction error, we expand the bounding box appropriately. The formulas to judge whether point P belong to p objects is as follows:

$$
\begin{aligned}
sin(\theta) * (P_y - B_y) + cos(\theta) * (P_x - B_x) &> B_y - \lambda * cov_y^{1/2} - l/2 \\
sin(\theta) * (P_y - B_y) + cos(\theta) * (P_x - B_x) &< B_y + \lambda * cov_y^{1/2} + l/2 \\
cos(\theta) * (P_x - B_x) - sin(\theta) * (P_y - B_y &> B_x - \lambda * cov_x^{1/2} - w/2 \\
cos(\theta) * (P_x - B_x) - sin(\theta) * (P_y - B_y &< B_x + \lambda * cov_x^{1/2} + w/2
\end{aligned}
\tag{1}
$$

Here P_x, P_y represent coordinates of the point, B_x, B_y, l, w, θ represent the center of the box, length, width and heading angle, cov_x, cov_y is the prediction error of box center got from covariance which is calucated in extended kalman filter when estimate the states of the box, λ is a parameter to control the influence from uncertainty of the center of the box. To get the 3D bounding box from the segmented object's points, we remove the points to the points center coordinate system by subtracting the x, y means of the object's points' position. Referring to Frustum PointNets [12], a preprocessed transformer network and box regression PointNets [13] are used to estimate object's amodal 3D bounding box. Since no tracking information for the first frame and the first observed objects, we apply Frustum PointNets to create 3D object bounding box.

3.2   Multi-object Tracking Management

The frustum point clouds f_i^t and the predicted 3D boxes $\tilde{x}_j^t$ are not always one-to-one correspondence. Therefore, the 2D detected objects c_i^t and the tracked 3D objects x_j need to be associated respectively. For stable tracked objects, we apply EKF to estimate the position and orientation of the objects' boxes in the current frame and use them as points segmentation input. In addition, we distinguish disappearance and appearance of objects with false alarms and missing alarms to deal with the latter two cases.

Objects Association. For some reasons, such as occlusion, two or more frustum areas projected by 2D boxes may have intersection areas, or two predicted boxes may intersect with the same frustum point cloud. We need to match 2D detection boxes $c_i^t \in C^t$ with 3D objects $x_j \in X$ which they are not one-to-one associated. Because the motion of objects in video has continuity, the 2D bounding box of the same object in two adjacent frames will have a similar position and size. Besides, the objects which are occluded and further having little 2D bounding boxes. Thus we calculate the IoU between 2D box c_i^t in current frame and 2D box c_j^{t-1} that has been associated to the 3D object x_j in the last frame. We match the 2D object box c_i^t which has larger IoU and tracked object x_j together. For the associated 2D object c_i^t and tracked 3D objects x_j, we save the 2D bounding box, category, 3D bounding box and current frame ID in the queue of the tracking management's objects x_j.

Tracked Object Prediction. We apply extended kalman filter to predict the state of stably tracked objects in current frame, and update the entire state of each object based on its corresponding 3D box state. We predict objects' state on the world coordinate system to make no effect on the movement of vehicle. In order to predict the state of objects more accurately, we use a constant acceleration and constant angular velocity model. We formulate the state of the 3D objects as an 10-dimensional vector $(x, y, z, \theta, vx, vy, vz, ax, ay, w)$, the variables vx, vy, vz, ax, ay represent the velocity and acceleration, w represents the angular velocity of the object, Δt represents time interval between two frames. The extended kalman filter updating model formulas are as follows:

$$x_{estimate} = x + \Delta t * vx + \Delta t^2 * ax/2 \qquad vx_{estimate} = vx + \Delta t * ax \qquad (2)$$

$$y_{estimate} = y + \Delta t * vy + \Delta t^2 * ay/2 \qquad vy_{estimate} = vy + \Delta t * ay \qquad (3)$$

$$z_{estimate} = z + \Delta t * vz \qquad (4)$$

$$\theta_{estimate} = \theta + \Delta t * w \qquad (5)$$

As a result, the element x, y, z, w of the predicted state and 3D bounding box size estimated in the last frame will be used in points segmentation as the input.

Miss and False Detection Handling. As the existing objects will disappear from the field of view and new objects will enter the detection area, we set status for objects to manage the tracked objects. For the objects tracked more than five frames, we consider the objects' status is stably tracked. We continuously track the stably tracked objects, record their trajectory, and predict their position and heading. If the stably tracked object loses less than three frames, we keep predicting its position and giving a hypothetical trajectory. If the objects are associated again, we consider there is a missing alarm for detection. The missing alarm may be caused by false negative 2D detection or there are no points in the area that predict object box intersecting with frustum. First, we apply RTS algorithm to smooth object's trajectory and get more accurate location in lost frame. Let's assume object X^{ST} is tracked in frame $(0, N)$ while lost in frame $j \in (0, N)$, $[Z_1, \cdots, Z_{j-1}, Z_{j+1}, \cdots, Z_N]$ is the object detection input in other frame, we need to calculate the optimal state estimate of object $\tilde{X}_j$ in frame j. The RTS smoothing algorithm is mainly reflected in the backward filtering process, we save state vector and variance matrix estimates and predictions value $\tilde{X}_F(k \mid k), \tilde{X}_F(k \mid k-1), \tilde{P}_F(k \mid k), \tilde{P}_F(k \mid k-1)$ (k means for frame k) calculated in extended kalman filter for every frame. We initialize the smoother, let:

$$\tilde{X}_S(N \mid N) = \tilde{X}_F(N \mid N) \qquad \tilde{P}_S(N \mid N) = \tilde{P}_F(N \mid N) \qquad (6)$$

Subscript S indicates optimal smoothing and subscript F indicates kalman filter. In frame j the smoothing gain of RTS smoothing algorithm is as follows:

$$\overline{K_S(j)} = P_F(j \mid j)\Phi_{j+1,j}^T P_F^{-1}(j+1 \mid j) \qquad (7)$$

In the formula (7), $\Phi_{j+1,j}$ is the jacobian matrix in extended kalman filter. And the smooth state vector and variance matrix in frame j are updated as:

$$\tilde{X}_S(j \mid N) = \tilde{X}_F(j \mid j) + \overline{K_S(j)}[\tilde{X}_S(j+1 \mid N) - \tilde{X}_F(j+1 \mid j)] \qquad (8)$$

$$\tilde{P}_S(j \mid N) = \tilde{P}_F(j \mid j) + \overline{K_S(j)}[\tilde{P}_S(j+1 \mid N) - \tilde{P}_F(j+1 \mid j)]\overline{K_S(j)}^T \qquad (9)$$

We can get more accurate position and heading of objects in frame from the smooth state estimate vector $X_S(j \mid N)$, and take it as a prediction input to segment objects point cloud and generate 3D box. If there is no segmented point, we use the predicted value as a result. In addition, if the stably tracked objects miss more than five frames, we suppose the objects are out of range and stop tracking, thereby we distinguish objects disappearance and missing detection, and supply the detection.

For new detected objects and unassociated objects, we set their status as trackable. If the object misses after being detected only one frame, we consider that is a false alarm and discard that. For objects which are continuously detected, the status is updated to tracked.

4   Experiments

In this section, we present experiments we have performed and analyze the results. We evaluate our methods and compare with the other multi-object tracking methods. Then we will show examples to prove that our method works.

4.1   Qualitative Evaluation

Our method is tested on the challenging KITTI Benchmark [1]. We choose Recurrent Rolling Convolution [15] as 2D detection input, and train the F-PointNets network and box estimation network in our framework on KITTI 3D detection dataset. Our method need continuous frame information, so, we evaluate the proposed detection and multi-object tracking framework on tracking dataset. To evaluate the performance of our method, we adopt the MOT metrics. We compare our approach with three MOT methods which also use LiDAR. Table 1 shows the multi-object tracking evaluation results of our methods and other methods on test set. Our method has close performance with FANTrack on MOTA,MT,PT and ML. It indicates that our method has a similar performance on tracking accuracy and lost targets with FANTrack. Our method have a lower FRG value, which shows that our method has effect on distinguishing and supplying missing alarm. The reason for the low MOTP value may be that sometimes the tracked segment objects' points only based on predict boxes are not precise enough.

In addition to reduce the impact of the detector on performance we compare our methods with AB3DMOT using F-PointNets as 3D detectors on KITTI train set. Table 2 shows the results of AB3DMOT using PointRCNN as input, AB3DMOT using F-PointNets as input and our methods. We can see that the detector has a great impact on the tracking results. Our method performs better than AB3DMOT on MOTA and MOTP if using a similar detector.

Table 1. Results on the KITTI tracking test set

Method	MOTA	MOTP	Recall	Precision	MT	PT	ML	IDS	FRG
Complexer-YOLO [19]	75.70	78.46	85.32	95.18	58.00	36.92	**5.08**	1186	2092
FANTrack [4]	77.72	82.33	83.66	96.15	62.62	28.62	8.77	150	812
AB3DMOT	**83.84**	**85.24**	**88.32**	96.98	**66.92**	21.69	11.38	**9**	224
Ours	77.22	79.00	82.91	**97.40**	62.16	29.19	8.65	145	**205**

Table 2. Compared with AB3DMOT(2) on the KITTI tracking train set

Method	MOTA	MOTP	Recall	Precision	MT	PT	ML	IDS	FRG
AB3DMOT-PointRCNN	83.35	78.43	92.17	93.86	75.68	20.54	3.24	0	30
AB3DMOT-FPointNets	76.04	78.36	80.54	**97.60**	56.22	35.14	8.65	**6**	**50**
Ours	**77.47**	**78.84**	**84.16**	96.79	**65.41**	**30.81**	**3.78**	104	193

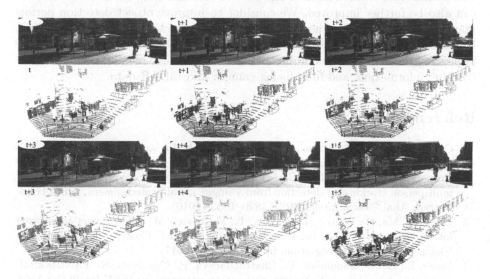

Fig. 3. Instance of where missing alarm and false alarm has been detected and correctly handled. The green 2D boxes are true positive results, red 2D boxes in frame $t, t + 1$ and $t + 3$ are discarded, because no points belong to objects in the frustum. The green 3D boxes are true positive 3D detection obtained from 2D detection and predicted 3D boxes. The purple 3D boxes in frame $t + 3$ and $t + 4$ are supplied objects of which 2D detection is missing. The blue 3D boxes in frame $t + 3$ and $t + 4$ are ground truth. (Color figure online)

4.2 Performance

Figure 3 shows an example about the handle of missing alarm and false alarm in our framework. The object with ID 3 is tracked stably before frame $t+3$, but lost 2D detection in frame $t + 3$ and $t + 4$. We keep tracking the object and predict its position and trajectory. It is associated again in frame $t + 5$ and frames after $t + 5$, so we consider missing alarm of the object with ID 3 happend. We smooth the predicted location in frame $t + 3$ and $t + 4$ by object's location after frame $t + 5$ and supply its 3D detection in frame $t + 3$ and $t + 4$. The purple 3D boxes are the results and the blue 3D boxes are ground truth. In frame $t, t + 1$ and $t + 3$,the red 2D boxes are discarded as false alarm for lacking LiDAR points.

5 Conclusion

In this paper, we present a 3D detection and tracking coupling framework. We achieve 3D detection and multi-object tracking through a 2D detection result. And our framework can effectively reduce miss alarm and false alarm in a single frame. Our method still has a long way to go. Though our method and 2D detector are independent, more precise 2D detector can bring superior performance. The points segmentation method and 3D box estimate network in our framework can also be further improved. We consider to improve object detection performance and focus more on the connection between tracking and object detection, because both are important to autonomous driving. Our future work will include improving the 2D detector and box estimation network, making better using of tracking information, and fully fusing camera and LiDAR data.

References

1. Geiger, A., Lenz, P., Urtasun, R.: Are we ready for autonomous driving? The KITTI vision benchmark suite. In: CVPR (2012)
2. Breitenstein, M.D., Reichlin, F., Leibe, B., Koller-Meier, E., Van Gool, L.: Online multiperson tracking-by-detection from a single, uncalibrated camera. IEEE Trans. Pattern Anal. Mach. Intell. **33**(9), 1820–1833 (2010)
3. Chen, X., Ma, H., Wan, J., Li, B., Xia, T.: Multi-view 3D object detection network for autonomous driving. In: Proceedings of the IEEE Conference on Computer Vision and Pattern Recognition, pp. 1907–1915 (2017)
4. Baser, E., Balasubramanian, V., Bhattacharyya, P., Czarnecki, K.: FANtrack: 3D multi-object tracking with feature association network. In: IEEE Intelligent Vehicles Symposium (2019)
5. Frossard, D., Urtasun, R.: End-to-end learning of multi-sensor 3D tracking by detection. In: 2018 IEEE International Conference on Robotics and Automation (ICRA), pp. 635–642. IEEE (2018)
6. Girshick, R.: Fast R-CNN. In: Proceedings of the IEEE international conference on computer vision, pp. 1440–1448 (2015)
7. Himmelsbach, M., Hundelshausen, F.V., Wuensche, H.J.: Fast segmentation of 3D point clouds for ground vehicles. In: IEEE Intelligent Vehicles Symposium, Proceedings, pp. 560–565 (2010)
8. Himmelsbach, M., Wuensche, H.J.: Tracking and classification of arbitrary objects with bottom-up/top-down detection. In: 2012 IEEE Intelligent Vehicles Symposium, pp. 577–582. IEEE (2012)
9. Ku, J., Mozifian, M., Lee, J., Harakeh, A., Waslander, S.L.: Joint 3D proposal generation and object detection from view aggregation. In: 2018 IEEE/RSJ International Conference on Intelligent Robots and Systems (IROS), pp. 1–8. IEEE (2018)

10. Lang, A.H., Vora, S., Caesar, H., Zhou, L., Yang, J., Beijbom, O.: PointPillars: fast encoders for object detection from point clouds. In: Proceedings of the IEEE Conference on Computer Vision and Pattern Recognition, pp. 12697–12705 (2019)
11. Lenz, P., Geiger, A., Urtasun, R.: FollowMe: efficient online min-cost flow tracking with bounded memory and computation. In: Proceedings of the IEEE International Conference on Computer Vision, pp. 4364–4372 (2015)
12. Qi, C.R., Liu, W., Wu, C., Su, H., Guibas, L.J.: Frustum pointnets for 3D object detection from RGB-D data. In: Proceedings of the IEEE Conference on Computer Vision and Pattern Recognition, pp. 918–927 (2018)
13. Qi, C.R., Su, H., Mo, K., Guibas, L.J.: PointNet: deep learning on point sets for 3D classification and segmentation. In: Proceedings of the IEEE conference on computer vision and pattern recognition, pp. 652–660 (2017)
14. Qi, C.R., Yi, L., Su, H., Guibas, L.J.: Pointnet++: deep hierarchical feature learning on point sets in a metric space. In: Advances in neural information processing systems, pp. 5099–5108 (2017)
15. Ren, J., et al.: Accurate single stage detector using recurrent rolling convolution. In: Proceedings of the IEEE Conference on Computer Vision and Pattern Recognition (2017)
16. Ren, S., He, K., Girshick, R., Sun, J.: Faster R-CNN: towards real-time object detection with region proposal networks. In: Advances in neural information processing systems, pp. 91–99 (2015)
17. Sharma, S., Ansari, J.A., Murthy, J.K., Krishna, K.M.: Beyond pixels: leveraging geometry and shape cues for online multi-object tracking. In: 2018 IEEE International Conference on Robotics and Automation (ICRA), pp. 3508–3515. IEEE (2018)
18. Shi, S., Wang, X., Li, H.: PointRCNN: 3D object proposal generation and detection from point cloud. In: Proceedings of the IEEE Conference on Computer Vision and Pattern Recognition, pp. 770–779 (2019)
19. Simon, M., et al.: Complexer-YOLO: real-time 3D object detection and tracking on semantic point clouds. In: Proceedings of the IEEE Conference on Computer Vision and Pattern Recognition Workshops, pp. 770–779 (2019)
20. Xiang, Y., Alahi, A., Savarese, S.: Learning to track: online multi-object tracking by decision making. In: Proceedings of the IEEE international conference on computer vision, pp. 4705–4713 (2015)
21. Yan, Y., Mao, Y., Li, B.: Second: sparsely embedded convolutional detection. Sensors **18**(10), 3337 (2018)
22. Zhou, Y., Tuzel, O.: VoxelNet: end-to-end learning for point cloud based 3D object detection. In: Proceedings of the IEEE Conference on Computer Vision and Pattern Recognition, pp. 4490–4499 (2018)

Ontologies/AI

Automated MeSH Indexing of Biomedical Literature Using Contextualized Word Representations

Dimitrios A. Koutsomitropoulos(✉) and Andreas D. Andriopoulos

Computer Engineering and Informatics Department, School of Engineering,
University of Patras, 26504 Patras, Greece
koutsomi@ceid.upatras.gr, a.andriopoulos@upatras.gr

Abstract. Appropriate indexing of resources is necessary for their efficient search, discovery and utilization. Relying solely on manual effort is time-consuming, costly and error prone. On the other hand, the special nature, volume and broadness of biomedical literature pose barriers for automated methods. We argue that current word embedding algorithms can be efficiently used to support the task of biomedical text classification. Both deep- and shallow network approaches are implemented and evaluated. Large datasets of biomedical citations and full texts are harvested for their metadata and used for training and testing. The ontology representation of Medical Subject Headings provides machine-readable labels and specifies the dimensionality of the problem space. These automated approaches are still far from entirely substituting human experts, yet they can be useful as a mechanism for validation and recommendation. Dataset balancing, distributed processing and training parallelization in GPUs, all play an important part regarding the effectiveness and performance of proposed methods.

Keywords: Classification · Indexing · Word embeddings · Thesauri · Ontologies · Doc2Vec · ELMo · MeSH · Deep learning

1 Introduction

Digital biomedical assets include a variety of information ranging from medical records to equipment measurements to clinical trials and research outcomes. The digitization and availability of biomedical literature is important at least in two aspects: first, this information is a valuable source for Open Education Resources (OERs) that can be used in distance training and e-learning scenarios; second, future research advancements can stem from the careful examination and synthesis of past results.

For both these directions to take effect, it is critical to consider automatic classification and indexing as a means to enable efficient knowledge management and discovery for these assets. In addition, the sheer volume of biomedical literature is continuously increasing and puts excessive strain on manual cataloguing processes: For example, the

© IFIP International Federation for Information Processing 2020
Published by Springer Nature Switzerland AG 2020
I. Maglogiannis et al. (Eds.): AIAI 2020, IFIP AICT 583, pp. 343–354, 2020.
https://doi.org/10.1007/978-3-030-49161-1_29

US National Library of Medicine experiences daily a workload of approximately 7,000 articles for processing [12].

Research in the automatic indexing of literature is constantly advancing and various approaches are recently proposed, a fact that indicates this is still an open problem. These approaches include multi-label classification using machine learning techniques, training methods and models from large lexical corpora as well as semantic classification approaches using existing thematic vocabularies. To this end, the Medical Subject Headings (MeSH) is the de-facto standard for thematically annotating biomedical resources [18].

In this paper we propose and evaluate an approach for automatically annotating biomedical articles with MeSH terms. While such efforts have been investigated before, in this work we are interested in the performance of current state-of-the-art algorithms based on contextualized word representations or word embeddings. We suggest producing vectorized word and paragraph representations of articles based on context and existing thematic annotations (labels). Consequently, we seek to infer the most similar terms stored by the model without the need and overhead of a separate classifier. Moreover, we combine these algorithms with structured semantic representations in Web Ontology Language format (OWL), such as the implementation of the MeSH thesaurus in OWL Simple Knowledge Organization Systems (SKOS) [20]. Finally, we investigate the effect and feasibility of employing distributed data manipulation and file system techniques for dataset preprocessing and training.

The rest of this paper is organized as follows: in Sect. 2 we summarize current word embedding approaches as the main background and identify the problem of automated indexing; in Sect. 3 we review relevant literature in the field of biomedical text classification; Sect. 4 presents our methodology and approach, by outlining the indexing procedure designed, describing the algorithms used and discussing optimizations regarding dataset balancing, distributed processing and training parallelization. Section 5 contains the results of the various experiments and their analysis, while Sect. 6 outlines our conclusions and future work.

2 Background

Word embedding techniques [9] convert words into word vectors. The following approaches have emerged in recent years with the performance of text recognition as the primary objective. The start has been made with the Word2Vec algorithm [11] where unique vector word representations are generated by means of shallow neural networks and the prediction method as well as by the explicit extension Doc2Vec (Document to Vector) [8], where a unique vector representation can also be given for whole texts. Next, the Global Vectors (GloVe) algorithm [14] manages to transfer the words into a vector space by making use of the enumeration method. Then, the FastText algorithm [4] achieves not only the management of a large bulk of data in optimal time but also better word embedding due to the use of syllables. In addition, the ELMo algorithm [15] uses deep neural networks, LSTMs, and a different vector representation for a word the meaning of which differentiates. Lastly, the BERT algorithm [2] also generates different representations of a word according to its meaning, but instead of LSTMs it uses transformer elements [21].

Assigning a topic to text data is a demanding process. Nevertheless, if approached correctly, it ensures easier and more accurate access for the end user. A typical subcase of the topic assignment problem is the attempt to create MeSH indexes in repositories with biomedical publications. This particular task, which improves the time and the quality of information retrieved from the repositories, is usually undertaken by field experts. However, this manual approach is a time consuming process (it takes two to three months to incorporate new articles), but also a costly one (the cost for each article is approximately $10) [10].

3   Related Work

A plethora of research approaches have attempted to tackle with the problem of indexing. To do this, they use word embeddings in combination with classifiers. Typical cases are discussed below.

MeSH Now [10] classifies the candidate terms based on the relevance of the target-article and selects the one with the highest ranking, thus achieving a 0.61 F-score. To do this, researchers are using k-NN and Support Vector Machine (SVM) algorithms. Another approach, named DeepMeSH [13], deals with two challenges, that is, it attempts to examine both the frequency characteristics of the MeSH tags and the semantics of the references themselves (citations). For the first it proposes a deep semantic representation called D2V-TFIDF and for the latter a classification framework. A k-NN classifier is used to rate the candidate MeSH headings. This system achieves an F-score of 0.63. Another study [5] uses the Word2Vec algorithm on all the abstracts of the PubMed repository, thereby generating a complete dictionary of 1,701,632 unique words. The use of these vectors as a method to reduce dimensionality is examined by allowing greater scaling in hierarchical text classification algorithms, such as k-NN. By selecting a skip-gram neural network model (FastText) vectors of size 300 are generated with different windows from 2 to 25, which the authors call MeSH-gram with an F-score of 0.64 [1].

Moreover, taking into consideration the assumption that similar documents are classified under similar MeSH terms, the cosine similarity metric and a representation of the thesaurus as a graph database, scientists proceed to an implementation with an F-score of 0.69 [16]. A problem-solving approach is considered starting from converting texts into vectors with the use of Elastic Search and identifying the most similar texts with the help of the cosine similarity metric. Then, by deriving the tags from these texts and calculating the frequency of occurrence in conjunction with similarity, an evaluation function is defined which classifies documents.

BioWordVec [22] is an open set of biomedical word vectors/embeddings which combines subword information from unlabeled biomedical text with MeSH. There are two steps in this method: first, constructing MeSH term graph based on its RDF data and sampling the MeSH term sequences and, second, employing the FastText subword embedding model to learn the distributed word embeddings based on text sequences and MeSH term sequences. In this way, the value of the F-score metric is improved to 0.69 and 0.72 for CNN and RNN models respectively.

4 Methodology

4.1 Automated Indexing Procedure

The proposed approach for the indexing of biomedical resources starts with assembling the datasets to be used for training. We then proceed by evaluating and reporting on two prominent embedding algorithms, namely Doc2Vcc and ELMo. The models constructed with these algorithms, once trained, can be used to suggest thematic classification terms from the MeSH vocabulary.

In an earlier work we have shown how to glean together resources from various open repositories, including biomedical ones, in a federated manner. User query terms can be reverse-engineered to provide additional MeSH recommendations, based on query expansion [7]. Finally, we can combine and assess these semantic recommendations by virtue of a trained embeddings model [6].

Each item's metadata is scraped for the title and abstract of the item (*body of text*). This body of text is next fed into the model and its vector similarity score is computed against the list of MeSH terms available in the vocabulary. Training datasets comprise biomedical literature from open access repositories including PubMed [19], EuropePMC [3] and ClinicalTrials [17] along with their handpicked MeSH terms. For those terms that may not occur at all within the datasets, we fall back to their scopeNote annotations within the MeSH ontology. As a result, the model comes up with a set of suggestions together with their similarity score (Fig. 1).

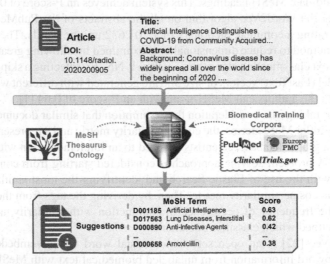

Fig. 1. A specific item gets subject annotations and their similarity scores are computed.

4.2 Datasets

For the application of the Doc2Vec and ELMo methods, a dataset from the PubMed repository with records of biomedical citations and abstracts was used. In December of every year, the core PubMed dataset integrates any updates that have occurred in the field. Each day, the National Library of Medicine produces updated files that include new, revised and deleted citations. About 30 M records, which are collected annually, can be accessed by researchers as of December 2018. Another source is EuropePMC, which is a European-based database that mirrors PubMed abstracts but also provides free access to full-texts and an additional 5 M other relevant resources.

Each entry in the dataset contains information, such as the title and abstract of the article, and the journal which the article was published in. It also includes a list of subject headings that follow the MeSH thesaurus. These headings are selected and inserted after manual reading of the publication by human indexers. Indexers typically select 10-12 MeSH terms to describe every indexed paper. MeSH is a specialized solution for achieving a uniform and consistent indexing of biomedical literature. In addition, it has already been implemented in SKOS [20]. It is a large and dense thesaurus consisting of 23,883 concepts.

The ClinicalTrials repository also provides medical data, that is, records with scientific research studies in XML. ClinicalTrials contains a total of 316,342 records and each record is also MeSH indexed. We include these records with the ones obtained from PubMed and EuropePMC for the purposes of variability and dataset diversity.

4.3 Using a Distributed File System

Our initial methodology for collecting data for training followed a serial approach [6]. Access to PubMed can be done easily via File Transfer Protocol (FTP). The baseline folder includes 972 zip files, up to a certain date (December 2018). Each file is managed individually with the information being serially extracted, thus rendering the completion of the entire effort a time costly process. In addition to delays, this problem makes the entire process susceptible to the need for constant internet connection to the repositories and any interruptions that may occur.

Algorithm 1: Dataset preparation procedure	
	input: XML files from repository
	output: two CSV files
Step 1.	**for each** file ∈ repository **do**
Step 2.	connect to FTP server
Step 3.	get file to local disk
Step 4.	store file as line in RDD
Step 5.	delete file from local disk
Step 6.	**end for**
Step 7.	parse file - useful information is extracted
Step 8.	convert RDD to DataFrame
Step 9.	write useful information to CSV files

To solve this particular problem, we investigate the use of a distributed infrastructure at the initial stage of data collection. For this purpose, Apache Spark[1], a framework for parallel data management, was used. XML files are now stored as a whole in a DataFrame. In detail, all the information in an XML file is read as a line and then converted into Resilient Distributed Dataset (RDD). The useful information is then extracted as in the

---

[1] https://spark.apache.org/.

previous procedure. Finally, the RDD is converted into a DataFrame, from which the information is easily extracted, for example, into CSV files. With this process, although extracting data from repositories is still not avoided, parsing can be now performed on a distributed infrastructure.

4.4   Balancing

An essential part of the dataset preparation process is to cover the whole thesaurus as thoroughly as possible. Thus, apart from the full coverage of the thesaurus terms, the model must also learn each term with an adequate number of examples (term annotations) and this number should be similar between the terms. Otherwise, there will be a bias towards specific terms that happen to have several training samples vs. others which might have only a few. Therefore, to achieve a balanced dataset for training, when a term is incorporated into the set, its number of samples is restricted by an upper limit (Algorithm 2). If there are fewer samples the term is ignored; if there are more, exceeding annotations are cut off. The final dataset, as shown in Table 1, does not fully cover the thesaurus, but the terms it contains are represented by an adequate and uniform number of samples.

Algorithm 2: Dataset balancing procedure
input: two CSV files, thesaurus terms
output: two CSV files with balanced dataset
Step 1.   **for each** term ∈ thesaurus **do**
Step 2.   compute number of term occurrences in CSV file
Step 3.   **if** number >= 100 **then**
Step 4.   collect first 100 samples
Step 5.   add samples in two lists
Step 6.   **end if**
Step 7.   **end for**
Step 8.   write lists to CSV files

Table 1. Details of dataset

	Balanced dataset	Test dataeset
Total items	100,000	10,000
Total annotations	140,000	14,000
Average # terms per item	1.4	1.4
Thesaurus terms coverage rate (%)	4%	4%

4.5   Experiments

Doc2Vec Model. To create the model, the input is formed from the "one-hot" vectors of the fixed-size body of text, which is equal to the dictionary size. The hidden plane includes 100 nodes with linear triggering functions and with the same size as the resulting vector dimensions. At the output level there will also be a vector equal to the dictionary size, while the activation function will be *softmax*.

The training of the Doc2Vec model, with the help of the Gensim library[2], is performed by using the following parameters: *train epochs* 100, *size vector* 100, *learning parameter* 0.025 and *min count* 10. A variety of tests were performed to estimate these values. Tests have shown that when there are only a few samples per term, a larger number of epochs can compensate for the sparsity of training samples. In addition, removing words with less than 10 occurrences also creates better and faster vector representations for thesaurus terms. The created model is stored so that it can be called directly when needed. This model, with the adopted weights of the synapses that have emerged, is in fact nothing more than a dictionary. The content of this dictionary is the set of words used in the training along with their vector representations as well as a vector representation for each complete body of text.

ELMo Model. Initially, all the vectors are extracted through a URL connection[3]. Then, all the words related to the topic and mentioned in labels, are converted into classes, that is, numeric values, e.g. 0, 1, 2, etc., which in turn become "one-hot" vectors. To create the model, we have used the Keras library[4]. The input layer receives one body of text (title and abstract) at a time. The next layer is the *Lambda*, which is supplied by the input layer, and uses the ELMo embeddings, with an output count of 1024. To create the ELMo embeddings, the vectors derived from the URL are used and the full body of text is converted into string format and compressed at the same time with the help of the tensorflow package. Selecting the default parameter in this process ensures that the vector representation of the body of text will be the average of the word vectors. The next layer will be the dense one, which in turn is powered by the Lambda layer and contains 256 nodes and the *ReLU* activation function. Finally, we have the output layer, which is also a dense layer with as many nodes as classes, and in this case the activation function will be *softmax*. In the final stage of the system (compile), the loss parameter is *categorical_crossentropy*, because of the categorization being attempted. The *ADAM* optimization, and the accuracy metric are chosen.

The training of the ELMo model is done by starting a tensorflow session with parameters: *epochs* 10 and *batch_size* 10. This ensures that training will not take longer than 10 epochs and the data will be transferred in packages of 10 samples. Upon completion, the finalized weights are stored, for the purpose of immediately creating a model either for evaluation or any other process required.

---

[2] https://radimrehurek.com/gensim/.

[3] https://tfhub.dev/google/elmo/3.

[4] https://keras.io/.

4.6 Scalability

In the context of the implementation of the above two models, the most significant stage is the execution of the algorithm responsible for their training. This process, depending on the architecture of the model, is particularly demanding on computing resources.

Most experiments are conducted on average commodity hardware (Intel i7, 2.6 GHz, 4-cores CPU with 16 GB of RAM). In a shallow neural network, such as Doc2Vec, execution can be completed relatively quickly without significant processing power demands. Our test configuration appears sufficient for at least 100 epochs to be achieved, so that the model can be trained properly. However, this is not the case for deep neural networks, such as ELMo. In these models, the multiple layers with many connections add complexity and increase computational power requirements in order to be adequately trained. Therefore, finding a way to parallelize the whole process is considered necessary, not only for the optimization of the training time, but also for the completion of the entire effort.

Specifically, options associated with increasing the size of the training set in combination with the number of epochs can well lead either to a collapse of the algorithm execution or to prohibitive completion times. As an example, training ELMo on a dataset of 1,000 samples (10 labels with 100 items each) took about 20 *min* in the above hardware configuration.

Based on this concern, we have conducted experiments on infrastructures with a large number of Graphics Processing Units (GPUs). GPUs owe their speed to high bandwidth and generally to hardware which enables them to perform calculations at a much higher rate than conventional CPUs. We experimented on high-performance hardware with two Intel Xeon CPUs including 12-cores, 32 GB of RAM and a Nvidia V100 GPU. The V100 has 32 GB of memory and 5,120 cores and supports CUDA v. 10.1, an API that allows the parallel use of GPUs by machine learning algorithms. In this configuration, training with the 1,000 samples dataset took only 30 s for the same model. This faster execution (40x) can help run experiments with larger datasets that would otherwise be prohibitive to achieve.

5   Results

5.1   Doc2Vec

To evaluate the Doc2Vec approach, a total of 10,000 samples was used. This test set is balanced with the same procedure as the 100 K dataset we have used previously for training. Therefore, each MeSH label occurring in the test set appears within the annotations of 10 bibliographic items vs. 100 in the training set. In both sets, a total of 1,000 distinct MeSH labels, out of the available 22 K, are considered. A body of text (title and abstract) is given as input to the model, which in turn generates a vector. The model then searches through the set of vectors, already incorporated from training, to find those that are close to the generated one. The process is quite difficult due to the plethora of labels in total, but also individually per sample as each one contains a finite non-constant number of labels. The threshold value reported is the similarity score above which suggestions are considered. At most 10 suggestions are produced by the model.

The following Fig. 2 plots precision (*P*) and recall (*R*) for various threshold values. For our purposes, a single information need is defined as a single-term match with an item (1-1) and we report the mean value of these metrics over all these matches, i.e. they are micro-averaged. Precision takes into account how many of these 1-1 suggestions are correct out of the total suggestions made, while recall considers correct suggestions out of the total number of suggestions contained in the ground truth.

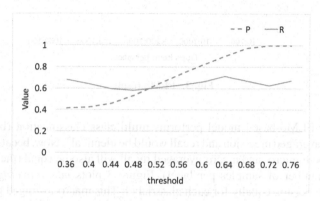

Fig. 2. Doc2Vec results

Other than the standard tradeoff between precision and recall, we first notice that the higher the threshold, the better the quality of results is for *P*. However, an increase in the threshold causes recall to drop. This makes sense, because there may be considerably fewer than 10 suggestions for higher thresholds, i.e. only few of the terms pass the similarity threshold, thus leaving out some relevant terms. For middle threshold values, more that 60% of predictions are correct and also cover over 60% of the ground truth.

5.2 ELMo

Similarly balanced test sets were also used to evaluate the ELMo model, with a varying number of labels and samples per label. The results obtained are shown in Fig. 3.

Initially, equally comparable or better results to previous related work are observed for precision and recall, when a small number of labels is selected. Given that the dataset is balanced, these considerably improve when allowing 10x more samples for each label, thus allowing for better embeddings to be learned by the model. Raising the number of labels to 100 increases the dimensionality of the problem; now, each item must be classified among 100 classes rather than simply 10. Consequently, we notice a decrease in both metrics which, however, is ameliorated with the use of more samples (1,000 per label), as expected. Nonetheless, further increasing the complexity of the dataset, by allowing 1,000 labels, causes absolute values to decrease considerably. Even with the better performing hardware, any further efforts to improve the results in the case of 1,000 labels, notably by increasing the number of samples, do not succeed, as the execution of the algorithm is interrupted each time due to the need for additional memory.

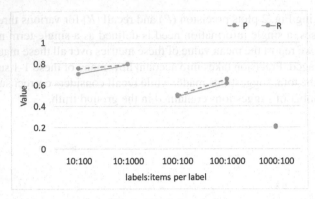

Fig. 3. ELMo results

Since our ELMo-based model performs multi-class classification (but not multi-label), micro-averaged precision and recall would be identical[5]. Now, because our test set is balanced, both micro- and macro-averaged recall will also be equal (the denominator is the fixed number of samples per label). Figure 3 plots *macro-averaged* precision i.e. averages precision equally for each class/label. This macro-averaged precision will always be higher than the micro averaged one, thus explaining why P appears better than R, but only slightly, because of the absence of imbalance. P and R are virtually the same.

5.3 Discussion

The Doc2Vec model is capable of making matching suggestions on its own, by employing similarity scores. A threshold value of 0.64 is where precision and recall acquire their maximum values. Certainly, fewer than 10 suggestions may pass this limit, but these are more accurate, apparently because of their higher score; they also occur more frequently within the terms suggested by experts in the ground truth: on average, each biomedical item in the test set hardly contains 2 MeSH terms, let alone 10 (see Table 1). This observation serves as validation of the fact that the similarity measure produced by the model is highly relevant and correlated with the quality of suggestions.

The ELMo model, in turn, has allowed us to construct a multi-class classification pipeline that is built around the ELMo embeddings. This manages to achieve very good results when dimensionality is kept low, i.e. a relatively small number of labels or classes is selected. It also seems to outperform the Doc2Vec approach, even though classes are fewer and the recommendation problem is reduced to multi-class classification. However, Doc2Vec surpasses the ELMo classification pipeline when there is a need to choose labels from a broader space. One the other hand, the threshold existence favors Doc2Vec, in the sense that fewer ground truth annotations pass its mark and are, therefore, considered when computing retrieval metrics. Further attempts for improvement where not possible for ELMo, a fact that confirms it is a very computationally expensive module compared to word embedding modules that only perform embedding lookups.

---

[5] https://scikit-learn.org/stable/modules/model_evaluation.html Sect. 3.3.2.8.2.

6   Conclusions and Future Work

Contextualized word representations have revolutionized the way traditional NLP used to operate and perform. Neural networks and deep learning techniques combined with evolving hardware configurations can offer efficient solutions to text processing tasks that would be otherwise impossible to perform on a large scale. We have shown that word embeddings can be a critical part and deserve careful consideration when approaching the problem of automated text indexing. Especially in the biomedical domain, the shifting nature of research trends and the complexity of authoritative controlled vocabularies still pose challenges for fully automated classification of biomedical literature.

To this end, we have investigated both deep- and shallow learning approaches and attempted comparison in terms of performance. Dimensionality reduction, either implied by setting a threshold or directly posing a hard cap over available classification choices, appears necessary for automated recommendations to be feasible and of any practical value. Still, a careful dataset balancing as well as the capability of deep networks to leverage distributed GPU architectures are demonstrated beneficial and should be exercised whenever possible.

As a next step, we intend to further evaluate our ELMo implementation and design a model that would perform multi-label classification using the ELMo embeddings. In addition, we are planning to release a web-based service offering access to the recommendations provided by the two models. This would facilitate interoperability, for example, with learning management systems and other repositories. Finally, we see improvements in the way classification suggestions are being offered, especially in view of the density of the thesaurus used: other ontological relations, such as generalization or specialization of concepts can be taken into account in order to discover and prune hierarchy trees appearing in the recommendations list.

References

1. Abdeddaïm, S., Vimard, S., Soualmia, L.F.: The MeSH-gram Neural Network Model: Extending Word Embedding Vectors with MeSH Concepts for UMLS Semantic Similarity and Relatedness in the Biomedical Domain, arXiv:1812.02309v1 [cs.CL] (2018)
2. Devlin, J., Chang, M. W., Lee, K., Toutanova, K.: BERT: Pre-training of Deep Bidirectional Transformers for Language Understanding, arXiv:04805v2 [cs.CL] (2019)
3. Europe PMC Consortium. Metadata of all Full-Text Europe PMC articles. europepmc.org/ftp/pmclitemetadata/
4. Joulin, A., Grave, E., Bojanowski, P., Mikolov, T.: Bag of Tricks for Efficient Text Classification, arXiv:1607.01759v3 [cs.CL] (2016)
5. Kosmopoulos, A., Androutsopoulos, I., Paliouras, G.: Biomedical semantic indexing using dense word vectors. In: BioASQ (2015)
6. Koutsomitropoulos, D., Andriopoulos, A., Likothanassis, S.: Subject classification of learning resources using word embeddings and semantic thesauri. In: IEEE Innovations in Intelligent Systems and Applications (INISTA), Sofia, Bulgaria (2019)
7. Koutsomitropoulos, D.: Semantic annotation and harvesting of federated scholarly data using ontologies. Digit. Libr. Perspect. 35(3–4), 157–171 (2019)
8. Le, Q.V., Mikolov, T.: Distributed representations of sentences and documents. In: 31st International Conference on Machine Learning, ICML, Beijing, China (2014)

9. Li, Y., Yang, T.: Word embedding for understanding natural language: a survey. In: Srinivasan, S. (ed.) Guide to Big Data Applications. SBD, vol. 26, pp. 83–104. Springer, Cham (2018). https://doi.org/10.1007/978-3-319-53817-4_4

10. Mao, Y., Lu, Z.: MeSH now: automatic MeSH indexing at PubMed scale via learning to rank. J. Biomed. Semant. **8**(1), 15 (2017). https://doi.org/10.1186/s13326-017-0123-3

11. Mikolov, T., Chen, K., Corrado, G., Dean, J.: Efficient estimation of word representations in vector space. In: ICLR Workshop (2013)

12. Mork, J.G., Jimeno-Yepes, A., Aronson, A.R.: The NLM medical text indexer system for indexing biomedical literature. In: Conference and Labs of the Evaluation Forum 2013 (CLEF 2013), Valencia, Spain (2013)

13. Peng, S., You, R., Wang, H., Zhai, C., Mamitsuka, H., Zhu, S.: DeepMeSH: deep semantic representation for improving large-scale MeSH indexing. Bioinform. **32**(12), i70–i79 (2016). https://doi.org/10.1093/bioinformatics/btw294

14. Pennington, J., Socher, R., Manning, C.D.: GloVe: global vectors for word representation. In: Conference on Empirical Methods in Natural Language Processing (EMNLP), Doha, Qatar, pp. 1532–1543 (2014)

15. Peters, M.E., Neumann, M., Iyyer, M., Gardner, M., Clark, C., Lee, K., Zettlemoyer, L.: Deep contextualized word representations, arXiv:1802.05365v2 [cs.CL], NAACL (2018)

16. Segura, B., Martínez, P., Carruan, M.A.: Search and graph database technologies for biomedical semantic Indexing: experimental analysis. JMIR Med. Inform. **5**(4), e48 (2017). https://doi.org/10.2196/medinform.7059

17. U.S. National Library of Medicine. ClinicalTrials.gov. https://clinicaltrials.gov

18. U.S. National Library of Medicine. Medical Subject Headings, 2019. https://www.nlm.nih.gov/mesh/meshhome.html

19. U.S. National Library of Medicine. PubMed.gov. https://www.nlm.nih.gov/databases/download/pubmed_medline.html

20. van Assem, M., Malaisé, V., Miles, A., Schreiber, G.: A method to convert thesauri to SKOS. In: Sure, Y., Domingue, J. (eds.) ESWC 2006. LNCS, vol. 4011, pp. 95–109. Springer, Heidelberg (2006). https://doi.org/10.1007/11762256_10

21. Vaswani, A., et al.: Attention is all you need. In: Advances in Neural Information Processing Systems, pp. 6000–6010 (2017)

22. Zhang, Y., Chen, Q., Yang, Z., et al.: BioWordVec, improving biomedical word embeddings with subword information and MeSH. Sci. Data **6**, 52 (2019). https://doi.org/10.1038/s41597-019-0055-0

Knowledge-Based Management and Reasoning on Cultural and Natural Touristic Routes

Evangelos A. Stathopoulos[✉], Alexandros Kokkalas, Eirini E. Mitsopoulou,
Athanasios T. Patenidis, Georgios Meditskos, Sotiris Diplaris,
Ioannis Paliokas, Stefanos Vrochidis, Konstantinos Votis, Dimitrios Tzovaras,
and Ioannis Kompatsiaris

Information Technologies Institute, Centre for Research & Technology - Hellas,
Thessaloniki, Greece
{estathop,akokkalas,emitsopou,apatenidis,gmeditsk,diplaris,ipaliokas,
stefanos,kvotis,dimitrios.tzovaras,ikom}@iti.gr

Abstract. There is great potential in interdisciplinary traveling platforms mingling knowledge about cultural heritage aspects, such as places with schedules providing visits or even containing augmented reality features also, along with environmental concerns to enhance personalized tourist experience and tripping avocation. For an ontological framework to support and nominate trip detours of targeted interests according to end-users, it should incorporate and unify as much heterogeneous information, deriving either from web sources or wherever there are ubiquitously available such as sensors or open databases. A plethora of qualitatively diverse data along with adequate quantities of them escalate the contingent results in terms of conferring a plurality of relevant options which can be utterly manifested through involving axioms with rule-based reasoning functionalities upon properties considered to be irrelevant to each other at first glance. Thus, managing to import predefined concepts from other ontologies, such as temporality or spatiality, and combine them with new defined concepts to tourist assets, such as points of interest, results in novel meaningful relationships never established before. Apart from the utilization of pre-existent resources and logic towards automatic detouring suggestions, a wide-spectrum modeling enables a suitable problem statement relevant to the e-Tracer framework and comprehension of the issues, providing the opportunity of statistical analysis of knowledge when adequate amounts amassed.

Keywords: Ontologies · Reasoning · Semantically enriched geodata ·
Data homogenization · Route recommendation subsystem

1 Introduction

Vast amounts of data are effusive throughout every ecosystem. Gradually, the ability to effectively capture data for knowledge extraction has increased.

© IFIP International Federation for Information Processing 2020
Published by Springer Nature Switzerland AG 2020
I. Maglogiannis et al. (Eds.): AIAI 2020, IFIP AICT 583, pp. 355–367, 2020.
https://doi.org/10.1007/978-3-030-49161-1_30

Digital agents tend to pseudomimic mental processes, such as deductive reasoning, intricate decision making, inferring general assumptions and so on. The chasm between data and knowledge is bridged by semantics, inserting and fusing contexts into otherwise meaningless data. Ultimately, the interest is not on the value itself but on its representation and meaning inside a system and its exploitation. Today, there are many methodologies followed towards knowledge design and manipulation; in that aspect, one can amalgamate disparate conceptions into a unified model. To achieve the wished level of homogeneity, the affinities of entities must become firmly established, the capitalization of which is the rapidity and deftness of knowledge elaboration to infer with logic as a basis.

In this paper we describe an ontology-based framework for capturing and interlinking assets of cultural and natural substance facilitating the formation of routes via the utilization of spatio-temporal rule-based reasoning. The ontological model encompasses and depicts every unique data genre present in the workflow of the platform: genres related to weather, topological formats expressing geometry and geospatiality (such as points in 3D space and routes), time formats expressing temporality, user profiling, Augmented Reality (AR) and hierarchical content categorization of places along with other information. That way, places might be discarded or included in a final route recommendation for the end-user. To complete this task, a systematic evaluation was performed to assess the abundance of our approach.

Our methodology starts by identifying each relevant entity tagged as coherent with the e-Tracer objectives, as described above. Thereinafter, an extensive research on pre-existing ontologies encapsulating relative knowledge concepts was conducted. Upon the dilemma emerged concerning either the entire imports of concepts (where the majority of structures were needed) or manufacturing our own, we concluding in building custom concepts targeting precisely our objectives and linking afterwards where applicable. At the same time, we paid effort to keep a minimalistic approach in the design of the overall system. Based on the ontology already created, the aggregation of web content or content from official databases was populated in the knowledge base. Finally, rules that should diminish significantly the offered number of selections based on property constraints were implemented and incorporated into web Application Programming Interfaces (APIs).

The rest of the paper is structured as follows: Sect. 2 presents work which at some extent seems coherent to this paper. Sections 3 provides an overview of the framework and bestows the overall vision and motivation. In Sect. 4 elaboration on the inference, validation and consistency capabilities are exhibited, while in Sect. 5 fundamental reasoning functionalities relying on time and geo-spatial properties are showcased. In Sect. 6 quantitative and qualitative evaluation is displayed and, finally, Sect. 7 concludes our work.

2   Related Work

Data about Cultural and Natural places are bound with their location information and time parameters. With the increasing amount of geospatial data being
published online and the geographic information taking a crucial part in several
central hubs on the Linked Data Web; geospatial semantics, geo-ontologies and
semantic interoperability can have a key role in supporting publishing, retrieving,
and reusing data while reducing the risks of misinterpretations [1,2]. Moreover,
various semantic web technologies have been adapted for geospatial data, with
progress in the effectiveness of the methods used in [3] with space and time having
a key role for definition, organization and mutual interaction between concepts
for knowledge engineering [4]. In similar approach, RDF models regarding space-
time events have been designed, that integrate spatial, temporal and semantic
relations for capturing factors behind certain geographic changes [5].

Aggregating geospatial datasets into a single one is a challenging task. To
tackle this problem semantic technologies were deployed so as to automate the
geospatial data conflation process. By using ontology, RDF data conversion and
a set of SWRL rules, one can produce a *Points of Interest (POIs)* dataset with
reduced duplicates and improved accuracy [6]. Furthermore, the semantic ontological network graph (SONET) [7] is an ontological network to match categories
across multiple heterogeneous sources of *POIs* of Volunteered Geographic Information (VGI) data. This ontological network advances the study of VGI data by
enabling cross-platform analysis while it supports the use of *POI* data in land
use mapping and population modeling applications. Deepening in the issue of
heterogeneity since particularly cultural heritage consists of multiple resources
which might include entities such as places, events, availability, and others that
have special characteristics and might be connected with each other. Cultural-
ON is a suite of ontology modules, to model the principal elements identified
in the cultural heritage data type classification. The result is a knowledge base
consisting of semantic interconnections with also other data available in the Web
to be exploited according to different tasks and users preferences [8].

Taking advantage of cultural assents and towards touristic recommendations,
there are several ontology-based systems available. They provide personalized
suggestions to users, based on user profiles and information concerning the suggested locations. The results are ranked based on profile assignment, content
filtering and user feedback [9] or by ontology-based content analyzer, ontology-
based profile learner, and ontology-based filtering component [10] or in the case
of STAAR (Semantic Tourist information Access and Recommending) where
algorithms take into consideration itinerary length and user interests [11].

3   e-Tracer Framework

In a world with liberty and convenience of locomotion, data can be aggregated in
abundance, enabling the sector of personalized tourism to attribute with further
enhancements to provide a more pleasant experience to the end-user, an effort
also facilitated by e-Tracer [12], a national funded project.

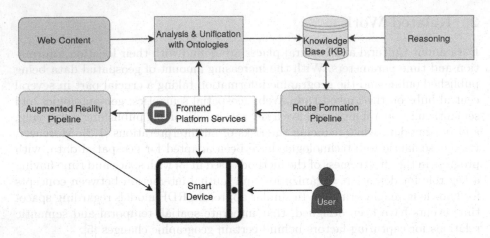

Fig. 1. The conceptual framework of e-Tracer

3.1   Key Concepts and Vision

The conceptual architecture of the platform is depicted in Fig. 1. Briefly, the *Web Content* involves data derived from official websites and open governmental databases. For the former the easIE framework [13] was used to scrap content, while for the later massive file exports sufficed. Furthermore, the *Route Formation Pipeline* expresses a complex algorithm to create routes, containing not only reasoning functionalities but also personalization based on similarity measures and graph algorithms such as the traveling salesman approaches, Dijkstra's shortest path and so on. Moreover, the *Augmented Reality Pipeline* encompasses search, identification, retrieval and display of objects on a *Smart Device*. Concluding, for aggregating information and presenting it to the *User* the *Frontend Platform Services* are responsible for.

Our work is focused on the semantics and aims to enrich e-Tracer with such capabilities. It acts as a semantic middleware, capturing, interlinking and serving results. This is materialized in the *"Analysis & Unification with Ontologies"* component where dissimilar content is homogenized and stored locally in compliance with Resource Description Framework (RDF) triplets inside a *Knowledge Base (KB)*. Additionally, it provides the semantic infrastructure to retrieve stored assets in an asynchronous on-demand manner to fulfil dynamic querying requirements. Furthermore, the reasoning occurs either automatically on each update of the *KB* or by calling it as a service where data are reciprocated amphidromously. Concluding, it is obvious from Fig. 1 that the use of semantics stands in the core of the platform adding extra value and coordinating a majority of processes.

3.2   e-Tracer Ontological Core Model

Figure 2 illustrates the upper-level concepts of the e-Tracer hierarchical model where each differentiated arrow line is depicting a distinct type of connection

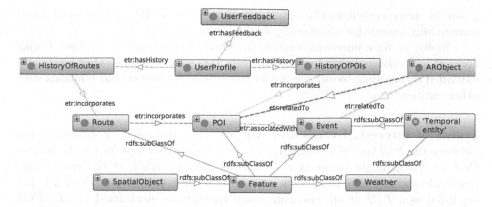

Fig. 2. The upper-level core concepts of the e-Tracer model and the semantic conjugation

and dotted lines implying customness. The conceptual model revolves around the notions of *point of interest (POI)*, *event, augmented reality object, spatial object, temporal entity, route, user profile, user feedback* and *weather*. Each distinct connotation is potentially intertwined with others in manifold degrees, both at root and lower modeled levels, in a way that there should not be perceived in the overview graph such a phenomenon as an orphaned node when data increment efficiently and satisfactorily. Intuitively, entities will be linked with pre-existing online entities as soon as the core model reaches near its terminal configuration after numerous successive iterations. In the succeeding subsections we circumscribe extensively each key concept along with its semantic correlation with the rest.

Re-used Concepts. The concept of the *Event* is described both accurately and sufficiently by the ontology for linking open descriptions of events, LODE [14]. In its simplicity, it is explained more thoroughly in the next subsection along with how supplementary content was formed on top of this work to extend it.

The GeoSPARQL ontology [15], abiding by strictly to the Open Geospatial Consortium's standards, was shaped to represent and provide functionalities for objects possessing physical extent. In that aspect, as a "Feature" can be characterized anything from points in space to more intricate corporeal spatioformations such as polylines and acanonic, convex or non-convex, polygons. Noticeably, this supplementary property can be exploited to typify e-Tracer's concepts such as *routes, POIs, events* and furthermore the *weather*.

In conjunction with efforts towards the aforementioned localization, OWL-Time [16] instils the nature of temporality into the e-Tracer ontology. Periodic and sporadic chronic intervals can be analytically specified and represented, spanning from time-limited events to recurring schedules of points of particular interest, as well as unfolding iterative events such as weather phenomena. Consequently, the fusion of space and time capacitates a thorough designation for when an entity is shifting through those dimensions, facilitating perpetual and unremitting knowledge monitoring of its evolutionary existence.

Finally, we took into consideration previous work regarding the *User Profile* concept and collected very few content from [17] as described further below and extended it according to our needs with additional properties and relations with other entities.

Points of Interest, Events and Routes. The necessity for differentiation between the *POI* and the *Event* relies on the logical assumption which dictates that an *Event* might occur at a *POI*, coinciding spatially at the exact same longitude and latitude, thus being associated with it or at a place which is not regarded as a *POI* at all, also containing coordinates unclaimed by any *POI*. On the contrary, a *POI* might or might not ever hold an event, without being self-defined by it in any case. In terms of OWL 2 semantics, this is defined as:

$$Route \equiv Feature \sqcap \forall includes.(POI \sqcap \forall isAssociatedWith.Event)$$

Regarding temporal discrepancies, we take into consideration that a *POI* is permanently established, mayhaps withholding a somewhat fixed schedule, whereas an *Event* is strictly time-limited and even if periodicity is witnessed, each resumption will be considered a distinct instance. Apart from the dimensional perspective, a qualitatively extensive context-based research, based on local governmental resolutions[1,2,3,4,5], has been conducted so as to conclude to a class-based hierarchical categorization of eventual supported types of *POIs* and *Events* of interest.

Ultimately, the existence of an instance of the class *Route* solely depends on it incorporating either an instance of a *POI* or that of an *Event* at least, in a ranked manner which mandates the order of visit each *Route* suggests. *Routes* are being generated dynamically in a more compound integrated algorithm, which is out of the scope of this paper, due to the humongous population of potential options, thus not pre-processed or stored locally where the later occurs only after an end-user has completed successfully a proposed detour, on several occasions accompanied by an evaluation of his overall experience in the form of *User Feedback*.

User Profile, User Feedback and History of Usage. The *User Profile* construction encapsulates basic demographic information plus personal interests complemented by the end-user himself or potential impairments. Each instance is bound to possess a *History of Routes* and a *History of POIs* archiving each *Route*, and within, each *POI* the user has indeed paid a visit to, which is cross

---

[1] http://odysseus.culture.gr.
[2] http://listedmonuments.culture.gr.
[3] http://estia.minenv.gr/.
[4] http://www.minagric.gr.
[5] http://www.opengov.gr/.

validated via the end-user's smart device geolocalization. Furthermore, the sub-classes of *User Feedback* are related at a lower level with the historical assets of the *POI* and the *Route* of the ontology, rendering infeasible to submit a personal standardized evaluation about a spatial entity without a priori ratified attendance by the appropriate digital agent of the platform.

$$UserProfile \equiv \forall hasFeedback.UserFeedback \; \sqcap \; \forall hasRouteHistory$$
$$.(HistoryOfRoutes \sqcap \exists hasRoute.Route)$$

Augmented Reality and Weather Features. The *Augmented Reality Object* intends to enclose a profusion of essential properties and describe thoroughly in a semantic way 2D, 2.5D and 3D objects. Unfortunately, there is a scarcity in modelled digital assets to the applied interests of e-Tracer, nevertheless, due to the purposeful adaptability of the system beyond fixed use cases, it was deemed vital to patronize such features. Pragmatically, such an object might only be related to spatial entities, potentially deriving implicit inferences from them according to each scenario. Moreover, the *Weather* in its simplicity is considered to hold, apart from self-explanatory data properties, both temporal and spatial attributes, achieving to monitor evolutionary weather phenomena across regions pertaining *POIs* or *Events*, eventually served to the end-user as plain information or taken into consideration in rule-based reasoning upon nominating *routes*.

4 Inference and Validation

4.1 Implicit Relationships

Extra logical assumptions are the outcome of blending native OWL 2 RL reasoning and manually constructed custom rules, where the prior relies on the OWL 2 RL profile semantics [18]. Sadly, OWL 2 is limited as it serves modeling only for instances related in a tree-like approach [19]. Our framework implements domain rules on top of the standard graphs in order to enunciate richer relations by the utilization of CONSTRUCT graph motifs, thus enabling the identification of valid inferences. For example, an *Augmented Reality Object* instance never contains information about its geolocalization but in our ontology is always attached to a spatial entity, which in turns contains coordinates that can be bequeathed to it via the suitable SPARQL CONSTRUCT query shown below:

```
CONSTRUCT {
    ?arobject geo:asWKT ?coordinates.
} WHERE {
    ?arobject etr:relatedTo ?a2.
    ?a2 geo:hasGeometry ?a3.
    ?a3 geo:asWKT ?coordinates.
}
```

4.2   Consistency and Validation Check

The validation procedure guarantees the consistency of the framework along with the quality, both morphologic and syntactic. This scope is fulfilled through the usage of both custom SHACL [20] validation rules and native ontology consistency checking, always adhering to the closed-world paradigm. The latter manages validation by considering the semantics at TBOX, such as class disjointness, whereas the first discerns constraint contraventions like imperfect information or cardinality contradictions. For example, a SHACL shape representing a constraint which forces all *POIs* to contain exactly a single ID as a data property of type string is shown below:

```
etr:POIIDShape
    a sh:NodeShape;
    sh:targetClass etr:POI;
    sh:property [
        sh:path etr:hasID;
        sh:datatype xsd:string;
        sh:minCount 1;
        sh:maxCount 1;
    ].
```

5   Spatio-Temporal Rule-Based Reasoning

Entire concepts were imported from well-known ontologies and were combined so as to administer especial properties to specific instances, thus conferring additional capabilities on ruled-based reasoning in order to succeed in a significant diminish in the pool of recommended selections which flow to posterior in-chain services with ultimate objective to deliver delightful route suggestions to the end-user.

The very essence of reasoning in e-Tracer relies on the meaningful restriction of proffered choices. It has already been stated explicitly that each place of interest withholds formal standardized coordinates nearby a major traffic network. On top of those coordinates, functions and APIs based on [15] were developed and utilized so as to estimate euclidean distances between interchanges of the initial route and the places of interest, realistically serving as a lower distance bound where at best case the euclidean distance equals the actual driving distance. In addition, the fixed traffic network speed limits facilitated the development of an algorithm about approximate calculation of the time needed to arrive from one place to another, also serving as a lower estimation bound.

The algorithms described above were fused into dynamic hybrid rules expressed in complex SPARQL queries, combining time and space dimensions and an additional boolean variable of accessibility impairment to showcase the true potential of complex rule-based reasoning. Furthermore, it is not obligatory to set all limitations at once for the API to function, e.g. sometimes we only mind for distance and not at all for time or accessibility. Consequently, by applying limitations when retrieving places of interest correlated to each interchange within the main route of the end-user, the options stand fewer than before based on logic and necessarily satisfy either default constraints or constraints set by the end-user himself. A sample SPARQL pseudocode applying time and space constraints is provided in Algorithm 1 where if input variables are set to zero the algorithm does not consider that variable for filtering at all:

Algorithm 1: Spatio-temporal Reasoning SPARQL Pseudocode

Data: Interchange, POI_Coordinates, Interchange_Coordinates
Input   : DistanceOfTravel, TimeOfTravel $\in \mathbb{N}$
Output: A list of POIs
initialization;
foreach *Interchange* **do**
    GET each *POI_coordinates and the Interchange_coordinates;*
    foreach *POI_Coordinates* **do**
        if $DistanceOfTravel \neq 0$ **then**
            $X=euclidean\_distance($ *POI_Coordinates,*
            Interchange_Coordinates);
        end
        FILTER *(X ≤ DistanceOfTravel) ;*
        if $TimeOfTravel \neq 0$ **and** $DistanceOfTravel \neq 0$ **then**
            $Y = (X \times 60) \div 90000;$
        end
        FILTER *(Y ≤ TimeOfTravel) ;*
        if $TimeOfTravel \neq 0$ **and** $DistanceOfTravel = 0$ **then**
            $Z=euclidean\_distance(POI\_Coordinates,$
            $Interchange\_Coordinates) \times 60 \div 90000;$
        end
        FILTER *(Z ≤ TimeOfTravel);*
    end
end

---

6   Evaluation

Currently, a user-centered evaluation stands infeasible as the pilots are due to commence in subsequent months, ipso facto we focalized in system-wise benchmarking. In that aspect, we demonstrate the population of the stored entities, shown in detail in Table 1. The triple store at our disposal is a GraphDB 9.1.1 Free Edition with currently stored 15286 triples which was deployed at a server with Ubuntu 18.04.4 LTS (Bionic Beaver) 64-bit operating system, an Intel Xeon(R) Silver 4108 CPU @ 1.80 GHz × 32, 62.5 GB of RAM and a Hard Disk Drive of 3.6 TB capacity.

Table 1. The number of *POIs* and *Events* with (average) sum of properties for each, inside the knowledge base

#*POIs* & *Events*	#Properties	Avg. Properties per *POI*/*Event*
257	11627	≈45

Unfortunately, it is only anticipated to incline the evaluation towards the engineering response times in a manner where the bias is eliminated. All but one methods were manufactured as dynamic RESTful, thereby we ensured upon summoning that the variables on call conform to a uniformly distributed pseudogenerator with their range values spanning with equal probability of selection to all meaningful and valid content. All of them gratify the competency questions which were documented formerly of the creation of the e-Tracer ontology, a subset of which is showcased in Table 2, along with mean response times and standard deviations, elicited from 1000 executions for each.

Table 2. Exemplary competency questions

#	Question	Mean (SD) in msec
Q1	Retrieve all registered *POI* names with their respective IDs	27 ± 8
Q2	Retrieve all related properties to a pseudorandom *POI*	19 ± 10
Q3	Retrieve all *POI* names with their IDs registered to a pseudorandom interchange bound to pseudorandom time & distance constraints	22 ± 11
Q4	Retrieve all related properties to *POIs* registered to a pseudorandom interchange bound to pseudorandom time & distance constraints	114 ± 139
Q5	Retrieve all related properties from multiple *POIs* registered to pseudorandom multiple interchanges bound to pseudorandom time & distance constraints	3277 ± 4491

A simulation example of our approach is displayed in Fig. 3, while moving from point A to point B, where all *POIs* retrieved from the interchange without any reasoning occurring stand 43. On the contrary, it is conceived that the number of 5 *POIs* is noticeably less when the constraint of time is set to 30 min and that of the distance to 20 km.

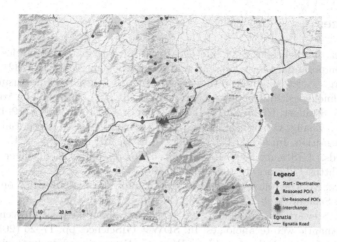

Fig. 3. Geographic map of simulation

7   Conclusion and Future Work

In this paper we presented an ontology-based framework for encapsulating and interlinking resources of cultural and natural nature towards the construction of suggestive enhanced routes. On top of the structured knowledge we practised rule-based reasoning based on spatial and temporal properties of the assets.

At the moment, the work featured is part of a synthesis of services, where dynamic routes are formed based only on the distinct unary level. Consequently, looking to the future, the next step is to exploit knowledge at a more aggregated level, such as applying reasoning at route level. Apart from reasoning, at final stages the resources of e-Tracer ought to be openly linked to other efforts, following the principles of *Open Data & Linked Data*.

The evaluation plan at a cultural level will be orchestrated by *Piraeus Bank Group Cultural Foundation* whereas at environmental level *Axios - Loudias - Aliakmonas Delta, Koronia-Volvi and Pamvotis lakes* protected area management bodies are responsible for. The platform encapsulates the *Egnatia Motorway* axis for pilots and content provided by the prior organizations. Finally, the assessment of the prototypes will be conducted during the pilot tests applied to 2 collaborating museums (*the silversmithing museum* and *the silk museum*) and at least 3 areas of environmental interest.

Acknowledgements. This work is co-financed by the European Union and Greek national funds via the Operational Program Competitiveness, Entrepreneurship and Innovation, under the call RESEARCH-CREATE-INNOVATE (project code: T1EΔK-00410).

References

1. Janowicz, K., Scheider, S., Pehle, T., Hart, G.: Geospatial semantics and linked spatiotemporal data-Past, present, and future. Semant. Web **3**(4), 321–332 (2012)
2. Homburg, T., et al.: Interpreting heterogeneous geospatial data using semantic web technologies. In: Gervasi, O., et al. (eds.) ICCSA 2016. LNCS, vol. 9788, pp. 240–255. Springer, Cham (2016). https://doi.org/10.1007/978-3-319-42111-7_19
3. Zhao, T., Zhang, C., Wei, M., Peng, Z.-R.: Ontology-based geospatial data query and integration. In: Cova, T.J., Miller, H.J., Beard, K., Frank, A.U., Goodchild, M.F. (eds.) GIScience 2008. LNCS, vol. 5266, pp. 370–392. Springer, Heidelberg (2008). https://doi.org/10.1007/978-3-540-87473-7_24
4. Janowicz, K.: The role of space and time for knowledge organization on the semantic web. Semant. Web **1**(1, 2), 25–32 (2010)
5. Fan, J., Stewart, K.: Modeling and reasoning about geospatial event dynamics using semantic web technologies. In: SDW@ GIScience, pp. 17–25 (2016)
6. Yu, F., McMeekin, D.A., Arnold, L., West, G.: Semantic web technologies automate geospatial data conflation: conflating points of interest data for emergency response services. In: Kiefer, P., Huang, H., Van de Weghe, N., Raubal, M. (eds.) LBS 2018. LNGC, pp. 111–131. Springer, Cham (2018). https://doi.org/10.1007/978-3-319-71470-7_6
7. Palumbo, R., Thompson, L., Thakur, G.: SONET: a semantic ontological network graph for managing points of interest data heterogeneity. In: Proceedings of the 3rd ACM SIGSPATIAL International Workshop on Geospatial Humanities, pp. 1–6, November 2019
8. Lodi, G., et al.: Semantic web for cultural heritage valorisation. In: Hai-Jew, S. (ed.) Data Analytics in Digital Humanities. MSA, pp. 3–37. Springer, Cham (2017). https://doi.org/10.1007/978-3-319-54499-1_1
9. Alonso, K., et al.: Ontology-based tourism for all recommender and information retrieval system for interactive community displays. In: 2012 8th International Conference on Information Science and Digital Content Technology (ICIDT 2012), vol. 3, pp. 650–655. IEEE, June 2012
10. Bahramian, Z., Abbaspour, R.A.: An ontology-based tourism recommender system based on spreading activation model. International Archives of the Photogrammetry, Remote Sensing & Spatial Information Sciences, 40 (2015)
11. Cao, T.D., Phan, T.H., Nguyen, A.D.: An ontology based approach to data representation and information search in smart tourist guide system. In: 2011 Third International Conference on Knowledge and Systems Engineering, pp. 171–175. IEEE, October 2011
12. Stathopoulos, E.A., et al.: Smart discovery of cultural and natural tourist routes. In: IEEE/WIC/ACM International Conference on Web Intelligence-Companion Volume, pp. 208–214, October 2019. https://doi.org/10.1145/3358695.3361105
13. Gkatziaki, V., Papadopoulos, S., Mills, R., Diplaris, S., Tsampoulatidis, I., Kompatsiaris, I.: easIE: easy-to-use information extraction for constructing CSR databases from the web. ACM Trans. Internet Technol. (TOIT) **18**(4), 1–21 (2018)
14. Shaw, R., Troncy, R., Hardman, L.: LODE: linking open descriptions of events. In: Gómez-Pérez, A., Yu, Y., Ding, Y. (eds.) ASWC 2009. LNCS, vol. 5926, pp. 153–167. Springer, Heidelberg (2009). https://doi.org/10.1007/978-3-642-10871-6_11
15. Battle, R., Kolas, D.: Geosparql: enabling a geospatial semantic web. Semant. Web J. **3**(4), 355–370 (2011)

16. Hobbs, J.R., Pan, F.: Time ontology in OWL. W3C working draft, 27, 133 (2006)
17. Maria, G., Akrivi, K., Costas, V., George, L., Constantin, H.: Creating an ontology for the user profile: method and applications. In: Proceedings AI* AI Workshop RCIS (2007)
18. Motik, B., Grau, B.C., Horrocks, I., Wu, Z., Fokoue, A., Lutz, C.: OWL 2 web ontology language profiles. W3C recommendation, 27, 61 (2009)
19. Motik, B., Cuenca Grau, B., Sattler, U.: Structured objects in OWL: representation and reasoning. In: Proceedings of the 17th International Conference on World Wide Web, pp. 555–564, April 2008
20. Knublauch, H., Kontokostas, D.: Shapes Constraint Language (SHACL), W3C Recommendation. World Wide Web Consortium (2017)

Ontological Foundations of Modelling Security Policies for Logical Analytics

Karolina Bataityte[1]([✉]), Vassil Vassilev[2], and Olivia Jo Gill[1]

[1] School of Computing, London Metropolitan University, London, UK
{k.bataityte,v.vassilev,o.gill}@londonmet.ac.uk
[2] Cyber Security Research Centre, London Metropolitan University, London, UK

Abstract. Modelling of knowledge and actions in AI has advanced over the years but it is still a challenging topic due to the infamous frame problem, the inadequate formalization and the lack of automation. Some problems in cyber security such as logical vulnerability, risk assessment, policy validation etc. still require formal approach. In this paper we present the foundations of a new formal framework to address these challenges. Our approach is based on three-level formalisation: ontological, logical and analytical levels. Here we are presenting the first two levels which allow to model the security policies and provide a practical solution to the frame problem by efficient utilization of parameters as side effects. Key concepts are the situations, actions, events and rules. Our framework has potential use for analysis of a wide range of transactional systems within the financial, commercial and business domains and further work will include analytical level where we can perform vulnerability analysis of the model.

Keywords: Security policies · Modelling · Ontologies · Knowledge representation · Situations and actions · Frame problem

1 Introduction

In recent years there has been an increase in the interest of analysing the logical vulnerability and the security policies of cyber systems. The security policies cover a wide range of situations: how to prevent unauthorized access to the information, secure the operations, control the transactions, neutralize malicious activities, etc. Any gaps or inconsistencies in the security policies can open the door for logical vulnerabilities and leave the system exposed [3]. Logical analysis of the vulnerability requires modelling of the online operations with sufficient background information to cover the security - user credentials and

This research is partially funded by Lloyds Banking Group in London, UK. However, no actual data from the bank has been used, the results and the opinions formulated in the paper are the author's and the examples are for illustration purpose only, without any resemblance to the actual banking policies and practices.

© IFIP International Federation for Information Processing 2020
Published by Springer Nature Switzerland AG 2020
I. Maglogiannis et al. (Eds.): AIAI 2020, IFIP AICT 583, pp. 368–380, 2020.
https://doi.org/10.1007/978-3-030-49161-1_31

profiles, needed for identification, authentication and authorisation, communication channels, physical connections and logical sessions for operations and transaction control, threat intelligence for security protection, etc. Our approach for addressing it is to represent the domain knowledge in the ontological model and to formulate the security policies as a system of rules, so that we can analyse them *formally*. For this purpose, we developed a theory of Situations and Actions in Description Logic (DL) and modelled the Security Policies in Clausal Logic (CL), which can be implemented using the standard languages of Semantic Web - Ontology Web Language (OWL) [5] and Semantic Web Rule Language (SWRL)[6]. We can model dynamic changes and synchronous actions with different security events asynchronously.

The paper is organized as follows. In Sect. 2 we will present the overall methodology which we follow. In Sect. 3 we will present logical foundations. In Sect. 4 we will introduce the ontological level. Section 5 will consider the security policies as rules on logical level. Section 6 we will conclude the paper and comment on the security policy analysis on Analytical Level.

2   Methodology

There are a number research projects being conducted that are developing ontological models for the different security purposes. They each use their own vocabulary, however, they use the same semantic web technologies (e.g., [4,9]). We separate the model of the world (*ontological level*) from the model of the policies which govern the changes in the world (*logical level*) and the model of the dynamic changes as a result of decisions (*analytical level*) (see Fig. 1). For each of the three levels we will use different formal systems, suitable for modelling of an aspect of the problem in a manner, similar to the infamous "layered cake" of the Semantic Web [8].

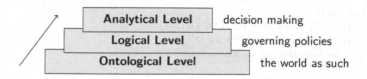

Fig. 1. Multi-level model for analysis

The Ontological Level models the world using the vocabulary presented in Sect. 4.1. The conceptualization is similar to the famous *situation calculus (Sit-Calc)* [7], but formulating it using the language of DL makes it more "object-oriented" and allows for a new solution of the *frame problem* [10]. Using DL on this level allows us to implement the model entirely using OWL.

The Logical Level models the policies, captures constraints and completeness. It reflects the expert knowledge in the domain, which can be formulated as logical rules in CL and can be represented in computer format using SWRL.

The Analytical Level will deal with the analysis of the policies on a directed graph, considering the situations as nodes and the actions as edges however it is beyond the scope of this paper and is left for the next publication.

3 Logical Foundations

For developing of the theory of situations and actions we consider DL called $\mathcal{ALC}$ [13] which is not the most expressive but is expressive enough to support our needs without being too complicated beyond the necessity. More constructors can be added to extend $\mathcal{ALC}$ if the modelling requires it. As we choose DL for comfortable implementation in OWL, similarly we choose CL as we can implement rules in SWRL. The following two logics can be glued together for modelling the domain ontology and the policies within that domain.

3.1 Description Logic $\mathcal{ALC}$ as a Modelling Language

The syntax and the semantic interpretation is shown in Table 1. The interpretation I is a pair $I = (\triangle^I, \cdot^I)$, where $\triangle^I$ is a non-empty set (domain) and $\cdot^I$ is a mapping function [12].

Table 1. Syntax and semantics

Concepts		Roles	
Syntax	Semantics	Syntax	Semantics
$\top$	$\triangle^I$	R	$R^I \subseteq \triangle^I \times \triangle^I$
$\bot$	$\emptyset$	$Domain(R,C)$	$<a,b> \in R^I \rightarrow a \in C^I$
A	$A^I \subseteq \triangle^I$	$Range(R,C)$	$<a,b> \in R^I \rightarrow b \in C^I$
$\neg C$	$\triangle^I \backslash C^I$		
$C \sqcap D$	$C^I \cap D^I$		
$C \sqcup D$	$C^I \cup D^I$		
$\forall R.C$	$\{a \in \triangle^I | \forall b.(<a,b> \in R^I \rightarrow b \in C^I)\}$		
$\exists R.C$	$\{a \in \triangle^I | \exists b.(<a,b> \in R^I \wedge b \in C^I)\}$		

where C, D are concepts, A is an atomic concept, R is a role.

Given interpretation I in model M with axiom α, we say that M is a model of α under I if M satisfies α, written $I \models \alpha$. We will be expressing the domain restrictions as $\exists R.\top \sqsubseteq C$ and the range restrictions as $\top \sqsubseteq \forall R.C$ [13]. By adding domain and range axioms we are able to have a fixed structure of the real world we are modelling without the necessity to use more expressive language or nonstandard semantics.

3.2   Clausal Logic and SWRL

In most logical languages it is possible to formulate rules, which are necessary for modelling structural constraints and dynamic changes. We have chosen a version of the first order clausal logic similar to the horn-clause predicate logic because its serialized version SWRL refers directly to the terms of OWL.

SWRL Knowledge Base (K) is defined as follows: $K = (\Sigma, R)$ where Σ is KB of $\mathcal{ALC}$ and R is set of rules. The rule is composed of *body* and *head* which is represented as following: *body* $\rightarrow$ *head*. It consists of a conjunctions of atoms which are classes $C(i)$ (concepts in $\mathcal{ALC}$) and object properties $R(i, j)$ (roles in $\mathcal{ALC}$) [6].

4   Ontological Level: The Domain Model

The term *ontology* in a narrow logical sense provides the *terminology*, which can be used for building the domain model, together with its *interpretation* in the *semantic domain* [11]. The cyber security operations require accounting of both static and dynamic semantic considerations, in order to have an adequate and semantically rich ontology for the analysis.

4.1   Terminological Vocabulary

In our ontology the semantic domain, $\triangle$, is a non-empty set, split into three disjoint subdomains. **Entities, Events** and **Situations** (plural) as $\triangle_{Entities}$, $\triangle_{Events}$ and $\triangle_{Situations}$ respectively. In our theory we will use three terms with predefined meaning: *Entity, Event* and *Situation* (singular), which will be three separate taxonomies representing the static model of the world. The interpretation of $\mathcal{ALC}$ concepts in the domain are as follows: $Entity^I \subseteq \triangle_{Entities}^I$, $Event^I \subseteq \triangle_{Events}^I$ and $Situation^I \subseteq \triangle_{Situations}^I$. Our terminology (Table 2) will also include some predefined roles, one of them is *Action* ($Action^I \subseteq \triangle_{Situations}^I \times \triangle_{Situations}^I$), which can be used as a top of the hierarchy of actions. The ontology can have as many specific *named concepts* and *named roles* as needed, (noted as $Entity_x$, $Situation_y$, $Event_e$, $Action_z$), with the intended meaning and interpretations in the semantic subdomains introduced above in accordance with the syntax and semantics of $\mathcal{ALC}$ as presented in Sect. 3.1. Concepts from three subdomains must be disjoint as follows:

$$Situation \sqcap Event \sqsubseteq \bot, Situation \sqcap Entity \sqsubseteq \bot, Entity \sqcap Event \sqsubseteq \bot. \quad (1)$$

On the ontological level we are using the $\mathcal{ALC}$ TBox for formulating the terminological axioms and the RBox for the relational axioms, while the ABox will incorporate the assertions later on.

Table 2. Vocabulary of the domain ontology

Term	DL category	Use in modelling	Condition
Situation	Concept	Partial static description of the world	axiom 1
Event	Concept	Asynchronous activity	axiom 1
Entity	Concept	Qualitative descriptor	axiom 1
Action	Role	Synchronous activity	axiom 2
occur-in	Role	Event occurrence	axiom 4
present-at	Role	Situation description	axiom 6
part-of	Role	Event description	axiom 5
describe	Role	Describing entities and specifying dependencies	axiom 7
chain	Role	Connecting events causally	axiom 3

4.2   Static Modelling of the World

Here we are defining a fix static structure of the modelling world using terms above. A *Situation* is a concept, which represents a partial description of the world in a specific moment of time. Two *Situation* concepts can be connected via *Action* roles to model the potential change:

$$\exists Action.\top \sqsubseteq Situation, \top \sqsubseteq \forall Action.Situation \tag{2}$$

The events are asynchronous activities which are modelled using *Event* concepts, linked through the predefined role *chain* in a causal chain (axiom 3). The intended meaning of *Event* is to represent a real-world events which can occur in the situations through the predefined role *occur-in* with domain *Event* and range *Situation* (axiom 4). This way we can formulate security policies with regard to planned and unexpected activities (events), which may or may not happen in the situations.

$$\exists chain.\top \sqsubseteq Event, \top \sqsubseteq \forall chain.Event \tag{3}$$

$$\exists occur-in.\top \sqsubseteq Event, \top \sqsubseteq \forall occur-in.Situation \tag{4}$$

The *Entity* concepts are used to describe situations and events using the predifined roles from the vocabulary: *part-of* with domain *Entity* and range *Event* (axiom 5); *present-at* with domain *Entity* and range *Situation* (axiom 6); *describe* with domain *Entity* and range *Entity* (axiom 7).

$$\exists part-of.\top \sqsubseteq Entity, \top \sqsubseteq \forall part-of.Event \tag{5}$$

$$\exists present-at.\top \sqsubseteq Entity, \top \sqsubseteq \forall present-at.Situation \tag{6}$$

$$\exists describe.\top \sqsubseteq Entity, \top \sqsubseteq \forall describe.Entity \tag{7}$$

It is important to note that the events do not change the situations in our theory, they can only occur in them; the changes can be caused only by actions. So that events are described as asynchronous activities while actions are purely synchronous activities.

4.3   World Dynamics

In state-based dynamic theories which uses DL, the actions are represented as ⟨*precondition, occlusion, post-condition*⟩ triplets [1,2]. Unfortunately, there is no easy implementation of such a formalism since it has additional syntactic structure.

We have adopted the view that the dynamic changes are possible only through actions, similar to the original SitCalc from the early days of AI [10]. This logic formalism encounters the infamous frame problem, caused by the propositional treatment of the situations which require them to incorporate their parameters as arguments.

However, in our approach the definition of the actions (as relations between the situations) looks almost identical to SitCalc approach. The partitioning of our ontology has interesting and unexpected characteristics with practical importance for applications. We define the parameters of the actions contextually. In our approach the actions can change the situations only through their parameters, which are entities, but the action parameters are no longer attributed to the actions – they are attributed to the situations which the actions relate instead. This completely eliminates the need for heavy "frame axioms" because the complete absence of any "side effect" of the actions.

If we have TBox T with situations and entities as follows:

$$T := \{\text{Entity}_x \sqsubseteq Entity, \text{Situation}_y \sqsubseteq Situation\} \tag{8}$$

and Entity$_x$ describe Situation$_y$, T is extended as follows:

$$T' := T \cup \{\text{Entity}_x \sqsubseteq \exists present\text{-}at.\text{Situation}_y\}. \tag{9}$$

Example 1. Let's consider the situation *LoggedIn* and the entity *User*. For this scenario the TBox T is as follows:

$$T := \{User \sqsubseteq Entity, LoggedIn \sqsubseteq Situation, User \sqsubseteq \exists present\text{-}at.LoggedIn\}$$

Each situation can be described by a number of entities. Since the actions change the situations, they will affect these entities but not directly. So, we can consider the entities which describe all situations in which a given action applies as its *input parameters* and similarly, entities which describe the situations to which the action leads as its *output parameters*. NB: not all entities are input and/or output parameters, some of them just describe the situation without being needed for an action. To specify the parameters of all actions, we can create a GBox G as follows:

Definition 1. *A GBox* $G = \{\langle \text{Entity}_x, \text{Action}_z\rangle, \langle \text{Action}_z, \text{Entity}_y\rangle\}$ *is a set of pairs of actions and entities, representing the action parameters where pair* ⟨Entity$_x$, Action$_z$⟩ *is for input parameters and pair* ⟨Action$_z$, Entity$_y$⟩ *is for output parameters.*

The action parameters will be important on the Analytical Level since the input parameters are binding the actions, making them executable, while the output parameters are producing the effect, determining the changes in the situations.

In order for an entity to be an input parameter, it must meet the following conditions:

1. $\exists \text{Action}_z.\top \sqsubseteq \text{Situation}_x$,
2. Entity$_e \sqsubseteq \exists present\text{-}at.\text{Situation}_x$.

If both conditions hold, we can say GBox $G = \{\langle \text{Entity}_e, \text{Action}_z \rangle\}$. It can be formalized as the following axiom:

$$\text{Entity}_e \sqsubseteq \exists present\text{-}at.(\text{Situation}_x \sqcap \exists \text{Action}_z.\top) \tag{10}$$

which says that Entity_e is connected to a Situation_x via *present-at* and there is an Action_z starting at Situation_x and leading to another unknown *Situation*. This gives us the first criteria for analysing the descriptive completeness of the security policies with respect to the possibility of binding the input parameters of the applicable actions to the descriptions of the situations in which they apply.

In order for an entity to be an output parameter, it must meet the following conditions:

1. $\top \sqsubseteq \forall \text{Action}_z.\text{Situation}_y$,
2. $\text{Entity}_e \sqsubseteq \exists present\text{-}at.\text{Situation}_y$.

If both conditions hold, we can say GBox $G = \{\langle \text{Action}_z, \text{Entity}_e \rangle\}$. It can be formalized as follows:

$$\text{Entity}_e \sqsubseteq \exists present\text{-}at.\exists \text{Action}_z.\text{Situation}_y \tag{11}$$

which says that Entity_e describes Situation_y via *present-at* and Action_z leads to Situation_y after it executes.

Example 2. In Fig. 2 we have a scenario which starts in situation Situation_1 and finishes in Situation_3 after executing Action_1 and Action_2. The two actions have parameters amongst the entities which are present in the corresponding situations. In this case $G = \{\langle \text{Entity}_2, \text{Action}_1 \rangle, \langle \text{Action}_1, \text{Entity}_3 \rangle, \langle \text{Action}_1, \text{Entity}_4 \rangle, \langle \text{Entity}_4, \text{Action}_2 \rangle\}$. Amongst the parameters Entity_4 is both input and output parameter of Action_2. Entity_1 and Entity_5 simply describe the situations without being needed for actions.

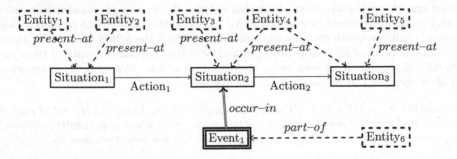

Fig. 2. A graphical representation of two-step journey

The ontological considerations we have presented so far can be constructed in any variation of DL. Since such a theory can be serialized directly in OWL, the process of developing the ontology can be done entirely interactively using any standard ontology editor, such as **Protégé**.

5   Logical Level: Constraints, Dependencies and Domain Policies

In order to describe the logical characteristics of the model, as well as to represent adequately the domain policies controlling the execution of the actions, we can use axioms, rules of inference and heuristic rules. Although DL and CL, as theoretical base of our framework, have well-defined inference mechanisms for practical purposes, it is more convenient to work with derived inference rules rather than the rules of inference within the underlying logic. In this section we will discuss some derived rules of our framework which allow us to automate this process.

5.1   Parameter Binding and Entity Completion

To make sure that our KB is descriptively complete, we need to guarantee that it contains all needed information in the TBox (the ontology model) to match the SWRL rules (the policies) so that the policy rules which prescribe actions actually lead to executable actions. In practice this means that all parameters of the actions in the head of the rules must be bound to the situations in which the rules apply. This can be implemented using an algorithm which uses the ontology in the TBox to check if the parameters of the actions prescribed by the rules are defined.

The following derived rule captures the parameters of various events in the situations to prevent the loss of bindings. It is used to implement a "reasoner" which performs a secondary logical inference according to the following schema:

$$\frac{Entity \sqsubseteq \exists part{-}of.Event}{Event \sqsubseteq \exists occur{-}in.Situation}$$
$$\therefore Entity \sqsubseteq \exists present{-}at.Situation$$

Derived Inference Rule 1 (Entity Triangulation). Let the following TBox T be given:

$$T := \{Entity \sqsubseteq \exists part{-}of.Event, \qquad (12a)$$

$$Event \sqsubseteq \exists occur\ in.Situation\} \qquad (12b)$$

Then the following holds:

$$T' := T \cup \{Entity \sqsubseteq \exists present{-}at.Situation\}. \qquad (13)$$

Proof. The TBox T holds since it states the domain and range of $part{-}of$ (12a) and $occur{-}in$ (12b) roles which satisfy the axioms 5 and 4 respectively. The same concept $Event$ is used as range of $part{-}of$ (12a) and as a domain of $occur{-}in$ (12b). Therefore, we can substitute $Event$ in 12a by the right-hand side of 12b to derive $Entity \sqsubseteq \exists part{-}of.\exists occur{-}in.Situation$. As we can see, $Entity$ is connected to $Situation$ via two roles. We know from Sect. 4.2, this can be done via $present{-}at$ (axiom 6), therefore, it can be expressed as $Entity \sqsubseteq \exists present{-}at.Situation$ (13).

□

5.2   Transitivity of the Roles and Entity Propagation

The next derived rule reflects the abstract "transitivity" of the logical descriptions within one and the same situation. It can be accounted by another "reasoner" which performs secondary inference according to the following schemas against concept *Situation* or *Event*:

$$\frac{\text{Entity}_y \sqsubseteq \exists describe.\text{Entity}_x \qquad \text{Entity}_y \sqsubseteq \exists describe.\text{Entity}_x}{\text{Entity}_x \sqsubseteq \exists present\text{-}at.\text{Situation}_x \qquad \text{Entity}_x \sqsubseteq \exists present\text{-}at.\text{Event}_z}$$
$$\therefore \text{Entity}_y \sqsubseteq \exists present\text{-}at.\text{Situation}_x \quad \therefore \text{Entity}_y \sqsubseteq \exists present\text{-}at.\text{Event}_z$$

Derived Inference Rule 2 (Entity Transitivity). Let the following TBox T be given:

$$T := \{\text{Entity}_y \sqsubseteq \exists describe.\text{Entity}_x, \tag{14a}$$

$$\text{Entity}_x \sqsubseteq \exists present\text{-}at.\text{Situation}_x\} \tag{14b}$$

Then the following holds:

$$T' := T \cup \{\text{Entity}_y \sqsubseteq \exists present\text{-}at.\text{Situation}_x\}. \tag{15}$$

Proof. The TBox T holds since it states the domain and range of *describe* (14a) and *present-at* (14b) roles which satisfy the axioms 7 and 6 respectively. The same concept Entity$_x$ is used as range of *describe* (14a) and domain of *present-at* (14b). Therefore, we can substitute Entity$_x$ in 14a by the right-hand side of 14b to derive Entity$_y$ $\sqsubseteq$ $\exists describe.\exists present\text{-}at.Situation_x$. As we can see, Entity$_y$ is connected to Situation$_x$ via two roles. Therefore Entity$_y$ is connected to Situation$_x$ and we can simply rewrite it as Entity$_y$ $\sqsubseteq$ $\exists present\text{-}at$.Situation$_x$ (15).   □

5.3   Conceptual Taxonomies and Entity Inheritance

Although the DL allows to automate the subsumption of concepts, we can extend our framework with additional inheritance mechanisms to allow full "parameter inheritance" in the style of object-oriented programming. This is possible because the entities, which are connected to situations or to events, are like the class attributes in object-oriented parlance. It is relatively straightforward to construct algorithmic reasoners which tackle more complex inheritance of entities, along the taxonomic hierarchies of situations and events.

$$\frac{\text{Situation}_y \sqsubseteq \exists \text{Situation}_x \qquad \text{Event}_y \sqsubseteq \exists \text{Event}_x}{\text{Entity}_x \sqsubseteq \exists present\text{-}at.\text{Situation}_x \qquad \text{Entity}_x \sqsubseteq \exists present\text{-}at.\text{Event}_x}$$
$$\therefore \text{Entity}_x \sqsubseteq \exists present\text{-}at.\text{Situation}_y \quad \therefore \text{Entity}_x \sqsubseteq \exists present\text{-}at.\text{Situation}_y$$

Derived Inference Rule 3 (Entity Inheritance). Let the following TBox T be given:

$$T := \{\text{Situation}_y \sqsubseteq \text{Situation}_x, \tag{16a}$$

$$\text{Entity}_x \sqsubseteq \exists present\text{-}at.\text{Situation}_x\} \tag{16b}$$

Then the following holds:

$$T' := T \cup \{\text{Entity}_x \sqsubseteq \exists present\text{-}at.\text{Situation}_y\}. \tag{17}$$

Proof. The TBox T holds since it states that Situation$_y$ is a sub-concept of Situation$_x$ (16a) and Entity$_x$ is related to Situation$_x$ via *present-at* (16b). Therefore, Entity$_x$ is also related to sub-concept of Situation$_x$, which is Situation$_y$ (17).   □

5.4   Frame Problem

In our dynamical model the situations change as a result of the actions. The only way the change from one situation to another situation can affect the descriptions (entities) of the latter situation, is through the output parameters of the actions causing the transition. The specific changes caused by the actions must be specified by the corresponding rules of the security policy. The following two principles allow us to avoid the frame problem by formulating rules according to those principles. They will also guide the changes on the Logical Level.

Principle of Preservation: Any description of the situations within the domain of the action in terms of input parameters remains unchanged.

Principle of Propagation: Any description of the situations within the range of the action in terms of output parameters may change as a result of the action.

5.5   Policy Rules

The policies on the Logical Level are rules which link the concepts and roles from the Ontological Level. Such rules have clausal form and can be represented as SWRL expressions (Sect. 3.2). This makes possible the use of the ontological editors like **Protégé** for modelling of the policies as well.

The policy rules can be modelled as SWRL rules using different templates which combine *Situation*, *Event*, *Entity* and *Action* atoms in the body and the head of the rule to serve different purposes - for analysis of the situations, making decisions for continuation of the journey, or responding to events. Two such templates are shown below, which can be finely tuned to the particular need of the analysis.

1. $situation_i(?sa) \wedge entity_k(?ia) \wedge present{-}at(?ia, ?sa) \wedge ... \wedge action_n(?sa, ?sb) \rightarrow$ $situation_j(?sb) \wedge entity_l(?ib) \wedge present{-}at(?ib, ?sb) \wedge ...$
2. $situation_i(?sa) \wedge entity_k(?ia) \wedge present{-}at(?ia, ?sa) \wedge event_l(?ea) \wedge occur{-}in(?ea,$ $?sa) \wedge ... \wedge entity_k(?ib) \wedge part{-}of(?ib, ?ea) \wedge ... \wedge action_n(?sa, ?sb) \rightarrow situation_j(?sb) \wedge$ $entity_k(?ic) \wedge present{-}at(?ic, ?sb)...$

In the templates above $situation_i(?s)$, $entity_k(?en)$, $event_l(?ev)$ are SWRL classes (which correspond to $\mathcal{ALC}$ concepts), $action_n(?sa, ?sb)$ is an SWRL object property (which corresponds to $\mathcal{ALC}$ role) and they have to be adopted to the specific scenario. Other classes/concepts and object properties/roles do not have to be adopted to the scenario and can be used as it is ($present{-}at(?ib, ?sb)$, $occur{-}in(?ea, ?sa)$, etc.).

5.6   Detailed Example

In this section we will present a more detailed example of the use of our framework for the analysis of a typical online banking transaction. The fragment was built in Protégé 5.1.0 with FaCT++ 1.6.5 reasoner. WebVOWL 1.1.7 was used for visualization of Fig. 3 and Fig. 4 was created using a drawing tool since software to generate these graphs is still in the development stage. Some specifications such as TBox, RBox and some of the named concepts are omitted for the sake of clarity and brevity. The purpose of this example is to illustrate our framework as well as show the interpretation and understanding of it.

Lets consider the case when transaction is requested: we start in the initial situation *S_TransactionRequested*; then there are three possible events which may or may not happen: *E_AccountIn Overdraft*, *E_MaxOverdraftReached*, *E_AccountOverloaded*. Reaching the final situation will depend on the policy rules expressed in SWRL. Figure 3 shows the interpretation fragment of ontological vocabulary from Ontological Level where yellow arrows represents some of the derived inference rules from Logical Level. Figure 4 visualises SWRL rules on the Logical Level which will be used on the Analytical Level for the analysis. Some of the SWRL rules are as follows:

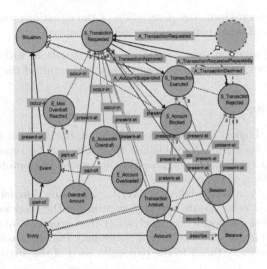

Fig. 3. Example visualization of ontological and logical levels

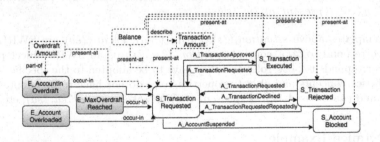

Fig. 4. Example visualization of logical and analytical levels

- $S\_TransactionRequested(?sa) \wedge Balance(?ix) \wedge present-at(?ix, ?sa) \wedge$
  $TransactionAmount(?it) \wedge present-at(?it, ?sa) \wedge E\_AccountInOverdraft(?ea) \wedge$
  $occur - in(?ea, ?sa) \wedge OverdraftAmount(?io) \wedge part-of(?io, ?ea) \wedge$
  $present-at(?io, ?sa) \wedge A\_TransactionApproved(?sa, ?sb) \rightarrow Balance(?iy) \wedge$
  $S\_TransactionExecuted(?sb) \wedge present-at(?iy, ?sb)$

- $S\_TransactionRequested(?sa) \wedge Balance(?ix) \wedge TransactionAmount(?it) \wedge$
  $present\text{-}at(?ix, ?sa) \wedge present\text{-}at(?it, ?sa) \wedge E\_MaxOverdraftReached(?eb) \wedge$
  $occur\text{-}in(?eb, ?sa) \wedge A\_TransactionDeclined(?sa, ?sb) \rightarrow$
  $S\_TransactionRejected(?sb)$
- $S\_TransactionRequested(?sa) \wedge Balance(?ix) \wedge present\text{-}at(?ib, ?sa) \wedge$
  $TransactionAmount(?it) \wedge present\text{-}at(?it, ?sa) \wedge E\_AccountOverloaded(?eb) \wedge$
  $occur\text{-}in(?eb, ?sa) \wedge A\_AccountSuspended(?sa, ?sb) \rightarrow S\_AccountBlocked(?sb)$

Although some of the rules may look too complex, most of the literals in it are type checking conditions which can be eliminated from the formulation by adopting a separate type checking algorithm.

6 Conclusion and Further Work

In this paper we presented ontological and logical considerations of knowledge representation for security analysis of the cyber systems operating in a workflow manner. As well as the processing of transactions in dynamic systems, which involve synchronous and asynchronous activities such as events and actions have been described. We outlined a multi-level framework for representing the ontology and modelling the security policies which enables analysing of some logical problems such as vulnerability analysis and risk assessment. It is entirely based on the use of standard modelling languages of the Semantic Web, which greatly simplifies the implementation, makes it transparent and efficient. Our framework provides a theoretical basis for solving some of the hard problems in modelling dynamic behaviour such as the infamous frame problem. We utilize the concept of state, to provide a proper distinction between the static characteristics of the situations and the possible side effect of the actions on them. We have a pilot implementation, of the framework, written in Java, which makes use of the APIs for OWL and SWRL available in Jena for processing the ontological representation and the security policies in symbolic form [14]. It allows us to perform various logical analytics related to logical vulnerability, risk assessment and policy validation. We are currently use this framework to cross-channel transaction processing, in digital banking, for preventing social engineering fraud.

The semantic and logical considerations discussed above provide the formal ground for formalizing the concepts of accessibility, logical vulnerability and risks. Within our framework this can be done by simulating different scenarios for execution of the actions, under the conditions imposed on the situations and with possibility for events happening in them. Although such an analysis is beyond the scope of this paper, the experiments we conducted using our prototype implementation, have demonstrated that this approach is both transparent and convenient to be used for practical purposes [14].

Currently, we are working on an extension of the framework (to equip it) with risk analysis capabilities, based on the naive Bayesian theory. We are also exploring the potential use of the same framework in other areas, related to workflow control such as production line fault recovery and safety management, like evacuation in the event of fire or other disastrous situations.

References

1. Baader, F., Lutz, C., Miličic, M., Sattler, U., Wolter, F.: Integrating description logics and action formalisms: First results. In: Proceedings of the 20th National Conference on Artificial Intelligence, AAAI 2005, pp. 572–577. AAAI Press (2005)
2. Chang, L., Lin, F., Shi, Z.: A dynamic description logic for representation and reasoning about actions. In: KSEM (2007)
3. Jizba, E., Chen, Y., Sun, F., Pellegrino, G.: Web logic vulnerability. https://users.cs.northwestern.edu/~ychen/classes/cs450-s14/lectures/Web. Accessed January 2020
4. Granadillo, G., Ben Mustapha, Y., Hachem, N., Debar, H.: An ontology-driven approach to model siem information and operations using the SWRL formalism. Int. J. Electron. Secur. Digit. Forensics **4**, 104–123 (2012)
5. Hitzler, P., Krötzsch, M., Rudolph, S.: Foundations of Semantic Web Technologies. Chapman & Hall/CRC, London (2009)
6. Lawan, A., Rakib, A.: The semantic web rule language expressiveness extensions-a survey, March 2019
7. McCarthy, J., Hayes, P.: Some philisophical problems from the standpoint of artificial intelligence. In: Machine Intelligence, vol. 4, pp. 463–502. Edinburgh University Press, Edinburgh (1969)
8. Passin, T.B.: The Explorer's Guide to the Semantic Web. Manning Publications, Shelter Island (2004)
9. Pereira, T., Santos, H.: An ontology based approach to information security. In: Sartori, F., Sicilia, M.Á., Manouselis, N. (eds.) MTSR 2009. CCIS, vol. 46, pp. 183–192. Springer, Heidelberg (2009). https://doi.org/10.1007/978-3-642-04590-5_17
10. Reiter, R.: Knowledge in Action: Logical Foundations for Specifying and Implementing Dynamical Systems. MIT Press, Cambridge (2001)
11. Sånchez, D., Cavero, J.M., Marcos MartÁnez, E.: The road toward ontologies. In: Sharman, R., Kishore, R., Ramesh, R. (eds.) Ontologies. Integrated Series in Information Systems, vol. 14, pp. 3–20. Springer, Boston (2007). https://doi.org/10.1007/978-0-387-37022-4_1
12. Szeredi, P., Lukácsy, G., Benkő, T.: The Semantic Web Explained: The Technology and Mathematics Behind Web 3.0. Cambridge University Press, New York (2014)
13. Tsarkov, D., Horrocks, I.: Efficient reasoning with range and domain constraints. In: Proceedings of the 2004 International Workshop on Description Logics (DL2004) (2004)
14. Vassilev, V., Sowinski-Mydlarz, V., Gasiorowski, P., et al.: Intelligence graphs for threat intelligence and security policy validation of cyber systems. In: Bansal, P., Tushir, M., Balas, V., Srivastava, R. (eds.) Proceedings of International Conference on Artificial Intelligence and Applications, ICAIA 2020, Janakpuri, India. Springer (2020, in print)

RDF Reasoning on Large Ontologies: A Study on Cultural Heritage and Wikidata

Nuno Freire[1]([✉]) [iD] and Diogo Proença[1,2] [iD]

[1] INESC-ID, Rua Alves Redol 9, 1000-029 Lisbon, Portugal
{nuno.freire,diogo.proenca}@tecnico.ulisboa.pt
[2] IST, Universidade de Lisboa, Avenida Rovisco Pais, 2, 1049-001 Lisbon, Portugal

Abstract. Large ontologies are available as linked data, and they are used across many domains, but to process them considerable resources are required. RDF provides automation possibilities for semantic interpretation, which can lower the effort. We address the usage of RDF reasoning in large ontologies, and we test approaches for solving reasoning problems, having in mind use cases of low availability of computational resources. In our experiment, we designed and evaluated a method based on a reasoning problem of inferring Schema.org statements from cultural objects described in Wikidata. The method defines two intermediate tasks that reduce the volume of data used during the execution of the RDF reasoner, resulting in an efficient execution taking on average 10.3 ± 7.6 ms per RDF resource. The inferences obtained in the Wikidata test were analysed and found to be correct, and the computational resource requirements for reasoning were significantly reduced. Schema.org inference resulted in at least one *rdf:type* statement for each cultural resource, but the inference of Schema.org predicates was below expectations. Our experiment on cultural data has shown that Wikidata contains alignment statements to other ontologies used in the cultural domain, which with the application of RDF and OWL reasoning can be used to infer views of Wikidata expressed in cultural domain's data models.

Keywords: Data volume · Reasoning · Wikidata · Schema.org · Semantic web

1 Introduction

Nowadays, large ontologies are available as linked data and with open licenses that allow for their reuse in a wide variety of applications across all domains of knowledge. Some examples are DBpedia[1] and Wikidata[2]. The usefulness of these ontologies is clearly acknowledged in many domains, but due to their high data volume, their reuse requires the commitment of considerable human resources for acquiring the knowledge about the ontologies' data models and for the development of the information systems for their processing.

---

[1] https://wiki.dbpedia.org/.
[2] https://www.wikidata.org/wiki/Wikidata:Main_Page.

I. Maglogiannis et al. (Eds.): AIAI 2020, IFIP AICT 583, pp. 381–393, 2020.
https://doi.org/10.1007/978-3-030-49161-1_32

The Resource Description Framework (RDF) provides possibilities for automation of data processing and semantic interpretation that can lower the effort for reusing large ontologies and for the development of general-purpose data processing tools. In our work, we address the usage of RDF reasoning for automated discovery of new facts about the ontologies' data.

RDF reasoning processes are based on a set of rules. Both the RDF Schema (RDF(S)) and Web Ontology Language (OWL) specifications include entailment rules to derive new statements from known ones. RDF(S) entailments are included in its semantics specification [1]. OWL has specifications for its RDF-based semantics [2] and its direct semantics [3], which include an extensive set of entailment rules.

In cultural heritage, the domain where we conduct our research, semantic data is highly valued and applied for descriptions of cultural objects. Reasoning is often mentioned and implemented but is mostly put into practice with pragmatic implementations that use ad-hoc data processing and querying of triple stores or SPARQL endpoints.

Our research focuses on defining a method for reasoning on large ontologies that can be systematically applied to varied reasoning contexts. We studied the problem using Wikidata as the target ontology for reasoning, and a reasoning problem with potential application in culture domain. Our approach aims to be lightweight, due to the lower capacity of information technologies employed by most cultural institutions, in contrast to other domains.

Our work provides three main scientific contributions:

- It identifies some limiting computational aspects of applying RDF reasoning to large volumes of data;
- It defines, tests and evaluates a method for RDF reasoning in large ontologies;
- It provides observations and evidence of Wikidata's potential to provide alignments from its data modes to other ontologies which can, with the use of RDF and OWL reasoning, be used to infer views of Wikidata expressed in cultural domain's ontologies.

We follow, in Sect. 2, by describing related work on linked data and reasoning, and also the research on reasoning in the cultural domain. Section 3 presents our proposed method for reasoning in large ontologies. The setup for the evaluation on Wikidata of our method is presented in Sect. 4. Section 5 presents the results from the evaluations and their analysis. Section 5 finalizes by summarizing the method, highlights the conclusions of the study and describes future work.

2 Related Work

Linked data has a large diversity of research topics related to our work. Scalability is one of the most addressed topics, with many facets such as indexing, federated querying, aggregation and reasoning. The reuse of published linked data by third parties has revealed data quality to be a challenge as well, both at the level of semantics and at the level of syntax [4–6]. Reuse of linked data is one of the concerns of our work, in which data quality is a relevant aspect. In this area, significant work has been done to facilitate the reuse of linked data by aggregation and data cleaning [7, 8].

Reasoning on linked data is also an active research topic. A comprehensive analysis and description of techniques has been published in [9]. Regarding the particular aspect of scalable reasoning, the related work employs techniques based on high computational capacity [10–13], which are beyond the capacity of most cultural institutions. The application of reasoning in large volumes of RDF data was addressed by [14], but with a different target use case – data streams, which is a problem with characteristics that differ from those of reasoning in large ontologies.

Regarding cultural heritage, although the use of linked data has been the focus of research, most of the published work addresses mainly the aspect of the publication of linked data [15–17]. Large ontologies have been published and are maintained by cultural organizations, mostly by national libraries that have built and maintained these ontologies to support the information needs from bibliographic data. This ontology development has been a long-term practice, and started much earlier than when the semantic web emerged. However, the scalability of the application of RDF(S) or OWL reasoning on cultural ontologies has not been studied.

3 Method

We are addressing reasoning problems where an ontology is used to make inferences about a target dataset. To cope with the large amount of data for reasoning, we tested the reduction of data volume used for reasoning (we will mention in Sect. 4 the approaches we attempted but were unable to run).

Subsection 3.1 describes the general process of our approach, including the main tasks, software components and data flow. Subsection 3.2presents how we applied and evaluated the process in an experiment on Wikidata.

3.1 The Process

Figure 1 shows an overview of the process followed in our method. It consists of the following tasks:

1. A domain expert defines the reasoning problem. The expert makes the following specifications:

    1.1. Specification of the triple patterns required from the ontology for the target reasoning ruleset. This specification allows the RDF reasoner to reduce the number of statements used during the reasoning process;

    1.2. Specification of the SPARQL endpoint(s) of the ontology(ies). The endpoints are used to collect the triple patterns specified in the previous item;

    1.3. Specification of the RDF reasoning ruleset to be applied;

    1.4. Define partitions of the target dataset into sub datasets where the reasoning problem can be applied independently from the rest of the dataset. This specification allows the execution of the RDF reasoner to be done on less data by executing it independently on each of the dataset partitions.

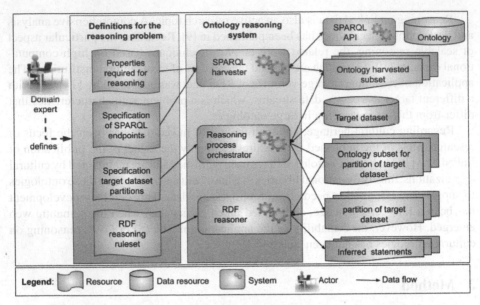

Fig. 1. Overview of the applied method for RDF reasoning in large ontologies.

2. RDF reasoning software, adapted to this process, executes the reasoning process following the specifications prepared by the domain expert:

2.1. A SPARQL harvester collects all triples from the ontology that use the specified triple patterns, storing them in a local triple store;

2.2. The RDF reasoner applies the reasoning ruleset to each sub dataset:

2.2.1. Before applying the reasoning rules, this RDF reasoner further reduces the subset of harvested statements from the ontology, selecting all resources about the specified properties, and all subjects and objects of statements using these properties as predicates. This selection of statements is applied recursively to all referred resources;

2.2.2. The reasoner executes using the ontology as supporting data, and the target subset as the main model for reasoning where all inferred statements are added;

3.2 Evaluation on Wikidata

We applied the reasoning method using Wikidata as the target ontology. The tested reasoning problem is for a use case of the culture domain. Wikidata contains RDF statements that align its classes and properties to several other ontologies. Alignments are not available for all classes and properties but their availability is high and has shown potential to support automatic interpretation of Wikidata [18]. One of these ontologies is Schema.org, which is also applied for cultural heritage objects.

For evaluating our method, we defined a use case that can be solved by RDF reasoning on Wikidata and Schema.org. We formulate it as follows: *"as a data re-user, I would like to obtain RDF data about a Wikidata entity represented as Schema.org"*.

Although Wikidata makes some use of Schema.org for its RDF output, it is only used for a limited set of properties that include only human-readable labels [18]. Wikidata's RDF output predominantly uses Wikidata's properties and classes, and Wikibase classes (Wikibase is the software on which Wikidata runs). The reasoning rules defined by RDF Semantics [1] and OWL [2, 3], enable the inference of Schema.org statements. By reasoning on the alignment statements that exist in Wikidata's RDF resources for its classes and properties, combined with the statements in RDF resources of Schema.org and Wikidata classes and properties, it is possible to infer the Schema.org properties and also the *rdf:type* properties using Schema.org classes as object.

The subset of RDF(S) and OWL reasoning rules that is required for our use case, and which we have setup in our RDF reasoner, are listed in Table 1[3]. To solve the reasoning problem, the reasoner requires that the statements match any of the triple patterns that appear in the rule trigger conditions, therefore for our use case, the reasoner requires data from Wikidata and the definition of OWL itself. The reasoner also requires these triple patterns from Schema.org in order to fulfil our use case. Note that RDF resources from RDF(S) do not have to be included. This is because RDF(S) is the foundational semantics for reasoning, therefore the base implementations of reasoners along with the RDF(S) inference rules have all the implicit meaning required for reasoning.

Our SPARQL harvester collected all statements using the required triple patterns from Wikidata and stored them locally in a triple store. Similarly, for Schema.org we collected the statements according to the triple patterns but given the much smaller size of Schema.org we simply harvested them from Schema.org's OWL definition file[4]. For the applied ruleset, we collected statements with the following properties: *rdfs:subclassOf, rdfs:subPropertyOf, owl:equivalentProperty, owl:equivalentClass* and *owl:sameAs*. For all the resources appearing as subjects in the harvested statements, we also harvest their *rdf:type* statements.

It is important to point out that Wikidata's RDF output is using almost exclusively Wikidata's properties and classes. In an earlier study, where we analysed Wikidata's RDF output about cultural heritage resources [18], we have observed only two properties from RDF in use: *rdf:type* and *rdf:label*. In the form that Wikidata's RDF is available, the application of the reasoning rules of RDF(S) and OWL would not be triggered. But Wikidata defines equivalent properties for all the necessary properties to perform the required reasoning, therefore, we harvested these equivalent properties instead of the RDF(S) and OWL ones.

The RDF resources of Wikidata's properties do not state their equivalence to RDF(S) and OWL. To allow the reasoning rules to trigger on the Wikidata properties, we added

---

[3] For readability purposes, in this text we abbreviate namespaces as follows: rdf for http://www.w3.org/1999/02/22-rdf-syntax-ns#; rdfs for http://www.w3.org/2000/01/rdf-schema#; owl for http://www.w3.org/2002/07/owl#; schema for http://schema.org/; wdt for http://www.wikidata.org/prop/direct/ .

[4] Schema.org OWL definition is made available at https://schema.org/docs/developers.html . Its web page states that currently, it is in an experimental stage.

Table 1. The subset of RDFS and OWL reasoning rules required for our use case.

Source of rule	Rule trigger condition	Rule entailment
RDFS	(? A ?P ?B), (?P rdfs:subPropertyOf ?Q)	(?A ?Q ?B)
RDFS	(?X rdfs:subClassOf ?Y), (?A rdf:type ?X)	(?A rdf:type ?Y)
OWL	(?P owl:equivalentProperty ?Q)	(?P rdfs:subPropertyOf ?Q), (?Q rdfs:subPropertyOf ?P)
OWL	(?P rdfs:subPropertyOf ?Q), (?Q rdfs:subPropertyOf ?P)	(?P owl:equivalentProperty ?Q)
OWL	(?P owl:sameAs ?Q), (?P rdf:type rdf:Property), (?Q rdf:type rdf:Property)	(?P owl:equivalentProperty ?Q)
OWL	(?P owl:equivalentClass ?Q)	(?P rdfs:subClassOf ?Q), (?Q rdfs:subClassOf ?P)
OWL	(?P owl:sameAs ?Q), (?P rdf:type rdfs:Class), (?Q rdf:type rdfs:Class)	(?P owl:equivalentClass ?Q)
OWL	(?A owl:sameAs ?B) (?B owl:sameAs ?C)	(?A owl:sameAs ?C)
OWL	(?X owl:sameAs ?Y), (?X rdf:type owl:Class)	(?X owl:equivalentClass ?Y)
OWL	(?X owl:sameAs ?Y), (?X rdf:type rdf:Property)	(?X owl:equivalentProperty ?Y)

owl:equivalentProperty statements to the harvested dataset. Table 2 lists the equivalent properties and the respective alignment statements. With the alignment statements in the data available for reasoning, the RDFS and OWL rules make complete inferences.

Once we reach this step, the reasoner's setup is complete and the reasoner is ready to be executed on Wikidata entities. For this evaluation, we identified Wikidata resources about cultural heritage objects by querying its SPARQL API, and checking for Wikidata entities containing the property *wdt:P727* (Europeana ID[5]). This property stands for the identifier assigned by Europeana to cultural heritage objects described in its dataset, therefore we consider it a reliable form of identifying cultural heritage objects in Wikidata. We collected a total of 11,928 resources from Wikidata in our sample and executed the reasoner.

We partitioned the sample by individual RDF resources that represent a cultural heritage object. We applied the final tasks of our reasoning to each partition. All the data from the ontologies required for the reasoning problem is collected first. For each partition, we selected from the ontologies all RDF resources on the predicates present in the

---

[5] https://www.wikidata.org/wiki/Property:P727.

Table 2. The property alignment statements we applied to allow the reasoning rules to trigger on the Wikidata properties.

Wikidata property	Alignment statements
Equivalent class (*wdt:P1709*)	`(wdt:P1709 owl:equivalentProperty` `owl:equivalentClass)`
Equivalent property (*wdt:P1628*)	`(wdt:P1628 owl:equivalentProperty` `owl:equivalentProperty)`
Subclass of (*wdt:P279*)	`(wdt:P279 owl:equivalentProperty` `rdfs:subClassOf)`
Subproperty of (*wdt:P1647*)	`(wdt:P1647 owl:equivalentProperty` `rdfs:subPropertyOf)`
Instance of (*wdt:P31*)	`(wdt:P31 owl:equivalentProperty` `rdf:type))`
External superpropery (*wdt:P2235*)	`(wdt:P2235 owl:equivalentProperty` `rdfs:subPropertyOf)`

partition, and all RDF resources of subjects and objects of the selected statements. This selection of statements is applied recursively to all referred RDF resources ensuring that the reasoner will execute with all necessary statements from Wikidata and Schema.org properties, which are required for an individual partition.

Finally, the reasoner was executed. We collected all the inferred statements and logged the running time of the reasoner. We create a data profile of the inferred statements for analysis. The results and their analysis are presented in Sect. 4.

4   Evaluation Results

For evaluating our method, we measured the number of statements used for reasoning at three stages: (1) the original ontologies; (2) after selecting triples necessary for the reasoning rules; and (3) for the statements about the cultural heritage objects. The reasoner execution time was also measured at the same three stages. Our third evaluation measured the number and characterization of the statements inferred in the final result.

Our experiments applied the same execution environment for all tests. It ran in a server with a Intel(R) Core (TM) i7-3770 CPU at 3.40 GHz. We have not applied any parallel processing, therefore the experiments run with one thread only, and the Java runtime environment was set for a limit of 16 GB memory usage. The software we implemented to support the experiment was a Java application that used Apache Jena[6] for the required RDF processing components: the RDF reasoner, triple store and RDF programming interface.

Table 3 summarizes the results we obtained on the number of statements for reasoning at the three stages. It breaks down the result by the three ontologies we used for the study on Wikidata. We cannot estimate the reduction of statements as a percentage of the

---

[6] https://jena.apache.org/.

original collection because the number of statements for Wikidata is unknown to us, but clearly it is a very small fraction of the original size, judging by its final average obtained for the final selection of triples and from the measurements obtained from Schema.org and OWL.

Table 3. The results in number of statements used for reasoning at three stages of the analysis: the original ontologies, after selecting triples necessary for the reasoning rules, and for the statements about the cultural heritage objects.

Ontology	Original size	Subset for rules	Subset for reasoning on an RDF resource (average per RDF resource)
Wikidata	unknown (latest dump file in Turtle format has 16 GB compressed)	3,010,212 (3,009,062 about classes plus 1,150 about properties)	256.9 ± 73.4
Schema.org	29.715	984	12.8 ± 5.3
OWL	450	6	2.0 ± 0.1
Total RDF statements	Unknown	3,011,202	271.8 ± 77.4

We attempted to execute the reasoner at all the three stages of the experiment and measured the execution time for each one. With the available computational resources, it was not possible to successfully execute it whenever too many statements from the ontologies were used. Using the complete ontologies exceeded the memory capacity. After reducing the reasoning data to the necessary statements for the reasoning rules, the reasoning time was too long for real-world applicability.

Reasoning ran successfully when executed at the final stage of our method where the ontologies statements used were the specific ones required for the reasoning rules and the RDF resource that was the target of reasoning. Table 4 presents the results for the time taken to execute the RDF reasoner, broken down in two operations: (1) the selection of statements from the ontologies; and (2) the execution of the RDF reasoner. For our complete sample of 11,928 Wikidata resources, the total runtime was of approximately two minutes.

Our third measurement was performed on the statements inferred in the final stage. We measured the amount of statements by predicates and their namespaces. Since our evaluation included the inference of *rdf:type* statements based on the transitivity property of *rdf:subClassOf*, we have measured the number of *rdf:type* statements inferred as well, grouping by the namespaces of the objects of statements.

From the sample of 11,928 Wikidata resources, the reasoning inferred 1,785,227 statements, averaging approximately 150 statements per resource. These statements contained predicates from 43 different namespaces, and the most frequently found are listed in Table 5. Most of the top ones where expected since they were related to those found in the reasoning rules applied. It was surprising, however, the large amount of statements

Table 4. The Results of execution time.

Operation	Execution on ontologies original size	Execution on the statement subset required by the rules	Execution on the statement subset for reasoning on a specific RDF resource
Ontology subset selection	N/A	N/A	6.8 ± 5.0 ms
Reasoning time	Reasoner failure (high memory requirements)	Reasoning not successful (too long - several hours)	3.5 ± 5.1 ms
Total time - ontology subset selection plus reasoning time	Unknown	Unknown	10.3 ± 7.6 ms

inferred from other namespaces, which amounted to 36% of the inferred statements. These results make it evident that Wikidata's alignments to properties and classes of other ontologies are frequently available, and that they can support the automatic semantics processing by general purpose RDF processing tools.

Table 5. The average number of inferred statements from a Wikidata RDF resource.

Namespace of predicates	Statements
http://d-nb.info/standards/elementset/gnd#	473,076
http://www.w3.org/1999/02/22-rdf-syntax -ns#	427,361
http://schema.org/	114,782
http://www.w3.org/2000/01/rdf-schema#	33,037
http://xmlns.com/foaf/0.1/	32,188
http://purl.org/dc/terms/	23,343
http://id.loc.gov/ontologies/bibframe/	20,019
http://www.cidoc-crm.org/cidoc-crm/	16,894
http://www.w3.org/2006/vcard/ns#	15,184
http://purl.obolibrary.org/obo/	14,224

Regarding the inference of *rdf:type* statements, we observed that 12 namespaces had at least 1 inference of *rdf:type* for more than 99% of the Wikidata RDF resources. The details for these namespaces are shown in Table 6. Schema.org had the highest average of inferences per resource, which is not surprising since the complete class structure of Schema.org was included in the source data for reasoning. These results support the conclusion that Wikidata's alignments to classes of other ontologies are frequently available.

Table 6. The namespaces with *rdf:type* inferences for at least 99% of the Wikidata RDF resources, and their respective averages per resource.

Namespace of rdf:type object	RDF resources with 1+ inferences	Average inferences per RDF resource
http://schema.org/	11,928	4.0 ± 0.5
http://www.cidoc-crm.org/Ent ity/e1-crm-entity/	11,928	1.0 ± 0.0
http://id.loc.gov/ontologies/bib frame/	11,927	3.1 ± 0.4
http://www.cidoc-crm.org/ent ity/e25-man-made-feature/	11,927	1.0 ± 0.0
http://purl.org/dc/dcmitype/	11,926	1.0 ± 0.1
https://d-nb.info/standards/ele mentset/gnd#	11,925	1.0 ± 0.0
http://dbpedia.org/ontology/	11,925	1.0 ± 0.3
http://www.cidoc-crm.org/Ent ity/e2-temporal-entity/	11,918	1.0 ± 0.0
http://def.seegrid.csiro.au/iso tc211/iso19108/2002/temporal#	11,918	1.0 ± 0.0
http://www.cidoc-crm.org/ent ity/e18-physical-thing/	11,908	1.0 ± 0.0
http://purl.org/dc/terms/	11,908	1.0 ± 0.1
http://www.cidoc-crm.org/ent ity/e24-physical-man-made-thing/	11,894	1.0 ± 0.1

We have further analysed the inferred statements containing predicates or objects with Schema.org URIs. Table 7 focuses on the inference results of Schema.org. Regarding the inference of *rdf:type* properties with Schema.org classes, at least one inference was always made from each Wikidata resource, and on average, 4.0 ± 0.5 *rdf:type* statements were inferred per resource. Regarding the inference of statements having Schema.org predicates, an average of 9.6 ± 2.5 statements were inferred. Altogether, an average of 13.6 ± 2.5 Schema.org statements were inferred.

5 Conclusion and Future Work

We have tested several approaches for solving RDF reasoning problems in large ontologies with limited computational resources. We identified that with a high volume of input data for reasoning, the memory requirements of the RDF reasoner become very demanding leading to an extremely long runtime or even to the impossibility of a successful execution.

Table 7. The average amount of inferred statements from a Wikidata RDF resource and the breakdown for those that contain Schema.org predicates or in objects of *rdf:type* properties.

Statements in Wikidata source resources	Total inferred statements	Inferred statements with Schema.org		
		In predicates	In *rdf:type* objects	Total
271.8 ± 77.4	149.7 ± 33.5	9.6 ± 2.5	4.0 ± 0.5	13.6 ± 2.5

Our method defines two intermediate tasks that reduce the volume of data used during reasoning. The first task is executed in advance of the reasoning, and it creates a subset of the ontology that contains only the statements that match the triple patterns included in at least one of the reasoning rules. The second reduction is performed when a reasoning request is invoked for a fragment of the target dataset, and it selects the statements of the ontology that are needed for the reasoning. The RDF reasoner runs efficiently when using only the resulting ontology subset and the target data.

Besides the evaluation of our reasoning method, it was also possible to evaluate Wikidata's potential for automatic semantics processing by general purpose RDF tools. Our conclusions pertain mainly to the context of cultural heritage data. We found that Wikidata's classes and properties frequently contain alignments to other ontologies, which are nowadays in use by the cultural domain.

We tested the inferences from Wikidata's Schema.org alignments. The inference of *rdf:type* statements was very positive, with at least one statement inferred for each cultural resource. However, the inference of statements with Schema.org predicates were not as high as we initially expected. We believe the difference between the different results is explained by the extensive class hierarchy of Wikidata along with the reasoning rules defined for *rdf:type*. These rules infer statements using all the super classes of the original statement, which makes an equivalence to at least one Schema.org class to be available in most cases. The reasoning rule for *owl:equivalentProperty*, however, does not infer the *owl:equivalentProperty* from super properties, it is inferred only from the particular Wikidata's property in the predicate leading to fewer alignments being available.

Our experiment on cultural data from Wikidata provided evidence supporting that the need for high resources may be mitigated since its data model contains alignments statements to several other ontologies in use by the cultural domain. Along with the application of RDF and OWL reasoning these alignments can be used to infer views of Wikidata expressed in cultural domain's data models. In addition, our method allows to lower the computational resources required for reasoning on Wikidata.

The positive results obtained with Wikidata motivate further work for maturing our method into a generic software framework to solve reasoning problems in large volumes of RDF data. The prototype should be redesigned into a framework supporting machine-readable definitions of large-scale reasoning problems. This definition of a reasoning problem must allow the configuration of all the subtasks of our method: the data source(s) for the ontology(ies); the triple patterns, or fragments, from the ontology for the reasoning problem; the reasoning rules; and triple fragments for the target dataset. We will start by an investigation of available standard vocabularies that address some of these configuration requirements. We expect that DCAT [19] and VoID [20] might support the configuration of data sources, and SHACL [21] or ShEx [22] might support the configuration of triple fragments.

Regarding Wikidata, our experiment supported that its Schema.org view may fulfil the requirements of some applications in culture. The use of alignment statements in Wikidata for ontologies of other domains should also be investigated.

Acknowledgments. We would like to acknowledge Antoine Isaac from the Europeana Foundation for his contribution to the preliminary discussion of our work regarding RDF reasoning and Wikidata. We also acknowledge the contribution of João Cardoso from INESC-ID for his review of the article.

This work was partly supported by Portuguese national funds through Fundação para a Ciência e a Tecnologia (FCT) with reference UIDB/50021/2020 and by the European Commission under contract number 30-CE-0885387/00-80.

References

1. Hayes, P., Patel-Schneider, P.F (eds.): RDF 1.1 Semantics. W3C Recommendation, W3C (2014)
2. Schneider, M. (eds.): OWL 2 Web Ontology Language RDF-Based Semantics, 2nd edn. W3C Recommendation, W3C (2012)
3. Motik, B., Patel-Schneider, P.F., Grau, B.C. (eds.).: OWL 2 Web Ontology Language Direct Semantics, 2nd edn. W3C Recommendation, W3C (2012)
4. Rietveld, L.: Publishing and Consuming linked data: optimizing for the unknown. In: Studies on the Semantic Web, vol. 21. IOS Press (2016)
5. Radulovic, F., Mihindukulasooriya, N., García-Castro, R., Gomez-Pérez, A.: A comprehensive quality model for Linked Data. In: Semantic Web, vol. 9, no. 1. IOS Press (2018)
6. Beek, W., Rietveld, L., Ilievski, F., Schlobach, S.: LOD lab: scalable linked data processing. In: Pan, J.Z., et al. (eds.) Reasoning Web 2016. LNCS, vol. 9885, pp. 124–155. Springer, Cham (2017). https://doi.org/10.1007/978-3-319-49493-7_4
7. Beek, W., Rietveld, L., Schlobach, S., van Harmelen, F.: LOD laundromat: why the semantic web needs centralization (even if we don't like it). In: IEEE Internet Computing, vol. 20, no. 2. IEEE (2016)
8. Fernández, J.D., Beek, W., Martínez-Prieto, M.A., Arias, M.: LOD-a-lot. In: d'Amato, C., et al. (eds.) ISWC 2017. LNCS, vol. 10588, pp. 75–83. Springer, Cham (2017). https://doi.org/10.1007/978-3-319-68204-4_7
9. Hogan, A.; Reasoning techniques for the web of data. In: Studies on the Semantic Web, vol, 19. IOS Press (2014)
10. Oren, E., Kotoulas, S., Anadiotis, G., Siebes, R., ten Teije, A., van Harmelen, F.: Marvin: distributed reasoning over large-scale Semantic Web data. J. Web Semant. 4(7), 305–316 (2009). https://doi.org/10.1016/j.websem.2009.09.002
11. Stuckenschmidt, H., Broekstra, J.: Time – space trade-offs in scaling up RDF schema reasoning. In: Dean, M., et al. (eds.) WISE 2005. LNCS, vol. 3807, pp. 172–181. Springer, Heidelberg (2005). https://doi.org/10.1007/11581116_18
12. Gu, R., Wang, S., Wang, F., Yuan, C., Huang, Y.: Cichlid: efficient large scale RDFS/OWL reasoning with spark. In: 2015 IEEE International Parallel and Distributed Processing Symposium, pp. 700–709 (2015). https://doi.org/10.1109/ipdps.2015.14

13. Ravindra, P., Deshpande, V.V., Anyanwu, K.: Towards scalable RDF graph analytics on MapReduce. In: Proceedings of the 2010 Workshop on Massive Data Analytics on the Cloud (MDAC 2010). ACM (2010). https://doi.org/10.1145/1779599.1779604

14. Komazec, S., Cerri, D.: Valle, E.D., Horrocks, I., Bozzon, A. (eds.) 1st International Workshop on Ordering and Reasoning (OrdRing 2011). CEUR-WS (2011)

15. Simou, N., Chortaras, A., Stamou, G., Kollias, S.: Enriching and publishing cultural heritage as linked open data. In: Ioannides, M., Magnenat-Thalmann, N., Papagiannakis, G. (eds.) Mixed Reality and Gamification for Cultural Heritage, pp. 201–223. Springer, Cham (2017). https://doi.org/10.1007/978-3-319-49607-8_7

16. Hyvönen, E.: Publishing and using cultural heritage linked data on the semantic web. In: Synthesis Lectures on the Semantic Web: Theory and Technology. Morgan & Claypool (2012)

17. Jones, E., Seikel, M. (eds.): Linked Data for Cultural Heritage. Facet Publishing (2016)

18. Freire, N., Isaac, A.: Technical usability of Wikidata's linked data. In: Abramowicz, W., Corchuelo, R. (eds.) BIS 2019. LNBIP, vol. 373, pp. 556–567. Springer, Cham (2019). https://doi.org/10.1007/978-3-030-36691-9_47

19. Maali, F., Reikson, J.: Data Catalog Vocabulary (DCAT). W3C Recommendation, W3C (2019)

20. Alexander, K., Cyganiak, R., Hausenblas, M., Zhao, J.: Describing Linked Datasets with the VoID Vocabulary. W3C Interest Group Note, W3C (2011)

21. Knublauch, H., Kontokostas, D. (eds.): Shapes Constraint Language (SHACL). W3C Recommendation, W3C (2017)

22. Prud'hommeaux, E., Boneva, I., Gayo, J.E.L., Kellog, G. (eds.): Shape Expressions Language 2.1. W3C Draft Community Group Report, W3C (2018)

Sentiment Analysis/Recommender Systems

A Deep Learning Approach to Aspect-Based Sentiment Prediction

Georgios Alexandridis[1]($\boxtimes$)(iD), Konstantinos Michalakis[1](iD), John Aliprantis[1](iD), Pavlos Polydoras[2], Panagiotis Tsantilas[2], and George Caridakis[1](iD)

[1] Intelligent Interaction Research Group, Cultural Technology Department, University of the Aegean, University Hill, 81100 Mytilene, Lesvos, Greece
{gealexandri,kmichalak,jalip,gcari}@aegean.gr
[2] Palo Services, 9, Chavriou Street, 10562 Athens, Greece
{pp,pt}@paloservices.com
https://ii.ct.aegean.gr/, https://www.paloservices.com/

Abstract. Sentiment analysis is a vigorous research area, with many application domains. In this work, aspect-based sentiment prediction is examined as a component of a larger architecture that crawls, indexes and stores documents from a wide variety of online sources, including the most popular social networks. The textual part of the collected information is processed by a hybrid bi-directional long short-term memory architecture, coupled with convolutional layers along with an attention mechanism. The extracted textual features are then combined with other characteristics, such as the number of repetitions, the type and frequency of emoji ideograms in a fully-connected, feed-forward artificial neural network that performs the final prediction task. The obtained results, especially for the negative sentiment class, which is of particular importance in certain cases, are encouraging, underlying the robustness of the proposed approach.

Keywords: Aspect-based sentiment analysis · Bi-directional long short-term memory units · Convolutional neural networks · Attention mechanism · Deep learning

1 Introduction

Sentiment analysis or *opinion mining* has become a vigorous research area, especially in recent years, with the vast expansion of the *world-wide web* and the proliferation of *online social networks* (OSNs), like *Facebook, Twitter* and *Instagram*. Indeed, people discuss, voice opinions, share digital content and generally engage in activities, in a large public space. This reality has caught the attention of businesses and organizations, whose objective is to study and analyze public opinion with respect to the products and services they offer. Ideally, the aforementioned parties need not conduct surveys or opinion polls any more, as there is an abundance of relevant information available online.

© IFIP International Federation for Information Processing 2020
Published by Springer Nature Switzerland AG 2020
I. Maglogiannis et al. (Eds.): AIAI 2020, IFIP AICT 583, pp. 397–408, 2020.
https://doi.org/10.1007/978-3-030-49161-1_33

However, locating and extracting user opinion from online sources (social media sites, blog posts, forums, etc.) is a rather cumbersome task. Apart from the huge volume of information that needs to be processed, one has to be familiarized with the specifics of each service (e.g. API calls) and with the sentiment annotation processes. Therefore, it is not uncommon for companies to resort to specialized analysts that offer content services for consumers and brands.

From the business analyst perspective, sentiment analysis is a multi-faceted task. In [8], three distinct levels of analysis are identified; (i) document, (ii) sentence, (iii) entity and aspect. At the first level, a single sentiment is assigned on the whole document (e.g. positive or negative). This is practical for sources like news agencies, that usually discuss only one entity. At the second level of analysis, sentiment is extracted on a per sentence basis, having application on documents discussing more than one entities or on micro-blogging platforms like Twitter, where documents commonly consist of a few sentences.

The third level of analysis is the most demanding task, as instead of examining language constructs, the emphasis is placed on the entity or the aspect level. For example, a tweet stating *"Company X offers a great service, Thank God I switched over from Company Y"* can be classified as positive, w.r.t. Company's X service, negative w.r.t to Company Y and neutral, w.r.t. other similar companies. Therefore, the same text excerpt may have different interpretations. Additionally, subjective criteria may arise when deciding upon opinion or sentiment; for instance, a business may consider the reproduction of one of its press releases by a news agency a positive event, while another may view this event as neutral.

In this work, sentiment prediction is modelled as a supervised classification problem and is addressed using a deep learning architecture, based on *Bidirectional Long Short-Term Memory* (BiLSTM) units [5], combined with convolutional attention layers [10]. More specifically, Sect. 2 discusses related work and Sect. 3 presents the overall system architecture. Section 4 describes the data collected by the system, while Sect. 5 presents the feature extraction procedure, the implemented model and the obtained results. Finally, the work concludes in Sect. 6.

2   Related Work

Even though the research areas of sentiment analysis and opinion mining firstly appeared in 2003 [2], a multitude of works have been published on the subject ever since [8], based on various methodologies [11]. Nevertheless, in recent years, most state-of-the-art approaches are related to deep learning techniques. For example, the key element of the proposed system in [3] (studying the domain adaptation problem for sentiment classification) is a *stacked denoising autoencoder* that performs unsupervised feature extraction using both labeled and unlabeled samples. In [17], a neural network consisting of convolutional and LSTM layers is being presented, that learns document representations by considering sentence relationships. Other works combine the use of LSTMs with *attention mechanisms*; for instance, in [20], the document-level aspect-sentiment rating prediction task is formulated as a comprehension problem that is being addressed by a hierarchical interactive attention-based model.

LSTMs have also been used in aspect level sentiment classification. In [16], the *target-dependent* and *target-connection* extensions to LSTM are proposed. The target is considered as another input dimension and is subsequently concatenated with the other features. A similar approach is followed in the current work; however, in the proposed methodology bi-directional LSTMs are employed instead. Bi-directional LSTMs with word embeddings at their input are used in [13] for aspect level sentiment classification, without an attention mechanism though.

An attention-based LSTM methodology for aspect-based sentiment analysis is described in [18], where the attention mechanism has been found to be effective in enforcing the model to focus on the important parts of each sentence, with respect to a specific aspect. Finally, in [19], two attention-based bidirectional LSTMs are proposed; however, unlike our approach, no other input features are considered apart from text.

3   System Architecture

The overall system architecture is depicted in Fig. 1. It begins with *data crawling*, the process of systematically accessing a disparate set of online sources, gathering data that satisfy certain filtering criteria and forwarding them into a data repository. The set of data sources being accessed includes traditional web sources such as news sites, blogs, forums, as well as OSNs. There is a constantly updating registry of specific access points per medium, limiting search space only to relevant sources.

Each data source is correlated with potentially multiple data formats, depending on the information granularity. For instance, YouTube exposes hierarchical information, starting from a channel, drilling into metrics (followers and likes), its videos and finally video comments and reactions. The crawling process of this hierarchy needs to be addressed by the corresponding crawler, both in terms of data navigation and crawling policy. The latter is strongly related to data "freshness" (i.e. breaking news need to be crawled as fast as possible), as well as importance from a business perspective (e.g. more popular Instagram accounts need to be crawled more often). The crawling process, directed by these factors, pushes the accessed data sets to the ingestion services, in a streaming manner and re-iterates.

The subsequent step is data cleansing & homogenization. During this process, data are being stripped off inconsistencies attributed to major errors (e.g. missing article date), garbage information injection (e.g. ads in articles) and even erroneous semantics, such as out-of-scope or inappropriate content. Finally, data are stored in the *data lake* [12], a logical database which is the major hub of information exchange among the services. At its final version, the original raw data unit is upscaled into a mention, supplemented by derived information including named entities, image/pattern recognition measures, as well as sentiment

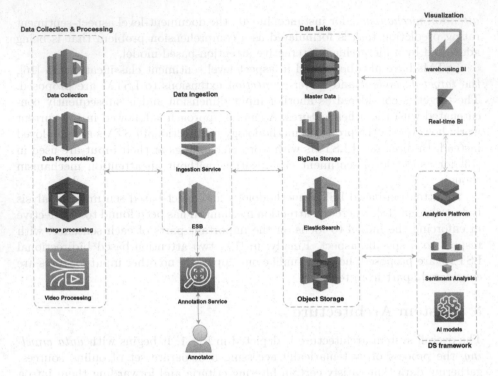

Fig. 1. Overall system architecture

indication. Mention semantics extend to a broader definition and usage context which includes a specific domain (e.g. telecommunications) and intended usage (e.g. competition analysis). Naturally, one document may correspond to multiple mentions, each associated with a different semantic context.

3.1   Annotation

Sentiment annotation [8] is the process of assigning specific sentiment values to a given mention. Currently, this process considers three values; namely *positive*, *negative* and *neutral*, but it can be generalized to a more extensive set. Orthogonal to sentiment assignment per se, however, is the knowledge base according to which a specific sentiment value can be extracted from a given mention. These rules can be arbitrarily chosen, based on specific criteria, driving sentiment analysis outcome accordingly.

In principle, sentiment annotation criteria are either *global* or *specific*, with the former referring to commonly agreed criteria, such as association of negative sentiment and lists of insulting words or phrases. The latter refer to specific rules, which may contradict the global ones and in those cases, they take precedence. The unit of focus is the aspect of the given mention being examined by the rule. Aspects include the text of the mention and its metadata, which in turn consist of the respective data source information (e.g. news site and related category), entity type (e.g. Facebook post or comment), generation time (e.g. twitter comments posted after midnight), author, etc. In all cases, the aspect is well-defined prior to the annotating process and it is uniquely identified by a respective identifier (aspect id).

As stated above, one raw data record is associated with potentially multiple mentions, each corresponding to a different perspective. This results to possibly multiple sentiment values for the same record (remember the discussion about the different interpretations of the same tweet in Sect. 1), which are determined manually by a human annotator, studying the defined rules and applying them by assigning sentiment values to automatically selected samples. Sample selection follows the stratified random sampling methodology [14], with subgroups defined by the respective data sources (sites, blogs, social media, etc). Annotation is software-assisted and forms the respective data set (Sect. 4).

4   Data

Based on the procedure discussed above, 343,956 Greek language documents have been crawled and annotated over a period spanning nearly 2 years (September 2017 to June 2019). Table 1 outlines the distribution of their sources. As it is evident, the various sources are not evenly represented in this dataset because of the compliance of the crawling procedure to data protection regulations, as discussed in Sect. 3. For this reason, only public and business accounts are being processed and since Twitter is the most popular OSN in which information is disseminated predominately publicly, it is over-represented. The same reasoning is applied to news sources as well, as the vast majority of Greek news agencies and outlets are being monitored and indexed on a daily basis.

Table 1. Source medium distribution

Source	Entries	Percentage
Tweets	160,905	46.78%
News articles	87,750	25.51%
Facebook posts	39,591	11.51%
Facebook comments	25,250	7.34%
Blog posts	14,975	4.35%
Instagram	10,784	3.14%
Other	4,701	1.37%
Total	343,956	100.00%

Table 2 displays the frequency of appearance of specific domains within the crawled data. More than half of the collected information is about telecommunication businesses (mobile phone operators, Internet service providers, etc.), followed by tobacco companies (about 20%). Another interesting observation is the relatively large number of documents related to political parties and politicians, attributed to the fact that 2019 has been an election year in Greece. It should also be noted that the appearance of each specific domain is not evenly spread across all sources (that is, according to the distribution of Table 1). For example, information about the banking sector is predominately collected from news outlets (∼ 70%, when news articles constitute a quarter of the dataset), while politics appear evenly on Twitter and on news articles (∼ 50% and ∼ 40%, respectively).

Table 2. Domain distribution

Domain	Entries	Percentage
Telecom	178,739	51.97%
Tobacco	72,822	21.17%
Banks	36,582	10.64%
Politics	30,677	8.92%
Retail	11,756	3.42%
Transport	8,079	2.35%
Misc	5,301	1.54%
Total	343,956	100.00%

Finally, Table 3 summarizes the distribution of the three categories of annotated sentiment (Sect. 3.1) over the whole dataset. In total, five persons participated in the annotation task, all of whom had received special training on annotation guidelines. Additionally, to further eliminate bias in the labels, well-defined annotation rules, such as cross-validation, irregular intervals and random data distribution (to all available annotators) across the dataset, have also been adopted.

As it is evident, it is highly imbalanced, since the *neutral* class is assigned to the overwhelming majority of the cases, while the other two (and especially the *positive* class) are clearly underrepresented. If sentiment distribution is further analyzed on a per source basis, the most negative content (∼ 30%) appears on Twitter, a medium offering relatively anonymity (and thus, more "freedom") to its users. On the other hand, the least polarized opinions and at the same time the most "neutral" ones (more than 90%) appear on news articles. The latter are written by journalists who, most of the time, use a professional, unbiased language. Lastly, the most positive sentiment is expressed on Facebook comments (∼ 12%), which is three times more than the average.

A similar analysis on a per domain basis is also interesting. By far, the most negative (and the least neutral or positive) feelings are expressed when politics are discussed, indicating that this is a highly polarized topic. The most neutral content, on the other hand, is again related to the banking sector, since most

Table 3. Annotated sentiment distribution

Sentiment class	Percentage
Positive	4.03%
Neutral	78.53%
Negative	17.44%

of the relevant content in the collected dataset originates from news articles, as it has been already argued. Finally, the transportation sector has received the most positive comments (around 10%).

4.1 Preprocessing

Prior to performing the sentiment prediction task, a number of data preprocessing steps are necessary. Initially, the textual part of each record is cleaned; that is, extra white space, non printable characters and other artifacts (e.g. HTML tags) are removed. Subsequently, the words that comprise the text are mapped to an *embedding space*, using *fastText* [1], a *natural language processing* methodology. In the end, the text of each document, is represented by the embeddings (vectors) of its words.

Among the non-printable characters that are extracted in the cleaning phase are *emojis* [9], a short of ideograms used in electronic communications to express feeling and emotions that are directly related to the sentimental state of the author of the document (e.g. smileys, sad or angry faces, etc). Since emojis do carry sentiment information, they are expected to positively contribute to the opinion mining task. A common methodology of including emojis in the prediction task would be to map them to a continuous vector space, usually consisting of two dimensions (sentiment score and neutrality) [9]. However, a different approach has been followed in this work; instead of using emoji embeddings, a vector designating the frequency of appearance of each emoji has been constructed for each record.

Another important characteristic that might be related to the sentiment value of a record is the number of its repetitions (retweets, shares, reposts, etc). The intuition behind this type of reasoning is that widely-spread content may carry significant emotional weight and therefore a correlation might exist between the number of times a text excerpt appears and its content. This characteristic follows a power law distribution in the dataset; the overwhelming majority of documents appear only once, while less than a $1,000$ records have been repeated more than 10 times. For this reason, in the model of Sect. 5.2, the logarithm of the number of repetitions is considered.

5 Experiments

The experiments that follow have been performed on the collected corpus presented above. In order to maintain temporal consistency, the dataset has been chronologically split into a training set (63.75% of the samples, earliest in time), a validation set (11.25% of the samples, subsequent in time) and a test set (25% of the samples, latest in time).

5.1 Feature Extraction

The predominant feature extraction activity involves the textual parts of each record in the collection. It is achieved by a stacked, two-layered BiLSTM network (Fig. 2), which is considered to be among the state-of-the-art in capturing the spatial relationship between words and the order they appear in a text sequence [21]. The neural embeddings of the words are provided to the network in the order they appear in text, with a small amount of Gaussian noise ($\mu = 0, \sigma = 1$) added to them, as a regularization effect that reduces overfitting. After extensive experimentation, the optimal number of units for each layer have been determined to be 150, with *dropout* layers applied in-between them ($p = 0.3$) [15].

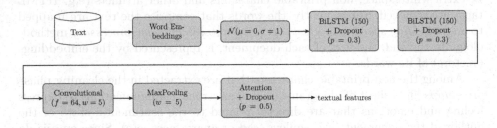

Fig. 2. Textual feature extraction procedure

After the BiLSTM layers, an one-dimensional convolutional layer follows, with 64 filters and a window size of 5. Again, both of the aforementioned hyper-parameters have been determined after experimentation. Subsequently, a max-pooling layer of an equal window size downsamples the output of the convolutional layer. The textual feature extraction is finalized with an attention layer, whose addition counterbalances the decline in performance when dealing with long sentences. Lastly, the feature extraction procedure concludes with a Dropout layer ($p = 0.5$).

The other three features to be considered do not require such an extensive feature extraction procedure. Aspect is incorporated through aspect id, an one-hot encoded variable (Sect. 3.1), while the presence of emojis is quantified as a frequency vector. Finally, the number of repetitions of each document is provided to the model via its logarithm (Sect. 4.1).

5.2 Model Selection

After experimenting with various techniques and architectures, the optimal model has been determined to be a fully-connected feed-forward artificial neural network, consisting of two hidden layers (Fig. 3). The first hidden layer is comprised of 1024 neurons and the second of 128. Their activation function is the *rectified linear unit* [6] (in contrast to the output layer, where *softmax* activation is used instead [4]). Network training has been based on the Adam optimization algorithm [7], with a learning rate of 10^{-3} and hyperparameters β_1, β_2 being fixed at 0.9 and 0.999, respectively. Finally, the *fastText* word embedding vectors used in the experiments have been pretrained on a corpus of more than $2,000,000$ words.

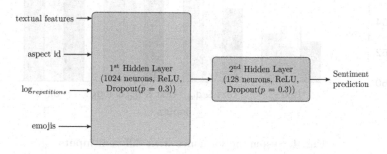

Fig. 3. Fully-connected model

5.3   Results

Figures 4 and 5 examine system performance regarding various aspects on *Precision, Recall* and their harmonic mean (*F1-score*); a set of popular information-retrieval metrics, widely used in sentiment analysis tasks [8]. The former is equal to the ratio of the correctly classified documents to a given class over the total number of classified documents to that class, while the latter is equal to the ratio of the correctly classified documents to a given class over the total number of documents that belong to that class. The values displayed in Fig. 4 are averaged over all classes, while the results in Fig. 5 are given for each class separately.

Figure 4 summarizes system performance with respect to the different inputs. When only the textual part of each record is considered, the efficiency of the proposed approach is limited, an indication that in the environment described in Sect. 3, text alone is not a sufficient indicator for the prediction task. When aspect-related information is considered, Recall increases by more than 6% (followed by a smaller boost in Precision), meaning that the system can better discriminate in-between the classes. The addition of the logarithm of the number of repetitions marginally affects Recall, but further enhances Precision, with the F1-score in this case being slightly better than the previous one. Finally, when all four inputs are provided (text, aspect, logarithm of the number of repetitions, emojis), system throughput is further enhanced, with all three metrics being above (70%), adding more than 8% to the overall system performance.

Figure 5 displays the per-class performance of the examined metrics. Even though class labels are highly imbalanced (Table 3), the system achieves very

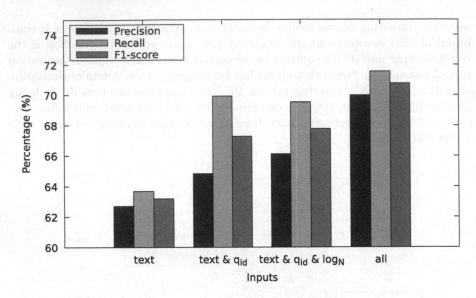

Fig. 4. System performance w.r.t different inputs

good results for the negative class, a characteristic that is of significant importance, as the main concern of many businesses is to be able to timely identify and respond to unpleasant content. On the other hand, the predictions on the

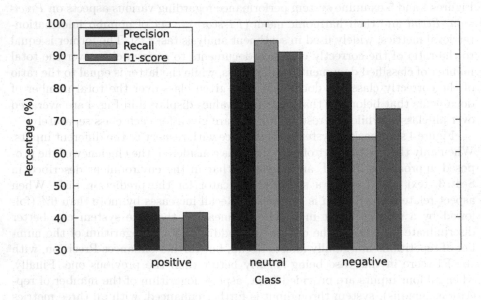

Fig. 5. System performance w.r.t different classes

positive class are clearly below average and therefore more effort should be put in the direction of improving system efficacy for this particular case, as well.

6 Conclusions

In this work, a novel hybrid bi-directional LSTM/CNN feature extraction architecture has been presented, as part of a broader system that performs aspect-based sentiment analysis. The obtained results, on a corpus selected from Greek-language content from OSNs and other sources, are encouraging, especially on the negative class that is of particular interest to businesses. Nevertheless, the outlined architecture needs to be further fine-tuned and reasoned upon, as the system demonstrates sub-optimal performance in identifying positive sentiment.

The proposed architecture may be extended in a number of ways. An obvious direction would be to consider additional textual characteristics that convey aspect and sentiment-based information. For instance, feature extraction from hashtags, which are quite popular on OSNs, is expected to further aid the desired task. Additionally, the number of repetitions could be leveraged by examining frequency patterns in-between the users that are either mentioned on or just redistribute content.

Finally, the quality of the already extracted features may be farther enhanced. For example, the application of dimensionality reduction techniques, such as principal component analysis, on the emoji frequency matrix can help determine which of the available emojis have the greatest impact on the sentiment analysis task.

Acknowledgements. This research has been co-financed by the European Regional Development Fund of the European Union and Greek national funds through the Operational Program Competitiveness, Entrepreneurship and Innovation, under the call RESEARCH - CREATE - INNOVATE (project code: T1EDK-03470).

References

1. Bojanowski, P., Grave, E., Joulin, A., Mikolov, T.: Enriching word vectors with subword information. arXiv preprint arXiv:1607.04606 (2016)
2. Dave, K., Lawrence, S., Pennock, D.M.: Mining the peanut gallery: opinion extraction and semantic classification of product reviews. In: Proceedings of the 12th International Conference on World Wide Web. WWW 2003, pp. 519–528. ACM, New York (2003). https://doi.org/10.1145/775152.775226, https://doi.acm.org/10.1145/775152.775226
3. Glorot, X., Bordes, A., Bengio, Y.: Domain adaptation for large-scale sentiment classification: a deep learning approach. In: Proceedings of the 28th International Conference on International Conference on Machine Learning. ICML 2011, pp. 513–520. Omnipress, USA (2011), http://dl.acm.org/citation.cfm?id=3104482.3104547
4. Goodfellow, I., Bengio, Y., Courville, A.: Deep Learning. MIT Press, Cambridge (2016). http://www.deeplearningbook.org

5. Graves, A., Fernández, S., Schmidhuber, J.: Bidirectional LSTM networks for improved phoneme classification and recognition. In: Duch, W., Kacprzyk, J., Oja, E., Zadrożny, S. (eds.) ICANN 2005. LNCS, vol. 3697, pp. 799–804. Springer, Heidelberg (2005). https://doi.org/10.1007/11550907_126
6. Hahnloser, R.H., Sarpeshkar, R., Mahowald, M.A., Douglas, R.J., Seung, H.S.: Digital selection and analogue amplification coexist in a cortex-inspired silicon circuit. Nature **405**(6789), 947 (2000)
7. Kingma, D.P., Ba, J.: Adam: a method for stochastic optimization. In: 3rd International Conference on Learning Representations. ICLR 2015, San Diego, CA, USA, 7–9 May 2015. Conference Track Proceedings (2015). http://arxiv.org/abs/1412.6980
8. Liu, B.: Sentiment Analysis and Opinion Mining. Morgan & Claypool Publishers, San Rafael (2012)
9. Novak, P.K., Smailović, J., Sluban, B., Mozetič, I.: Sentiment of emojis. PloS One **10**(12), e0144296 (2015)
10. Raffel, C., Ellis, D.P.W.: Feed-forward networks with attention can solve some long-term memory problems. CoRR abs/1512.08756 (2015). http://arxiv.org/abs/1512.08756
11. Ravi, K., Ravi, V.: A survey on opinion mining and sentiment analysis: tasks, approaches and applications. Knowl.-Based Syst. **89**, 14–46 (2015). https://doi.org/10.1016/j.knosys.2015.06.015. http://www.sciencedirect.com/science/article/pii/S0950705115002336
12. CITO Research: Putting the data lake to work: a guide to best practices. Technical report, Teradata (2014)
13. Ruder, S., Ghaffari, P., Breslin, J.G.: A hierarchical model of reviews for aspect-based sentiment analysis. arXiv preprint arXiv:1609.02745 (2016)
14. Särndal, C.E., Swensson, B., Wretman, J.: Model Assisted Survey Sampling. Springer, Heidelberg (2003)
15. Srivastava, N., Hinton, G., Krizhevsky, A., Sutskever, I., Salakhutdinov, R.: Dropout: a simple way to prevent neural networks from overfitting. J. Mach. Learn. Res. **15**, 1929–1958 (2014). http://jmlr.org/papers/v15/srivastava14a.html
16. Tang, D., Qin, B., Feng, X., Liu, T.: Effective LSTMs for target-dependent sentiment classification. arXiv preprint arXiv:1512.01100 (2015)
17. Tang, D., Qin, B., Liu, T.: Document modeling with gated recurrent neural network for sentiment classification. In: Proceedings of the 2015 Conference on Empirical Methods in Natural Language Processing, pp. 1422–1432 (2015)
18. Wang, Y., Huang, M., Zhao, L., et al.: Attention-based LSTM for aspect-level sentiment classification. In: Proceedings of the 2016 Conference on Empirical Methods in Natural Language Processing, pp. 606–615 (2016)
19. Yang, M., Tu, W., Wang, J., Xu, F., Chen, X.: Attention based LSTM for target dependent sentiment classification (2017). https://aaai.org/ocs/index.php/AAAI/AAAI17/paper/view/14151
20. Yin, Y., Song, Y., Zhang, M.: Document-level multi-aspect sentiment classification as machine comprehension. In: Proceedings of the 2017 Conference on Empirical Methods in Natural Language Processing, pp. 2044–2054 (2017)
21. Zhou, P., Qi, Z., Zheng, S., Xu, J., Bao, H., Xu, B.: Text classification improved by integrating bidirectional LSTM with two-dimensional max pooling. In: COLING 2016, 26th International Conference on Computational Linguistics, Proceedings of the Conference: Technical Papers, Osaka, Japan, 11–16 December 2016, pp. 3485–3495 (2016). http://aclweb.org/anthology/C/C16/C16-1329.pdf

On the Reusability of Sentiment Analysis Datasets in Applications with Dissimilar Contexts

S. Sarlis and I. Maglogiannis(✉) 🆔

Department of Digital Systems, University of Piraeus, Piraeus, Greece
stevensarlis@gmail.com, imaglo@unipi.gr

Abstract. The main goal of this paper is to evaluate the usability of several algorithms on various sentiment-labeled datasets. The process of creating good semantic vector representations for textual data is considered a very demanding task for the research community. The first and most important step of a Natural Language Processing (NLP) system, is text preprocessing, which greatly affects the overall accuracy of the classification algorithms. In this work, two vector space models are created, and a study consisting of a variety of algorithms, is performed on them. The work is based on the IMDb dataset which contains movie reviews along with their associated labels (positive or negative). The goal is to obtain the model with the highest accuracy and the best generalization. To measure how well these models generalize in other domains, several datasets, which are further analyzed later, are used.

1 Introduction

Sentiment analysis, also known as opinion mining, refers to the subjective analysis of a text [1]. Its basic task is predicting the overall sentiment polarity of a given sentence. In more recent works [2], sentiment analysis expands this task on trying to identify the emotional status of a sentence (e.g., sadness, excitement, joy, anger). It mainly applies to reviews, social media posts and survey responses, helping businesses understand their customer's opinions about their products and services, improving their marketing strategies and decision policies [3]. This practice of extracting insights from data is widely adopted by organizations and enterprises, affecting many aspects of human life.

Every day, many positive and negative comments are shared on the Internet. The impact of these comments on the economy is palpable and has led to the implementation of techniques that take advantage and extract knowledge from them. The sentiment of a movie review is usually associated with a distinct rating, which can be used for classification problems. From a user's perspective, it can be used as a recommendation tool for movie selection [4].

The IMDb dataset contains 50,000 reviews and is evenly divided into 25,000 train reviews and 25,000 test reviews. Movie reviews are labeled as positive or negative, and the challenge is to predict the sentiment of an unseen review. In this work, the

I. Maglogiannis et al. (Eds.): AIAI 2020, IFIP AICT 583, pp. 409–418, 2020.
https://doi.org/10.1007/978-3-030-49161-1_34

following classifiers are used: Logistic Regression, Bernoulli Naïve Bayes, Multinomial Naïve Bayes, Linear Support Vector Machine and two Neural Networks. Different
techniques, including Count Vectorizer, TF-IDF (Term Frequency – Inverse Document
Frequency) Vectorizer, lemmatization, stopwords, n-grams, minimum-maximum number of words and max features are implemented. The experiments conducted show that
Logistic Regression, Linear Support Vector Machine and Neural Network classifiers
perform noticeably better than Bernoulli Naïve Bayes and Multinomial Naïve Bayes.

The study aims to find answers to the following three questions:

1. How various classification algorithms perform on the IMDb sentiment analysis
   classification problem?
2. How much pre-processing improves the performance of these classifiers?
3. Do these classifiers generalize well to other datasets and can they therefore be used
   in different real-life problems?

The remainder of this paper is structured in four sections, as follows: Sect. 2 presents
the background information and related research work, while Sect. 3 proposes the sentiment analysis system architecture. Section 4 describes the system in practice and reports
the experiments conducted and the corresponding results. Finally, Sect. 5 concludes the
paper and provides future work ideas.

2   Background and Related Work

Several approaches have been proposed for the extraction of features from text and the
exploitation of them in forming appropriate classifying models. Initial attempts in order
to represent a document used Bag of Words (BOW) representations, where the document was represented as the sum of one-hot vectors of its words. However, in the field of
NLP, state-of-the-art Word Embeddings outperform BOW [5], as BOW representations
lose some of the semantic meaning associated with text [6]. Therefore, more recent
works create vector representations for both words and documents, to retain as much
information as possible. Furthermore, word embeddings can boost the generalization
capabilities of neural systems [7, 8]. However, in this work, BOW representations are
used, as they require just a few lines of code to build the models and they work just fine in
these specific datasets. The scikit-learn library, incorporated in Python [9], provides different schemes that can be used, including word counts with Count Vectorizer and word
frequencies with TF-IDF (Term Frequency – Inverse Document Frequency) Vectorizer,
which are further analyzed in this work. Moreover, [10] presented the Delta TF-IDF
technique, which is considered to be an improvement of the TF-IDF. The difference is
that in regular TF-IDF, each word or combination of words, is associated with a TF value
(word counts in the document) and an IDF value (word counts in all documents), while
delta TF-IDF gives more weight to words that occur more often in that text, and are rarer
in oppositely labeled documents. It is shown that the Delta TF-IDF produces noticeably
better results than the regular TF-IDF. As mentioned earlier, text preprocessing is considered the most important step for NLP tasks. In an NLP pipeline, many preprocessing
techniques are used including but not limited to lowercasing, lemmatization, character

removal. These techniques have already been studied in [3] and they are further analyzed later in this work.

3   Proposed Methodology for Sentiment Analysis

The process steps and the workflow diagram used for the sentiment analysis is shown in Fig. 1. Firstly, the IMDb dataset is collected and imported. Subsequently, in order to remove noise and clean the data, pre-processing is performed. The next step is feature extraction and feature selection, where each document is represented as a document-term matrix. In this matrix, rows refer to documents in the collection and columns correspond to terms. In the final step, the classification process is performed, and the used algorithms are evaluated.

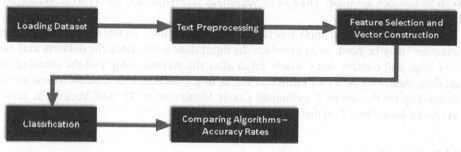

Fig. 1. Workflow diagram

3.1   Text Preprocessing

Before performing any classification algorithm, it is important to clean up the data. Finding and removing noise, greatly affects the accuracy of the classification algorithms. The following preprocessing techniques are applied: lowercasing, lemmatizing and removing noise characters and stop words. Lowercasing is probably the simplest and most effective preprocessing technique, where words map to the same lowercase form. This technique applies to all characters of the input text. Furthermore, the reviews are cleaned by removing HTML tags, non-word characters, punctuation and accents, which do not serve any purpose for detecting sentiment. Subsequently, a set of commonly used, low information words (i.e., stopwords), which do not carry any sentiment, is removed. During the stopwords process, the words included in the "NLTK list of English stopwords" are removed, except for the words "no" and "not" which are excluded, as they affect the value of the label. This initiative stems from the idea that, for example, the expression "not good" expresses a negative feeling, whereas if the word "not" was removed, the expression would become "good", which has a positive connotation, thus distorting the result. The process of lemmatization replaces the various forms a word can have, with its corresponding lemma (i.e., cutting out endings). This process is similar to stemming but it is preferred, as it doesn't just chop things off, but it actually transforms words to their actual root. Lemmatization is a standard preprocessing technique for linear text classification systems and has been shown to improve classification accuracy in sentiment analysis tasks [11]. Finally, n-grams and more specifically bi-grams are used in some models. N-grams are a contiguous sequence of n items and consequently bi-grams are a sequence of 2 words [12].

3.2 Feature Selection and Feature Extraction (Vectorization)

Common machine learning algorithms expect a two-dimensional table as an input, where rows are instances and columns are attributes. In this context, documents must be converted into vector representations. This process is called "feature extraction" or "vectorization" and is an important step when analyzing text. Words are coded as integers or as floating-point values, depending on which technique is implemented and are then used as an input to machine learning algorithms. In this work, two of the most well-known such techniques are used: The Count Vectorizer technique and the TF-IDF Vectorizer technique [13].

Moreover, in order to apply machine learning algorithms on text segments, "feature selection" may be need, so as to reduce the input dimension, since the datasets may be very large and contain many words. Even after the preprocessing and the cleaning of the data, there may be a lot of different words in the dataset, which create a dictionary. Depending on the selected technique, Count Vectorizer or TF-IDF Vectorizer, these words are sorted based on their importance.

3.2.1 Count Vectorizer

Count Vectorizer is one of the simplest and relatively satisfactory techniques. It counts the number of times a word appears in a document and uses this value as its weight. The Count Vectorizer technique provides tokenization of the text documents and builds a vocabulary of words. Count Vectorizer counts words, so it returns integers, while TF-IDF Vectorizer assigns a score to words, so it returns floats.

3.2.2 Term Frequency and Inverse Document Frequency (TF-IDF) Vectorizer

In the TF-IDF Vectorizer technique, the weight corresponding to each word not only depends on its frequency in the document, but also on how repetitive this word is in all documents. TF-IDF is a statistic that aims to reflect a word's importance in a document within a collection of documents [14]. This technique is mainly used in information retrieval in order to rank how important a keyword is to a given document in a corpus and is a common approach for sentiment analysis [15]. Term Frequency (TF) is the frequency of the term within a document and indicates how many times a word is used within this document [14]. TF describes how important a word is in the document, but it doesn't consider the possibility of that word occurring to other documents too, thus a balance term is needed. The Inverse Document Frequency (IDF) refers to the specificity of a term, which can be quantified as an inverse function of the number of documents in which it occurs. The more documents that have a specific word, the lower its IDF score is.

3.3 Classification and Comparison

To solve the classification problem, the following algorithms are used: Logistic Regression (LR), Bernoulli Naïve Bayes (BNB), Multinomial Naïve Bayes (MNB), Linear Support Vector Machine (Linear SVM) and two Neural Networks (NN1 & NN2). All the parameters of the classifiers use the default values provided by the scikit-learn package, except for the Logistic Regression and the Linear SVM, where in the Count Vectorizer technique, the value of the max interactions parameter is set to 1,000 and 10,000 respectively. Furthermore, the Linear SVM classifier uses the hinge loss function, which is the standard SVM loss. Regarding the architecture of the Neural Networks, they are Sequential Neural Networks with four Dense layers. The first three Dense layers use the "relu" activation function and the last uses the "sigmoid". The Networks use two Dropout layers to avoid overfitting, "killing" random nodes each time, so that the network can recalibrate its weights. The optimizer is "adam" and the loss function is binary cross entropy. The batch size is set to 500 and the epochs used are only two, as the model overfits if trained longer. The two networks need to know what input shape they should expect. In this work, they have different input shapes. More specifically, the NN1 has an input shape equal to 57,210 and the NN2 has an input shape equal to 50,000. This results from the fact that in the NN1 the two vectorization techniques (Count and TF-IDF Vectorizer) are implemented with the first approach, which does not consider max features, while in the NN2 the two techniques are implemented with the second approach, which considers max features (50,000) and bi-grams. Max features refer to the number of words or combinations of words required, to construct the feature vectors. The metric used to evaluate these classifiers is accuracy, as there is an equal number of samples belonging to each class [16]. Moreover, the models are trained on the IMDb dataset and tested on a variety of domains including food, clothing, tweet and hotel datasets. The IMDb is selected as the training dataset, as it contains more than 83 million registered users [4], providing a rich and diverse source of human sentiments.

4 Experimentation and Results

This section reports the experiments conducted and the corresponding results. More specifically, in this section, the evaluation of the performance of these algorithms and their ability to generalize to other sentiment recognition problems is explored.

4.1 Description of Problems and Datasets

The IMDb dataset contains 50,000 reviews and is divided into 25,000 train and 25,000 test reviews. Movie reviews are labeled as positive or negative. The first problem is a classification problem, in which a variety of classification algorithms is being evaluated. After preprocessing and cleaning the data, there are 57,210 different words in the train set (25,000 reviews). Since there is no limitation on the run-time of the algorithms and the memory size, a large set of words was chosen, which leads to better accuracy rates. Therefore, for every classifier, the Count Vectorizer and the TF-IDF Vectorizer techniques are implemented. The main contribution of this work is to examine the generalization of the trained with the IMDb dataset classifiers on other sentiment recognition problems. A large set of words is chosen, in order to see if the classification models can generalize well to other datasets, since with larger sets of words (i.e., dictionaries) it is more likely to achieve better generalization results. More specifically, two different approaches are implemented. The first, takes all the 57,210 words for constructing the feature vectors. The second one, takes the top 50,000 words or combinations of words (i.e., bi-grams) for constructing the feature vectors. The datasets used to measure how well these models generalize include Women's E-Commerce Clothing Reviews Dataset [17], Amazon Fine Food Reviews Dataset [18], Datafiniti's Business Hotel Reviews Dataset [19], Consumer Reviews of Amazon Products Dataset [20] and two Twitter Datasets [21, 22]. The first four datasets have a score column with values between 1 to 5, therefore in order to determine if a sentiment is positive or negative, reviews with score ≥ 4 are mapped to label 1 (positive review) and reviews with score ≤ 2 are mapped to label 0 (negative review). The Twitter datasets are ready to be used directly in binary classification models (0 for negative and 1 for positive sentiment) (Table 1).

Table 1. Datasets utilized in this work and corresponding samples.

Datasets	Test samples
IMDb large movie review dataset v1.0	25,000
Women's e-commerce clothing reviews	20,615
Amazon fine food reviews	36,813
Datafiniti's business hotel reviews	8,438
Consumer reviews of Amazon products	33,128
Tweets dataset - Ayoub Benaissa	99,989
Sentiment140 dataset with 1.6 million tweets	100,000

4.2 Results

As already mentioned, the IMDb dataset contains 50,000 reviews and the 25,000 of them are used to train the classification models. The other four datasets consisted of more test samples but since their label was mapped to 1 and 0 if they had score ≥ 4 or score ≤ 2 respectively, records with score equal to 3 were ignored, as they did not serve any purpose. The first Twitter dataset was fully used and from the second Twitter dataset 1/16 of the data was randomly selected. The results are shown in Tables 2 and 3 and they are derived from the following classification algorithms: Logistic Regression (LR), Bernoulli Naïve Bayes (BNB), Multinomial Naïve Bayes (MNB), Linear Support Vector Machine (Linear SVM), Neural Network (NN1) and the following vectorization techniques: Count Vectorizer, TF-IDF Vectorizer. These two vectorization techniques are also implemented with the second approach, which considers max features (50,000) and bi-grams and is tested on the second Neural Network (NN2). The results on the IMDb Dataset concern the 25,000 reviews used for testing.

Table 2. The accuracy of the models using the **Count Vectorizer technique**.

Classifier	Experiments						
	Trained on this set (50% split)	IMDb trained classifiers evaluated on various datasets					
	IMDb dataset	Clothing dataset	Food dataset	Hotel dataset	Product dataset	1st Twitter dataset	2nd Twitter dataset
Logistic Regression	0.85824	0.80392	**0.75470**	**0.77281**	**0.84876**	**0.59829**	0.58368
Bernoulli Naïve Bayes	0.8196	0.45651	0.40895	0.57703	0.29890	0.45524	0.51936
Multinomial Naïve Bayes	0.82076	0.70972	0.5827	**0.77447**	0.75066	0.57522	0.58787
Linear SVM	0.8318	0.69876	0.70211	0.71284	0.80919	0.57536	0.56601
Neural Network I	**0.8769**	**0.8385**	0.7287	**0.7817**	**0.8405**	**0.6043**	**0.5979**
Neural Network II	**0.8865**	**0.8432**	**0.7659**	**0.7687**	0.8156	0.5774	**0.5907**

Table 3. The accuracy of the models using the **TF-IDF Vectorizer technique**.

Classifier	Experiments						
	Trained on this set (50% split)	IMDb trained classifiers evaluated on various datasets					
	IMDb dataset	Clothing dataset	Food dataset	Hotel dataset	Product dataset	1st Twitter dataset	2nd Twitter dataset
Logistic Regression	**0.87996**	**0.84826**	**0.78792**	**0.82223**	**0.84964**	**0.61008**	**0.6053**
Bernoulli Naïve Bayes	0.8196	0.45651	0.40895	0.57703	0.29890	0.45524	0.51936
Multinomial Naïve Bayes	0.82744	0.69721	0.57618	0.73631	0.73025	0.56942	0.58517
Linear SVM	**0.87516**	**0.83371**	0.74568	0.78916	0.81010	**0.59899**	**0.59454**
Neural Network I	**0.8786**	0.7392	0.6747	0.7234	0.7479	0.5800	**0.5952**
Neural Network II	**0.8888**	**0.8488**	**0.7776**	**0.8072**	0.8310	0.5997	**0.5944**

4.3 Discussion on the Results

From the experimental results listed above, the following conclusions arise:

1. In each model, TF-IDF technique provides higher accuracy rates, except for Multinomial Naïve Bayes and Neural Network I, where Count Vectorizer technique gives higher percentages. More specifically, in these two models, TF-IDF is better on the IMDb test set, resulting from the dataset in which the models have been trained and worse in other datasets. This means that it does not generalize as good as the other technique.

   a. The Multinomial classifier is used when features have discrete values. In practice, fractional counts such as TF-IDF also work [23], however they lose some accuracy, as shown in the results.
   b. As far as the Neural Network I is concerned, which approaches the task by taking all the 57,210 words to construct the feature vectors, Count Vectorizer outperforms TF-IDF Vectorizer. This approach is followed by all the models except for the Neural Network II, which takes the max features (50,000) and bi-grams.

2. The Bernoulli Naïve Bayes classifier gives the same accuracy rates for both techniques, in each dataset. Bernoulli Naïve Bayes is used for features with binary or boolean values. Therefore, TF-IDF values are converted to 0 or 1 (e.g., float values equal to zero are mapped to zero and float values greater than zero are mapped to one), hence, to Count Vectorizer values. The comparison of Bernoulli and Multinomial models was conducted, since training Naïve Bayes models is fairly cheap.

3. Lastly, neither of the classifiers performs good on the two Twitter datasets. This is due to the fact that when analyzing tweets, different preprocessing techniques should be used. In particular, punctuation marks and emoticons may need to be preserved. Tweets preprocessing may involve converting many types of emoticons into tags that express their sentiment (i.e. :(→ **unhappy, sad**) [24]. Therefore, emoticons play an essential role in tweets analysis as they serve the purpose of detecting emotion. Moreover, tweets are often too short in length, which makes the sentiment analysis process difficult.

Generally, as far as the Count Vectorizer technique is concerned, the models with the highest accuracy are the two Neural Networks and the Logistic Regression classifier. Regarding the TF-IDF Vectorizer technique, the models with the highest accuracy are the Neural Network II and the Logistic Regression classifier. The Neural Network I and the Linear SVM only performed good accuracy rates in some datasets.

Finally, comparing all the mentioned models, those with the best accuracy rates are the Neural Network II and the Logistic Regression classifiers using the TF-IDF Vectorizer technique. These models achieve accuracies between 0,78 to 0,88, excluding the two Twitter datasets.

5 Conclusion

In this work, several classification algorithms were implemented and evaluated on sentiment-labeled datasets. The current state-of-the-art accuracy on IMDb [24], scores 97.42%, using document embeddings trained with cosine similarity. The scope of this work is different. The goal is not only to achieve good accuracy rates but also to get good generalization results on various datasets, by using traditional machine learning techniques. Logistic Regression and the second implementation of the Neural Network worked best on the IMDb dataset. More specifically, an accuracy score of 0,8888 was achieved on this dataset. Furthermore, the impact of constructing the feature vectors, on the performance of sentiment analysis classifiers, was analyzed. The Neural Network II with the second approach which takes the top 50,000 words or bi-grams to construct the feature vectors, generalizes better that the Neural Network I on every dataset. Common preprocessing techniques were used such as lowercasing, lemmatization, stopwords and non-word removal. All techniques provided significant improvements to the classifiers performances. Some of the techniques simply removed noise from the data, while others increased the importance of specific words or combinations of words. None of the classifiers performed well in the Twitter datasets. These moderate percentages may also be due to the misspellings occurring in data [25]. It is obvious that analyzing social media data is hard, as many of them are composed of misspelled words and special words or expressions. These are difficult for researchers to understand, as they are not used in everyday speech.

However, sentiment analysis is not only useful for classification problems. An important perspective for future work would be to use sentiment analysis as a recommendation tool. Finally, datasets with discrete rating scores of each review (e.g. from 1 to 5) could potentially lead to better performance of the models used.

References

1. Angiani, G., et al.: A comparison between preprocessing techniques for sentiment analysis in Twitter. In: KDWeb (2016)
2. Cambria, E., Olsher, D., Rajagopal, D.: SenticNet 3: a common and common-sense knowledge base for cognition-driven sentiment analysis. In: Twenty-Eighth AAAI Conference on Artificial Intelligence (2014)
3. Younis, E.M.G.: Sentiment analysis and text mining for social media microblogs using open source tools: an empirical study. Int. J. Comput. Appl. 112(5), 44–48 (2015)
4. Lopez, B., Minh, A.N., Xavier, S.: IMDb sentiment analysis. COMP 551 - Group 17 (2019)
5. Hirose, A., Ozawa, S., Doya, K., Ikeda, K., Lee, M., Liu, D. (eds.): ICONIP 2016. LNCS, vol. 9950. Springer, Cham (2016). https://doi.org/10.1007/978-3-319-46681-1
6. Pasi, G., Piwowarski, B., Azzopardi, L., Hanbury, A. (eds.): ECIR 2018. LNCS, vol. 10772. Springer, Cham (2018). https://doi.org/10.1007/978-3-319-76941-7
7. Camacho-Collados, J., Pilehvar, M.T.: On the role of text preprocessing in neural network architectures: an evaluation study on text categorization and sentiment analysis. arXiv preprint arXiv:1707.01780 (2017)
8. Camacho-Collados, J., Pilehvar, M.T.: From word to sense embeddings: a survey on vector representations of meaning. J. Artif. Intell. Res. 63, 743–788 (2018)

9. Wagh, B., Shinde, J.V., Kale, P.A.: A Twitter sentiment analysis using NLTK and machine learning techniques. Int. J. Emerg. Res. Manag. Technol. **6**(12), 37–44 (2017)
10. Martineau, J.C., Finin, T.: Delta TFIDF: an improved feature space for sentiment analysis. In: Third International AAAI Conference on Weblogs and Social Media (2009)
11. Toman, M., Tesar, R., Jezek, K.: Influence of word normalization on text classification. Proc. InSciT **4**, 354–358 (2006)
12. Aisopos, F., Papadakis, G., Varvarigou, T.: Sentiment analysis of social media content using N-Gram graphs. In: Proceedings of the 3rd ACM SIGMM International Workshop on Social Media (2011)
13. Avinash, M., Sivasankar, E.: A study of feature extraction techniques for sentiment analysis. In: Abraham, A., Dutta, P., Mandal, J., Bhattacharya, A., Dutta, S. (eds.) Emerging Technologies in Data Mining and Information Security, vol. 814, pp. 475–486. Springer, Singapore (2019). https://doi.org/10.1007/978-981-13-1501-5_41
14. Tarimer, İ., Çoban, A., Kocaman, A.E.: Sentiment analysis on IMDB movie comments and Twitter data by machine learning and vector space techniques. arXiv preprint arXiv:1903. 11983 (2019)
15. Pelaez, A., Talal, A., Mohsen, G.: Sentiment analysis of IMDb movie. Mach. Learn. **198**(536) (2015)
16. Oswal, N.: Predicting rainfall using machine learning techniques. arXiv preprint arXiv:1910. 13827 (2019)
17. Nicapotato: Women's E-Commerce Clothing Reviews. Kaggle, 3 February 2018. www.kaggle.com/nicapotato/womens-ecommerce-clothing-reviews
18. Stanford Network Analysis Project: Amazon Fine Food Reviews. Kaggle, 1 May 2017. www.kaggle.com/snap/amazon-fine-food-reviews
19. Datafiniti: Hotel Reviews. Kaggle, 24 June 2019. www.kaggle.com/datafiniti/hotel-reviews
20. Datafiniti: Consumer Reviews of Amazon Products. Kaggle, 20 May 2019. www.kaggle.com/datafiniti/consumer-reviews-of-amazon-products
21. Youben: Twitter Sentiment Analysis. Kaggle, 21 October 2018. www.kaggle.com/youben/twitter-sentiment-analysis/data
22. mistryjimit26: Twitter Sentiment Analysis Basic. Kaggle, 21 September 2018. www.kaggle.com/mistryjimit26/twitter-sentiment-analysis-basic/data
23. Kibriya, A.M., Frank, E., Pfahringer, B., Holmes, G.: Multinomial Naive Bayes for text categorization revisited. In: Webb, G.I., Yu, X. (eds.) AI 2004. LNCS (LNAI), vol. 3339, pp. 488–499. Springer, Heidelberg (2004). https://doi.org/10.1007/978-3-540-30549-1_43
24. Thongtan, T., Phienthrakul, T.: Sentiment classification using document embeddings trained with cosine similarity. In: Proceedings of the 57th Annual Meeting of the Association for Computational Linguistics: Student Research Workshop (2019)
25. Agarwal, A., et al.: Sentiment analysis of Twitter data. In: Proceedings of the Workshop on Language in Social Media (LSM 2011) (2011)

Opinion Mining of Consumer Reviews Using Deep Neural Networks with Word-Sentiment Associations

Petr Hajek[1]([⊠]) [iD], Aliaksandr Barushka[1] [iD], and Michal Munk[1,2] [iD]

[1] Institute of System Engineering and Informatics, Faculty of Economics and Administration,
University of Pardubice, Studentska 84, 532 10 Pardubice, Czech Republic
`petr.hajek@upce.cz, aliaksandr.barushka@student.upce.cz,`
`mmunk@ukf.sk`
[2] Department of Computer Science, Constantine the Philosopher University in Nitra, 949 74
Nitra, Slovakia

Abstract. Automated opinion mining of consumer reviews is becoming increasingly important due to the rising influence of reviews on online retail shopping. Existing approaches to automated opinion classification rely either on sentiment lexicons or supervised machine learning. Deep neural networks perform this classification task particularly well by utilizing dense document representation in terms of word embeddings. However, this representation model does not consider the sentiment polarity or sentiment intensity of the words. To overcome this problem, we propose a novel model of deep neural network with word-sentiment associations. This model produces richer document representation that incorporates both word context and word sentiment. Specifically, our model utilizes pre-trained word embeddings and lexicon-based sentiment indicators to provide inputs to a deep feed-forward neural network. To verify the effectiveness of the proposed model, a benchmark dataset of Amazon reviews is used. Our results strongly support integrated document representation, which shows that the proposed model outperforms other existing machine learning approaches to opinion mining of consumer reviews.

Keywords: Opinion mining · Consumer review · Word embedding · Lexicon · Sentiment · Deep neural network

1 Introduction

Opinion mining (sentiment analysis) of consumer reviews studies consumers' opinions on products and services [1]. The increasing number of users on online platforms produces a huge number of online product reviews. In the last two decades, opinion mining has become one of the most important text classification tasks because consumers' opinions affect the purchase decisions of other consumers. In addition, consumers' opinions in online reviews provide invaluable insights into consumer behavior and are thus

I. Maglogiannis et al. (Eds.): AIAI 2020, IFIP AICT 583, pp. 419–429, 2020.
https://doi.org/10.1007/978-3-030-49161-1_35

central to companies. The large number of consumer reviews available across diverse online sources has led to the necessity of employing automated opinion mining systems. Numerous machine learning methods have been used for this task, including methods with supervised learning and methods exploiting sentiment lexicons [1]. Recently, deep neural networks have emerged as an effective tool. Multiple layers enable learning complex representations of features [2]. Many deep neural networks in this domain use word embeddings as input features. Thus, words are transformed from a high-dimensional sparse space to lower-dimensional dense vectors, representing latent features and word context.

Opinion mining has been investigated at three levels of granularity, namely the document, sentence and aspect levels. For example, product reviews can be represented as documents classified into positive or negative opinion categories. Note that in this task, it is assumed that the review concerns a single product entity. In sentence-level categorization, only opinionated sentences must be first selected. Aspect-level opinion mining requires the identification of a product's aspect (target). In other words, this approach comprises several subtasks such as aspect extraction and aspect opinion classification.

Concerning the features used for opinion mining of consumer reviews, the bag-of-words model represents a traditional document representation in which word frequencies are calculated for each word (phrase) in the vocabulary [3]. However, this approach results in high-dimensional sparse document representation. Moreover, this representation ignores word order. In the case of using n-grams instead of single words, a short context is considered. To overcome these problems, word embeddings were introduced to produce low-dimensional dense word representation [4–7]. Compared with bag-of-words, word embeddings are also more effective in modeling word context and word meaning. After the appropriate document representation is generated, various neural network models can be employed for opinion classification. Alternatively, neural networks can be used to produce word embeddings; then other machine learning methods, such as support vector machines, can be used for the classification task [8].

The core problem of word embedding representations in existing studies is that the sentiment polarity and intensity of the words are ignored. As a result, a word embedding may comprise words with opposite sentiment polarity. This study aims to overcome this problem by developing a deep neural network model integrating word embeddings with their sentiment associations obtained from a wide range of lexicons. To further improve the performance of the opinion classifier in domain-specific context of reviews on different products, bag-of-words features are incorporated into the model.

The rest of this paper is structured as follows. Section 2 briefly reviews the recent advances in deep learning for opinion mining of consumer reviews. Section 3 outlines the proposed model. In Sect. 4, the benchmark dataset is introduced. Section 5 presents the results of the experiments in comparison with existing approaches. Section 6 presents future research directions and concludes the paper.

2   Deep Learning for Opinion Mining of Consumer Reviews – A Literature Review

This section reviews existing deep neural network (DNN)-based approaches to opinion mining of consumer reviews. As demonstrated in earlier studies, NNs outperform other traditional machine learning methods such as support vector machine (SVM) and Naïve Bayes (NB) in this task, irrespective of the context of balanced/unbalanced datasets [9]. However, the initial efforts in this domain relied on a traditional bag-of-words model that produced high-dimensional and sparse datasets. It should be noted that shallow NNs are not effective in handling sparse datasets [10]. By contrast, DNNs have the capacity to overcome this problem by capturing more complex features from the data. A DNN unsupervised learning approach was developed in [11] to show that word representation can be effectively learned by a stacked denoising autoencoder and that this representation can also be easily adapted to different review domains. To address the problem of scalability with the high-dimensional bag-of-words representation of the traditional autoencoders, a semisupervised autoencoder was developed for sentiment analysis in [12]. Supervision is introduced into the model via the loss function obtained from a linear classifier. Convolutional NNs (CNNs) were also employed to use the bag-of-words representation [3], which was also one of the first attempts to effectively use word order for opinion classification.

To further improve the performance of DNNs in opinion classification, vector representation models such as Word2Vec [13, 14] and Glove [15] were used to generate dense documents by reconstructing the linguistic context of the words. As a result, words that share a common context are located close to each other in the vector space, and the dimensionality of the space is reduced to several hundred word embeddings. CNNs and long short-term memory (LSTM) NNs were used to learn sentiment representation from word embeddings by [4]. In the next step, document representation was learned using gated recurrent units (GRUs). Different approaches for generating word embeddings were combined in a CNN model that outperformed SVM and NB. Another proposed CNN model integrates word embeddings with the representation of user text, thus incorporating user preferences [5]. Similarly, user and product information were utilized in an LSTM model with word and sentence attention [6]. To overcome the problem of the memory unit with long texts, a cached LSTM model was developed to capture the overall semantic representation [7]. Cross-domain sentiment classification represents another challenge in related literature. To learn a document representation that can be shared across domains, an end-to-end adversarial memory network was introduced in [16].

Recently, a cross-modality consistent regression model was employed to utilize three different CNN models with attention mechanisms, namely semantic, lexicon and sentiment representations. It was shown that sentiment and lexicon representations overcome the disadvantages of semantic embeddings in Twitter sentiment analysis [17]. Indeed, word embeddings used in previous studies ignore the sentiment polarity and sentiment intensity of the words and, hence, often combine words with different sentiment polarity. This may lead to misrepresentation of the documents in the context of sentiment analysis. Moreover, the hybrid representation models combining word embeddings with the traditional bag-of-words representation may further improve the classification performance in related tasks due to highly domain-specific context [18, 19]. Product reviews from different domains is exactly such a task. Inspired by these observations, the original contribution of this study is the proposal of a DNN model integrating word embeddings, bag-of-words and a wide range of sentiment polarity and sentiment intensity features to overcome the problems of the above approaches. Notably, word-sentiment associations enable to obtain both the meaning and sentiment intensity of the words in the review representation. Deep feed-forward neural network (DFFNN) was employed in this integrated model to effectively handle the high-dimensional sparse bag-of-words representation [10].

3   DNN Model with Word-Sentiment Associations

The architecture of the proposed DNN with word-sentiment associations (DNN-WSA) model for opinion mining of consumer reviews is presented in Fig. 1. The DFFNN with two dense hidden layers was used to process the variety in the input features, including both the word-sentiment representation and the n-gram representation.

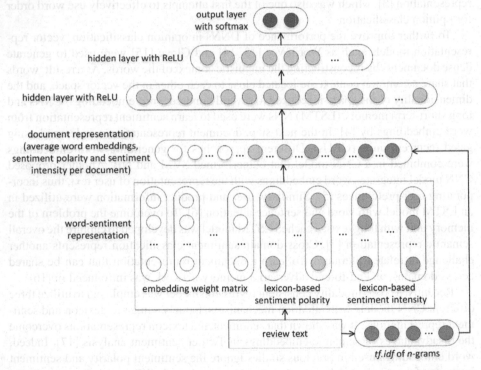

Fig. 1. The proposed DNN-WSA architecture for opinion mining of consumer reviews.

The word-sentiment representation is generated in two steps. First, word embeddings are trained using the Skip-Gram model because it is reportedly more effective than its competitors in exploiting the word context [13]. Second, the vocabulary obtained from the corpus of reviews is compared with several lexicons to append lexicon-based sentiment polarity and intensity.

To calculate the embedding weight matrix, the adapted embedding function is applied to each word w_t in the vocabulary. The embedding function is adapted for the sequence

$W = \{w_1, w_2, ..., w_t, ..., w_T\}$ of training words so that the following objective function is maximized

$$E = \frac{1}{T}\sum_{t=1}^{T}\sum_{-c \leq j \leq c} \log p(w_{t+j}|w_t),\qquad(1)$$

where c represents the context window radius (how many surrounding words are considered), and $p(w_{t+1}|w_t)$ is the probability of the output word given the input words calculated using the hierarchical softmax algorithm

$$p(w_O|w_I) = \prod_{j=1}^{L(w)-1} \sigma\left(\llbracket n(w, j+1) = \mathrm{ch}(n(w,j))\rrbracket v_{n(w_O,j)}^T v_{w_I}\right),\qquad(2)$$

where w_I and w_O are input and output words, respectively; v_w and v'_w denote the vector representations of the input and output words, respectively; $n(w, j)$ is the j-th node in the binary tree; $L(w)$ is the path length in the tree; $\mathrm{ch}(n)$ represents a child node; and $\sigma(x)$ denotes a sigmoidal function, where if x is true, then $\llbracket x \rrbracket = 1$; otherwise $\llbracket x \rrbracket = -1$. To obtain the document representation for the next layer in the DNN-WSA architecture, the mean values of the vectors from the embedding weight matrix were calculated.

To complement the word-sentiment representation with the sentiment polarity and intensity, we used several predefined sentiment lexicons. To obtain a reliable sentiment assessment, it is suggested not to rely on a single lexicon [20]. Moreover, the combination of lexicon-based sentiment indicators overcomes the problem of susceptibility to indirect opinions typically present in the machine learning models. To calculate sentiment polarity, we used two handcrafted lexicons of positive and negative words: Bing Liu's opinion lexicon [21] and OpinionFinder [20]. One shortcoming of these lexicons is that equal weight is assigned to all words regardless of their sentiment intensity. To address this issue, we incorporated the sentiment intensity indicators obtained from the following lexicons with pre-trained sentiment strengths [20, 22]: S140, NRC Hashtag, AFINN and SentiWordNet. Thus, the overall positive and negative scores can be calculated for each lexicon. In addition, the combination of several lexicons ensures higher lexical coverage [20].

To obtain the n-gram representation, the weight of each n-gram is calculated as follows

$$\omega_{ij} = \left(1 + \log(tf_{ij})\right) \times \log(N/df_i),\qquad(3)$$

where ω_{ij} denotes the weight of the i-th n-gram in the j-th document (review); $j = 1, 2, ...$, N; and tf_{ij} and df_i represent term and document frequency, respectively. Thus, review length is considered, and a relatively higher weight is assigned to rare n-grams. For further processing, the n-grams are ranked according to their weights, and top n-grams are selected to enter the document representation layer in the DNN architecture.

The next two hidden layers are used to process the complex relationship between the document representation and output sentiment positive/negative classes. To avoid overfitting and to make the training more effective, we used dropout regularization (dropout rate of 0.2 and 0.5 for the input and the two hidden layers, respectively) and ReLU (rectified linear units), respectively. The mini-batch gradient descent algorithm with $b = 100$ mini-batches, a learning rate of 0.1 and 1,000 iterations provided us with good and stable convergence behavior. Different numbers n_{h1} and n_{h2} of ReLU in the two hidden layers $= \{2^4, 2^5, 2^6, 2^7\}$ were tested to obtain the optimal architecture. As presented below, the best results were obtained for $n_{h1} = 2^5$ and $n_{h2} = 2^4$ neurons. Note that we also experimented with one hidden layer but without improvement. The objective function was represented by cross-entropy loss. The overall complexity of the proposed model can be expressed as $O(b \times I \times (m \times n_{h1} + n_{h1} \times n_{h2} + n_{h2} \times n_O))$, where I is the number of iterations; m denotes the number of features in the document representation layer; and n_{h1}, n_{h2} and n_O represent the numbers of neurons in the first and second hidden layers and the output layer, respectively.

4 Data and Preprocessing

For the experiments, a large enough Amazon dataset that is openly accessible at Kaggle[1] was used. The dataset, provided by Xiang Zhang, was originally used in [23] to classify opinions in consumer reviews using temporal CNNs with character-level features. The dataset was collected from the Stanford Network Analysis Project since 1994 [24], resulting in ~34 million reviews from ~6.6 million users on ~2.4 million products. The mean character length of the reviews was 764 (90.9 words). Extremely short and long reviews were discarded, and duplicates were removed. Users' rating scores were used to categorize the consumer reviews into positive and negative classes. More precisely, labels 1 and 2 were converted to negative opinion, and the scores of 4 and 5 were transformed to positive opinion. We used the testing data from the original dataset, represented by 130,000 samples from each score category. Overall, the dataset comprised 400,000 reviews evenly distributed into positive and negative opinion classes. Review title and review content were used in the dataset.

In the data pre-processing step, we performed tokenization (using the following delimiters: ".,;:'"()?!"), removal of stopwords (using the Rainbow list for noise reduction), and transformation to lowercase letters.

5 Experimental Results

The experiments were conducted on the Amazon dataset of 400,000 reviews. To learn word embeddings, we used the Skip-Gram model trained on the Amazon dataset. As shown in Fig. 2, we experimented with different settings of the model; the best performance was achieved with 200 word embeddings and context window radius $c = 5$. The Skip-Gram model was trained in the Deeplearning4j environment (distributed, open-source DNN library written for Java, compatible with Scala or Clojure and integrated with distributed computing frameworks Apache Spark and Hadoop). Regarding the bag-of-words representation, the top 1,000 n-grams (unigrams, bigrams and trigrams) were generated according to their $tf.idf$ (term frequency – inverse document frequency) weights in agreement with the previous literature [25]. To obtain the word-sentiment associations, the AffectiveTweets package was employed.

[1] https://www.kaggle.com/bittlingmayer/amazonreviews.

In our experiments, three evaluation measures were considered: accuracy (Acc), area under receiver operating characteristic curve (AUC), and F-score. To evaluate the performance of the proposed model, stratified 5-fold cross-validation was performed. The mean values and standard deviations are presented.

In a further set of experiments, we examined the effects of the used word representations. Figure 3 shows that the DNN model using lexicon-based sentiment features had the worst performance. More precisely, the DNNs with n-gram and Skip-Gram features increased accuracy by 2.7% and 3.0%, respectively, compared with DNN-LexSent. DNN-BoW and DNN-SkipGram performed similarly in terms of all the evaluation measures. The DNN-WSA model performed best with a 3.8% increase in accuracy compared with the DNN-SkipGram model. Overall, the combination of the three word representations performed significantly better than the baseline models at the 5% significance level using the Wilcoxon signed rank test.

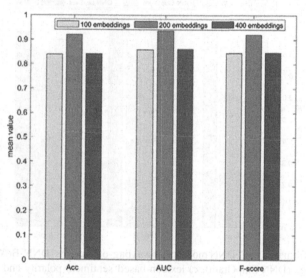

Fig. 2. The effect of the number of word embeddings on the performance of the DFFNN model with two hidden layers of $n_{h1} = 2^5$ and $n_{h2} = 2^4$ neurons.

To comprehensively evaluate the effectiveness of the DNN-WSA model, we compared its performance against the following existing models:

- Improved Naïve Bayes (INB-1) [26] accommodates the sentiment word using the SentiWordNet lexicon in the feature extraction component. Following [26], we extracted the unigrams, bigrams and sentiment patterns.
- Support vector machine with word sense disambiguation (SVM-WSD) [27] uses adverbs scored using the SentiWordNet lexicon as input features. Thus, positive and negative scores were assigned to adverbs, and SVM was trained using the LibLINEAR library. L2-regularized L2-loss SVM type was employed with cost parameter $C = 1$.

- A multiple classifier model combining three baseline classifiers, namely NB, SVM and bagging (NB+SVM+Bagging) [28]. In agreement with the original study, we used unigrams as features and voting as the meta-classifier.
- LSTM [4] and CNN [4] were used to obtain the semantic sentence-level representation. Following [6], the dimension of hidden/cell states was set to 200, corresponding to the number of word embeddings. The CNN architecture comprised the convolutional layer with five filters of size 5 and a max pooling layer of size 4. For both models, the sentence representation was fixed and the number of words in the sentence corresponded to the review with maximum length. Document representation for both models was produced as the composition of sentence representation using GRUs. Stochastic gradient descent with Adam optimizer was used to train both models in the Deeplearning4j environment.

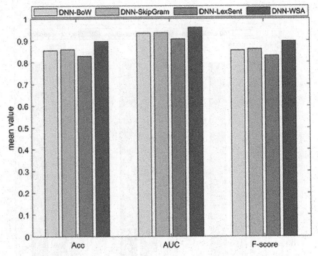

Fig. 3. The performance of FFDNN models using a) bag-of-n-grams (DNN-BoW), b) Skip-Gram word embeddings (DNN-SkipGram), c) lexicon-based sentiment polarity and intensity (DNN-LexSent), and d) all the word representations together (DNN-WSA). All the models were trained using two hidden layers with $n_{h1} = 2^5$ and $n_{h2} = 2^4$ neurons.

Table 1 shows the results of DNN-WSA in comparison with the above models. Note that the proposed model not only performed best in terms of all the used evaluation measures, but its performance was also significantly better at the 5% significance level using the Wilcoxon signed rank test, which demonstrates the effectiveness of the proposed model. SVM-WSD also performed well in terms of accuracy, especially when considering the computational time.

In this study, we adopted the testing time criterion - as suggested in related studies [19] - to show the real-time capacity of consumer review classifiers. The proposed DNN-WSA model performed the worst regarding time efficiency, but it can still be considered time efficient with approximately 7,700 reviews classified per second. Recall that the key determinants of the overall complexity are the numbers of iterations and features in the DNN model. Therefore, better time efficiency can be expected with the decrease in the number of *n*-grams. Overall, the DNN-WSA model performed well for both opinion categories, as indicated by the high value of AUC. The other two DNN models, LSTM and CNN, also performed well regarding AUC. Additionally, the high value of the *F*-score for DNN-WSA indicates a balanced performance in terms of precision (0.896 on average) and recall (0.899).

Table 1. Results of the experiments.

Model	Acc [%]	AUC	*F*-score	Testing time [s]
INB–1 [26]	78.84 ± 0.70	0.833 ± 0.008	0.796 ± 0.005	5.862 ± 0.363
SVM-WSD [27]	85.61 ± 2.67	0.856 ± 0.027	0.862 ± 0.022	**0.070 ± 0.011**
NB+SVM+Bagging [28]	83.10 ± 0.68	0.906 ± 0.006	0.833 ± 0.006	0.105 ± 0.011
LSTM [4]	84.05 ± 0.28	0.917 ± 0.003	0.841 ± 0.002	2.042 ± 0.128
CNN [4]	84.29 ± 0.17	0.921 ± 0.001	0.844 ± 0.003	8.139 ± 0.286
DNN–WSA (this study)	**89.70 ± 0.92**	**0.959 ± 0.011**	**0.897 ± 0.009**	10.398 ± 0.226

Notes: The best results are in bold. The experiments were conducted using AMD Opteron 6180 SE 2.50 GHz with twelve cores/threads and 256 GB RAM.

6 Conclusion

In this study, we proposed an efficient DNN model integrating word-sentiment associations for the opinion mining of consumer reviews. We proved the model's performance improvement compared with baseline word representations by conducting extensive experiments on the Amazon dataset. We compared the proposed DNN-WSA model with several existing approaches, including both DNNs and other machine learning methods. Hence, the effectiveness of the proposed model was demonstrated. The results of the experiments suggest that word-sentiment associations might be more effective than word representation based on word embeddings only. Integrating the word-sentiment associations with *n*-gram representation provides further improvement. However, such a word representation model leads to a partly sparse dataset, which necessitates further requirements for the opinion mining machine learning methods. We showed that the proposed DNN model can handle such a word representation model.

In future research, a more thorough analysis can be performed by investigating the word-sentiment associations at the entity/aspect level. One of the limitations of the proposed model is that only local features were captured. Therefore, alternative DNN models with attention mechanisms could be considered to overcome this limitation. A cross-domain modification of the model is another problem that needs to be addressed. The *n*-gram feature extraction used in this study does not consider the semantic similarity or the discriminative ability of words. Therefore, enhanced *n*-gram representations [29] are recommended to reduce the dimensionality and sparsity of the data. The application of an effective feature selection method may also lead to lower computational complexity and improved time efficiency [30]. Alternative embedding-based schemes can also be utilized [31].

Acknowledgments. This article was supported by the scientific research project of the Czech Sciences Foundation Grant No. 19-15498S and by the Student Grant Competition SGS_2020.

References

1. Liu, B.: Sentiment Analysis: Mining Opinions, Sentiments, and Emotions. The Cambridge University Press, Cambridge (2015)
2. Zhang, L., Shuai, W., Liu, B.: Deep learning for sentiment analysis: a survey. Wiley Interdisc. Rev.: Data Min. Knowl. Discov. **8**(4), e1253 (2018)
3. Johnson, R., Zhang, T.: Effective use of word order for text categorization with convolutional neural networks. In: Proceedings of the Conference of the North American Chapter of the Association for Computational Linguistics: Human Language Technologies, pp. 103–112 (2015)
4. Tang, D., Qin, B., Liu, T.: Document modelling with gated recurrent neural network for sentiment classification. In: Proceedings of the Conference on Empirical Methods in Natural Language Processing, pp. 1422–1432 (2015)
5. Tang, D., Qin, B., Liu, T.: Learning semantic representations of users and products for document level sentiment classification. In: Proceedings of the Annual Meeting of the Association for Computational Linguistics, pp. 1014–1023 (2015)
6. Chen, H., Sun, M., Tu, C., Lin, Y., Liu, Z.: Neural sentiment classification with user and product attention. In: Proceedings of the Conference on Empirical Methods in Natural Language Processing, pp. 1650–1659 (2016)
7. Xu, J., Chen, D., Qiu, X., Huang, X.: Cached long short-term memory neural networks for document-level sentiment classification. In: Proceedings of the Conference on Empirical Methods in Natural Language Processing, pp. 1660–1669 (2016)
8. Do, H.H., Prasad, P.W.C., Maag, A., Alsadoon, A.: Deep learning for aspect-based sentiment analysis: a comparative review. Expert Syst. Appl. **118**, 272–299 (2019)
9. Moraes, R., Valiati, J.F., Neto, W.P.: Document-level sentiment classification: an empirical comparison between SVM and ANN. Expert Syst. Appl. **40**, 621–633 (2013)
10. Barushka, A., Hajek, P.: Spam filtering using integrated distribution-based balancing approach and regularized deep neural networks. Appl. Intell. **48**(10), 3538–3556 (2018)
11. Glorot, X., Bordes, A., Bengio, Y.: Domain adaption for large-scale sentiment classification: a deep learning approach. In: Proceedings of the 28th International Conference on Machine Learning, ICML, pp. 513–520 (2011)
12. Zhai, S., Zhang, Z. M.: Semisupervised autoencoder for sentiment analysis. In: Proceedings of AAAI Conference on Artificial Intelligence, AAAI, pp. 1394–1400 (2016)
13. Mikolov, T., Sutskever, I., Chen, K., Corrado, G. S., Dean, J.: Distributed representations of words and phrases and their compositionality. In: Advances in Neural Information Processing Systems, NIPS, vol. 26, pp. 3111–3119 (2013)
14. Le, Q., Mikolov, T.: Distributed representations of sentences and documents. In: International Conference on Machine Learning, JMLR, vol. 32, pp. 1188–1196 (2014)
15. Pennington, J., Socher, R., Manning, C.D.: GloVe: global vectors for word representation. In: Proceedings of the Conference on Empirical Methods on Natural Language Processing, pp. 1532–1543 (2014)

16. Li, Z., Zhang, Y., Wei, Y., Wu, Y., Yang, Q.: End-to-end adversarial memory network for cross-domain sentiment classification. In: Proceedings of the International Joint Conference on Artificial Intelligence, pp. 2237–2243 (2017)
17. Zhang, Z., Zou, Y., Gan, C.: Textual sentiment analysis via three different attention convolutional neural networks and cross-modality consistent regression. Neurocomputing **275**, 1407–1415 (2018)
18. Sun, C., Du, Q., Tian, G.: Exploiting product related review features for fake review detection. Math. Probl. Eng. 1–7 (2016)
19. Hajek, P., Barushka, A., Munk, M.: Fake consumer review detection using deep neural networks integrating word embeddings and emotion mining. Neural Comput. Appl. 1–16 (2020)
20. Bravo-Marquez, F., Frank, E., Mohammad, S. M., Pfahringer, B.: Determining word-emotion associations from tweets by multi-label classification. In: 2016 IEEE/WIC/ACM International Conference on Web Intelligence, pp. 536–539. IEEE (2016)
21. Hu, M., Liu, B.: Mining and summarizing customer reviews. In: Proceedings of the 10th ACM SIGKDD International Conference on Knowledge Discovery and Data Mining, pp. 168–177. ACM (2004)
22. Kiritchenko, S., Zhu, X., Mohammad, S.M.: Sentiment analysis of short informal texts. J. Artif. Intell. Res. **50**, 723–762 (2014)
23. Zhang, X., LeCun, Y.: Text understanding from scratch. arXiv preprint arXiv:1502.01710 (2015)
24. McAuley, J., Leskovec, J.: Hidden factors and hidden topics: understanding rating dimensions with review text. In: Proceedings of the 7th ACM Conference on Recommender Systems, pp. 165–172 (2013)
25. Kouloumpis, E., Wilson, T., Moore, J.: Twitter sentiment analysis: the good the bad and the omg!. In: Fifth International AAAI Conference on Weblogs and Social Media, pp. 538–541 (2011)
26. Kang, H., Yoo, S.J., Han, D.: Senti-lexicon and improved Naïve Bayes algorithms for sentiment analysis of restaurant reviews. Expert Syst. Appl. **39**(5), 6000–6010 (2012)
27. Kausar, S., Huahu, X., Shabir, M.Y., Ahmad, W.: A sentiment polarity categorization technique for online product reviews. IEEE Access **8**, 3594–3605 (2019)
28. Catal, C., Nangir, M.: A sentiment classification model based on multiple classifiers. Appl. Soft Comput. **50**, 135–141 (2017)
29. Chen, X., Xue, Y., Zhao, H., Lu, X., Hu, X., Ma, Z.: A novel feature extraction methodology for sentiment analysis of product reviews. Neural Comput. Appl. **31**(10), 6625–6642 (2019)
30. Barushka, A., Hajek, P.: Spam detection on social networks using cost-sensitive feature selection and ensemble-based regularized deep neural networks. Neural Comput. Appl. 1–19 (2020)
31. Onan, A.: Deep learning based sentiment analysis on product reviews on Twitter. In: Younas, M., Awan, I., Benbernou, S. (eds.) Innovate-Data 2019. CCIS, vol. 1054, pp. 80–91. Springer, Cham (2019). https://doi.org/10.1007/978-3-030-27355-2_6

Sentiment Analysis on Movie Scripts and Reviews
Utilizing Sentiment Scores in Rating Prediction

Paschalis Frangidis [ID], Konstantinos Georgiou[✉] [ID], and Stefanos Papadopoulos [ID]

School of Informatics, Aristotle University of Thessaloniki, Thessaloniki, Greece
{frangidis,georgiouka,stefanospi}@csd.auth.gr

Abstract. In recent years, many models for predicting movie ratings have been proposed, focusing on utilizing movie reviews combined with sentiment analysis tools. In this study, we offer a different approach based on the emotionally analyzed concatenation of movie script and their respective reviews. The rationale behind this model is that if the emotional experience described by the reviewer corresponds with or diverges from the emotions expressed in the movie script, then this correlation will be reflected in the particular rating of the movie. We collected a dataset consisting of 747 movie scripts and 78.000 reviews and recreated many conventional approaches for movie rating prediction, including Vector Semantics and Sentiment Analysis techniques ran with a variety of Machine Learning algorithms, in order to more accurately evaluate the performance of our model and the validity of our hypothesis. The results indicate that our proposed combination of features achieves a notable performance, similar to conventional approaches.

Keywords: Natural Language Processing · Sentiment analysis · Machine learning · Prediction

1 Introduction

Movie scripts are an interesting source of text, due to the diverse display of sentiments expressed in them. In principle, they are a storytelling device where the screenwriter is trying to convey something meaningful. The script's "emotional charge" is usually a tool to achieve the aforementioned goal or a byproduct of the process- and indeed a very important and powerful one. It is probably the most immediate point of resonance and communication with the audience. Movies usually contain scenes where emotions alter dynamically, between happiness and sadness, calmness and anger in order to aid the narrative progression, while some works are characterized by an overarching emotional 'weight', such as sadness in a tragedy-based film. In order to achieve this goal, the script needs to be written in a manner that captures the appropriate sentiments and allows the actors to portray it in their performances. It is very common nowadays, for the audience and critics alike to express their enjoyment of a movie or lack thereof, in the form of online reviews. In sites like "Metacritic" or "Rotten Tomatoes" reviews, analyses and criticisms are collected to create an average estimation of how popular and well-received

© IFIP International Federation for Information Processing 2020
Published by Springer Nature Switzerland AG 2020
I. Maglogiannis et al. (Eds.): AIAI 2020, IFIP AICT 583, pp. 430–438, 2020.
https://doi.org/10.1007/978-3-030-49161-1_36

a movie is. This information has already been exploited in many applications including movie recommendations and genre classification systems. However, most reviewers do not simply state their like or dislike of a movie and instead tend to focus on the different feelings that the movie evoked in them. It would be an interesting inquiry to analyze the intended emotion of a movie's script, compare it to the received emotional response of the reviewer and study whether the unison or dissonance between the two are correlated or not. The primary goal of this study is to accumulate movie scripts and their respective reviews in order to examine the validity of the aforementioned statement and examine whether the relationship between the intended emotional weight of the movie and the received emotion of the reviewer can help in accurately predicting movie ratings. Our proposed model was tested in conjunction with more conventional approaches by running multiple experiments with different combinations of features like Vector Semantics, typical Sentiment Analysis and using different Machine Learning algorithms.

The remaining chapters are structured as follows: in Sect. 2, we present some related work and pinpoint our contribution to the topic. In Sect. 3, we analyze the dataset and present our key features, and in Sect. 4 the methodology applied is explained. In Sect. 5, the basic results of our study are presented and discussed, while in Sect. 6 we provide our main conclusions, accompanied by some interesting future work suggestions.

2 Background and Related Work

Sentiment analysis, as a field of study, has met a significant increase in its applicability in various sectors, as it can easily become cross disciplinary and facilitate different processes [8, 15, 16]. Its simplicity and straightforward approach have established it as a respected factor of text analysis. Indicative sectors where sentiment analysis is implemented are business systems, marketing campaigns and recommender systems [3].

In recent years, sentiment analysis in conjunction with NLP and ML techniques have been used in an array of different applications concerning movie scripts and their reviews. They have been used in order to identify patterns in movie structures [2] and learn to predict the following emotional state based on the previous [6], showing respectively that 'successful' movies follow specific narrative progressions and have a certain "flow" and consistency in the way that emotional states unfold. Additionally, it has been observed that existing binary ("positive/negative") sentiment analysis techniques have a 'positivity bias', favoring the learning of positive emotions over negative ones, hence "underestimating" the latter's existence. This can create a significant discrepancy in accuracy of around 10 to 30%. It is shown that taking into consideration meta-features (capital letters, punctuation and parts of speech) helps mitigate this problem [4].

Other approaches have tried to classify reviews based on semantically similar words [3] in order to detect communities of reviewers via clustering. Some interesting alternatives propose the identification of the driving aspects of a movie, them being mainly associated with the screenplay, the acting and the plot or some more particular characteristics (e.g. music, effects) [7], and how they are reflected on the reviews in sentiments. Results indicate that the acting and the plot are usually the most important factors that influence a review. However, variations in these studies which examine the same aspects in different movie genres may alter the driving factors in respective reviews. Typically, the methodology implemented to perform sentiment analysis, directly involves classic text mining solutions such as N-Gram extraction or Part-of-Speech tagging in combination with Naïve Bayes (NB) or Recurrent Neural Networks (RNN) [9]. However, different techniques, such as the use of a Gini Index or Support Vector Machine (SVM) approaches [10, 11] have increased the accuracy of results. The Skip Gram and Continuous Bag of Words (CBOW) models [3] have also shown promising results.

A recent research venture [1] has tried a mixture of techniques to extract the emotion out of reviews. Using a combination of emotion lexicon and word embeddings in order to extract reviews' sentiment, they acquire a satisfactory level of prediction to the reviewers' binary score of the movie. Moreover, it provides solutions for both English and Greek datasets. This particular approach is going to be extended in the current research. Regarding the classification of movie scripts based on NLP, there have been attempts in using subtitle files of different movies to ultimately predict their genre [8]. The premise is that via sentiment analysis of the subtitles' context, sentiment can be extracted and thus, the genre of the movie can be identified. Initial findings show that these techniques yield better results when applied to action, romance or horror movies. However, further analysis could strengthen the prediction margin.

The arising importance of sentiment analysis in movie and review evaluation has also prompted well known and respected online competition networks like Kaggle [11] to organize initiatives in order to suggest innovative solutions to the problem. It is evident, though, that in order to produce a well-documented solution, features have to be carefully selected, taking into consideration the variance of terms, the handling of negation and the treatment of opinion words [10].

Our paper contributes to the current scientific framework by exploiting data both from movie scripts and their corresponding reviews and by applying two specialized lexicons in order to conduct sentiment and emotion analysis. We believe our research to be of notable importance as it can indicate a new approach to the problem which treats the review text and the movie script as coexisting entities and merges them for optimal results.

3 Data Collection and Preprocessing

In order to gather data for movie scripts and movie reviews, we used simple scrapers in Python to crawl specific web pages. For the movie scripts, we used the IMSDB website, which contains more than 1100 movie scripts and drafts. Our reviews were gathered from the Rotten Tomatoes website, a well-known source for reviewing films, also by using a web scraper. The scripts gathered were in.txt format and the reviews were saved in a CSV file. Before continuing to the preprocessing phase, we noticed a significant imbalance in our review dataset which contained 55.886 fresh/positive movies and 22.193 rotten/negative. This imbalance would create inaccurate predictions in our model and in order to correct it, we performed undersampling, keeping all of the "rotten" and 25.000 of the "fresh" reviews. "Fresh/Rotten" values were transformed into 0 and 1 and were used as the model's target variable.

The scraper utilized for obtaining the movie scripts has a quite simple and easy to understand structure. It uses the Beautiful Soup package and redirects to the IMSDB website, locating all the script titles, categorized in alphabetical order. The scraper's first operation is to store all the <a> tags, corresponding to movie titles, in a list. Afterwards, it iterates the list of titles and redirects to the corresponding URLs, where it loads scripts in.txt files and creates a directory to store them. The scraper keeps only data present in <body> tags and opts to ignore tags like <head> or <footer>.

To obtain the corresponding movie reviews, we modified a Beautiful Soup based scraper which collected review texts and stored them to a shared CSV file. Our data were limited due to the reason that Rotten Tomatoes only contains an excerpt of the original review. During the collection phase, we filtered reviews which contained no text, as we deemed them unsuitable for our model.

In the preprocessing phase, we applied sentence splitting and word tokenization. Punctuation marks and stop-words were removed but only after experimenting and keeping track of the results with their inclusion. We then created a punctuation-removal list which consisted of all the punctuation marks we aimed to remove. At first, only exclamation marks were excluded from the punctuation list and the words "not" and "but" from the stopword-removal list in order to study their effects on the VADER sentiment analysis, as discussed in the next section. Their effect on the predictive accuracy was considered negligible, hence all stop words and punctuation marks were eventually removed. Finally, we applied part-of-speech tagging and lemmatization because we noticed that the lexicon-based sentiment analysis tools we selected lacked many derivatives and would lose out on a significant segment of our dataset.

4 Methodology

Our proposed model is based on calculating the emotional weight of a movie's script and combining it with the emotion expressed by the reviewer. For Sentiment and Emotion Analysis, we considered many different tools but ended up selecting VADER and NRC. VADER (or Valence Aware Dictionary and sEntiment Reasoner) is a binary Sentiment Analysis tool using a dictionary approach, containing 7.518 uni-grams including punctuation, slang words, initialisms, acronyms and emoticons. VADER receives a sentence as input and returns 4 values, negative, neutral, positive and compound which is the 'normalized weighted composite score'. Each output is ranging from −1,5 to +1,5, from 'Very Negative' to 'Very Positive', but we normalized them into a range of {0 to 1}. VADER is widely used [12, 13] and preferred as a sentiment analysis tool because of its advanced heuristics. It includes 5 built-in and pre-trained heuristics taking into consideration punctuation marks (especially exclamation points), capitalization, degree modifiers (boosters and dampeners) Shift Polarity (with words like "but") and Negation Handling using tri-grams.

NRC, on the other hand, is an emotion analysis tool, categorizing a sentence into eight emotions (anger, fear, anticipation, trust, surprise, sadness, joy, disgust) and two sentiments (positive, negative). Regarding the selected emotions in the NRC lexicon, there is a consensus among Psychology Researchers about Joy, Sadness, Anger, Disgust, Surprise and Fear being categorized as 'Basic Human emotions' showing their importance and universal nature [14]. In addition to these six emotions, the NRC contains words about Trust and Anticipation and although they may not be considered as 'basic emotions' by everyone, we believe them to be important especially for analyzing texts of movie scripts and their reviews. This lexicon contains more than 14000 words, with each one being scored on every emotion. We had to calculate the average of these ten emotion scores for every review, after having summed the scores for every word. We hypothesize that NRC is better suited for the model we are trying to build due to the complexity of movie scripts but both sentiment analysis tools were tested empirically.

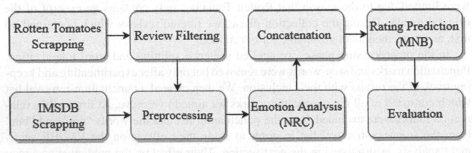

Fig. 1. Applied workflow

Our model requires the concatenation of the emotionally analyzed of both the movie scripts and their reviews using the NRC lexicon (Fig. 1). In order to evaluate our model more reliably, we reproduced many different conventional methods for movie rating predictions. As features we used Vector Semantics like CountVectorizer and TF-IDF in combination with NRC and VADER. Indicatively, the experimental combinations that were tested on the movie reviews alone were:

1. TF-IDF.
2. CountVectorizer.
3. VADER.
4. VADER combined with NRC.
5. VADER (including stop words 'not and but') combined with NRC.
6. CountVectorizer (CV) combined with VADER.
7. CV combined with NRC.
8. CV combined with VADER and NRC.

The dataset was split into train and test data, by 75% and 25% respectively. Finally, regarding the selection of machine learning algorithms we experimented with the following algorithms: Multinomial Naive Bayes (MNB), Logistic Regression (LR), SVM and Multilayer Perceptron (MLP). All algorithms were able to return satisfactory results but MNB was selected because it consistently resulted in higher predictive accuracy (Table 1) and was significantly faster computationally. We selected to test the four machine learning models against two feature combinations, CV as a representative of Vector Semantic approaches and CV in combination with NRC to add information about the emotions into the equation.

Table 1. F1-scores of the four compared ML models.

	MNB	LR	MLP	SVM
CV	**0.807**	0.791	0.789	0.565
CV with NRC	**0.768**	0.747	0.702	0.422

5   Results

The evaluation of the aforementioned experiments was conducted with the use of Accuracy, Precision, Recall and F1 as the selected metrics. The bar plot below (Fig. 2), presents an indicative selection of experiments, from the total that were performed, which we considered to be the most important. Based on the visualized results, it seems that the CountVectorizer method produces slightly better results than TF-IDF method, possibly because of better document representation. The VADER lexicon showed the worst performance when used on its own. Additionally, it appears that its heuristics, despite increasing the intensity of the sentiment, did not play a crucial role in improving the predictive accuracy. VADER's low performance was expected, as we initially hypothesized that binary sentiment analysis (positive/negative) was too simplistic to capture the complexity of a movie's script. We believe that the problem lies in the interpretive ambiguity of binary sentiment approaches. A negative sentiment score for a review can have two very different meanings. It may show legitimate dislike for the movie or it may express the reviewer's experience of sadness or horror which can technically be considered as "negative" emotions but are expected in an effective and well-made drama or horror movie respectively. Furthermore, confirming our hypothesis, the NRC lexicon greatly outperforms VADER by 10%, proving the importance of taking into consideration a multitude of emotions.

An unexpected finding is that the combination of Vector Semantics approaches with sentiment lexicons yields significantly different results based on which lexicon is applied. The VADER lexicon slightly enhances the performance of the CountVectorizer - if only by 0.1% - while the NRC lexicon reduces the performance. This can possibly be attributed to the relative increase in complexity of the model when combined with the multiple emotions of the NRC lexicon, which led to constrained results. Finally, our proposed model, which took into consideration both the review and the script emotions without any other NLP technique, managed to reach impressive precision percentages, only 0.8% lower than the best performing model but didn't quite reach the same level of accuracy. While not being the best performing model, it indicates the potential and validity of our initial hypothesis which dictated that the relationship between the expressed emotion of the movie and the received emotion by the reviewer can be a potent predictor for movie's rating.

Generally, Vector Semantics approaches (TF-IDF and CountVectorizer) performed much better than simply using sentiment and emotional lexicons. They seem better at identifying the importance of each word inside the text which helps the machine learning model to comprehend its structure and any possible hidden patterns. Sentiment analysis tools are already being used in a variety of cases, but they are still in a transitional stage.

NRC, especially, doesn't take at all into consideration the context of a word - it simply returns its pre-classified score - which we believe to be the most important reason for its lacking performance. Another factor that probably contributed to our model's restrained performance was the absence of the totality of each review. We only had a small excerpt from the whole review provided by Rotten Tomatoes. By analyzing the review in its totality it's possible that our model's predictive accuracy would have improved.

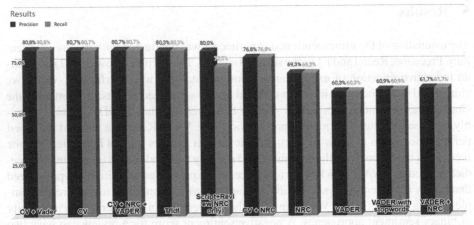

Fig. 2. Evaluation scores of the proposed model and the baseline methods. The proposed model is referred as "Script + Review (NRC only)".

6   Conclusions – Future Work

This paper explored the effect of incorporating emotion and sentiment analysis in predicting movie ratings and more specifically, the combinations of the movie script's intended emotion and the emotional response of the reviewer. Data were collected from relevant sources and contained a fair amount of scripts and reviews. We tried several machine learning models and experimented with the usage of sentiment and emotion analysis with two well-known lexicons, one intended for sentiment analysis and one for emotion analysis. We examined their performance on their own and in combination with Vector Semantics tools. The performance of our proposed model was actually quite impressive, almost reaching the best performing model in precision but we believe that the results were limited by the lack of negation handling and the fact that we used only excerpts of the reviews, provided by Rotten Tomatoes, and not the whole text. We believe that the accuracy of the model would improve if we were able to obtain the whole body of the reviews and thus have a larger corpus for analysis. However, the results may still end up being unsatisfying, due to the absence of negation handling and polarity shift which should also be tested. Of course, such enhancements would require different preprocessing of the data by keeping appropriate stop words or applying different representation methods.

Regarding suggestions for improving similar models in the future, we thought it would be useful to incorporate character and contextual analysis. Although sentiment can indeed be extracted from simple words, there should be additional weight to some character names. For example, the names of some famous villains can easily be associated with a specific sentiment, like fear, and thus should be considered non neutral words when conducting an analysis of sentiment. This can lead to better results in many cases and better characterization of the script's context as a whole.

Another possible expansion of the model would be the implementation of different document representation methods, such as Word2Vec embeddings or the use of N-Grams. It would be interesting to explore how different methods affect the quality of results and whether the combinations of lexicons and approaches would have similar outputs. Finally, we considered that a possible practical use of our model would be to expand its use not only in reviews but other forms of criticism like YouTube Comments or Tweets and apply the same methodology to give an early estimation of a movie's box office success or failure. It is obvious that such predictions will only be vague, as a movie's gross depends on a variety of factors (marketing, production, cast, merchandise etc.) but this model can certainly serve as a supplementary tool for further assurance of success.

Acknowledgements. This paper was directly based on an assignment we completed during our "Text Mining and Natural Language Processing" Course in the Data and Web Science MSc program of the School of Informatics of the Aristotle University of Thessaloniki. We would like to personally thank Dr. Grigorios Tsoumakas, our supervising professor, who provided direction, reviewed the project and encouraged us to submit it as a conference paper.

References

1. Giatsoglou, M., Vozalis, M., Diamantaras, K., Vakali, A., Sarigiannidis, G., Chatzisavvas, K.: Sentiment analysis leveraging emotions and word embeddings. Exp. Syst. Appl. **69**, 214–224 (2017)
2. Lee, S., Yu, H., Cheong, Y.: Analyzing movie scripts as unstructured text. In: Proceedings of IEEE Third International Conference on Big Data Computing Service and Applications 2017 (BigDataService), pp. 249–254. IEEE, San Fransisco (2017)
3. Chakraborty, K., Bhattacharyya, S., Bag, R., Hassanien, A.E.: Comparative sentiment analysis on a set of movie reviews using deep learning approach. In: Hassanien, A.E., Tolba, Mohamed F., Elhoseny, M., Mostafa, M. (eds.) AMLTA 2018. AISC, vol. 723, pp. 311–318. Springer, Cham (2018). https://doi.org/10.1007/978-3-319-74690-6_31
4. Kim, J., Ha, Y., Kang, S., Lim, H., Cha, M.: Detecting multiclass emotions from labeled movie scripts. In: IEEE International Conference on Big Data and Smart Computing (BigComp), 2018, pp. 590–594. IEEE, Shanghai (2018)
5. Kim, D., Lee, S., Cheong, Y.: Predicting emotion in movie scripts using deep learning. In: IEEE International Conference on Big Data and Smart Computing (Bigcomp) 2018, pp. 530–532. IEEE, Shanghai (2018)
6. Sahu, T., Ahuja, S.: Sentiment analysis of movie reviews: a study on feature selection & classification algorithms. In: International Conference on Microelectronics, Computing and Communications (MicroCom) 2016, pp. 1–6. IEEE, Durgapur (2016)

7. Parkhe, V., Biswas, B.: Sentiment analysis of movie reviews: finding most important movie aspects using driving factors. Soft Comput. **20**, 3373–3379 (2015). https://doi.org/10.1007/s00500-015-1779-1

8. Mesnil, G., Mikolov, T., Ranzato, M., Bengio, Y.: Ensemble of Generative and Discriminative Techniques for Sentiment Analysis of Movie Reviews. CoRR (2014)

9. Sureja, N., Sherasiya, F.: Using sentimental analysis approach review on classification of movie script. Int. J. Eng. Dev. Res. **5**, 616–620 (2017)

10. Manek, A., Shenoy, P., Mohan, M., Venugopal, K.R.: Aspect term extraction for sentiment analysis in large movie reviews using Gini Index feature selection method and SVM classifier. World Wide Web **20**, 135–154 (2016)

11. Rotten Tomatoes Movie Database. https://www.kaggle.com/ayushkalla1/rotten-tomatoes-movie-database. Accessed 07 Jan 2020

12. Park, C., Seo, D.: Sentiment analysis of twitter corpus related to artificial intelligence assistants. In: 5th International Conference on Industrial Engineering and Applications (ICIEA). 2018, pp. 495–498. IEEE, Singapore (2018)

13. Newman, H., Joyner, D.: Sentiment analysis of student evaluations of teaching. In: Penstein Rosé, C., et al. (eds.) AIED 2018. LNCS (LNAI), vol. 10948, pp. 246–250. Springer, Cham (2018). https://doi.org/10.1007/978-3-319-93846-2_45

14. Kowalska, M., Wróbel, M.: Basic Emotions. In: Zeigler-Hill, V., Shackelford, T. (eds.) Encyclopedia of Personality and Individual Differences. Springer, Cham (2017)

15. Pang, B., Lee, L.: Opinion mining and sentiment analysis. Found. Trends® Inf. Retriev. **2**, 1–135 (2008)

16. Medhat, W., Hassan, A., Korashy, H.: Sentiment analysis algorithms and applications: a survey. Ain Shams Eng. J. **5**, 1093–1113 (2014)

The MuseLearn Platform: Personalized Content for Museum Visitors Assisted by Vision-Based Recognition and 3D Pose Estimation of Exhibits

G. Styliaras[1], C. Constantinopoulos[1], P. Panteleris[2], D. Michel[2], N. Pantzou[1],
K. Papavasileiou[1], K. Tzortzi[1], A. Argyros[2], and D. Kosmopoulos[1(✉)]

[1] University of Patras, Patras, Greece
{gstyl,kkonstantino,cpapavas,ktzortzi,dkosmo}@upatras.gr,
padeler@ics.forth.gr
[2] Foundation for Research and Technology – Hellas (FORTH), Patras, Greece
npantzou@upatras.gr, {michel,argyros}@ics.forth.gr

Abstract. MuseLearn is a platform that enhances the presentation of the exhibits
of a museum with multimedia-rich content that is adapted and recommended for
certain visitor profiles and playbacks on their mobile devices. The platform consists mainly of a content management system that stores and prepares multimedia
material for the presentation of exhibits; a recommender system that monitors
objectively the visitor's behavior so that it can further adapt the content to their
needs; and a pose estimation system that identifies an exhibit and links it to the
additional content that is prepared for it. We present the systems and the initial
results for a selected set of exhibits in Herakleidon Museum, a museum holding
temporary exhibitions mainly about ancient Greek technology. The initial evaluation that we presented is encouraging for all systems. Thus, the plan is to use the
developed systems for all museum exhibits as well as to enhance their functionality.

Keywords: Museum guide system · Recommender system · Pose estimation ·
Content management system

1 Introduction

Museums are perhaps the most important institutions for the preservation and promotion
of the world's cultural heritage and act as powerful learning environments as well,
contributing to social and economic development. Traditionally, museums used linear,
non-interactive ways of presenting exhibits and guide material, e.g. audio tours or QR
codes that offer static information. Multimedia technologies have recently begun to be
exploited, but they require considerable investment on the part of museums and often
have limitations on the extent of the presented material but also on the number of people
who can use them. Also, a key concern for museums has been the evaluation of their
exhibitions and activities, as it is critical to understand how satisfied visitors are about
the exhibits, the additional digital content and the way guidance material is presented.

I. Maglogiannis et al. (Eds.): AIAI 2020, IFIP AICT 583, pp. 439–451, 2020.
https://doi.org/10.1007/978-3-030-49161-1_37

In this paper, we present the initial results of the project MuseLearn, the main objective of which is the development of an innovative guide platform in a museum for mobile devices that will provide personalized additional multimedia content by using exhibit detection, as a basis for providing recommendations to visitors. A key aim of the project is to increase the number of museum visitors and their satisfaction.

The project is implemented in the Herakleidon Museum, Athens [1], which is a museum organising temporary exhibitions with focus on ancient Greek technology. The current exhibitions (2019) deal with the themes of ancient automata (such as the Antikythera Mechanism, and ancient war technology (such as a battering ram). The exhibits are of educational character, in the sense that they take the form of informative material, representations and reconstructions based on original artifacts.

Regarding the presentation of the initial results of MuseLearn, we first present the developed in Sect. 2. In Sect. 3 we discuss existing approaches and work related to these components. Section 4 presents the management of additional multimedia content that is offered to visitors. Section 5 describes the proposed recommender system along with visitor tracking techniques. Section 6 details the exhibit detection technique used in the project. Section 7 describes the implementation of pilot content and initial evaluation

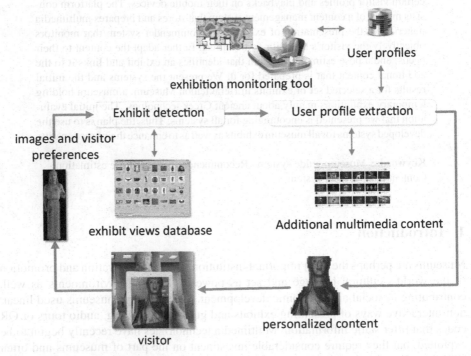

Fig. 1. Outline of the MuseLearn platform. The visitor can see the exhibit through a mobile device. The exhibit is then recognized by the system (based on the acquired image and the stored exhibits) and the appropriate digital material is suggested based on the user's profile. Concurrently the users' choices are recorded. Asynchronously, the curator may analyze the exhibition statistics.

results in all components of the project. Finally, Sect. 8 discusses the key conclusions of the paper and presents plans for future work.

2 Platform Architecture

The MuseLearn platform (see Fig. 1) consists of the following basic systems that are presented in the following sections:

- A **content management system** that stores multiply structured multimedia material that is offered to visitors through augmented reality for better understanding the exhibits of the museum.
- A **recommender system** that allows adaptation of the information to the needs and requirements of visitors.
- An **exhibit detection** and pose estimation system that can detect the exhibit the visitor is interested in and provides identification (exhibit name/id) and orientation information (3D pose) using the camera of a mobile device (phone/tablet).

According to the typical usage scenario, the visitor sees the exhibit via a portable device. The exhibit is then detected by the pose estimation system (based on the images of stored exhibits) and the recommender system suggests appropriate digital material stored in the CMS to the visitor according to their profile. At the same time, visitor choices are recorded for further refining their profile.

3 Related Work

Content Management Systems (CMS): Regarding content management systems (CMS) for museums should offer administrators the ability to structure content in various multimedia formats for artifacts, concepts and supplementary material that is necessary for their presentation [2]. A research of commercial and research CMS that are mainly or indirectly targeted for museums is found in [3]. Argus [4] is a web-based Collections Management Solution (CMS) that can be used for managing and presenting artifacts and objects. Argus provides a tabular visual interface organized in rows for accessing and editing content. The retail oriented RetailPro platform [5] supports responsive design, portable content, reports, transformation of content parameters, reorganization of collections and organization of virtual collections. Content is mainly presented in semantically colored tables. The Museum System (TMS) [6] is a CMS that supports planning and managing of exhibitions and generates reports; it contains a digital asset management module and offers administrative support. MuseumPlus [7] supports managing collections and exhibitions and is in use in over 900 museum sites worldwide. Unlimited images and other multimedia content can be linked to objects, artists, addresses and other record types. CollectiveAccess [8] is a web-based open source CMS that embraces Dublin Core, PBCore, VRA Core etc. The embedded media viewer allows to enlarge and inspect uploaded images, video and audio playback with time-based annotations and PDF viewers. Museum Anywhere [9] offers collection management and tile-based presentation through mobile devices and supports integrating content from other museum management software. The proposed CMS aims to overcome limitations of existing museum CMS by introducing a simple but flexible and expandable structure for organizing content about exhibits and concepts, which can be visualized by using modern web technologies and exploiting the spatial and interface capabilities of mobile devices.

Museum Recommendation Systems: So far, museum recommendation systems are mainly used for supporting personalized museum tours. There are examples of recommendation systems in museums that offer unique tours based on specific interests [10, 11]. The traditional recommendation approaches are (a) Content-based filtering (e.g., [12]) gives recommendations based on similar content that was of interest to the user in the past and (b) Collaborative filtering (e.g., [13]) that are based on content that was of interest to the users that are most similar to the current one based on their selections. Combination of these leads to hybrid approaches. The latent factor models (e.g. [14]) are more recent and try to factorize the matrix that associates users and items to be selected via SVD, or sparse methods. Therefore, hidden relations between users and items may be identified. The Herakleidon Museum can provide a great deal of information to the visitors, in a variety of formats, such as texts, pictures, sounds, videos, games, etc., instead of a comprehensive description of the exhibits. Thus, arises the need for filtering, hierarchy and effective delivery of this information. Our goal is to create a recommender system to manage the large amount of multimedia information accompanying the exhibits of and to recommend visitors to explore the museum on the basis of their interest.

Exhibit Detection and 3D Pose Estimation: Regarding exhibit detection and 3D pose estimation modules, there is a large number of relevant works. Some recent relevant publications address some of the requirements of the MuseLearn pose estimation module. A very significant building block of the overall approach is the capability to localize a camera relative to the scene. Kinect Fusion [15] proposed by Newcomb et al., fuses all the data from a depth camera into a voxel space, that represents a dense 3D model of the scene to be reconstructed. ICP is used with it to track the camera position. More elaborated versions [16] allows it to operate in larger scenes. Although promising, the method requires depth input to operate and can be sensitive to outliers and specular surfaces. These requirements make it a bad fit for the MuseLearn scenarios. Mur-Artal et al. [17, 18] proposed the ORBSlam method. OrbSlam2 performs simultaneous localization and mapping of the environment. It tracks the camera position and builds a map of the environment based on ORB features. The map contains a sparse 3D point cloud, each point being associated with an ORB descriptor, and some geometric information about the view corresponding to it. A set of keyframes (i.e., representative frames) is also maintained in time, and each converted to a bag of words (BoW) descriptor [19]. This approach operates with RGBD or monocular/stereo RGB input, can make use of a static map, and works in real time, making it a possible base solution for our problem. However, the method assumes a static scene and is not robust to moving objects and large numbers of outliers. More recently, deep-learning based methods [36, 37] solve the object pose estimation problem from RGB input using Convolutional Neural Networks (CNNs). The networks are trained to predict the 2D coordinates of the projections of known 3D landmarks on the objects of interest. Subsequently, they solve the PnP problem to acquire the 3D object pose. Although promising, these approaches do not scale well with multiple exhibits and suffer in dynamic environments and in the case of partially occluded exhibits. This renders them unsuitable for the MuseLearn use case. FORTH's proposed method for exhibit identification and pose estimation borrows many ideas from [18], especially for the training phase as it will be detailed in later sections. Lourakis and Zabulis [20] detect the pose of known rigid objects in monocular RGB images, using a 3D model for each object. The model consists of 3D points each associated with a SIFT descriptor. A hand-held camera is used to capture the appearance of the object from all view angles. Then, the camera motion together with the 3D point cloud describing the object is recovered using "Shape from Motion" techniques similar to [21]. The authors report a run time of 0.6 s per frame. While the work of Lourakis et al. is a good candidate for solving our problem, the dense object representation makes it too computationally expensive, and the lack of use of background features makes it harder to detect featureless objects. However, aspects of this work such as the F2P features matching strategy and the pose estimation method (posest library) are used in the proposed method.

4   Content Management System

Modern web technologies have made possible the visualization of content on a spatial area. For example, Scalable Vector Graphics (SVG) and Cascading Style Sheets (CSS) may be employed for implementing vector shapes and animations that are presented uniformly in devices of different sizes and interface options. On the other hand, in a smartphone world, content-editing operations should resemble natural operations by introducing spatial interfaces and gestures and not relying on providing a desktop experience on a smaller screen. The Content Management System (CMS) developed for MuseLearn embraces these ideas for displaying and interacting with content spatially.

Two on-site surveys have been conducted in order to record the information provided by the museum and all kinds of exhibits and concepts. Information about exhibits and their related multimedia informative material has provided the basis for populating the system's database. The database of the CMS stores all information about the museum's exhibits and concepts that they may be assigned to. Database design is flexible enough so exhibits may be assigned to multiple concepts along with other multimedia material linked to them. More specifically, the main entities in the DB are:

- **Concepts:** The concepts that museum exhibits may be assigned to. For example, the exhibit "battering ram" is assigned to the concept "combative weaponry". Concepts also describe all kinds of informative material within the museum.
- **Exhibits:** The main table holding information about every exhibit, such a title and description about the exhibit, the exhibit code within the museum, the museum room where the exhibit lays and the exhibit types.
- **Links:** This table is responsible for defining all possible relations among exhibits and concepts. There may be relations between exhibits; between concepts for defining the concepts hierarchy.
- **Multimedia:** Multimedia content that is linked to an exhibit or a concept.

The database has been implemented in MySQL and content has been inserted about the presentational needs of exhibits and their supplementary content for both museum buildings along with the hierarchical structure of concepts for the certain museum and all kinds of relations among them. More specifically, the database currently stores 98 concepts of all hierarchy levels; 82 exhibits; 152 links of all types and 180 multimedia items for concepts and exhibits.

PHP code has been employed for the hierarchical visualization of museum content starting from the museum level, continuing with the two main collections ("Automata" and "Ancient War Technology") and finishing at single concept and exhibit level. For every concept, its description and multimedia content are displayed along with links to subordinate concepts that analyze it and exhibits that are related directly to the concept. Visualization is both textual, as in a subject catalog, and graph-based. The latter is implemented in SVG, so that it is accessible by all device types. Treant.js [22] free to use, graph creation library has been used for drawing content hierarchies.

5 Recommender System

For the recommender system a trial mode is scheduled to initiate a few users. Gradually and based on data from more users, the system will be retrained. The preferences will arise indirectly from the viewing time of the accompanying material. Also, when entering the museum, the user will have a profile that matches his/her demographics and/or short quote when signing up. Progressively the profile will be completed during the visit and as long as its preferences are known. We currently use alternatively latent factor approaches. For a new item or a new visitor, the new object has no score and the new user has no history, which is known as the cold-start problem. So any recommendation result could be doubtful. We handle it by using demographic data, which has been an effective method for cold-start in the past [23]. Different versions of the same content are associated with some initial and grossly-defined profiles based on a few deterministic rules, which are tested against a questionnaire as described in the following subsection. After some views the system is able to fine-tune the user profile.

The question as to how to identify and distinguish visitor profiles has been a key issue in both the museum studies (e.g., [24, 25]), and information systems literatures (see [2]). For the pilot system, it was decided to create and organize the content, addressing three broad requirements related to visitors' prior knowledge, experience and expectations about the exhibition: 'general interest', 'specialized knowledge', and 'enthusiasm about new technology'. To identify these broad requirements, a group of basic questions forms the starting section of the pilot system. These focus on visitors' demographic profile (sex, age range, highest education level), and their familiarity both with the subject of the exhibition and the use of technology. A sixth key question asks for the driving motivation behind the visit. This question draws on fundamental distinction of five identity-related categories of visitors ('explorers', 'facilitators', 'professional/hobbyists', 'experience seekers', and 'rechargers') [25]. At the pilot stage of the project, visitors' responses to these questions will offer, in combinations, 'personalized' content at the most basic level. For example, a visitor aged 17 and familiar with new technologies will be offered 'general interest' content, enhanced by interactive educational games. Visitors' responses will provide the foundations on which to fine-tune later the recommender system. These six questions are also used for the primary evaluation procedure, as described in Sect. 7.2.

6 Pose Estimation System

The developed exhibit localization and pose estimation method is tailored to the specialized requirements of MuseLearn. Specifically, the inference is performed using RGB input; the solution is scalable to large exhibitions; it supports multiple levels of pose estimation detail; is associated with small (re)configuration overhead; and achieves real time performance. The method is using a "features and reference poses" database that is generated offline, in order to identify the pose of exhibits in query images captured by mobile devices.

6.1 Training Phase

Creating the features and reference poses database is a critical step for the correct operation of the pose estimation pipeline. In order to maximize the accuracy, we employed the use of RGB-D cameras in the training phase. This type of sensors (i.e. MS Kinect2, Asus Xtion, ORBEC) provide apart from the usual red, green and blue channels an additional channel for depth. This channel provides distance information for each imaged point of the scene.

Training Sequence Acquisition: For each exhibit we capture an RGB-D video sequence. The goal is to cover a wide range of possible visitor viewing directions.

Feature and Pose Extraction: We apply a SLAM method to extract features and key frames and camera poses. For the purposes of MuseLearn we chose ORBSlam2 [18] as the SLAM method and the ORB [26] features and descriptor for feature extraction.

Exhibit Annotation: An annotator selects a reference frame, pose and bounding volume for each exhibit in the museum coordinate frame. The common coordinate frame is required in order to report pose information to the other MuseLearn modules.

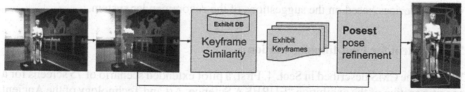

Fig. 2. The pipeline of the pose estimation method during inference. From left to right: (i) the query image captured using the camera on the mobile device (ii) ORB features are detected on the query image and their descriptors are computed (iii) The BoW representation of the ORB features is used to find the most similar (high similarity score) keyframes in the exhibit db (iv, v) Pose refinement using exhibit frames and matching ORB features from the query image (vi) the exhibit pose with the highest score is returned.

Exhibit DB: In the final step we automatically select frames from the training set where each exhibit is fully visible. The ORB features from these "keyframes" are converted into a bag of words (BoW) representation and are inserted in the exhibit database.

6.2 Exhibit Identification and Pose Estimation

The first step in the pose inference pipeline is to compute the 2D features $O(I)$ in the input image I. From these, a BoW representation q is created. We compute the similarity score S between the BoW of the query image and all the keyframes K in the exhibit database. The set C of keyframes that have a similarity score above a threshold S_i (q, K_i) > T_s are the initial "coarse" estimations. If the query from the mobile device requires only exhibit identification (no pose estimation), the exhibit ID of the keyframe in C with the highest similarity score is returned. The pipeline is depicted in Fig. 2.

The pose estimation is a refinement step over C using the features $O(I)$. Initially we compute C' that is the subset of keyframes in C that belong to the identified exhibit. Using the ORB descriptors and matching described in [26] we find the best correspondences $M(O(I), K_j)$ between $O(I)$ and the ORB features of each K_j keyframe in C'. Posest [20] is applied to all matches were $|M| < T_m$. T_m is the minimum number of features on the keyframe that belong to the exhibit (i.e., not in the background) and have a good match in $O(I)$. Posest uses a RANSAC scheme to iteratively select a subset of M and compute the rigid transformation that best explains the camera motion with respect to a given keyframe pose. The quality of the transformation is measured using the reprojection error of the matched ORB features from the query frame to the keyframe. The pose with the highest score is selected as the camera pose with respect to the exhibit.

7 Pilot Implementation and Primary Evaluation Procedure and Results

During the pilot implementation, we had the chance to test the various systems of the MuseLearn platform in the Herakleidon museum. Pilot multimedia-rich content has been developed for certain exhibits. When these exhibits are identified by the pose estimation system through visitors' mobile devices, the additional material will be displayed on visitors' devices based on the suggestions of the recommender system.

7.1 Content for Pilot Implementation

We used the CMS described in Sect. 4. First, a pilot extended scenario of 75 screens for a specific section of the exhibition "EUREKA Science, Art and Technology of the Ancient Greeks" was drafted. Focus was centered on exhibits related to armors and weapons from the Mycenean to the Hellenistic Period. We decided to fully deploy content for five exhibits belonging to the "Warrior and Armor/Equipment" section of the exhibition and for three assigned categories of users/testers (see Sect. 5). The five exhibits were selected on the basis of their associative and historical value, as well as their potency to enhance the storytelling, increase visitor engagement and unfold stories about individual artifacts or specific object groups.

Content creation decisions were taken by accounting for users of different needs, interests and knowledge (see [27]). Hence, two levels of content exist for each exhibit: specialized and basic content, enhanced occasionally by interactive applications. This content includes texts, images, videos, hyperlinks and games. More precisely, as in most guide systems, the user can access descriptions of the exhibits, information about their function and the historical and the social context of their use and production from a variety of sources, as well various interpretations and representations, both scientific and popular (see [2]) by using the content gathered as described in Sect. 4. The system also allows users/testers to zoom in and observe object details, consult the glossary for unknown scientific terms, place names and archaeological periods, and search for related content from the museum collection or from other physical or virtual spaces with the help of hyperlinks and object recommendation. Finally, interactive applications through matching, ordering and sequence games and object identification provide users with the opportunity to test their knowledge and acquire new, get acquainted with typologies, explore objects' correlation/interconnection and interact in a playful way.

7.2 Primary Evaluation Procedure and Results for Visitor Satisfaction

Nowadays, evaluation construes integral component of museum practice and research (i.e. [28–30]), but also of mobile application designing and multimedia development (i.e. [28, 31, 32]). To be beneficial, evaluation should be approached as "an ongoing process" [28]. The primary evaluation procedure was carried out by a group of six undergraduate students from the University of Patras. At this preliminary stage, both qualitative and quantitative methodologies were employed. First, a questionnaire was drafted both in English and Greek consisting of 15 questions and organized in two sections. The aim of this survey is to assess visitors' motivation, familiarity with the exhibition topic as well as information and communications technology, their level of satisfaction with information provided through various interpretive means, as well as their potential interest in using a mobile application during their visit.

The questionnaire is self-administered. Therefore, an introductory note informed the participants about the goals and context of this survey. As expected, the last section collected information about the age, gender, higher level of education and place of origin of the respondents. Initial feedback was encouraging. Visitors expressed their strong interest in using a mobile application to navigate around the exhibition space.

The primary evaluation procedure was complemented by observation of visitor behavior, which has become a key element in feedback studies on museum performance (see for example [24, 33–35]). The aim of this research was to observe how visitors move, explore and use space and display in the exhibition 'Technology of War in Ancient Greece', without the use of the mobile guide system. These data will then be used for a comparative study after the implementation of the platform and potentially lead to a deeper understanding of how the platform impacts on visitor experience.

Visitor behavior, in particular their patterns of moving and viewing, is analyzed using established techniques. First, the arrangement of the display is recorded on the building layout as the basis for designing the observation record sheet for mapping visitors' movement and interactions with the displays. Traces of the paths of visitors, who are randomly selected, spread across time periods and have consented to take part in the research, are recorded for their whole visit to the exhibition. When the visitor stops to look at a work, read a text, or watch a video, a stopping point is recorded on the plan of the exhibition by the observer. Other symbols are used to clarify where a visitor stops for longer periods of time. The tracking data are complemented by the recording of the total time visitors spent in the exhibition. The resulting ratings table which associates users with exhibits and their preferences will be used to initialize the system profiles and thus to address the cold-start problem in our system.

7.3   Primary Evaluation Results for the Pose Estimation System

Regarding the pose estimation system, we evaluated our baseline method using data acquired on two sites. **(i) The AMI facility:** The ambient intelligence (AMI) facility of FORTH hosts exhibition and demonstration areas. We trained our system on six (6) exhibits selected for their size and placement to be compatible with typical exhibits in a museum. Data were acquired with low- and high-quality cameras to investigate performance with different sensors. A training set and two test sets were acquired. The first test set uses the low quality RGB sensor of an Asus Xtion. The second uses a high-quality camera similar to the cameras found in modern phones and tablets. The exhibit locations and reference poses were annotated in the training dataset as well as in a test dataset. **(ii) Herakleidon museum:** We recorded a preliminary dataset of six exhibits. In both cases the exhibit locations/reference poses were annotated in the training dataset. The preliminary version of the pose estimation method focused on the exhibit identification and coarse pose estimation. Specifically, the system must correctly identify the quadrant from which the visitor is approaching the exhibit: Left, Right Front, Back.

The results of our experiments on both sites are given in Table 1 and Table 2.

Table 1. Average precision and recall on exhibit classification accuracy.

	Herakleidon	AMI (low quality)	AMI (high quality)
Average precision	0.947	0.980	0.971
Recall	0.771	0.844	0.810

7.4   Primary Evaluation Results for the Recommender System

We started to evaluate different approaches using public data, until we collected our proprietary ones for the Herakleidon museum. More specifically we evaluated the data provided in [11]. According to this, a dataset of visitors' times spent in each thematically organised exhibit area is created, which were recorded and collected by computer-supported methodology. The recommendations concern specific thematic areas. In order

Table 2. The coarse (quadrant) pose estimation accuracy. Two metrics are shown. Left is the error in degrees relative to the ground truth. On the right we quantize the same error using 3 thresholds. The quadrant approach corresponds to "<45°".

Error in degrees on the yaw axis		% of frames within correct quadrant		
Avg error (std dev)	Median error	<22.5°	<45°	<90°
8.98 (±7.94)	6.73	93.0%	99.8%	100%

to create appropriate recommendations, the exhibits were grouped into semantic and spatially coherent areas. With this in mind, 126 thematic and naturally organised areas were created at the museum's facilities. The time spent by 158 visitors on these areas of the museum was recorded and expressed implicitly the rating.

The viewing times are transformed to logarithmic with values between 1 and 8. We applied two methods: a simple collaborative filtering with KNN and a latent factor method with SVD. The root mean square error is calculated by ten-fold cross validation. With both methods the mean RMSE is around 1 with standard deviation between 0.02 and 0.03. There are similarities and differences with our system. We similarly use the viewing times as an implicit rating mechanism; the target users are museum visitors as well. However, we provide recommendations on supplementary digital material associated to museum artifacts and not directly on museum artifacts.

8 Future Work and Conclusions

We presented the MuseLearn platform that is constituted of three innovative systems (artifact detection, content, recommendation,). Some initial evaluation results were presented after the implementation of the platform for Herakleidon Museum.

In the future, subsystems of MuseLearn will be extended both in functionality and content for covering all exhibits of Herakleidon Museum and other museums. Content-wise, further support will be implemented for inserting and linking multimedia content, such as 3D representations and virtual reality material. Towards this direction, refined visitor profiles will be developed on the basis of the empirical findings, and their preference for certain themes, in conjunction with demographic data. We also plan the creation of more educational applications based on content that will have been prepared for the presentation of exhibits.

Acknowledgement. Co-financed by the EU and Greek national funds through the Operational Program Competitiveness, Entrepreneurship and Innovation, under the call RESEARCH-CREATE-INNOVATE (project: T1EDK-00502 - MuseLearn).

References

1. Herakleidon Museum. http://herakleidon-art.gr. Accessed 29 July 2019

2. Kosmopoulos, D.I., Styliaras, G.D.: A Survey on developing personalized content services in museums. Pervasive Mob. Comput. **47**, 54–77 (2018)
3. Lanir, J., Kuflik, T., Sheidin, J., Yavin, N., Leiderman, K., Segal, M.: Visualizing museum visitors' behavior: where do they go and what do they do there? Pers. Ubiquit. Comput. **21**(2), 313–326 (2017)
4. Lucidea, Argus, make your collection more visible and accessible than ever before. https://lucidea.com/argus/. Accessed 29 July 2019
5. RetailPro. http://www.retailpro.com/. Accessed 29 July 2019
6. GallerySystems, Collection management software, museum and art collections. https://www.gallerysystems.com/products-and-services/tms-suite/tms/. Accessed 29 July 2019
7. Zetcom, Museum plus. http://www.zetcom.com/en/productsen/. Accessed 29 July 2019
8. Collective Access. http://www.collectiveaccess.org/. Accessed 29 July 2019
9. Museum Anywhere. http://museumanywhere.com/. Accessed 29 July 2019
10. Keller, I., Viennet, E.: Recommender Systems for Museums: Evaluation on a Real Dataset, pp. 65–71. IARIA, Brussels (2015)
11. Bohnert, F., Zukerman, I.: Non-intrusive personalisation of the museum experience. In: Houben, G.-J., McCalla, G., Pianesi, F., Zancanaro, M. (eds.) UMAP 2009. LNCS, vol. 5535, pp. 197–209. Springer, Heidelberg (2009). https://doi.org/10.1007/978-3-642-02247-0_20
12. Pazzani, M.J., Billsus, D.: Content-based recommendation systems. In: Brusilovsky, P., Kobsa, A., Nejdl, W. (eds.) The Adaptive Web. LNCS, vol. 4321, pp. 325–341. Springer, Heidelberg (2007). https://doi.org/10.1007/978-3-540-72079-9_10
13. Herlocker, J.L., Konstan, J.A., Borchers, A., Riedl, J.: An algorithmic framework for performing collaborative filtering. In: ACM SIGIR Conference 1999, pp. 230–237 (1999)
14. Koren, Y., Bell, R., Volinsky, C.: Matrix factorization techniques for recommender systems. computer. IEEE Comput. Soc. **42**(8), 30–37 (2009)
15. Newcombe. R.A., et al.: KinectFusion: real-time dense surface mapping and tracking, In: IEEE International Symposium on Mixed and Augmented Reality (ISMAR) (2011)
16. Whelan, T., Kaess, M., Johannsson, H., Fallon, M., Leonard, J.J., McDonald, J.: Real-time large-scale dense RGB-D SLAM with volumetric fusion. Int. J. Robot. Res. **34**(4–5), 598–626 (2015)
17. Mur-Artal, R., Montiel, J.M., Tard'os, J.D.: ORB-SLAM: a versatile and accurate monocular SLAM system. IEEE Trans. on Robot. **31**(5), 1147–1163 (2015)
18. Mur-Artal, R., Montiel, J.M.M., Tard'os, J.D.: ORB-SLAM2: an open-source SLAM system for monocular, stereo and RGB-D cameras. IEEE Trans. Robot. **33**(5), 1255–1262 (2017)
19. Galvez-Lopez, D., Tard´os, J.D.: Bags of binary words for fast place recognition in image sequences. IEEE Trans. Robot. **28**(5), 1188–1197 (2012)
20. Lourakis, M., Zabulis, X.: Model-based pose estimation for rigid objects. Comput. Vis. Syst. **7963**, 83–92 (2013)
21. Snavely, N., Seitz, S., Szeliski, R.: Photo tourism: exploring photo collections in 3D. ACM Trans. Graph. **25**(3), 2006 (2006)
22. https://fperucic.github.io/treant-js. Accessed 29 July 2019
23. Solanki, S., Batra, S.: Recommender system using collaborative filtering and demographic characteristics of users. Int. J. Recent Innov. Trends Comput. Commun. **3**, 4735–4741 (2015)
24. Serrell, B.: Paying attention: Visitors and museum exhibitions. American Association of Museums 1998, Washington (1998)
25. Falk, J.: Identity and the Museum Visitor Experience. Routledge, Abingdon (2009)
26. Rublee, E., Rabaud, V., Konolige, K., Bradski, G.: ORB: an efficient alternative to SIFT or SURF. In: IEEE - ICCV, Barcelona, Spain, November 2011

27. Raptis, D., Tselios, N., Avouris, N.: Context-based design of mobile applications for museums: a survey of existing practices. In: International Conference on Human Computer Interaction with Mobile Devices & Services, pp. 153–160 (2005)
28. Economou, M.: Evaluation strategies in the cultural sector: the case of the Kelvingrove museum and art gallery in Glasgow. museum soc. **2**(1), 30–46 (2004)
29. Pontin, K., Lang, C., Reeve, J., Woolard, V.: Understanding museum evaluation. The Responsive Museum, Working with Audiences in the Twenty-First Century. Ashgate Publishing, Ltd., Farnham (2006)
30. Screven, C.G.: Exhibit evaluation: a goal-referenced approach. Curator **19**(4), 271–290 (1976)
31. Rasimah, C.M.Y., Ahmad, A., Zaman, H.B.: Evaluation of user acceptance of mixed reality technology. AJET **27**(8) (2011). https://ajet.org.au/index.php/AJET/article/view/899/176
32. Tost, L.P., Economou, M.: Exploring the suitability of virtual reality interactivity for exhibitions through an integrated evaluation: the case of the ename museum. J. Museol. **4**, 81–97 (2007)
33. Tzortzi, K.: Museum Space: Where Architecture Meets Museology. Routledge, London (2015)
34. Veron, E., Levasseur, M.: Ethnographie de l'Exposition: l'espace, le corps, le sens. Centre Georges Pompidou, Paris (1983)
35. Yalowitz, S.S., Bronnenkant, K.: Timing and tracking: unlocking visitor behavior. Visit. Stud. **12**(1), 47–64 (2009)
36. Tekin, B., Sinha, S.N., Fua, P.: Real-time seamless single shot 6D object pose prediction. In: Proceedings of the IEEE CVPR (2018)
37. Peng, S., et al.: Pvnet: Pixel-wise voting network for 6D of pose estimation. In: Proceedings of the IEEE CVPR (2019)

Promoting Diversity in Content Based Recommendation Using Feature Weighting and LSH

Dimosthenis Beleveslis and Christos Tjortjis(✉) ⓘ

The Data Mining and Analytics Research Group, School of Science and Technology,
International Hellenic University, 14th Km Thessaloniki – Moudania, 57001 Thermi, Greece
c.tjortjis@ihu.edu.gr

Abstract. This work proposes an efficient Content-Based (CB) product recommendation methodology that promotes diversity. A heuristic CB approach incorporating feature weighting and Locality-Sensitive Hashing (LSH) is used, along with the TF-IDF method and functionality of tuning the importance of product features to adjust its logic to the needs of various e-commerce sites. The problem of efficiently producing recommendations, without compromising similarity, is addressed by approximating product similarities via the LSH technique. The methodology is evaluated on two sets with real e-commerce data. The evaluation of the proposed methodology shows that the produced recommendations can help customers to continue browsing a site by providing them with the necessary "next step". Finally, it is demonstrated that the methodology incorporates recommendation diversity which can be adjusted by tuning the appropriate feature weights.

Keywords: Content-Based (CB) Recommendation Systems (RS) ·
eCommerce · Minhash · Locality-Sensitive Hashing (LSH) · Decision support

1 Introduction

A Recommendation System (RS) refers to an intelligent system that produces suggestions about items to users. Its role is to predict items that are likely to be of interest to the user. They provide personalized information by learning the user's interests through his/her interaction with items. The way that these recommendations are produced depends on the specific domain and the desired results. In general, they are based on item similarities, user preferences, past purchases and actions related to items and users.

Recommended items can potentially be anything that a user is looking for, such as products, movies, songs, services etc. [25–27]. E-commerce is a domain that RS are used extensively [1, 2]. Online stores need to offer customers a similar or even better shopping experience than that of an actual store.

© IFIP International Federation for Information Processing 2020
Published by Springer Nature Switzerland AG 2020
I. Maglogiannis et al. (Eds.): AIAI 2020, IFIP AICT 583, pp. 452–461, 2020.
https://doi.org/10.1007/978-3-030-49161-1_38

A customer can see product recommendations in several parts of an e-commerce site including the 'product page', showcasing a specific product, along with its detailed description and features [3]. The user can add the product to their cart or visit an alternative product page that better matches their preferences. The aim of the system is to display relevant items on such a page and help customers to continue browsing the site by providing them with the necessary "next step". The recommended products must be similar to the target product, by considering the importance of their features.

However, it is important to achieve diversity in recommendations because customers do not need a list of almost identical products. For example, when a user views a laptop of a specific brand and receives recommendations corresponding only to similar laptops of the same brand. In this case, the user will not be able to visit the product page of a laptop of an alternative brand by navigating through the recommendations.

Furthermore, the logic of such a RS relies on the idea of finding similar products by calculating similarities among them [4]. This is computationally heavy and time-consuming, for sites with thousands of products. Calculating the similarity between a product and other products, in order to find the most similar ones, is usually inefficient. So, it is important to apply more sophisticated methods that reduce the complexity and computational cost without compromising accuracy.

The scope of this research is to implement a Content-Based (CB) RS suitable to the 'product page' of an e-commerce site [5]. Specifically, we address the problem of product representation by incorporating a weighted method that offers the functionality to customize the importance of product features. In that way, the logic of the methodology is adjusted to the needs of the site. In addition, recommendation diversity is accomplished without compromising similarity. Furthermore, this work addresses the problem of efficient product similarity calculations for sites that dispose thousands of products. This problem is tackled by incorporating a Locality-Sensitive Hashing (LSH) method [6, 7]. Finally, our implemented heuristic approach is being evaluated based on real data from two e-commerce sites.

The remainder of the paper discusses related work and the data used in Sects. 2 and 3 respectively. The proposed methodology are detailed in Sect. 4. Section 5 presents results and evaluation. Section 6 concludes the paper.

2 Related Work

Zhang et al. addressed the problem of high dimensionality, traditionally faced by RS [8]. E-commerce platforms with very big number of items and users need to incorporate large volume of data in the recommendation process. This makes the generation of real time recommendations inefficient. They incorporated LSH techniques in a Collaborative Filtering (CF) approach in order to reduce time complexity. Minhash hashing method was used for binary data and simhash for real-valued data. Similar candidate pair identification was performed through LSH to increase efficiency of similarity computing, the most time-consuming task for traditional CF RS. By conducting experiments on synthetic and real-world datasets, it is shown that LSH can approximately preserve similarities of data whilst significantly reducing data dimensions.

Aytekin et al. use LSH to introduce a RS that can scale with increasing amounts of data [9]. They improved the LSH based recommender algorithms and systematically evaluated LSH in neighborhood-based CF. By experimenting with real datasets, they presented algorithms that have better execution time than standard LSH-based applications, whilst preserving prediction accuracy and producing recommendations with diversity.

Debnath et al. proposed a hybridization of CB and CF recommendation, which incorporates product attribute weighting [10]. They argued that human judgment of similarity between two items often gives different weights to different attributes and that RS need to consider this. The weights refer to the importance of each product attribute to customers and are estimated from a set of linear regression equations obtained from a social network graph, which captures human judgment about similarity of items. The proposed methodology is compared with CB methods that consider the importance of different products features as equal. The evaluation is based on IMDB recommendations as benchmark. The proposed method outperforms simple methods; hence, the effectiveness of feature weighting is demonstrated.

Barranco et al. proposed a feature weighting method to improve the CB filtering in cases of multi-valued item features [11]. They argued that a user considers some features as more important than others. This represents an implicit feature weighting, which can be subjective. The weight of each feature is computed according to: (i) the entropy or amount of information provided (the more entropy the higher the weight), and (ii) the correlation between items chosen by the user in the past and the values of some features of the set of items.

K. Bradley argued that recommendation diversity is important, and that traditional CB RS suffer from potentially poor diversity [23]. He also questioned the blind faith in the similarity assumption. A diversity preserving similarity-based retrieval algorithm is proposed. Multiple strategies for retrieving k items from a collection, each focusing on a different way of increasing diversity is presented. Finally, he concluded that an over emphasis on diversity can result in a corresponding drop in similarity.

3   Data

Two datasets were used to evaluate the proposed RS. The first one was created by scraping a real e-commerce site and comprises information for thousands of products. The second one was created by preprocessing an existing dataset that has been used for similar purposes in the past [27]. These datasets are described next.

3.1   Bestprice Dataset

Bestprice is a new dataset created for the purposes of this research. Data were gathered by scraping BestPrice.gr, a commercial site, where a customer can compare product prices across numerous e-shops [12]. A huge variety of product categories and subcategories is available. We used a subset consisting of data regarding 29.541 products that belong to 6 main technology categories, each comprising several subcategories, 22 in total. Additionally, there are 515 different product brands. A product title is also available for each item, along with a set of 10 product recommendations for each product, provided by the site in each product page, which were used for evaluation.

The dataset was preprocessed in order to assign a textual representation to each product. This text holds information that characterizes each product and is produced by concatenating its title, category, subcategory and brand. Each of these features were preprocessed before being added to the final text. First, the textual data of each feature were transformed to lowercase. Then categories, subcategories and brands were discriminated from other words by adding the suffixes '_cat', '_subcat' and '_brand' respectively. This helps to assign different weights to these specific words. Specific symbols that do not add any value were removed from the title. Finally, a string that consists of the above textual features was created for each product. This set of strings was used to create the TF-IDF matrix [13]. The information available and the final textual representation for a product is presented in Table 1.

Table 1. Product details for bestprice dataset.

Title	Category	Subcategory	Brand	Text
Omega Ice Box	Laptop_pc	Laptop_bases	Omega	omega_brand ice box laptop_pc_cat laptop_bases_subcat

3.2 Retailrocket Dataset

Retailrocket dataset is a dataset published by Retail Rocket that has been used in RS [14]. The data were collected from a real-world e-commerce website. We selected a subset of 28.241 products that belong to 6 main categories and 37 subcategories to produce a dataset with customer sessions. A session refers to a group of user interactions with an e-shop that take place within a given time frame. For example, a session may contain the information that a customer first viewed several products, added some to their cart and finally purchased them.

This session-based dataset was used to determine the product pages that users visit after viewing each product. An important aspect we considered was the number of consecutive events related with each other. We call this window size. Window size equal to 1 means that a product that a user viewed is linked only to the product that they visited next. Respectively, window size equal to 2 means that a product that a user viewed is linked to the next two products they visited. Furthermore, a representative text, that consists of the title, category and subcategory, has been assigned to each product. This set of texts was used in order to calculate the TF-IDF matrix.

4 Methodology

The heuristic approach for the proposed CB RS is based on feature weighting and LSH. The methodology consists of three parts. The first part refers to the method that is used in order to represent the set of products as a weighted matrix. The second is the weighted Minhash method that is used to create a compressed representation of each product by its Minhash signature and to approximate the Jaccard similarity of two sets. The last part is the efficient production of recommendations based on LSH. The implementation of the methodology based on the Bestprice dataset is presented in the following three subsections and in Fig. 1.

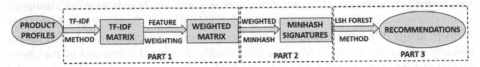

Fig. 1. Methodology diagram

4.1 Weighted Method

The first part involves creating a TF-IDF matrix for the set of 29.541 products. Typically, different features carry different amount of information. A word in a product title may be more informative than the rest, and this must be considered when calculating similarity. In our case, each product is represented by a text that is the concatenation of the product title and three specific features, as described in Sect. 3.1. The terms that correspond to the three specific product features are initially considered to be simple terms. These features are the category, subcategory and brand of each product. The total number of products is 29.541 and the respective corpus consists of 35.405 unique terms.

The target is to create a matrix that represents how important each term is for each product. A weight is assigned to each unique term for each product text. This weight corresponds to the significance of the term for each specific product. The more texts a term appears in, the less informative it is considered to be; thus it gets a smaller weight. Respectively, the significance of a term in a product text increases with the occurrences of it in that specific text. In particular, the TF-IDF matrix [13] has been created by using the corresponding functions of the 'scikit-learn' python library [15]. Furthermore, specific terms of the corpus have been given extra weight. These terms correspond to the three product features that are parts of each product text. The target is to have a functionality with which we can easily adjust the importance of each feature in the calculation of product similarities and enhance recommendation diversity, without compromising similarity or efficiency.

Assuming a products' textual representation that consists of N terms and that n of them correspond to specific product features, we assign a weight w_i ($w_1, w_2, ..., w_N$) to each of the N terms by applying the TF-IDF method. Then we adjust the importance of those n product features by multiplying their weights by small constants, while the rest of the weights remain the same. By multiplying a feature's weight with a constant greater than 1, this is considered as more important during the calculation of product similarities. The multiple scenarios that were implemented are presented in Sect. 5.1.

4.2 Weighted Minhash

The second part of the methodology aims to create a compressed representation for each product by applying the Minhash approach. Specifically, a method that incorporates real number weights has been used since the set of products is represented by a TF-IDF matrix. The weighted Minhash algorithm [16–19] that is available in 'datasketch' python library [20] is used. In practice, a Minhash signature has been created for each product based on the corresponding TF-IDF array. Hence, each product is represented by a much smaller array than before. Specifically, while the TF-IDF array consists of 35.405 elements, the Minhash signature consists of only 128 elements. The length of the signature corresponds to the 'sample_size' parameter that can be adjusted accordingly as by increasing the number of samples, a better accuracy is accomplished, at the expense of slower speed.

4.3 LSH Forest and Recommendations

In the third phase, the methodology uses the Minhash signatures in order to approximate the Jaccard similarity between products by applying the LSH approach. In particular, the LSH approach is used in order to find the most similar products to each product based on their Minhash signatures [21]. In particular, we search for the k-top similar products that correspond to k recommendations. For this reason, a variation of LSH that is known as LSH Forest [22] is used. Minhash LSH Forest, uses the Minhash representation of the query product and returns the k-top matching products that have the approximately highest Jaccard similarities with the query product. Hence, it is not necessary to pre-define a specific threshold for the Jaccard similarity score. In this way, we produced k recommendations for each of the 29.541 products in the Bestprice dataset.

5   Evaluation

The evaluation of the RS consists of two parts. The first part is based on the Bestprice dataset that was created by scraping a real e-commerce site and is presented in paragraph 3.1. The aim is to compare the recommendations of our methodology with those that were available on the site, concerning certain aspects. Specifically, we compare the diversity of the recommendations and we examine whether we can achieve different results by tuning the weights of the product features. The second part is based on the Retailrocket dataset that consists of real sessions as described in paragraph 3.2. The aim is to examine whether the recommendations that the proposed methodology produces would help users navigate an e-shop and prevent them from leaving the site without finding products that meet their preferences.

5.1   Recommendation Diversity

In this part of the evaluation process, the recommendation diversity of our methodology is examined. We evaluate the recommendations that our methodology produced for the Bestprice dataset. Finally, we compare the recommendations that our methodology produced with those that were scraped from the respective site. The dataset is presented in Sect. 3.1. A subset of 1.182 products has been used in the evaluation process.

Having applied the methodology that is presented in detail in Sect. 4, we have produced a set of ten recommendations for each product page. This number matches the number of recommendations that are available for each product in the Bestprice dataset. The experiment was conducted for 11 scenarios that are presented in Table 2. In each scenario we consider the importance of the product features (brand, category and subcategory) to be different, by tuning the corresponding weights (W_brand, W_cat and W_subcat). The average number of different brands, categories and subcategories in those sets of 10 recommendations is presented in Table 2.

Table 2. Average # of brands, categories, subcategories in the sets of 10 recommendations

	W_brand	W_cat	W_subcat	Brands	Categories	Subcategories
Scenario 1	0.5	1	1	3.07	1.16	1.30
Scenario 2	1	1	1	2.54	1.16	1.30
Scenario 3	1.5	1	1	2.16	1.16	1.31
Scenario 4	1	1	0.5	2.47	1.21	1.40
Scenario 5	1	1	1.5	2.58	1.14	1.25
Scenario 6	1	0.5	1	2.51	1.20	1.33
Scenario 7	1	1.5	1	2.63	1.12	1.29
Scenario 8	0.5	0.5	1	2.94	1.20	1.40
Scenario 9	1.5	1.5	1	2.19	1.14	1.26
Scenario 10	0.5	1	0.5	3.01	1.20	1.32
Scenario 11	1.5	1	1.5	2.23	1.13	1.29
BestPrice	–	–	–	5.16	1.0	1.0

In the first 3 scenarios, we see that the number of different brands decreases by increasing the corresponding weight. This means that considering the product brand as more important in the similarity calculations, results in recommendations that include more of the same brand for each product page. Respectively, in scenarios 4 and 5, the number of different subcategories decreases by the increase of the corresponding weight. The same happens for the product category feature in scenarios 6 and 7. Furthermore, in the rest of the scenarios, the weights of different combinations of the product features are tuned. The characteristics of the recommended products are affected by the weights that are assigned to each product feature.

In general, by observing Table 2, we see that the methodology incorporates diversity as it does not recommend products that are almost the same with each other. Specifically, products from more than one brand are present in each set of 10 recommendations. Also, by assigning the appropriate weights, we have product recommendations from more than one subcategory. There are also cases in which the recommended products belong to more than one category.

The recommendation diversity is very important for the quality of the system [23]. In CB RS, diversity can be as important as similarity [24]. Similarity assures that the recommended products are similar to the target product. Diversity means that the recommended products are not very similar to each other. The importance of the recommendation diversity can be explained through the following example. Assume that the target product is a Dell laptop and the system recommends 10 different Dell laptops. This might probably mean that the recommended products are very similar to the target product. However, the user will not have the option to move to a laptop of a different brand through the recommendations. Similar problems will occur in a case when all the recommended products are of the same subcategory or category.

In the last row of Table 2, we see the respective statistics in the product recommendations that were scraped from the site (BestPrice.gr). It is obvious that there is diversity in the recommendations concerning the brand feature. There is an average of more than 5 different brands in each set of 10 product recommendations. However, there is no diversity in the cases of subcategory and category. The system recommends products that belong only to the category and subcategory of the target product. As a result, a user does not have the option to move to a product of a similar subcategory through clicking one of the recommendations.

5.2 Session Based

By applying our methodology, we produced a set of ten recommendations for each product page in the Retailrocket dataset. The aim is to determine whether the recommendations would help customers navigate through the e-commerce site, by offering them the required next step in order to move from a product page to another.

For this purpose, we compare the recommendations with the product pages that users visited after viewing each single product. The test was conducted for multiple scenarios in which the importance of specific product features was tuned. These product features are the product category and subcategory. In addition, each scenario was tested for a window size up to 3. Table 3 presents the percentage of cases that at least one recommended product matches with one product view, for the scenarios where subcategory weights were tuned.

Table 3. Percentage of cases that at least one recommended product matches with views.

Window size	Subcategory weight			
	0.5	1	1.5	2.0
1	94.74%	94.79%	94.84%	94.86%
2	96.90%	97.01%	96.98%	97.03%
3	97.68%	97.73%	97.68%	97.73%

It turns out that the percentage of cases that at least one recommended product matches the real product views is between 94% and 97%, regardless the feature importance. The percentage increases with the increase of the window size. Similar results were accomplished by tuning the category weights. Hence, in most of the cases, the recommendations that our methodology produces would potentially help users move to a similar product page by clicking one of the recommended products.

6 Conclusions

In this paper we proposed a heuristic approach for CB RS based on feature weighting and LSH with the aim to promote recommendation diversity and make similarity calculations efficient. The proposed methodology creates a textual representation for each item based on its features. This is used in order to create a weighted representation of the set of items based on TF-IDF scheme. The importance of specific features is adjusted according to our needs by tuning the corresponding weights. Furthermore, the methodology incorporates the weighted Minhash algorithm in order to create a compressed representation of each item. Finally, the Minhash signatures are used by LSH forest in order to make the similarity calculations efficient and produce the recommendations.

Based on the evaluation of our proposed RS and the results that were produced by conducting several experiments, interesting conclusions were drawn. Regarding the recommendation diversity, results show that the characteristics of the produced recommendations can easily be adjusted to the desired results by tuning the product feature weights appropriately. The methodology incorporates recommendation diversity without compromising similarity. This aspect turns out to be very important because most of the CB approaches lack diversity [23]. Considering the product brand as less important in the similarity calculations, results in recommendations that include products of different brands. Similar results can be achieved by tuning other product features. In that way, customers see recommendations that are similar to the target product, but not very similar to each other.

Furthermore, our approach incorporates a hashing method that makes the product similarity calculations much faster and efficient than traditional systems. The representation of product profiles as compressed signatures, by applying the Minhash method, turns out to be effective. Finally, the approximation of product similarities by applying the LSH method results in a system that can handle thousands of products efficiently without compromising similarity. In that way, our proposed approach outperforms against others, as it combines recommendation diversity with calculation efficiency.

Regarding the quality of the produced recommendations, results show that in a very high percentage (94–97%) of cases, at least one recommended product matches with one product view that was generated based on real customer actions. Hence, our proposed approach was shown to offer recommendations that could potentially help users move to a similar product page instead of leaving the e-commerce site.

References

1. Schafer, J.B., Konstan, J.A., Riedl, J.: Recommender systems in e-commerce. In: ACM Conference on Electronic Commerce, pp. 158–166 (1999)
2. Karimova, F.: A survey of e-commerce recommender systems. Eur. Sci. J. **12**(34), 75–89 (2016). ISSN 1857-7881
3. Han, E.-H., Karypis, G.: Feature-based recommendation system. In: 14th ACM International Conference on Information and Knowledge Management, pp. 446–452 (2005)
4. Charikar, M.S.: Similarity estimation techniques from rounding algorithms. In: Annual ACM Symposium on Theory of Computing, pp. 380–388 (2002)

5. Lops, P., de Gemmis, M., Semeraro, G.: Content-based recommender systems: state of the art and trends. In: Ricci, F., Rokach, L., Shapira, B., Kantor, Paul B. (eds.) Recommender Systems Handbook, pp. 73–105. Springer, Boston, MA (2011). https://doi.org/10.1007/978-0-387-85820-3_3
6. Indyk, P., Motwani, R.: Approximate nearest neighbors: towards removing the curse of dimensionality. In: Annual ACM Symposium on Theory of Computing, pp. 604–613 (2000)
7. Gionis, A., Indyk, P., Motwani, R.: Similarity search in high dimensions via hashing. In: 25th International Conference on Very Large Data Bases, pp. 518–529 (1999)
8. Zhang, K., Fan, S., Wang, H.: An efficient recommender system using locality sensitive hashing. In: 51st Hawaii International Conference on System Sciences (2018)
9. Aytekin, A.M., Aytekin, T.: Real-time recommendation with locality sensitive hashing. J. Intell. Inf. Syst. **53**, 1–26 (2019)
10. Debnath, S., Ganguly, N., Mitra, P.: Feature weighting in content-based recommendation system using social network analysis. In: 17th International Conference on World Wide Web, pp. 1041–1042 (2008)
11. Barranco, Manuel J., Martínez, L.: A method for weighting multi-valued features in content-based filtering. In: García-Pedrajas, N., Herrera, F., Fyfe, C., Benítez, J.M., Ali, M. (eds.) IEA/AIE 2010. LNCS (LNAI), vol. 6098, pp. 409–418. Springer, Heidelberg (2010). https://doi.org/10.1007/978-3-642-13033-5_42
12. Bestprice Homepage. https://www.bestprice.gr/. Accessed 21 Feb 2020
13. Ramos, J.: Using TF-IDF to determine word relevance in document queries. Department of Computer Science, Rutgers University (2003)
14. Retailrocket dataset. https://www.kaggle.com/retailrocket/ecommerce-dataset. Accessed 21 Feb 2020
15. scikit-learn, Machine Learning in Python, scikit-learn.org
16. Wu, W., Li, B., Chen, L., Gao, J., Zhang, C.: A Review for Weighted Minhash Algorithms (2018)
17. Wu, W., Li, B., Chen, L., Zhang, C.: Consistent weighted sampling made more practical. In: 26th International Conference on World Wide Web, pp. 1035–1043 (2017)
18. Manasse, M., Mcsherry, F., Talwar, K.: (2007) Consistent weighted sampling. Technical Report (2010)
19. Ioffe, S.: Improved consistent sampling, weighted minhash and L1 sketching. In: IEEE International Conference on Data Mining, pp. 246–255 (2010)
20. datasketch 1.5.0. http://ekzhu.com/datasketch/. Accessed 21 Feb 2020
21. Rajaraman, A., Leskovec, J., Ullman, J.: Mining of Massive Datasets. Cambridge University Press, Cambridge (2014)
22. Bawa, M., Condie, T., Ganesan, P.: LSH forest: self-tuning indexes for similarity search. In: 14th International Conference on World Wide Web, pp. 651–660 (2005)
23. Bradley, K.: Improving recommendation diversity. In: National Conference in Artificial Intelligence and Cognitive Science, Maynooth, Ireland, pp. 75–84 (2001)
24. Smyth, B., McClave, P.: Similarity vs. diversity. In: Aha, D.W., Watson, I. (eds.) ICCBR 2001. LNCS (LNAI), vol. 2080, pp. 347–361. Springer, Heidelberg (2001). https://doi.org/10.1007/3-540-44593-5_25
25. Gerogiannis, V., Karageorgos, A., Liu, L., Tjortjis, C.: Personalised fuzzy recommendation for high involvement products. In: IEEE SMC, pp. 4884–4890 (2013)
26. Nalmpantis, O., Tjortjis, C.: The 50/50 recommender: a method incorporating personality into movie recommender systems. In: 8th EANN, pp. 498–507 (2017)
27. Schoinas, I., Tjortjis, C. MuSIF: a product recommendation system based on multi-source implicit feedback. In: 15th AIAI, pp. 660–672 (2019)

Author Index

Printed in the United States
by Baker & Taylor Publisher Services

Printed in the United States
by Baker & Taylor Publisher Services